ADVANCED TOLERANCING TECHNIQUES

WILEY SERIES IN ENGINEERING DESIGN AND AUTOMATION

Series Editor
HAMID R. PARSAEI

GENETIC ALGORITHMS AND ENGINEERING DESIGN
Mitsuo Gen and Runwei Cheng

ADVANCED TOLERANCING TECHNIQUES
Hong-Chao Zhang

ADVANCED TOLERANCING TECHNIQUES

HONG-CHAO ZHANG
Texas Technical University
Lubbock, Texas

A Wiley-Interscience Publication
JOHN WILEY & SONS, INC.
New York · Chichester · Weinheim · Brisbane · Singapore · Toronto

Library of Congress Cataloging in Publication Data:
Advanced tolerancing techniques / [edited by] Hong-Chao Zhang.
       p.   cm. — (Wiley series in engineering design and automation)
    Includes bibliographical references and index.
    ISBN 0-471-14594-7 (cloth : alk. paper)
    1. Tolerance (Engineering)   2. CAD/CAM systems. I. Zhang, Hong
-Chao. II. Series.
  TS172.A38   1997
  620′.0045—dc21   96-51775

Printed in the United States of America

10 9 8 7 6 5 4 3 2 1

FIFTY YEARS
OF DEDICATION:
A Tribute to Oliver R. Wade

Beginning during World War II in 1944 and sadly ending on D-day 1994, exactly fifty years later, one man's dedication to excellence in process planning left us. He left us with the fundamentals with which to build future process planning strategies. It was a unique privilege to have known, worked with, studied his work, and enjoyed life with Oliver R. Wade, this remarkable genius of process planning. Oliver never graduated from the prestigious college, Cooper Union for the Advancement of Science and Art, because his important developing skills were critical to the war effort. Yanked from the university and assigned to a secret operation along the U.S. east coast, he began his long career designing process plans for small, precise mechanisms which included optics. Following the war the "round up" occurred and all such draft-exempt scientists found themselves lined up, unceremoniously awaiting their military assignment. Oliver was sent to White Sands, New Mexico, where he found himself in a strangely bilingual community of Americans and recently relocated German engineers. They set about designing the product and process for producing the V-2 rocket, a recently acquired device of great interest at that time.

Having completed this assignment in the military, Oliver worked for Wright Aeronautical Division of Curtiss-Wright Corporation and continued planning processes for highly complex and precise aircraft engines and gearboxes. It was here that he developed the first computer program to assist with generating a tolerance chart. For the rest of his life he preached about tolerance charting, solved tolerance problems, and wrote about this methodology, which he believed in passionately. In 1967 the Industrial Press published his book, *Tolerance Control in Design and Manufacturing*. In 1983, the Society of Manufacturing Engineers devoted Chapter 2 of Volume I of their *Tool and Manufacturing Engineers Handbook* series to tolerance control, with a highly detailed chapter written by Oliver. A few companies have developed proprietary approaches to accomplish charting and a renewed interest in the subject is once again beginning to emerge, as evidenced by several chapters in this book. In short, Oliver evolved and documented a foundation for scientific process planning which brings together both product and process design engineers in a truly concurrent planning approach.

We are very indebted to Oliver for dedicating his life to developing and capturing this critical methodology—mostly for no significant gain other than that of promulgating these important fundamentals. When he was called upon to solve manufacturing problems, Oliver believed in the importance of each client's needs, to the extent that during each engagement, he would move on site, bring his dedicated wife, Jeannette, with him, and remain until solutions had been documented and in the hands of the client. Thus, Jeannette deserves a vote of thanks, also, for her support of Oliver's important teaching and problem solving. Fortunately for all of us, he has left us with the tools to dramatically improve the quality and timeliness of developing critical engineering solutions.

W. B. Krag

CONTENTS

PREFACE

Tolerancing techniques have been used for more than 100 years in machine shops and drafting rooms. The original concept of tolerancing, introduced early in the nineteenth century, enabled machine parts to be made precisely enough to be assembled interchangeably. With the requirements involved in mass production in the early twentieth century, tolerance analysis and control not only attracted a great deal of attention from design engineers, but were also a strong focus of manufacturing engineers. Tolerance assignment and satisfaction have had a significant impact on manufacturing cost and product quality. Tolerancing techniques were addressed adequately in the curricula of engineering colleges during the first half of the twentieth century, but since the 1960s, when engineering science replaced practice-oriented teaching in most U.S. engineering colleges, dimensional tolerancing has disappeared as a part of training programs for industrial design and manufacturing engineers. Manufacturers adjusted to this lack of training either by instituting in-house training programs or by subcontracting dimensional tolerancing work to specialty firms. When computer-aided design (CAD) prompted a return to a practice-oriented teaching approach in the mid-1980s, tolerancing was not restored as part of most curricula. Part of the reason for this has been the lack of a suitable advanced research/reference volume that treats recent improvements in tolerancing techniques in a manner compatible with CAD, computer-aided process planning (CAPP), computer-aided manufacturing (CAM), and coordinated measuring machine (CMM)-based inspection.

   In general, three categories of tolerancing schemes have been developed for industry use: parametric tolerancing, geometric tolerancing, and operational tolerancing. The first two categories are used primarily in product design, while the third category is used primarily in process design. Parametric tolerancing is based on ordinary dimensions (or vector proxies). There are three versions: worst-case limit tolerancing, statistical tolerancing, and vectorial tolerancing. These schemes are called parametric tolerancing because dimensions can be regarded as control parameters for an underlying mathematical representation. Therefore, in some of the literature, parametric tolerancing is also referred to as dimensional tolerancing. Dimension chain is one of the typical techniques used to deal with parametric tolerancing. The dimension chain can be analyzed by a conventional worst-case limit method or by a statistical approach. The advantage of vectorial tolerancing, a relatively new scheme, is that it parameterizes and tolerances pairs in terms that relate naturally to manufacturing and inspection processes. Since vectorial tolerancing emphasizes orientation by means of vectors, the scheme should have potential for supporting tolerancing control in process design.

Geometric tolerancing applies tolerances directly to attributes of features. A feature is defined as a generic shape or a piece that is part of the surface: a hole, slot, section face, or planar face. Attributes characterized by the feature include size, position, form, and spatial relation. The semantics of geometric tolerances are established primarily by a set of rules for implementing datum systems from physical part features and another set of rules for constructing spatial zones. The datum system is used essentially for traditional machine drawings, but it becomes the rule of thumb in selecting the setup position for machining parts. Although wire-frame geometrical models were used in CAD, the importance of datum was ignored initially by computer-aided graphics, thus wire-frame model techniques could not completely transform information to manufacturing engineering. The recent popular feature-based three-dimensional-solid geometrical model enabled datum to function via CAD and thus is considered as a possible way to integrate CAPP and CAM.

Operational tolerancing is different from both parametric tolerancing and geometric tolerancing in being used primarily in process design. When a product is designed with specific tolerances (either parametrical or geometrical, quite often both), manufacturing engineers must guarantee reasonable cost by means of manufacturing processes. A conventional tool used for operational tolerancing is dimension and tolerance charting, also referred to as dimension chain, dimension chart, or dimension and tolerance chain. With the application of computer technology, the traditional manual analysis of dimension and tolerance charting has created difficulties for computer-aided process planning. Selection of the correct clamping datum for machining is the rule of thumb for guaranteeing the tolerances specified for a part. To utilize computer-aided process planning in real manufacturing environments, efforts have been made to implement dimension and tolerance charting as computer techniques. Unfortunately, the differences between a datum used for design and a datum used for manufacturing have not been addressed adequately. Other areas that should be investigated include the relationship between a machining datum, operational tolerances, and the tolerances specified in design. The global relationship in process planning should also be examined.

As mentioned earlier, the ignorance involving tolerancing techniques in engineering colleges has created problems in today's manufacturing industry. The fact that design and manufacturing engineers possess insufficient knowledge of tolerancing techniques results in raising unnecessary production costs and jeopardizing product quality. With the rapid development of computer-aided technologies, the requirement to catch up with the appropriate techniques of advanced tolerancing becomes urgent for today's educational institutions, industrial training organizations, and national research laboratories. The objective of this book is to provide a comprehensive overview of current tolerancing techniques and to present updated research results relative to this area. We hope to bridge the gap between traditional tolerancing techniques and advanced tolerancing techniques that are compatible with computer-aided technologies.

All of the chapter contributors are active researchers in the area and many are well-known scholars whose research results have guided progress in the area. Other contributors are young and promising scholars. The book is divided into three parts, although there is some topic overlap. Part I provides a general review of fundamental knowledge of tolerance problems, such as tolerance chain and tolerance chart techniques and current statistical methods used in tolerancing. Part II focuses on new mathematical contributions and modeling techniques used in the area, such as optimization of tolerance design versus manufacturing cost. Part III emphasizes applications and industrial practices. For each part of the book, I have invited one or two keynote chapters from respected contributors.

In Part I, a brief historical background of tolerancing techniques is provided and some fundamental knowledge is addressed. For a history of more than 100 years, a general review and guideline is necessary for tomorrow's research direction. In Chapter 1, H. B. Voelcker addresses the problems of tolerancing techniques today and points out the general future direction. Tolerance analysis and control are important during manufacturing to ensure that mechanical components meet design specifications. To bridge the gap between traditional and advanced tolerancing techniques, tolerance charting is included in the book. K. Whybrew and G. A. Britton provide a comprehensive discussion of tolerance charting, divided between Chapter 2 and Chapter 17. Chapter 2 introduces the fundamentals of tolerance charting, with tolerance stackup discussed in detail. In Chapter 17 they describe CATCH, computer software for tolerance charting. In Chapter 3, W. B. Krag discusses how a tolerance chart is applied to process design. Mr. Krag is the successor to the intellectual properties of Oliver R. Wade, a pioneer in the use of tolerance charting in practical applications whom we have honored earlier in these pages. Chapter 4 is a comprehensive tutorial on tolerance analysis and synthesis by R. J. Gerth. Problems ranging from linear one-dimensional tolerance analysis to nonlinear two-dimensional tolerance analysis, from worst-case study to statistical approach, and from constant factor models to engineering redesigns are all addressed in this chapter.

Tolerance specifications must be defined mathematically to avoid ambiguous interpretation and to provide a sound basis for assessing the correctness of measurement techniques and algorithms. In Part II we present some theoretical studies of tolerancing that focus on improving tolerancing techniques through mathematical modeling and definitions, geometric tolerances, vectorial tolerancing, statistical tolerancing, tolerance formalization, and algorithm development. In Chapter 5, K. W. Chese, S. P. Magelby, and J. Gao present a new method for tolerance analysis of two- and three-dimensional mechanical assemblies in which the characteristic of tolerance is represented by means of vectors. Generalizing vector loop-based models to include small kinematic adjustments has the advantage over traditional tolerance analysis methods that it does not require an explicit function to describe the relationship between the resultant assembly dimension(s) and manufacturing components dimensions, which may be difficult or impossible to obtain for complex assemblies. In Chapter 6, V. Srinivasan, M. A. O'Connor, and F. W. Scholz provide mathematical and computational techniques for establishing statistical tolerance zones. Statistical tolerancing is a new technique whose use is strongly encouraged for practical applications. Ideal tolerance designs should not start from individual mechanical component specifications but should emphasize an assembly "budget" for allowable variation and distribute it to the constituent components. In reality, this problem is addressed by several iterations of tolerance analysis wherein component variations are composed to determine the assembly variation.

As an integral part of mechanical design, tolerance has a profound influence on the functional performance and manufacturing cost of a product. In this book, several contributions address that issue. New research activities to solve the problem of tight tolerance with cost are reviewed in Chapters 7 through 10. An alternative approach to concurrent tolerancing for accuracy and optimum cost of accuracy is presented by M. M. Sfantsikopoulos in Chapter 7. This consideration is based on a new cost-tolerance function that pursues an optimum solution for the explicit relationship between the size of a dimension, the tolerance assigned to it, and the related manufacturing cost. By the same token, in Chapter 8, G. Zhang addresses the issue of tolerance versus manufacturing cost and proposes various solutions. Z. Dong addresses the same problem in Chapter 9 but presents another way of using a joint evaluation and optimization of the functional performance and manufacturing cost at the product design stage. In Chapter 10, S. Lin, H. Wang, and C. Zhang address this problem again in terms of a concurrent optimization methodology based on a normal distribution assumption. A mixed-discrete nonlinear optimization model is formulated and solved by means of genetic algorithms.

The methodology presented in Chapter 11 by Y. Wang, S. Gupta, and S. Rao is a supplemental procedure used to enable tolerancing techniques to be applied efficiently. They also discuss the basic formulations of efficient facilitation of coordinate measuring machines (CMMs) for automated inspection and manufacturing process improvement. At the end of Part II, in Chapter 12, J. Dong and Y. Shi present an approach to tolerance evaluations and allocations. In mechanical assembly, variation in one design constraint often affects other constraints. The sensitivity of such a variation can be determined analytically using variational geometry theory.

Tolerancing techniques are typical practice-oriented subjects. A diverse spectrum of advanced tolerancing techniques are presented in the nine chapters of Part III. Although the emphasis is on computerized tolerancing techniques, the topics include tolerancing for function, fixture design, setup planning, and CMM inspection, as well as geometrical modeling and tolerance transforming. In Chapter 13, R. D. Weill addresses the issue of dimensioning and tolerancing for function (DTF). The concept is illustrated by numerous examples, including a complex assembly example that demonstrates the relative limitations of DTF caused by manufacturing and inspection constraints. In Chapter 14, A. Y. C. Nee and A. S. Kumar address an issue of computational geometry-based techniques to solve the tolerance accumulation problem of modular fixtures. Both parametric tolerancing and geometric tolerancing are studied. It is known that tolerancing has a significant impact on fixture design. Many different catalogs of fixtures are adopted in today's industrial practice. Modular fixtures are one of the most popular examples. In Chapter 15, Y. Rong addresses a similar problem but from a different viewpoint. Five basic dimension relationship models of locating datum and machining surfaces are analyzed for estimation of machining errors with various setup conditions. Locating errors that affect manufacturing accuracy is also addressed in this chapter. The information from the analysis is useful not only for fixture design but also for computer-aided process planning.

In Chapter 16, S. Huang and I discuss tolerance analysis for setup planning in computer-aided process planning. CAPP has been addressed widely for more than two decades. To date, however, few CAPP systems have been applied in the industrial manufacturing environment. One of the most critical problems is that tolerance analysis for process and setup planning has been somewhat overlooked in the CAPP research community. While many researchers focus their attention on tolerance chart analysis, the issue of tolerance analysis for setup planning remains relatively unexplored. This chapter represents an innovative approach to solve the problem by means of proactive tolerance control. Since tolerance charting is one of the most popular tolerance analysis tools for process control, the description by G. A. Britton and K. Whybrew of a computerized tolerance charting program is presented in Chapter 17.

Computer-aided representation and assessment of geometric tolerances in CAD/CAM systems are important topics because of their significant influence on downstream manufacturing activities. Due to lack of unique and unambiguous definitions of geometric tolerances and the implementation details of their assessment criteria, it is becoming increasingly difficult to assess the geometric characteristics of manufactured components from the voluminous CMM output data against the tolerances specified in the CAD database. In Chapter 18, U. Roy, X. Zang, and Y. Fang address the issues of (1) determining unique and unambiguous definitions of the tolerance of any feature, (2) establishing a definition of the measured sizes of all features, (3) determining the criteria for establishing datums from measured data points of the features measured, and (4) establishing computational techniques for assessing CMM data against the design specification stored in the CAD data model. Industrial practice is one of the keys to examining the techniques that have been developed by the research community. In Chapter 19, M. A. Buckingham presents the development of a new in-house standard to address the problem with traditional tolerancing methods using ANSI Y14.5 and relative guidance for inspection. In Chapter 20, M. M. Dowling, P. M. Griffin, K. Tsui, and C. Zhou present an innovative method for form error estimation by CMM inspection. The minimum zone and orthogonal least-squares methods are compared through the use of a simulation study for straightness and flatness form tolerances. The results indicate that the orthogonal least-squares method yields a smaller mean-squared error than that obtained using the minimum zone method, together with other advantages. In Chapter 21, S. Kumar and S. Raman introduce a simple method for transferring the tolerance and surface finish specifications from the design domain to the manufacturing domain by means of tolerance and surface finish relationships.

Tolerance standardization is always an important issue, especially in industrial applications. Although many chapters address issues through applications, readers might feel that one topic is missing: a comprehensive discussion concerning standard geometric dimensioning and tolerancing (GD&T). Standardization could not be included in this book because no contributor was available to focus on these issues. For those interested in this topic there are introductory-level publications available, one recommendation being ANSI Y14.5-M. Even without this topic, we feel that the book is one of the most comprehensive publications addressing advanced tolerancing techniques. A broad spectrum of currently active research results are covered. Whether you are interested in traditional worst-case analysis methodologies or in advanced statistical analysis methodologies, you will find relevant chapters in this book. Whether you are interested in the traditional manual tolerancing analysis approach or in a computerized approach, you will find chapters addressing these topics. Whether you are interested in product design tolerance allocations or in operational tolerance allocations, you will find the chapters you need. Whether you are interested in cost issues or optimization approaches, you will get answers from this book. The book can be considered a research reference book for readers interested either in theoretical research or in practical development of advanced tolerances techniques, including product design and manufacturing engineers, research and development project engineers, project directors, program monitors, system supervisors, shop floor foremen, and division and department leaders. The book should also be valuable for engineering professors who are researching and teaching advanced tolerancing techniques, as well as graduate and senior undergraduate students pursuing careers in mechanical, industrial, electrical and electronic, chemical, and manufacturing engineering. For those who are organizing short courses or preparing training programs, this book would be an excellent choice.

I would like to take this opportunity to express my gratitude to the contributors for their effort and patience in preparing and revising their manuscripts and to reviewers for providing thoughtful and constructive comments. Many friends, colleagues, and students made contributions to the book. I particularly appreciate the support of those colleagues who provided excellent reviews while knowing that their own contributions would not be included. Some people deserve special mention for their unselfish assistance. First, my thanks goes to Sam Hongdao Huang, my dear colleague, for his extensive editorial work on the book. Second, I thank Hamid K. Pasaei, the series editor of the book, and Robert L. Argentieri, executive editor at John Wiley & Sons, for their encouragement. My special thanks go to my family, who were very supportive during this period of time. I particularly want to thank my dearest mother, Li Shiaohui, whose immense love and support have helped me complete this work.

HONG-CHAO ZHANG

Lubbock, TX USA
Spring 1997

CONTRIBUTORS

G. A. Britton, Professor, Nanyang Technological University, Nanyang, Singapore

Mark Buckingham, New Mills Wotton-Under-Edge, Gloustershire, GL128JR, UK

Kenneth W. Chase, Professor, Brigham Young University, Provo, Utah

Jian (John) Dong, Professor, University of Connecticut, Storrs, Connecticut

Zuomin Dong, Professor, University of Victoria, Victoria, British Columbia, Canada

Mary M. Dowling, Professor, Georgia Institute of Technology, Atlanta, Georgia

Y. Fang, 2310 NE 48th Street, #722, Seattle, Washington

Jinsong Gao, Hewlett-Packard, San Diego, CA

Richard J. Gerth, Professor, Ohio University, Athens, Ohio

Paul M. Griffin, Associate Professor, Georgia Institute of Technology, Atlanta, Georgia

Shailendra Gupta, Structural Dynamics Research Corporation, Milford, Ohio

Samuel H. Huang, EDS/Unigraph Inc.

W. B. Krag, Consultant, Technical Memory Consulting, Inc.

Senthil Kumar, Lecturer, National University of Singapore, Republic of Singapore

Shui-Shun Lin, Professor, National Institute of Technology, Taichung, Taiwan

Spencer P. Magleby, Professor, Brigham Young University, Provo, Utah

A. Y. C. Nees, Professor, National University of Singapore, Republic of Singapore

Michael A. O'Connor, IBM T. J. Watson Research Center, Yorktown Heights, New York

S. Raman, Professor, School of Industrial Engineering, University of Oklahoma, Norman, Oklahoma

Srinavas Rao, Microcosm, Inc., Jessup, Maryland

Yiming (Kevin) Rong, Professor, Southern Illinois University of Carbondale, Carbondale, Illinois

Fritz W. Scholz, Boeing Information and Support Services, Seattle, Washington

M. M. Sfantsikopoulos, Professor National Technical University of Athens, Athens, Greece

Ying Shi, Doctoral student, University of Connecticut, Storrs, Connecticut

Vijay Srinivasan, IBM T. J. Watson Research Center, Yorktown Heights, New York, and Columbia University, New York, New York

Kwok-Leung Tsui, Associate Professor, Georgia Institute of Technology, Atlanta, Georgia

Roy Uptal, Professor, Department of Mechanical, Aerospace & Manufacturing Engineering, Syracuse University, Syracuse, New York

H. B. Voelcker, Professor, Department of Mechanical Engineering, Cornell University, Ithaca, New York

Hsu-Pin (Ben) Wang, Professor, Florida A&M University, Florida State University, Tallahassee, Florida

Yu Wang, Professor, University of Maryland, College Park, Maryland

Roland D. Weill, Professor, Israel Institute of Technology, Haifa, Israel

K. Whybrew, Professor, University of Canterbury, Christchurch, New Zealand

Chun (Chuck) Zhang, Professor, Department of Industrial Engineering FAMU-FSU College of Engineering, Tallahassee, Florida

G. Zhang, Professor, Chongqing University, Chongqing, China

Hong-Chao Zhang, Professor, Texas Tech University, Lubbock, Texas

Chen Zhou, Associate Professor, Georgia Institute of Technology, Atlanta, Georgia

I

TOLERANCE TECHNIQUES OVERVIEW AND FUNDAMENTAL TECHNOLOGY

1

DIMENSIONAL TOLERANCING TODAY, TOMORROW, AND BEYOND

H. B. VOELCKER

Cornell University
Ithaca, New York

1.1 PROLOGUE[1]

The teaching of dimensional tolerancing was abandoned in many American engineering colleges when drafting and similar practice-oriented skills fell victim to engineering science in the 1960s. Practice-oriented teaching began to return in the 1980s, and with it training in CAD (computer-aided design, the "new drafting"), but not training in tolerancing. Tolerancing practices in industry changed radically in the 20 years that engineering science reigned in the engineering colleges. Specifically, geometric and statistical tolerancing replaced old-style parametric limit tolerancing in many mechanical applications, but engineering educators have not brought the new methods into current curricula. I shall summarize some personal opinions on why this is so, why things should improve over the next decade, and why we should seek a more holistic role for geometry in mechanical engineering education.

---

[1]This essay, minus the Postscript (Section 1.8), was presented as a keynote talk in Session DE-8, Geometric Dimensioning and Tolerancing Education, of the 1995 ASME International Engineering Congress and Exposition (San Francisco, November 1995). It was published by the ASME (American Society of Mechanical Engineers) in pamphlet form as Paper 95-WA/DE-8. The Postscript was added in mid-1996 to summarize some significant changes in the standards communities. The modified version is published with permission of the ASME.

Advanced Tolerancing Techniques, Edited by Hong-Chao Zhang
ISBN 0-471-14594-7   © 1997 John Wiley & Sons, Inc.

1.2   GEOMETRIC TOLERANCING TODAY

Scalable drawings became the medium for defining mechanical parts and entire machines early in the nineteenth century. Dimensions were added to drawings later in the nineteenth century, when affordable steel scales and vernier instruments made precise shop-floor measurement feasible. Tolerances, expressed as plus/minus limits on dimensions, appeared early in the twentieth century. These three constructs, taken together, provided the first reasonably complete system for specifying symbolically the geometry of mechanical products. Note the elements:

- Form was handled by views on orthographic drawings.
- Dimension quantified form.
- Variation control, which is essential because human-made artifacts are inherently imprecise, was handled by a mixture of limits on dimensions (e.g., 1.500 ± 0.005) and process callouts (e.g., "Bore 1.875 thru").

An event—World War II, with its vast demands for multisource procurement—stimulated postwar changes in variation-control practice. Process specifications were prohibited to retain maximum manufacturing flexibility, and tolerances became the sole mechanism for variation control. More pervasively, dimensional limit tolerancing (now called *worst-case parametric tolerancing*), which suffers from several ambiguities, was replaced with *geometric tolerancing*. Geometric tolerancing is based on three central notions:

1. Conformance to a geometric tolerance requires that a surface feature, or an attribute of a feature (e.g., the axis of a hole), lie within a prescribed spatial zone. Note that this is a true *geometric* criterion, whereas conformance to a parametric tolerance is inherently numeric.
2. A geometric tolerance usually controls explicitly only one specified property of a feature, such as form (flatness, cylindricity) or position. However, subtle interactions between different tolerances on the same feature can complicate matters considerably.
3. Some containment zones (e.g., for form) can be positioned freely in space, whereas others (e.g., for position) are located on parts through reference features called *datums*. The use of containment zones deals directly with *imperfect form* and is the hallmark of geometric tolerancing.[2]

---

[2]Imperfect form can be handled in other ways: for example, by replacing imperfect planar and cylindrical features with fitted versions of perfect features. This approach is popular in some European circles.

These changes were codified in successive editions of the ASME Y14.5 standard (ASME, 1994a) and required a workforce generation—almost 30 years—for industry to digest. It is important to realize that geometric dimensioning and tolerancing (GD&T), as defined in Y14.5 (or alternatively, in a set of about 10 analogous ISO standards), is not based on explicit mathematical principles. It is a *codification of best practices* assembled by practical people with artisan backgrounds, who defined GD&T mainly through examples explained with prose and graphics.

A second event, the "Metrology crisis" of the 1980s, triggered two projects that may mark the start of a new era in tolerancing.[3] The first project sought to "mathematize" Y14.5; it has produced to date a new standard, Y14.5.1M-1994, *Mathematical Definition of Dimensioning and Tolerancing Principles* (ASME, 1994b). The second project's goal is a companion standard for Y14.5 (B89.3.2, now in draft) covering measurement methods. The title of Y14.5.1 is somewhat misleading because "principles" are covered only in the narrowest sense. Y14.5.1 is a literal mathematization: specifically, an enumerative redraft of Y14.5, with Y14.5's prose and graphics replaced by definitions in algebraic geometry of three characteristics for each tolerance class: the containment zone, the conformance criterion, and the actual value. [*Actual value* means, loosely, the smallest tolerance value (smallest zone) to which a particular physical feature can conform.] Y14.5.1 does not provide a generative reformulation of geometric tolerancing—that is, a sparse set of mathematical principles from which one can *construct* the various tolerance classes and their three characteristics.[4]

Given this history, it is easy to see why geometric tolerancing, as currently codified, is largely untaught in American engineering colleges.

- GD&T is complicated; while some rules are reasonably simple, others are complex or arbitrary, and there are subtleties and special cases.
- There is no set of underlying mathematical principles from which everything can be derived. (Y14.5.1 does yeoman service in resolving ambiguities in Y14.5, but it contributes little to the underlying-principles problem.)
- Although there are several (expensive) technician-level training manuals abroad, only the recent books by Liggett (1993) and Henzold (1995) can be considered as possible engineering texts.

---

[3]Industrial use of coordinate measuring machines (CMMs) for part inspection proliferated in the 1980s, but early CMM algorithms produced results different from those obtained with traditional methods (hard gages, open setup). This divergence was tagged as a "Metrology crisis" requiring attention by the responsible standards committees: B89 for metrology, Y14 for tolerances (Hook, 1993).

[4]I wish to emphasize, in fairness to its hardworking members, that the Y14.5.1 Committee *met its 1989 charter,* which was to "rephrase" Y14.5 in mathematics, not to reformulate it.

• Most engineering professors know nothing about GD&T, many view tolerancing as picky or boring or both, and there is no strong industrial pressure to teach GD&T.

The first three points can be summarized by saying that today's GD&T is *trainable* but not *teachable*. The final point—no strong industrial pressure—is a warning flag: Is the educational issue moot? I believe the issue is not moot and that the situation in industry is similar in some ways to that in the universities. To wit, a few companies "do it well" with strong in-house GD&T training programs.[5] Some other companies accept that GD&T is important, but because GD&T is complicated, they subcontract dimensional management to specialist firms. In industry at large, some design engineers have a superficial knowledge of GD&T, but few understand it well; many would rather not think about tolerancing. As a result, many parts and products almost certainly are overtoleranced or haphazardly toleranced, with predictable consequences.

In summary, variation control is obviously important, but the best current means to handle it—GD&T—is not taught in the universities and not used as effectively as it should be in industry. The problem lies not with practitioners and teachers, but with GD&T itself: It is genuinely *baroque*. But help is on the way. I believe that the era of practice-driven tolerancing is ending and that a new era of science-based tolerancing is opening. Some of the probable changes are summarized below. But first we must acknowledge statistical tolerancing.

1.3 STATISTICAL TOLERANCING TODAY

The assignment of parametric limit-tolerance values and geometric-tolerance values is usually governed by worst-case analysis. Specifically, if a part's dimensions are required to lie in specified intervals, or if a part's features must lie in specified spatial zones, one can design for interchangeable assembly by requiring that the intervals or zones of mating features be disjoint.

Statistical tolerancing is based on the following observation: When parts are made by properly controlled processes, the actual values of most dimensions lie near the centers of their tolerance regions. Therefore, one can exploit statistical averaging, especially when designing assemblies in which dimensions are chained and variability is cumulative. (In worst-case design, the cumulative variability grows linearly with the length of a dimension chain, whereas the statistical standard deviation grows on a root-mean-square basis if fluctuations in the component dimensions are uncorrelated.) This means

---

[5]Two specific examples: the Ford Motor Company (large) and Hutchinson Technology, Inc. (small).

that greater variability can be accepted in parts embedded in complex assemblies, because the fluctuations will usually average out. Total interchangeability cannot be guaranteed, but the probability of assembly failure can be made very small, at least in theory.

The ideas just summarized, with supporting design formulas based on normal statistics, were developed essentially independently in several organizations some 30 to 40 years ago (i.e., contemporaneously with geometric tolerancing). Statistical tolerancing has been used on a limited basis in several large organizations for 20 to 30 years, and today there is growing use of untidy (to the academic mind) mixtures of statistical tolerancing notions, statistical process control notions, and Taguchi quality-function notions.

The key point for present purposes is that statistical tolerancing, as presently formulated, is merely an extension of classical parametric tolerancing that provides an alternative to worst-case design (Bjorke, 1989). There is no accepted statistical interpretation of geometric tolerances. However, the latest version of Y14.5 contains a bizarre move in that direction,[6] and as noted below, a small but able band of researchers is tackling the problem head-on.[7]

1.4   GEOMETRIC TOLERANCING TOMORROW

I think it likely that at least three important advances in geometric tolerancing will be made over the next decade.

1. One or more generative formulations of geometric tolerancing will be produced. A generative formulation will be more general than current practice but should contain the current GD&T facilities as special cases (perhaps with a few changes for the sake of homogeneity). A generative formulation should be teachable in the engineering colleges because it will be based on a small set of underlying mathematical principles, and it should increase the broadscale comprehensibility of GD&T in industry.

2. Statistical extensions of geometric tolerancing will be developed. Progress toward this end is already visible (Srinivasan and O'Connor, 1994). The keys seem to lie in finding statistical characterizations for

---

[6]Section 2.16 of Y14.5M-1994 provides a new symbol to indicate a statistical tolerance and provides some stylistic rules for using the symbol. It does not provide a geometric or statistical explanation of what the symbol means and says only that features carrying the symbol "shall be produced with statistical process controls." This explanation may be inconsistent with Section 1.4e, which prohibits process specifications.

[7]There are scores of papers in operations research and industrial engineering journals on statistical tolerancing, because once a part or an assembly has been abstracted into a set of real numbers and intervals (classical dimensions and tolerances), it provides an "application domain" for wondrous exercises in nonlinear analysis and optimization. This literature is largely ignored by tolerancing practitioners.

the actual-value concept introduced in Y14.5.1 and in finding means to model and analyze tolerance accumulation.

3. GD&T's tight links to drafting practice will be generalized to enable tolerances to be associated with features in CAD systems based on solid modeling rather than being treated as annotation on projections. In principle this is not difficult (Guilford and Turner, 1993), and it may occur naturally if a generative formulation of GD&T is defined as a formal language (e.g., via techniques used to define programming languages) rather than as an extension of the current graphic syntax.

1.5   WILD CARDS

Two vigorous product-definition standardization projects may change the prospects just outlined. One is STEP (Standard for the Exchange of Product Data), which evolved from PDES (Product Data Exchange Specification) out of IGES [Initial Graphic Exchange Specification (now a standard)]. STEP is a very ambitious undertaking that can be viewed as a bottom-up approach to enterprise modeling, or alternatively, as a gigantic information structure designed to contain all product and production data. STEP is being developed mainly by computer and information system engineers, with support from major manufacturing companies, CAD vendors, and several federal agencies. It could muddy the waters considerably if it standardizes tolerance structures invented by computer scientists not acquainted with the nuances of mechanical design and manufacturing. Laurance (1994) provides a useful perspective.

The second project is GPS (Geometric Product Specification), a comprehensive set of European standards being developed by the Joint Harmonization Group (JHG) of the International Organization for Standaridzation (ISO). (The JHG is a coordinating committee chartered by the ISO Technical Committees responsible for drafting, tolerancing, and metrology.) Bennich (1994) summarizes some of the early GPS thinking, and a new technical report (ISO, 1995) summarizes the GPS master plan. GPS is focused on the semantics of tolerancing and metrology—that is, on the mathematical and physical *meanings* of tolerance specifications and measurement procedures—whereas STEP has a strong bias toward syntactics [e.g., structural definitions for frameworks (data structures) to contain tolerance and other data]. GPS and STEP would benefit from coordination and collaboration, but to date these two large projects are proceeding essentially independently.

1.6   TOLERANCING BEYOND TOMORROW

The improved versions of GD&T sketched above will not solve the tolerancing problem; they will merely provide tidier tools for attacking problems we really don't understand. In blunt terms, we have no global models for the "physics" of interchangeable assembly and for the geometric sensitivity of various kinds of mechanical "functionalism" (heat transfer, bearing hydrodynamics, and the like). In different words, we have no formal theories of tolerancing for assembly and for function. Without such models and theories, we are forced to tolerance on an enumerative, feature-by-feature basis, which is expensive, error-prone, and difficult to optimize.

Because research aimed at these major problems is showing progress, I shall hazard another prediction. At some time in the next 5 to 25 years, a proper theory of tolerancing for assembly will appear and will provide guidelines for designing efficient assembly tolerancing languages. (Tolerancing for function has subtleties I shall not attempt to address here.)

A final prediction, unrelated to the others: At some time in the next 5 to 15 years, dimensional tolerancing and surface tolerancing, and their associated bodies of measurement methods, will be unified, probably through spectral theory. There is already movement in this direction in ISO circles.

1.7 EPILOGUE: RESTORING PARITY

Some people believe that modern engineering was born two centuries ago, in 1795, when Gaspard Monge published the treatise on descriptive geometry that provided foundations for engineering drawing. Early in the nineteenth century, engineers' ability to define and describe forms, configurations, and spatial relations through drafting exceeded by a considerable margin their ability to calculate properties, control processes, synthesize materials, and make machines. As the nineteenth century unfolded, however, the various capabilities came into rough balance as advances were made in mathematical analysis, material science, and manufacturing processes and methods.

In this century our ability to calculate, control, and make materials and machines has outstripped our ability to deal effectively with geometry. For example, contrast the elegance of the analytical methods used to solve field problems with the crude methods used to handle the problems' spatial domains. For a second example, and to close the circle, consider dimensional tolerancing. Dimensional tolerancing is intrinsically geometric, yet we know little more about it in a fundamental sense than we did 50 years ago, and to date, CAD has contributed nothing.[8,9]

I raise these issues because mechanical engineering educators are the most obvious agents for restoring geometry to a position of parity in our pantheon.

---

[8]CAD is in some senses a disappointing development. It has brought some major efficiencies, but not conceptual advances or even conceptual clarity.

[9]Ferguson (1992) believes that traditionally educated engineers were better able to deal with geometry and spatial relations than those educated recently—since 1960, say. In his eyes, we are losing geometric competence while advancing on other fronts.

I am not arguing for a return to an old curriculum but for the synthesis and teaching of new, powerful, and elegant approaches to geometry. I don't know how to do what I am advocating, but I don't doubt its importance. Perhaps we must await the emergence of a second Monge.

1.8   POSTSCRIPT

This postscript updates my 1995 essay (see footnote 1) by summarizing some significant organizational changes in the standards communities that occurred early in 1996. The organization of the Western world's relevant standards communities may be summarized loosely as follows. In the United States, three major ASME committees—Y14 (dimensioning and tolerancing), B89 (dimensional metrology), and B46 (surface phenomena)—are the working surrogates of ANSI (American National Standards Institute). Other nations also address standardization through national organizations, and almost all (including the United States) are members of the ISO. In tolerancing and metrology, ISO had through 1995 three Technical Committees—TC10/SC5, TC3, and TC57—that were loosely analogous to the American Y14, B89, and B46 committees. The American committees communicated and collaborated with their ISO counterparts through TAGs (Technical Activity Groups—an ANSI formality) and U.S. membership in the ISO TCs.

In early 1996 European members of ISO, under Danish leadership, effected a major reorganization of the ISO TC structure whereby the heretofore ad hoc Joint Harmonization Group (JHG, see Section 1.5) was established as a single, new TC213 to replace the three traditional TCs, which were disbanded. TC213's charter spans the field—dimensional and surface tolerancing and metrology, plus relevant instrument technology and aspects of process control—and its working style is highly dynamic. Succinctly, TC213 intends to work on successive three-year standardization cycles, with the master plan (ISO, 1995) as its goal. TC213 concedes that mistakes are inevitable when forcing the pace in ill-understood fields, but "mistakes can be fixed in the next cycle."

The American standards community has reacted to this largely European initiative by creating a new ASME umbrella committee, H213, to interact with TC213. The traditional ASME committees—Y14, B89, and B46—remain in place, but their roles and working styles are likely to change in ways not predictable at this writing.

In summary, we really do seem to be in the midst of major changes in the ways in which we perceive, codify, teach, and apply technologies for controlling spatial variability. The newly aggressive TC213 community is attempting to write comprehensive standards, on accelerated schedules, for techniques and phenomena that are poorly understood. Perhaps this organizational freneticism will force the pace of scientific codification, but that cannot be taken for granted at this early date.

ACKNOWLEDGMENTS

My views on the matters discussed in this essay have been shaped by discussions with Vijay Srinivasan, Vadim Shapiro, and others too numerous to cite here. Research and teaching in the areas of dimensional tolerancing and metrology have been supported at Cornell by the National Science Foundation, the Alfred P. Sloan Foundation, AT&T Bell Laboratories, the Brown & Sharpe Manufacturing Company, the Ford Motor Company, and Sandia National Laboratories. The Department of Energy and the IBM Corporation have provided graduate fellowships.

REFERENCES

ASME, 1994a, *Dimensioning and Tolerancing,* ANSI Y14.5M-1994, American Society of Mechanical Engineers, New York.

ASME, 1994b. *Mathematical Definition of Dimensioning and Tolerancing Principles,* ANSI Y14.5.1M-94, American Society of Mechanical Engineers, New York.

Bennich, P., 1994, Chains of Standards: A New Concept in GPS Standards, *Manuf. Rev.,* Vol. 7, No. 1, pp. 29–38.

Bjorke, O., 1989, *Computer-Aided Tolerancing,* 2nd ed., ASME Press, New York.

Ferguson, E. S., 1992, *Engineering and the Mind's Eye,* MIT Press, Cambridge, MA.

Guilford, J., and Turner, J., 1993, Representation Primitives for Geometric Tolerancing, *Compt.-Aid. Des.* Vol. 25, No. 9, pp. 577–586.

Henzold, G., 1995, *Handbook of Geometrical Tolerancing,* Wiley International, New York.

Hook, R., 1993, Interaction of Dimensioning, Tolerancing, and Metrology, *Proceedings of the 1993 International Forum on Dimensional Tolerancing and Metrology,* V. Srinivasan and H. B. Voelcker (eds)., American Society of Mechanical Engineers, New York, Dearborn, MI, June 17–19, 1995, CRTD-27, pp. 1–4.

ISO, 1995, *Geometrical Product Specifications (GPS): Masterplan,* ISO/TR 14638: 1995(E), International Organization for Standardization, Geneva, Switzerland.

Laurance, N., 1994, A High Level View of STEP, *Manuf. Rev.* Vol. 7, No. 1, pp. 39–46.

Liggett, J. B., 1993, *Dimensional Variation Management Handbook,* Prentice Hall, Upper Saddle River, NJ.

Srinivasan, V., and O'Connor, M. A., 1994, On Interpreting Statistical Tolerancing, *Manuf. Rev.,* Vol. 7, No. 4, pp. 304–311.

2

TOLERANCE ANALYSIS IN MANUFACTURING AND TOLERANCE CHARTING

K. WHYBREW

University of Canterbury
Christchurch, New Zealand

G. A. BRITTON

Nanyang Technological University
Nanyang, Singapore

The Lord willin' and if the creek don't rise ...

Maybe, just maybe, someday most everyone in the manufacturing industry who has anything to do with tolerances will have some knowledge of tolerance charts.

Surely the Lord is willing, since so many would benefit ...

And it isn't the creek that is rising, it's things like the complexity of the parts, the severity of the tolerances, and greater demands for exact adherence to the dictates of the engineering drawing.

> Lee Gunderson,
> Sundstrand

2.1 INTRODUCTION

Tolerance analysis and control are important during manufacturing to ensure that parts meet design specifications. Analysis and control are very difficult

Advanced Tolerancing Techniques, Edited by Hong-Chao Zhang
ISBN 0-471-14594-7   © 1997 John Wiley & Sons, Inc.

when parts are complex, design tolerances are very tight, the number of operations needed to make the parts is large, and changes in machining datums are frequent. These conditions are common in the aerospace and automotive industries, and so, not surprisingly, these industries developed formal procedures to guarantee tolerance control. One of the procedures used during process planning is referred to as *tolerance charting*. A tolerance chart is a graphical tool for representing a manufacturing sequence and for checking that tolerance stackups meet the design specifications.

We have used the piece of home-spun philosophy in the quotation at the start of this chapter to emphasize the importance of tolerance charts in process planning. The quotation is from the introduction to Sundstrand's in-house manual on tolerance charting by Lee Gunderson. Wade (1983) gives further support for the tolerance charting technique. He emphasizes the "self-evident importance of doing it right first time" and presents tolerance charts as the only systematic way of ensuring tolerance control in manufacture. Just-in-time manufacture is only made possible by right-first-time manufacture! In this chapter we introduce the problems of tolerance control in manufacturing and review the role of tolerance charting in helping to control these problems. In Chapter 3, Krag explains and illustrates manual tolerance charting in detail. In Chapter 17, Britton and Whybrew review and discuss recent developments in computer-aided tolerance charting. There have been numerous publications on process planning, process control, and tolerance charting, many of which use different and often conflicting definitions and terminology. To avoid confusion, the terms used in this chapter and in Chapter 17 are defined in a glossary at the end of the chapter.

2.2 TOLERANCE STACKUP IN MANUFACTURING

Why does tolerance stackup occur in manufacturing, and why is it so difficult to control? For any given design specification dimension, good dimensional control is guaranteed only if a workpiece is located on one of the surfaces related by the dimension while the second surface is machined. In this circumstance, the final tolerance depends solely on the process tolerance of the machine making the part, control is not a problem, and tolerance analysis is not necessary. However, this is an ideal situation that is not often encountered in practice, and certainly not for complex parts. Instead, what normally happens is that the setups for locating and holding the workpiece require the use of locating surfaces not referred to directly by the design specification for the dimensions. When this occurs, tolerance stackup is inevitable.

Figure 2.1 illustrates a simple example of a manufacturing tolerance stack. (Note that a *tolerance stack* is a connected sequence of dimensions that contribute to the tolerance of a dimension.) The part is a cylindrical plug manufactured on a lathe. The workpiece is completed in one setup. In operation 1, surface 3 is machined with a nominal working dimension of 49.00 mm

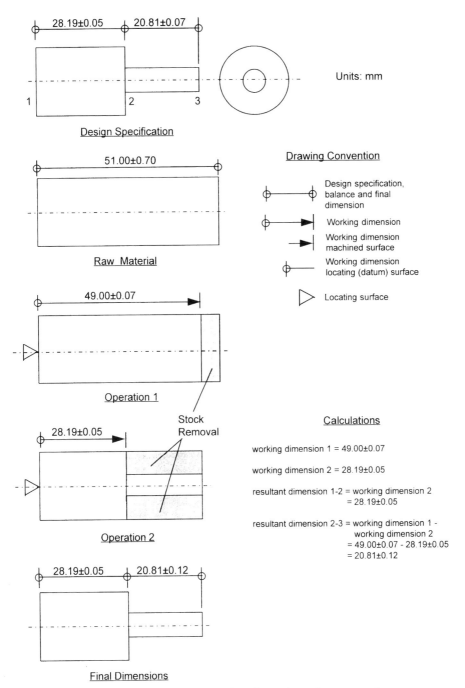

Figure 2.1. Design specification dimension tolerance stack.

from location surface 1, resulting in a working dimension of 49.00 ± 0.07 mm. Operation 2 machines surface 2 with a nominal working dimension of 28.19 mm from location surface 1, resulting in a working dimension of 28.19 ± 0.05 mm. The resultant (final) dimension between surfaces 1 and 2 (1–2) is the working dimension 1–2 and meets the design specification. The resultant dimension for 2–3 is a balance dimension and is obtained by subtracting working dimension 1–2 from working dimension 1–3. Its value is 20.81 ± 0.12 mm and it does not meet the design specification. The total tolerance for resultant dimension 2–3 is called a *tolerance stackup* and is the sum of the tolerances of the two other dimensions.

2.3   DIMENSIONS IN MANUFACTURING

In Section 2.2 a distinction has been made between different types of dimensions. This distinction is extremely important in tolerance analysis and control. The types of dimensions and other important concepts are defined below.

Dimension. A *dimension* is a closed interval of distances. It consists of two components: a nominal dimension and a tolerance. The nominal dimension is the mean value of the interval. The tolerance is the amount of deviation of the interval limits from the mean value. In manufacturing the nominal dimension is taken as the midpoint of the interval, and the tolerance as an equally disposed bilateral value. In this form tolerances are always added irrespective of whether a dimension is added to or subtracted from another. In addition, the tolerances are normally assumed to be at the extreme limits of the manufacturing processes and are referred to as *worst-case tolerances.* Figure 2.1 introduces a drawing convention used in tolerance charting to distinguish different types of dimension.

Design Specification Dimension. A *design specification dimension* is a dimension specified in a design drawing or model. It is sometimes referred to as a *blueprint dimension.* It is normally represented as a scalar dimension (with magnitude, but no direction), but can also be represented as a vector dimension (Johnson, 1954). In a vector representation of a design specification dimension, one of the design surfaces is taken as a location surface and the other as the end (finished) surface. Johnson introduced this concept as an aid for checking the correct input of design specification dimensions in manual charting.

Working Dimension. A *working dimension* is a dimension defined from a locating (datum) point to a machined surface. Strictly speaking, all working dimensions are vector dimensions; they have both magnitude and direction. The direction is defined from the location to the machined surface. However, a number of tolerance analysis techniques assume them to be scalar dimensions, having magnitude only. Working dimensions have different meaning in automatic and manual machining operations (Nee et al., 1995, p. 51). In automatic machining, such as computer numerical control (CNC) machining, the nominal component of the working dimension is equivalent to the programmed dimension using absolute measurement coordinates. It is the dimension from a specified machine zero to the surface being processed. If the workpiece is accurately located in a fixture, the origin for the dimension can be considered to be the location surface. In automatic machining process, the tolerance is the accuracy that is achievable by the machine for that particular operation. In manually controlled machining the working dimension has a different meaning. The nominal component is the distance measured from an existing surface to the surface being processed (the existing surface is not necessarily a location surface). The tolerance is an instruction to the machinist of the allowable deviation from the nominal dimension.

Balance Dimension. *Balance dimensions* are dimensions between workpiece surfaces that can be calculated from other dimensions. They are either intermediate dimensions that are used in manual tolerance analysis to facilitate tolerance calculations or final machined (resultant) dimensions.

Resultant Dimension. *Resultant dimensions* are the dimensions given by the finished surfaces of a workpiece. A resultant dimension is either a working dimension for the last operation producing the dimension, or a balance dimension when no further processing is done on the surfaces defining the dimension. The former situation occurs when the locating surface defines one end of the dimension and the machined surface is the other end. If the resultant dimension is not made in this way, a balance dimension can be used to calculate it. The upper and lower limits of a resultant dimension must be equal to or within the upper and lower limits of the equivalent design specification dimension. The upper and lower limits should be compared because quite often the nominal values of working dimensions are offset from the midpoint of the dimension interval to make it easier to machine the workpiece. However, notwithstanding this comment, it is also common practice to compare the resultant and design specification dimensions by equating their nominal dimensions and then checking that the resultant tolerances are less than or equal to the design specification tolerances.

Stock Removal Allowance. *Stock removal allowance* is the amount of material left after a process to allow for subsequent machining processes. Stock removal allowances for some different machining operations are illustrated in Figure 2.2.

Stock Removal Allowance Tolerance Stacks. A workpiece surface may be machined more than once. If this occurs, a tolerance stack can occur between each pair of machining cuts on the surface. The stackup from this stack affects

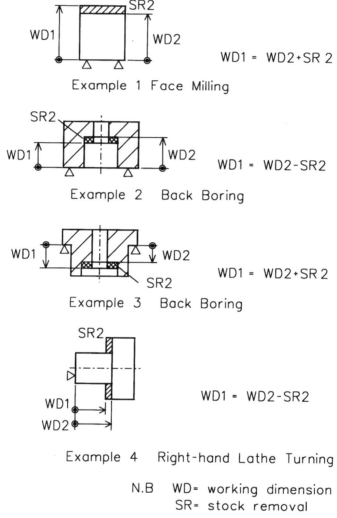

Figure 2.2. Some alternative stock removals. [From *Advanced Fixture Design for FMS* (Nee et al., 1995).]

the amount of material that is available for the second machining operation in each pair (the stock removal allowance). Figure 2.3 illustrates a simple tolerance stackup on a stock removal allowance of 2 mm. In this example tolerance stackup on the stock removal is 1.54 mm, or expressed bilaterally, ± 0.77 mm. The upper and lower limits for the stock removal allowance are therefore 2.77 mm and 1.23 mm. The lower limit is greater than zero and the stock removal provides a satisfactory amount of material for the second operation. In general, the stock removal allowance must be compared with half

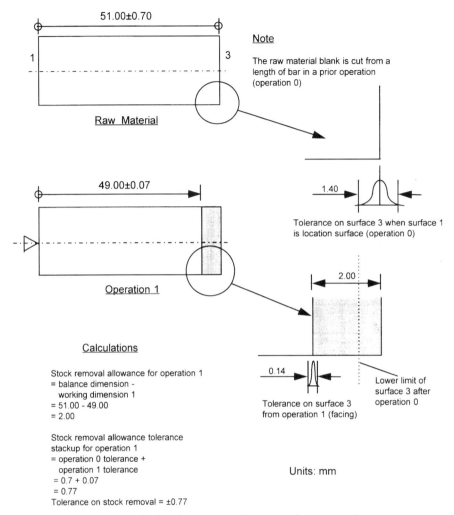

Figure 2.3. Stock removal allowance tolerance stack.

of the tolerance stackup. For the stock removal to be allowable it must be numerically greater than half the tolerance stackup (i.e., greater than the numerical value of the bilateral representation of tolerance stackup). (It should be noted that for many processes the minimum allowable stock removal is determined by the technological properties of the process and a minimum stock removal allowance that is greater than zero may be required.)

If the stock removal allowance is less than half the stackup, this does not mean that minimum stock removal is a negative quantity. It means that the high limit of the smaller dimension (second cut) and the lower limit of the larger dimension (first cut) could overlap. Should this occur there is insufficient material remaining for the second cut. It is possible to correct the situation by increasing the stock removal allowance. In some cases it may be possible to correct the situation simply by offsetting the nominal working dimension for operation 1. This second option is used for grinding allowances. When a grinding tolerance stackup is greater than the allowance, minimum and maximum values of stock allowance are calculated. The machinist is required to offset the grinding process as necessary to ensure that the actual allowance stays within these limits for each individual workpiece (Marks, 1953; Johnson, 1954).

2.4 TOLERANCE CONTROL IN MANUFACTURING

Tolerance control in manufacturing involves control of individual processes and control of tolerance stacks from a sequence of processes. These two kinds of tolerance control are discussed next. After this the relationships between tolerances and costs are discussed.

2.4.1 Control of Process Tolerances

Errors in a machining process can be caused by almost any aspect of the machining process. The sources of variation during the machining of a dimension of a workpiece are listed below.

- *Machine*
  - Clearance between moving parts of slideways and bearings
  - Geometric errors in slideways, bearings, and leadscrews
  - Dynamic stiffness
  - Resolution of the measuring system
  - Resolution of the positioning system
  - Thermal stability
- *Cutting Tool*
  - Tool wear
  - Variation of tool size and cutting geometry
  - Rigidity of the tool and support
  - Thermal stability
- *Fixture*
  - Variation between duplicate fixtures
  - Variation in location
  - Wear and contamination of locating surfaces
  - Deflection of locators and fixture
  - Thermal stability

- *Workpiece*
  - Variation in physical and chemical properties
  - Variation in workpiece size
  - Rigidity of workpiece
  - Thermal stability
  - Stress relaxation
- *Coolant*
  - Variation of flow
  - Variation of temperature
  - Contamination
  - Degradation
- *Operator.* Variations are particularly apt to occur if the finished size is under the direct control of the operator.
- *Environmental Conditions.* Changes in temperature affect the machine, fixture and tool geometry, and hysteresis in moving parts (e.g., slideways and bearings).
- *Process Variable.* Changes in process variables, such as feed and depth of cut, have a direct effect on workpiece size and geometric variation.

The dimensional variations can be classified into two groups: those that are random, unpredictable, and cannot be controlled, and those that are time dependent or capable of being controlled. Two examples of the first group are the effects of hysteresis and random variations of the chip-forming process. The second group includes the effects of tool and fixture wear over a period of time which can be measured and compensated for, or predicted by a tool wear management system (Torvinen et al., 1995).

A manufacturing process is said to be in a state of *control* if the statistical distribution of workpiece sizes does not vary over time; that is, there is no change in the mean value or standard deviation of the individual workpiece size distribution over time. The variability of a process in control is due only to random sources of variation and all assignable (knowable) sources of variation have been removed (BSI, 1985; Ford Motor Company, 1987; Juran, 1988). Statistical control charts can be used to ensure that the mean value and standard deviation are stable. These may control individual workpiece sizes (the population distribution) or the average values of small sets of samples (the sampling distribution).

Process capability is defined in Figure 2.4 as 6σ, where σ is the standard deviation of a process population distribution that is in a state of control (BSI, 1985; Ford Motor Company, 1987; Juran, 1988). It represents the best a manufacturing process can do under existing factory practices. For a machining process using cutting tools (e.g., turning or milling the process capability is defined by the statistical distribution produced by a newly sharpened tool.

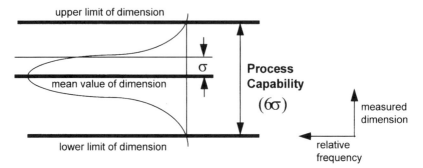

Figure 2.4. Process capability.

In practice, the variability of a manufacturing process will be greater than the process capability because of cutting tool wear, temperature variations during the day, size and geometric variations between fixtures, and machine wear (Bjorke, 1978). Cutting-tool wear has the most significant effect on variability. Figure 2.5 illustrates the changes in mean value and standard deviation produced by tool wear. We shall use the term *process tolerance* to refer to the actual overall variability of a process. The minimum value of a process tolerance is the process capability.

The maximum value of a process tolerance depends on the material being processed, the fixture repeatability, the cutting conditions, and the factory practice for controlling tool wear. The latter is controlled by controlling the time interval between tool offset adjustment or replacement. Three methods are commonly used:

1. Allow the tool to wear without adjustment, and replace the tool once the tool life or wearland reaches a predetermined value. The preset tool-

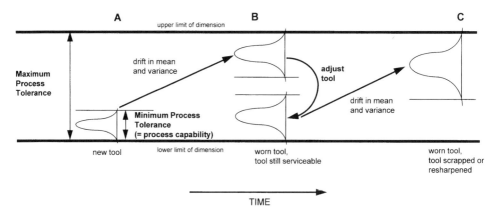

Figure 2.5. Process tolerance.

life value may be determined using tool wear or surface finish as the criterion. This option is represented in Figure 2.5 by the tool movement between points *A* and *B*. At point *B* the tool will be replaced or re-sharpened and the process reset to the condition shown at point *A*. The *A–B* cycle will then be repeated.

2. Allow the tool to wear with adjustment and replacement at predetermined time intervals using a PRE-control chart to adjust the process during these time intervals. This is represented in Figure 2.5 by the tool movement between points *A* and *C*. *A–B* represents the predetermined time interval between adjustments and *A–C* the time interval between tool replacement. The figure shows only one adjustment before the tool is replaced, but in practice several adjustments may be scheduled before replacement. Minor adjustments may also be made between points *A–B* and *B–C* due to the use of the PRE-control chart. For further details on the use of PRE-control charts, refer to Juran (1988).

3. Use a statistical control chart or a computerized tool management system to control the process and to determine when the tool needs to be adjusted or replaced; represented by the tool movement between points *A* and *C* in Figure 2.5. If a computerized tool management system is used, it is possible to correct for known sources of variation and hence to reduce the process tolerance (Torvinen et al., 1995).

Thus far, we have discussed process tolerances as if each process had a single tolerance value. In fact, this is not true. Modern machines are multi-functional (e.g., CNC machining centers). Different parts of a machine will wear at different rates and affect each function differently. The actual tolerance variation of a workpiece depends on the machine functions that are used to machine the workpiece (Torvinen et al., 1995). For example, wear in bearings supporting rotation of a machine spindle will affect tolerances during contour milling. However, there will be little or no effect during planar milling because the spindle is in a fixed position during the machining cycle. The tolerance variation for each function needs to be monitored and controlled. During process planning the appropriate tolerances on working dimensions need to be selected. This is almost always dependent on the detailed knowledge of the machine operator.

Finally, it is worth noting that some processes are surface roughness controlled rather than tolerance controlled (e.g., all finishing processes). In these cases the process tolerance is a result of the process conditions that are set to produce the required surface roughness. For processes that can produce different qualities of surface roughness there will be a set of tolerance values, with each value corresponding to a specified value of roughness. For example, consider finish turning. The surface roughness for turning can be controlled by varying the cutting speed and feedrate and/or by using a finishing tool. The final surface roughness depends on the process conditions, the finishing tool, and the control technique (Shirashi and Sato, 1990). The tolerance will also be affected by these factors.

Process control provides control of the tolerance(s) of each operation in a manufacturing sequence. Next we describe how to control tolerance stackups resulting from the sequence itself.

2.4.2   Control of Tolerance Stackup

Tolerance stackup is a direct result of the process plan and can occur for two reasons. The first and simplest type of stack occurs when two surfaces are machined from the same machining datum and the distance between these surfaces must be controlled, as illustrated in Figure 2.1. More complex tolerance stacks occur when machining datums are changed or are different from those specified by the designer. Datums are changed when indirect machining is used, a workpiece is relocated on a single machine, the measurement technique requires a new datum to be defined, or if a workpiece is machined on more than one machine.

Tolerance stackup needs to be controlled to ensure that resultant dimensions are within design specifications and that the stock removal allowances are adequate. Control of resultant dimensions is achieved as follows.

1. Identify the operations that contribute to the tolerance stackup for the resultant dimensions. Note that not all the operations in a process plan will contribute to tolerance stackup for resultant dimensions (e.g., some roughing and semiroughing cuts may not have an effect on the final dimension).
2. Sum the tolerances for these operations for each resultant dimension.
3. Compare the resultant dimensions with the design specification dimensions.
4. Change the manufacturing sequence if a design specification dimension cannot be met. A manufacturing sequence can be changed by selecting different machining datums or operations that can hold tighter tolerances. In some cases it may not be economical to meet a design specification dimension and the process planner may negotiate a change in the specification.

Each operation in a manufacturing sequence has its own requirements for stock removal allowance. If an operation cuts a surface that has already been cut, there may be a tolerance stack between the preceding and successive cuts. The tolerance stackup on the stock removal is the sum of the tolerances of the contributing dimensions between these two cuts. This stackup is not correlated in any way with the stock removal allowance required for the last cut. The allowance depends on the particular technological properties of the last operation, but the stackup depends on the sequence of operations between the previous and successive cuts on the surface.

Stock removal allowance is controlled as follows.

1. Identify all pairs of successive cuts in a manufacturing sequence.
2. Identify the operations contributing to the tolerance stackup for each pair.
3. Sum tolerances for these operations for each pair.
4. Compare the stackup with the recommended stock removal allowance for the operation producing the last cut in each pair.
5. Increase or offset the relevant working dimensions if the required stock removal allowances cannot be achieved.

Complex, high-precision parts have a relatively large number of surfaces to be machined and the machining sequence requires frequent change of machine datums. In these conditions it is impossible to achieve tolerance stackup control unless a formal technique is used. Tolerance charting is a formal technique for ensuring such control. It is based on one-dimensional chains of parallel part dimensions. Points are used to indicate the ends of each dimension in a chain. These points are critical control points that represent a workpiece's surfaces.

Ideally, a designer would like to control all points on surfaces that directly affect a part's design functionality. However, this is not practical during manufacturing because the cost of checking all points on these surfaces is prohibitively high. Instead, the designer specifies critical points that represent the surfaces. These points are considered necessary and sufficient to ensure that variation in size and form of the manufactured parts will not adversely affect design functionality.

During manufacture these designer-specified critical points, and intermediate manufacturing points, are used to define the manufacturing processes and are checked to ensure that the finished parts meet the design requirements. The dimension and tolerance checks, during manufacture, are always made between points. The check may be a single measurement (e.g., a micrometer measurement or a set of measurements (e.g., a runout check).

This critical-point technique works because it is based on well-developed theories of the mechanics of manufacturing processes, metrology, and process control. These theories guarantee that the critical points will actually provide the required degree of control during manufacture. Tolerance charting techniques use these critical points and theories to generate and check process plans prior to manufacture.

2.4.3   Cost–Tolerance Relationships

It is widely recognized that there is a strong correlation between the tolerance on a particular dimension and the cost of producing the feature to the specified dimension. Most design textbooks include diagrams similar to Figure 2.6 to illustrate this relationship, and there are many sources of data for

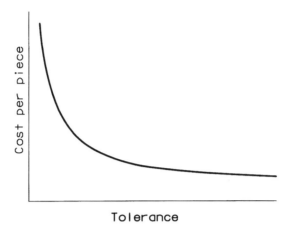

Figure 2.6. Cost–tolerance relationship used in design.

cost–tolerance relationships for finished design dimensions. This information is a valuable aid to designers in enabling them to design products that can be produced economically without unnecessarily high tolerances. However, the information provides little benefit to process planners who need to know the cost–tolerance relationship for each manufacturing operation.

Furthermore, it should not be inferred that there is a simple algebraic functional relationship between cost and tolerance. The cost of manufacture will not be reduced automatically by increasing a tolerance specification; the process planner must make some change to the process that will result in a cost reduction. Moreover, the process planner usually has several possible courses of action with which to effect a change of tolerance. If tolerance charts are to be used to optimize cost, the sources of cost in manufacture must be understood. The sources of variation in machining were itemized in Section 2.4.1. Most of these are capable of some measure of control or compensation, and there is invariably an associated cost in controlling the variation.

Some of the contributing costs of manufacture are itemized below.

Overhead Costs of the Machine. Machines of the same type do not necessarily have the same process capability. Machines capable of high accuracy are themselves manufactured to more exacting standards and are usually more expensive and hence attract higher overheads. If it is possible to schedule the work on a less accurate (lower cost) machine, production costs may be reduced.

Cost of Setup. In batch manufacture, a major contributor to the total cost of manufacturing a part is the cost of setting up for each operation. The cost comes from the time in handling the part, the waiting time, and the non-operation on the machine (Groover, 1987). Large cost savings are possible if the number of setups is reduced. If design tolerances are relaxed, the process planner can redesign the sequence of operations to eliminate finishing operations or by producing dimensioned features indirectly (i.e., increasing tolerance stacks).

Cost of Tooling. The cost of fixtures and special tools is also a major cost component in batch manufacture. Each setup usually requires additional tooling, and any reduction in the number of operations will result in reduced tooling costs. Fixtures and tooling contribute directly to the accuracy of a cutting process. Accuracy is usually a function of the stiffness of the fixture and the repeatability of location of the workpiece. For close-tolerance work, special-purpose fixtures with high rigidity are required. If the tolerance is relaxed, less rigid modular fixtures may be used at a greatly reduced cost (Nee et al., 1995).

Cost of the Process. The choice of process has the most significant effect on the process capability and on cost. There is usually an overlap between the capability of different processes and the cost tolerance relationship: for example, between grinding and lathework in Figure 2.7. Roughing as well as finishing operations are usually required to produce a finished dimension, and therefore the cost of additional set-ups must be taken into account:

$$\text{process cost} = \text{machine overhead} \times \text{operation time} + \text{tool service cost}$$

$$\text{operation time} = \frac{\text{volume of metal removed}}{\text{feed} \times \text{depth of cut} \times \text{cutting speed}}$$

Feed, depth of cut, and cutting speed also affect tool wear and hence tool service cost. In roughing operations these parameters are set to give minimum cost. For instance, in turning, depth of cut and feed are set to the maximum that the tools and machine will allow, and the speed is set to the optimum value giving the minimum total cost, as shown in Figure 2.8 (Boothroyd, 1981).

For finishing operations the overriding criteria are surface roughness and accuracy. Feed, depth of cut, and cutting speed affect the chip-forming process and surface roughness (Boothroyd, 1981; Shaw, 1986; Mital and Mehta, 1988; Shirashi and Sato, 1990) and must be selected to ensure that the specified finish is achieved. The criterion for effective tool life is usually deterioration of surface finish because of poor chip formation.

The cutting force is largely a function of feed and depth of cut:

$$\text{cutting force} = \text{specific cutting pressure} \times \text{feed} \times \text{depth of cut}$$

The deflection of the tool and workpiece and the magnitude of the random variables increase with increased force. Hence for accurate work, feed and depth of cut must be small. Process capability can be controlled by adjusting

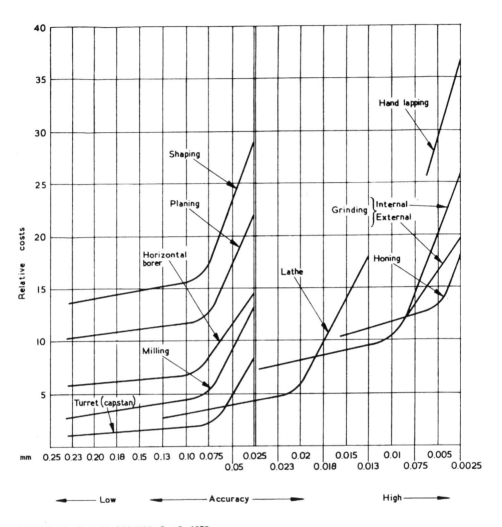

NOTE. See also figure 11 of BS 1134 : Part 2 : 1972.

Figure 2.7. Cost of various machine and hand processes for achieving set tolerances. [From *Manual of British Standards in Engineering and Design* (Parker, 1988).] Extracts from PD 6470:1981 are reproduced with the permission of BSI. Complete editions of the standards can be obtained by post from BSI Customer Services, 389 Chiswick High Road, London W4 4AL or through national standards bodies.

depth of cut and feed, but in practice, maintaining surface finish is usually an overriding consideration. Similar arguments apply to grinding and most conventional machining operations.

In batch manufacture the cost benefits possible from relaxing tolerances and changing the cutting parameters are small compared to possible savings

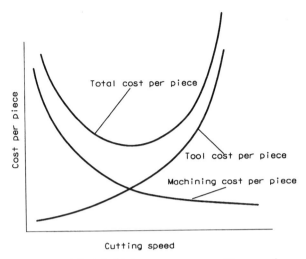

Figure 2.8. Machining cost versus cutting speed.

from the overhead, setup and fixture costs. However, in large volume (mass production) the cost reductions could be significant.

Cost of Inspection. The process tolerance is controlled by the period between sampling (Figure 2.5). With more frequent sampling of workpiece dimensions the process tolerance can be reduced by adjusting the mean dimension, by changing the tool setting, or by replacing a worn tool more frequently. There is a cost incurred by increasing the frequency of inspection. The process tolerance can be reduced to the process capability by employing 100% inspection or continuous in-process gaging, both of which will significantly increase the cost of the process.

Scrapping Costs. It is possible to maintain tolerances tighter than the process capability by means of 100% inspection and scrapping or reworking those components that fall outside the tolerance specified. This extremely expensive option will add considerable cost to the process.

2.5 TOLERANCE CHARTS

A tolerance chart is a graphical record of a process plan. It is valuable at all stages of the design–manufacturing cycle.

- *Design Function.* A tolerance chart provides formal verification that design specification dimensions can be met. It is a useful tool for satisfying the verification requirements of ISO 9001 with respect to part dimensions.

- *Process Planning Function.* A tolerance chart is a tool to plan and verify a manufacturing sequence from a dimensional and tolerancing point of view. It can also be used to negotiate changes in design specification dimensions when these cannot be met economically.
- *Manufacturing Function.* A tolerance chart is a set of instructions specifying the manufacturing sequence and the machine datums and allowable positional variations in machined dimensions at all stages in the manufacturing sequence.

The simplest form of tolerance chart layout is shown in Figure 2.10. This is the chart for the sequence of operations to manufacture the part shown in Figure 2.9. The example is adapted from an example in Wade (1967). A two-dimensional profile of the finished workpiece is drawn at the top of the chart. Lines are drawn down from selected points that represent surfaces, center-lines, or arc centers. We shall refer to these as *face lines*. These lines are used to indicate datum and machined (cut) surfaces. Design specification dimensions and the various types of operations are represented by symbols connecting two face lines. The symbols are defined in Figure 2.1.

Design specification dimensions are normally entered at the bottom of the chart. The nominal distance and bilateral tolerance are recorded on the chart. Resultant dimensions are also entered at the bottom of the chart. Machining operations are entered in the middle of the chart. Each machining operation is represented by an arrow. The datum surface of the operation is the tail of the arrow and is indicated by a small circle. The cut surface is the head of the arrow. The direction of the arrow is from the datum to the cut surface. This is an important feature of tolerance charting. A direction is assigned to all machined dimensions (i.e., the dimensions are vector and not scalar quantities). (Note that the direction of the arrow does not indicate the direction of tool movement during a cut.) Finished (last cut) surfaces are indicated by a small circle over the arrows.

Columns are drawn on the chart for recording the following manufacturing information:

- Operation number
- Working dimension (nominal dimension only)
- Operation tolerance
- Stock removal allowance
- Stock removal tolerance stackup

This is the minimum form of a tolerance chart. More complex forms suitable for manual charting calculations and the type of charts generated by our interactive computer program CATCH are presented in Chapters 3 and 17.

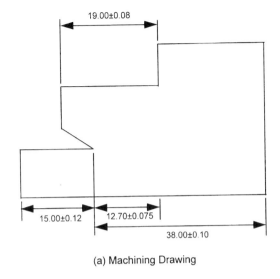

(a) Machining Drawing

Units: mm

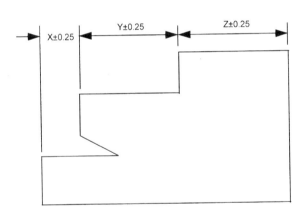

(b) Casting Drawing

Figure 2.9. Example part.

A tolerance chart is drawn in a two-dimensional plane and is used to check working and resultant dimensions in one direction only. For rotational parts a single chart per workpiece is sufficient to control tolerances along the axis of the workpiece. (There is no possibility of stackups occurring in the radial direction; therefore, a chart is not required.) For prismatic parts it is necessary to control tolerance stackups in at least three directions, so at least three charts are necessary for each workpiece. These charts will, in general, not be independent, as some surfaces, and hence tolerances, will appear on more than one chart. The charts must be linked together through the common surfaces.

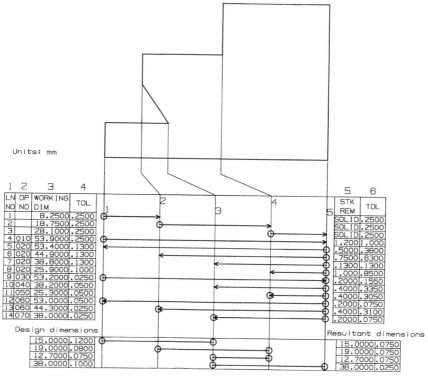

Units: mm

1 LN NO	2 OP NO	3 WORKING DIM	4 TOL	5 STK REM	6 TOL
1		8.2500	.2500	SOLID	.2500
2		18.7500	.2500	SOLID	.2500
3		28.1000	.2500	SOLID	.2500
4	010	53.9000	.2500	1.200	1.000
5	020	53.4000	.1300	.5000	.3800
6	020	44.9000	.1300	.7500	.6300
7	020	38.8000	.1300	.1300	.1300
8	020	25.9000	.1000	1.000	.8500
9	030	53.2000	.0250	.2000	.1550
10	040	38.2000	.0500	.4000	.3350
11	050	25.3000	.0500	.4000	.3050
12	060	53.0000	.0500	.2000	.0750
13	060	44.3000	.0250	.4000	.3100
14	070	38.0000	.0250	.2000	.0750

Design dimensions

15.0000	.1200
19.0000	.0800
12.7000	.0750
38.0000	.1000

Resultant dimensions

15.0000	.0750
19.0000	.0750
12.7000	.0750
38.0000	.0250

Figure 2.10. Tolerance chart for example part.

All tolerance charting procedures, whether manual or computer aided, must include:

- A means to generate, modify, and record a manufacturing sequence.
- A means to identify the operations contributing to the tolerance stacks of design specification dimensions, to calculate the resultant stackups, and to check these against the permissible design tolerances. This check is necessary to verify that the manufacturing sequence will produce dimensions that satisfy the design specification.
- A means to identify the operations contributing to the tolerance stackup between two successive cuts on the same surface, to calculate the stackup, and to check the stackup against the stock removal allowance. This must be performed for all surfaces that are cut more than once, and also for first cuts on forgings and castings to check the forging and casting allowances.
- A means to calculate the working dimensions.

A tolerance chart can represent the following (Marks, 1953; Gadzala, 1959; Wade, 1967, 1983; Britton et al., 1996):

- Normal metal cutting operations
- Heat treatment operations
- Plating operations
- Angular and diametrical cuts
- Surface finishing operations
- Welding operations
- Forming operations, such as forging and casting
- Geometric tolerances

2.5.1 Dimension Representation Schemes

The tolerance chart is only a means of recording and displaying the results of tolerance calculations. Some method of representing the tolerance stacks for both design specification dimensions and stock removal allowances is needed to calculate the tolerance stackups. Two main schemes are used: dimension chain and tree. The schemes are illustrated in Figure 2.11.

A dimension chain is a connected sequence of dimensions that contribute to the tolerance stackup of a resultant dimension. The dimensions in a chain may be scalar or vector dimensions. The chain for dimension 1–2 is a single dimension and is open. The chain for dimension 2–3 contains three dimensions: a balance dimension and two working dimensions that contribute to the tolerance stackup of the balance dimension. This chain must be closed. A technique called the *method of traces* is used to construct closed chains for manual charting (Gadzala, 1959; Wade, 1967, 1983). An explanation of tolerance stack calculations using a "schematic" based on the method of traces is provided in Chapter 3.

Several tree representation schemes have been proposed for tolerance analysis. Figure 2.11 shows a rooted tree scheme that was developed by the authors (Whybrew et al., 1990; Britton et al., 1992). Each cut surface on a workpiece is uniquely identified with a label and treated as a separate surface. These labels are the nodes on the tree. Operations are represented in the tree by lines joining the nodes representing datum and cut faces (this is a graphical representation of a vector dimension). Although the tree is a directed tree, for ease of use we do not use arrows to indicate the directions. Instead, the tree graph is read from top to bottom. This convention ensures that directionality is preserved. Each datum and cut surface appears once and once only in the tree. Successive cuts on the "same workpiece face" are differentiated by a number indicating the number of times that the face has been cut (e.g., the first cut is indicated by 01 and the second by 02). A tree has a unique starting point, the tree root, which is the datum surface for the first operation. The

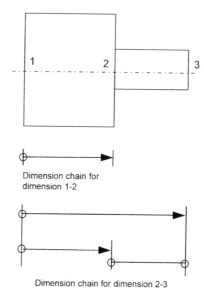

Dimension chain for
dimension 1-2

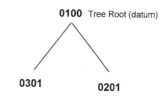

Dimension chain for dimension 2-3

(a) Dimension chain representation

0100 Tree Root (datum)

0301   **0201**

(b) Rooted tree representation

Figure 2.11. Dimension representation schemes.

root can be identified by the last two digits which are set to "00." Tolerance stackups and working dimensions can be calculated easily by using different rules to traverse the tree.

2.5.2   Tolerance Chart Calculations Using the Rooted Tree Algorithm

Our rooted tree algorithm for tolerance charting is introduced below using the example part in Figure 2.9. A full description of the computer implementation of this algorithm is given in Chapter 17. Although the algorithm can be implemented in different ways, the underlying logic and procedure are the same. Before the algorithm can be implemented a sketch of the workpiece must be drawn and the chart associated with it constructed as in Figure 2.10.

All known information is entered in the chart. The design nominal dimensions and tolerances are entered at the bottom of the chart and the resultant nominal dimensions are set the same as the design dimensions. The manufacturing sequence is entered line by line and the associated working dimensions are drawn using the convention defined in Figure 2.1. At this stage operation data are also entered in columns 1, 2, 4, and 5. The working dimensions and tolerance on stock removal are unknown at this stage, so columns 3 and 6 are left blank. The resultant tolerances are, of course, also unknown at this stage.

Concurrently with the development of the chart, the rooted tree shown in Figure 2.12 is constructed using the logic described above. The tree is related to the chart by numbers adjacent to the lines linking the nodes of the tree. These numbers are the line numbers of the tolerance chart for the operation represented by that line. Surface 1 has been chosen arbitrarily as the first datum surface, so 0100 is the root of the tree. The original casting dimensions are treated as working dimensions in lines 1, 2, and 3 of the chart and drawn as 0100–0201, 0201–0401, and 0401–0501 in the tree diagram. In the first operation, line 4, the workpiece is located on surface 1 and the new surface 5 is cut. This is represented in the tree diagram as the operation 0100–0502. In the next operation, line 5 in the chart, 0502 is the location surface and surface 0101 is cut. This is represented in the tree diagram as 0502–0101. The operations in lines 6, 7, and 8 also use surface 0502 for location and are represented as 0502–0202, 0502–0301, and 0502–0402. In line 9, surface 1 (now labeled 0101) is again used for location to finish machine surface 5. This is the final cut on surface 5 and is represented in the tree diagram as

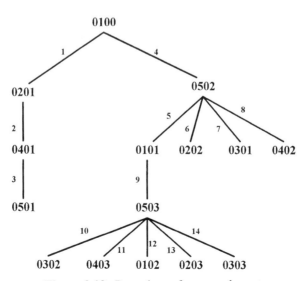

Figure 2.12. Rooted tree for example part.

0101–0503. This procedure is continued until the final cuts are made on all surfaces.

Once the manufacturing sequence has been entered and the tree diagram completed, it can be used for tolerance checking. Tolerance checking and calculation of working dimensions are performed in the following order.

1. *Check the resultant tolerances.* Each tolerance in the resultant column is calculated using the tree and tracing the path between the two surfaces defining the resultant nominal dimension. The tolerances of the working dimensions in this path are added to give the resultant tolerance. For example, the path between surfaces 0303 (line 14 on the chart) and 0403 (line 11) is 0303–0503–0403 and the resultant tolerance is the sum of the tolerances of the working dimension in the path (i.e., 0.025 + 0.050 = 0.075 mm).

2. *Calculate the tolerances for stock removal.* This is calculated using the tree and tracing the path between one cut on a surface and a prior cut. The tolerances of the working dimensions in this path are added to give the tolerance stackup for stock removal. For example, the stock removal tolerance stack for surface 0403 is the chain of working dimensions in the path between 0403 and 0402, which is 0403–0503–0101–0502–0402. From the tree diagram, the relevant contributing operations are line numbers 11, 9, 5, and 8. The tolerance stackup is 0.050 + 0.025 + 0.130 + 0.100 = 0.305 mm. This number is entered in column 6 for the operation cutting 0403 (line 11).

3. *Calculate the working nominal dimensions.* These are calculated by setting the finished nominal dimensions equal to the final dimensions of the part. Dimensions for other operations are calculated by adding or subtracting the appropriate stock removal allowance. For example, the working dimension defined by surfaces 0503 and 0102 is set to the finished length of the part, that is, 53.000 mm (line 12). The dimension defined by surfaces 0101 and 0503 equals the previous dimension plus the stock removal allowance for 0503, that is, 53.000 + 0.200 = 53.200 mm (line 9).

2.6 HISTORY OF TOLERANCE CHARTING

The first tolerance charts were produced manually. In the early days (1950s and 1960s), process planners were concerned with developing techniques for representing different manufacturing operations and speeding up the manual calculation of tolerance stackups and working dimensions. Marks (1953) briefly describes a manual method of charting in which the design specification and resultant dimensions are entered at the bottom of the chart. Most of the paper explains the need for stock removal allowances and how to calculate and check the stock removal tolerance stackups. Special emphasis is placed on calculating the stock allowance for grinding operations because:

- The stock removal tolerance stackups are often more than the desired allowances for these operations.
- Grinding is often performed after case hardening. When this happens it is important to ensure that the depth of cut is strictly controlled to prevent too much case-hardened material from being removed. It is normal practice to specify a minimum and a maximum stock removal allowance for grinding. The minimum value guarantees that there is sufficient material to clean up the surface and the maximum value ensures that the depth of cut is not too large.

Johnson (1954) outlines a complete method for manual charting using balance dimensions to reduce the amount of effort required to trace the tolerance stack chains for resultant dimensions and stock removal allowances. His general procedure is to calculate balance, resultant, and stock allowance tolerance stackups first. When these are satisfactory the working dimensions are calculated. The paper emphasizes the need to incorporate good machining practice on the chart. Johnson uses the same rule as Marks (1953) to calculate the minimum and maximum stock allowances for grinding operations. In addition, he outlines how to deal with polishing, lapping, plating, heat treatment, and machining operations in which a constant amount of material must be removed. A unique feature of Johnson's paper is that he uses an indexing technique to represent dimensions as vector quantities. It is very easy to check whether all the design specification dimensions have been entered correctly using the index numbers.

Mooney (1955) shows how to relax tolerances on working dimensions, taking into account practical manufacturing considerations. He also shows how to combine a number of single operations into one multitool operation. Anderson (1956) shows how to use tolerance charts for parts that are assembled during manufacture and machined after assembly. Wood (1957) discusses the relationship between an X-bar statistical process control chart and a tolerance chart. He shows that the X-bar chart controls a single operation (one line) on the tolerance chart. He argues that the tolerance chart provides the specification for control, while the X-bar chart provides the means for control.

Gadzala (1959) and Wade (1967, 1983) discuss, extensively, how to perform manual tolerance charting. They use the same technique, which they call the *method of traces,* for identifying tolerance chains. However, there are slight differences between their approaches. Wade bases his calculations on balance dimensions, whereas Gadzala avoids their use wherever possible. Gadzala uses a bottom-up technique for calculating working dimensions. Wade uses both bottom-up and top-down techniques. The latter can be used when the overall dimensions of the starting (initial) workpiece are known.

These two authors also show how to perform tolerance calculations for many different kinds of manufacturing processes and how to include geometric tolerances on charts.

Eary and Johnson (1962) give a cursory treatment of tolerance charting. However, they provide a detailed framework and rules for selecting machine datums and controlling workpieces during machining.

Manual charting is time consuming and tedious. Typically, it takes between 1 and 4 hours to produce a small chart. A large chart can take several days. Unless a process planner has considerable experience, mistakes are easily made and difficult to detect. For these reasons, and despite the seemingly obvious advantages, tolerance charts have only gained acceptance in companies where the cost of manufacturing errors and rework are exceptionally high (e.g., aerospace and automotive companies). The growth of interest in computer-aided process planning and computer-aided design of machining fixtures has placed emphasis on the importance of tolerance analysis in process planning (Sermsuti-Anuwat, 1992; Sermsuti-Anuwat et al., 1991, 1995; Sack, 1982). There is renewed interest in tolerance charting and in particular computer-aided tolerance charting.

This short history of tolerance charting is continued in Chapter 17, where the authors' computer-aided tolerance charting system CATCH is introduced.

ACKNOWLEDGMENTS

The authors gratefully acknowledge the support given by Sundstrand Pacific Aerospace Pte. Ltd. (Singapore), especially Eric Tan Hung Heng, section manager of manufacturing engineering, and the process planners in manufacturing engineering. We have gained considerable practical working knowledge of tolerance charting and process planning from our association with Sundstrand. We also wish to acknowledge Yongyooth Sermsuti-Anuwat, who collaborated with us at the University of Canterbury to develop the rooted tree graph technique. He also wrote the first computer program to use this technique for tolerance charting.

GLOSSARY

chain, dimension (or tolerance): a connected sequence of dimensions (or tolerances) that contribute to the tolerance of a dimension; a chain may be open or closed

dimension: a closed interval of distances normally specified by two components: a nominal dimension and a tolerance

dimension, balance: a dimension calculated from a working dimension chain that is used to speed up computations during manual tolerance charting

dimension, design specification: a part dimension specified by a designer, normally specified as scalar dimension but can also be considered as a vector dimension; sometimes referred to as a blueprint dimension

dimension, nominal: a mean value of a dimension

dimension, resultant: a finished dimension produced by a manufacturing sequence which must be within the design specification dimension

dimension, scalar: a dimension that has magnitude but no direction

dimension, vector: a dimension that has both magnitude and direction

dimension, working: a dimension produced by a manufacturing process; it is a vector dimension

face, cut: a surface of a workpiece that is cut or modified by a manufacturing process

face, datum: a surface of a workpiece that is used as a datum for controlling the dimension produced by a manufacturing process

face line: a line drawn on a tolerance chart to represent a surface of the workpiece

manufacturing sequence: an ordered sequence of manufacturing processes

operation: one or more manufacturing processes that are performed at a specified machine in a single setup; in tolerance charting the term *operation* is used both for single processes and combined processes

part: a finished product

process capability: a 6σ variation of workpiece dimensions produced by a manufacturing process that is under statistical process control

process tolerance: the actual variation of workpiece dimensions produced by a manufacturing process

stock removal allowance: the amount of material that is left on a workpiece to allow for further material removal processing

tolerance: the component of a dimension that specifies the amount of deviation from the nominal dimension; it is the difference between the upper and lower limits of the dimension and is normally expressed as an equal bilateral value

tolerance chart: a graphical record of a manufacturing sequence that facilitates checking of tolerance stacks

tolerance, resultant: the tolerance of a resultant dimension produced by the tolerance stack of all working dimensions contributing to the resultant dimension

tolerance stack: a chain of dimensions contributing to the total tolerance of a dimension

tolerance stackup: the total tolerance resulting from a tolerance stack

tolerance stackup, stock removal: the tolerance stackup from a tolerance stack between two machining operations on the same workpiece surface

tree, dimension: two or more dimension chains connected together; it may have branches

tree, directed dimension: a dimension tree composed solely of vector dimensions

tree, node: the intersection of two or more branches of a tree

tree, rooted dimension (rooted tree): a directed dimension tree that has a unique starting point, the root

tree, scalar dimension: a dimension tree composed solely of scalar dimensions

workpiece: a blank of material that is to be machined or processed

REFERENCES

Anderson, J. F., Jr., 1956, Assembly Tolerance Charts Save Time and Money, *Tool Eng.,* Vol. 37, No. 6, pp. 95–97.

Bjorke, O., 1978, *Computer-Aided Tolerancing,* Tapir, Trondheim, Norway.

Boothroyd, G., 1981, *Fundamentals of Metal Machining and Machine Tools,* McGraw-Hill International, Auckland, New Zealand.

Britton, G. A., Whybrew, K., and Sermsuti-Anuwat, Y., 1992, A Manual Graph Theoretic Method for Teaching Tolerance Charting, *Int. J Mech. Eng. Ed.* Vol. 20, No. 4, pp. 273–285.

Britton, G. A., Whybrew, K., and Tor, S. B., 1996, An Industrial Implementation of Computer-Aided Tolerance Charting, *Int. J. Adv. Manuf. Techol.,* Vol. 12, No. 2, pp. 122–131.

BSI, 1985, *Quality Management System: Quality Control,* BSI Handbook 24, British Standards Institution, London.

Eary, D. F., and Johnson, G. E., 1962, *Process Engineering for Manufacturing,* Prentice Hall, Upper Saddle River, NJ.

Ford Motor Company, 1987, *Continuing Process Control and Process Capability Improvement,* Ford Motor Company, Dearborn, MI.

Gadzala, J. L., 1959, *Dimensional Control in Precision Manufacturing,* McGraw-Hill, New York.

Groover, M. P., 1987, *Automation, Production Systems, and Computer Integrated Manufacturing,* Prentice Hall, Upper Saddle River, NJ.

Johnson, A., 1954, Index Tolerance Chart Simplifies Production, *Tool Eng.,* Vol. 32, No. 2, pp. 53–62.

Juran, J. M. (ed.), 1988, *Juran's Quality Control Handbook,* 4th ed., McGraw-Hill, New York.

Marks, C. J., 1953, Tolerance Charts Control Production Machining, *Am. Mach.,* Vol. 97, No. 5, pp. 114–116.

Mital, A., and Mehta, M., 1988, Surface Finish Prediction Models for Fine Turning, *Int. J. Prod. Res.,* Vol. 26, No. 12, pp. 1861–1876.

Mooney, C. T., 1955, How to Adjust Tolerance Charts, *Tool Eng.,* Vol. 35, No. 4, pp. 75–81.

Nee, A. Y. C., Whybrew, K., and Senthil kumar, A., 1995, *Advanced Fixture Design for FMS,* Springer-Verlag, London.

Parker, M. (ed.), 1988, *Manual of British Standards in Engineering and Design,* British Standards Institution in association with Hutchinson, London.

Sack, C. F., Jr., 1982, Computer Managed Process Planning: A Bridge Between CAD and CAM, *AUTOFACT 4,* Philadelphia, PA, November–December, pp. 7.15–7.31.

Sermsuti-Anuwat, Y., 1992, Computer-Aided Process Planning and Fixture Design (CAPPFD), Ph.D. thesis, University of Canterbury, Christchurch, New Zealand.

Sermsuit-Anuwat, Y., Whybrew, K., and Britton, G. A., 1991, Some Recent Developments in Tolerance Charting and Its Role in CIM, *Proceedings of the First International Conference on Computer Integrated Manufacturing,* Singapore, October 2–4, pp. 285–288.

Sermsuti-Anuwat, Y., Whybrew, K., and McCallion, H., 1995, "CAPPFD: A Tolerance Based Feature Sequencing CAPP System, *J. Syst. Eng.,* Vol. 5, No. 1, pp. 2–15.

Shaw, M. C., 1986, *Metal Cutting Principles,* Clarendon Press, Oxford.

Shirashi, M., and Sato, S., 1990, Dimensional and Surface Roughness Controls in a Turning Operation, *J. Eng. Ind.,* Vol. 112, pp. 78–83.

Torvinen, S., Andersson P. H., and Vihinen, J., 1995, Monitoring the Accuracy Characteristics of the Machinery by Using a Dynamic Measurement Approach, *Proceedings of the 3rd International Conference on Computer Integrated Manufacturing,* Singapore, July 11–14, pp. 1435–1442.

Wade, O. R., 1967, *Tolerance Control in Design and Manufacturing,* Industrial Press, New York.

Wade, O. R., 1983, Tolerance Control, Chapter 2 in *Tool and Manufacturing Engineers Handbook,* Vol. 1, *Machining,* T. J. Drozda and C. Wicks (eds)., Society of Manufacturing Engineers, Dearbon, MI.

Whybrew, K., Britton, G. A., Robinson, D. F., and Sermsuti-Anuwat, Y., 1990, A Graph-Theoretic Approach to Tolerance Charting, *Int. J. Adv. Manuf. Technol.,* Vol. 5, pp. 175–183.

Wood, W. K., 1957, Tolerance Charts Aid Dimensioning for Machining, *Am. Mach.* Vol. 101, No. 27, pp. 81–84.

3

MANUAL TOLERANCE CHARTING

W. B. KRAG

Technical Memory Consulting, Inc.

3.1 INTRODUCTION

The purpose of this chapter is to describe the steps necessary to manually construct a tolerance chart for planning a manufacturing process. Although this technique is not necessarily "advanced" (it has been adequately explained in the literature for some time), the previous publications are fairly complex and difficult to understand. It is hoped that this simplified explanation serves to give a person new to this technique a feeling for what is occurring and why this methodology is so critical for accurate process planning. For further detail about this technique, the reader is encouraged to study references Wade (1967, and 1983). Because the process designer must work closely with the product designer in developing a manufacturable solution as proven by a tolerance chart, this methodology is a key integrating approach for effective simultaneous engineering. In short, they both become integrated together as they collaborate on developing a single tolerance chart which proves that a part can be manufactured with specified equipment—right the first time.

3.2 DEFINITIONS AND APPLICATIONS

3.2.1 What Is Tolerance Charting?

Tolerance charting is a *precise mathematical technique* for planning a *manufacturing process* for *mechanical piece parts in process* having *successive*

Advanced Tolerancing Techniques, Edited by Hong-Chao Zhang
ISBN 0-471-14594-7   © 1997 John Wiley & Sons, Inc.

cuts and processes (etching, heat treating, plating, etc.). Let's take this definition and discuss it further.

Precise Mathematical Technique. Tolerance charting is a scientific technique for calculating (not guessing) in-process dimensions and tolerances for each step in a manufacturing process. Historically, process planners typically guess at in-process dimensions and tolerances and further, more likely than not, bluff the production environment by setting tighter tolerances than are scientifically required. This results in enormously costly launches, overqualified equipment, many potentially unnecessary engineering changes, and continual contention between product design and manufacturing engineering communities. The tolerance chart mathematically connects the finished dimensions and tolerances with all in-process dimensions and tolerances throughout the process back to the "as received" state (casting, forging, raw stock). Thus a tolerance chart provides the one document that both product designers and manufacturing engineers can utilize to solve tolerance problems analytically before production is launched. It can be used both as a planning tool and later as a diagnostic tool for solving very hidden problems that can have tremendous time, cost, and quality implications. This technique should be used as an important base planning tool for concurrent engineering initiatives.

Manufacturing Process. Manufacturing processes have been classified (Todd et al., 1994) into those that modify geometry and those that modify properties. Within geometric modification processes are those that conserve, reduce, or increase the mass of a piece part. These major classification categories have then been exploded into approximately 75 specific families of processes, including such operations as casting, molding, forging, forming, shearing, piercing, cutting, welding, soldering, and many chemical processes as well. As geometry is altered in these processes, most benefit from process planning using the tolerance charting methodology. Within the general classification of property modification processes are those that involve heat treating and surface finishing. As both these major categories can and most often either distort or add very small layers to a piece part and thereby influence part dimensions during the manufacturing process, they can also benefit from using the tolerance charting methodology to calculate in-process dimensions and tolerances. Thus there is a potentially very wide application of tolerance charting as a key methodology for designing most processes.

Mechanical Piece Parts. Mechanical piece parts include solid items that are successively processed until the desired finished state is achieved. This includes both the metallic and nonmetallic categories of parts, including plastics. Excluded from this universe are liquid products.

In-Process Dimensions. Most tolerancing analysis techniques allocate (worst-case or probabilistically) final assembly tolerances over the components within an assembly, resulting in tolerances assigned to each finished piece part within the assembly. This is all done at the product's finished state. (It is interesting to note that tolerance charting formats can be used for this allocation.) Now it is necessary to take each finished piece part and calculate the in-process dimensions, tolerances, and related stock removals for each step in the manufacturing process. This is the unique role of tolerance charting.

Successive Cuts and Processes.   A manufacturing process adds value. Some manufacturing processes are very short and form finished parts in one step (molding, stereolithography, one-step press forming, etc.) These typically do not need a tolerance chart for planning purposes unless following operations lead to significant dimensional change. Longer processes requiring successive process operations accumulate more setup variation and change datums as more dimensional varying steps are utilized. In these cases, tolerance charting technique provides a very essential analytical methodology which enables a planner to mathematically test the degree of excellence of any particular process design.

3.2.2   Applications of Tolerance Charting

With increases in product complexity, precision, and the number of process sequence steps tolerance charting becomes more and more essential for planning technically correct manufacturing processes. This is illustrated in Figure 3.1.

3.2.3   Brief History of Manual Tolerance Charting

Tolerance charting was utilized initially during World War II in designing processes for demanding mechanical products such as compact gearboxes,

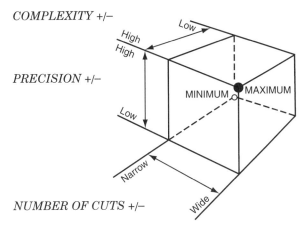

Figure 3.1. Tolerance charting cube. Tolerance chart application necessity grows as the location within the product/process cube approaches maximum.

aircraft reciprocating and jet engines, rapid-fire mechanisms, and high-volume machining lines. These application areas are still very valid. Industries that utilize tolerance charting vary from large corporations manufacturing aircraft engines, offroad vehicles, and automobiles, to NC-based machine shops.

Although knowledge about tolerance charting is in the public domain, tolerance charting is very often considered a proprietary methodology that provides a distinct advantage in being able to design world-class processes right the first time using in-place manufacturing equipment. As such, little is shared about this technique aside from the careful documentation provided by the late Oliver Wade (Wade, 1967) and his subsequent articles published by the Society of Manufacturing Engineering (SME), Dearborn, Michigan (e.g., Wade, 1983). A few other authors have also documented this technique and included tolerance charting or process charts in chapters in their books (Nicks, 1982; Curtis, 1983; Liggett, 1993) but have not devoted entire books to the subject. Several attempts have been made to computerize the technique, but none has resulted in commercially available products. A few companies have developed their own proprietary charting programs and are utilizing them in expert system-driven computer-aided process planning systems (CAPPs) for focused ranges of product types. The U.S. government has for many years been trying to evolve computer-based tolerance charting, working with companies that regularly utilize the technique, but the downsizing of the 1990s has curtailed these initiatives. Occasionally, the academic communities in the United States, Europe, and Far East (Tang et al., 1993) have initiated programs to computerize tolerance charting, but these also have typically resulted in efforts that have been fairly short lived and have had limited application.

3.3   BASIC RULES AND SYMBOLS USED IN DEVELOPING A TOLERANCE CHART

- *Bilateral Tolerances.* All dimensions and tolerances are adjusted and expressed as *equal, bilateral* tolerances; for example:

    1.500 ± 0.002   (not 1.499 + 0.003/−0.001)

- *Tolerances Are Always Added.* During the mathematical buildup of a tolerance chart, as is the case during a manufacturing process, tolerances continually stackup, that is, they are additive. Therefore, even though dimensions may be subtracted to achieve resultant lengths, their corresponding tolerances are always added. For example:

    1.500 ± 0.002 minus 1.000 ± 0.003 equals 0.500 ± 0.005

- *Bar End: -/---.* A line with a cutoff symbol indicates that the bar is faced off—normally used for chucked work.

- *Machining Dimension:* ●———▶. A line with an arrow at one end and a dot at the other is a machining dimension. The dot denotes the plane from which the cut is measured. The arrow denotes the surface that is being cut (or ground).
- *Location Plane:* ◆. The triangle whose vertex butts up against a plane indicates a datum for locating a part. Work done between centers is shown located by two triangles butting each end of the part.
- *Balance Dimensions:* ●———●. Resultant dimensions are balance dimensions. In tolerance chart form they are represented by a line with a dot at each end (barbell). In all cases in a tolerance chart the balance dimension is calculated by adding or subtracting known machining dimensions from *above* it in the chart (never below it). The three basic balance cases that form the foundation for most charting (see Figure 3.2) can be described as follows:

  Case A. Both cuts in this case begin from the same reference surface.

  Case B. The additive result of the first two cuts results in the overall length. The balance dimension represents the mathematical addition of the two cuts. Subtracting the final part width from this balance dimension at the bottom results in the stock removed.

  Case C. The reference surface is near the midpoint of the part. The balance dimension at the bottom is the additive result of the two cuts.
- *Altered Balance Dimension:* ▣———▣. The balance dimension symbol with a dot at each end enclosed within a box is used to highlight the fact that the dimension has been altered by some other operation such as heat treat or grinding angled surfaces when corner points move as stock is removed.
- *Plating:* ■. When plating is applied, this symbol is added to the "air" side of the surface to represent the thickness of the plate. Dashed lines

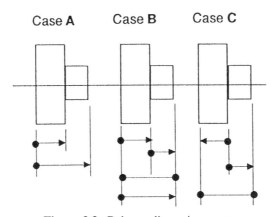

Figure 3.2. Balance dimension cases.

are dropped down from the "air" side of the rectangle to assist with altering calculated dimensions in the correct direction.

3.4   CONSTRUCTING A TOLERANCE CHART

3.4.1   Overall Approach

The overall approach for constructing a tolerance chart consists of four steps:

1. *Determine the process sequence.* Based on the available machinery, determine the basic sequence of operations, including the datums (see Figure 3.4).
2. *Initialize the chart.* Initialize the tolerance chart with all known information (see Figure 3.5).
3. *Balance the dimensions and stock removals.* Determine balance dimensions using skematics for close-looping previous cuts and determine stock removal dimensions and tolerances.
4. *Allocate the tolerances.* Determine in-process tolerances across specific manufacturing cuts.
5. *Calculate the dimensions, compare, and optimize.* Calculate the mean dimensions throughout the process and compare resulting dimensions and tolerances with final piece-part design requirements. Optimize the process as required.

The fact that dimensions are determined as a *final* step may frustrate process planners who are accustomed to working with dimensions first, treating tolerances as an add-on. Nevertheless, patience is important because in this case, working with the tolerances *first* results in mathematically correct process designs. Also, experience has shown that at the end of this procedure the in-process dimensions are relatively easy to calculate ("drop in") after all the prerequisite steps have been completed.

3.4.2   Determining the Process Sequence (Step 1)

Complete familiarity with the part (Figure 3.3) is the first task. Next, a rough process sequence is developed by the process designer (Figure 3.4). This is a very important step, and subtle differences in cut or operation sequence will result in significantly different results. Some sequences require more setups, and this causes more setup variation and additional datums to enter into the process. In this case, significantly tighter in-process tolerances may be required to be able to meet final piece-part requirements. By contrast, operation sequences requiring few setups and which utilize fewer datums result in simpler processes which can allocate wider tolerances to key manufacturing steps

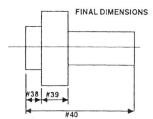

Figure 3.3. Basic schematic for the part to be manufactured.

and still meet final piece-part requirements. Thus the determining of process sequence is not a trivial exercise. Elsewhere in this publication you will find approaches to optimizing sequence such as the chain and graphical methods. Additionally, the tolerance chart is an excellent format for comparing two different process sequences to determine which is best. The chart can also be used in this way to justify the procurement of new equipment to assure capabilities are within requirements.

3.4.3   Initializing the Tolerance Chart (Step 2)

The initial tolerance chart (Figure 3.5) includes an exaggerated sketch of the part at the top of the chart. The exaggerated sketch is elongated axially to separate surfaces for easier charting and clarity. If the part is made from a casting or a forging, the forged/cast outline should be superimposed around the machined part with dashed lines. Below this part sketch vertical lines are drawn down from each part surface to be cut. At the bottom of the chart, resultant dimensions are indicated which match the final piece-part drawing requirements. On the left side of the chart, columns are used to number chart lines and operation cuts, reference to the revision level of any issued shop instruction sheets, machine names, and "machine to" dimensions and bilateral tolerances. To the right, columns are used to record balance dimensions and tolerances, stock removal dimensions and tolerances, and the chart line numbers utilized for each calculation. Although the charting examples used herein do not add further columns on the right, this can be a convenient place to indicate machine capability for each process step and resulting calculated CPK ratios. In this way, both product design and manufacturing engineering can easily spot problem areas in the emerging process design and make adjustments to the product design or the process design, improve the capability of the machine, or a combination of all of these—well before launch.

The tolerance chart, vertically, includes four regions as follows:

1. *Top Area:* piece-part exaggerated (elonagated if necessary) skematic showing each plane and, if applicable, showing the dashed outline of a

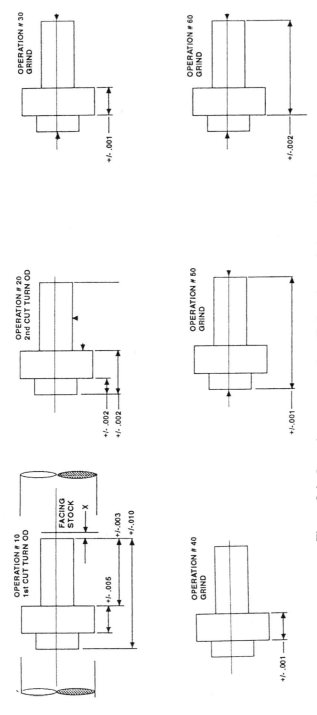

Figure 3.4. Operation sequences (usually based on existing machines).

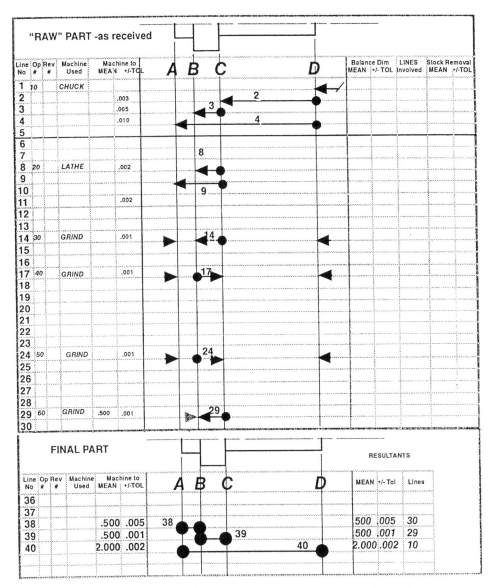

Figure 3.5. Initial tolerance chart showing datums, cuts, and final tolerance requirements.

casting or forging. Each plane is given a letter for reference (i.e., A, B, C, and D).

2. *Primary Area:* initial cuts that rough-form the piece part. In this region (lines 1 to 5) the part is being formed from a solid, so no stock removal dimensions/tolerances apply.

3. *Secondary Area:* for each secondary (or following) cut, stock removals must be determined. In some cases, minimum machining amounts are required by the process. In other cases the stock removals are the resultant of preceeding operations.

   (Balance dimensions are determined and located throughout both the primary and secondary areas.

4. *Final Product Area:* at the bottom of the chart (below a separation line) the final product piece-part dimensions and tolerances are noted at the left along with skematic line representations for the length of each finished dimension (barbell format). To the right is a summary table of the final dimensions and tolerances resulting from the tolerance chart calculations above. Each time a chart is completed, the "resultants" (below the line, on the right) are compared with the requirements issued by product design (below the line, on the left). If or as modifications are required either to eliminate negative tolerances or take full advantage of product design tolerances, the chart can be reworked from bottom to top.

3.4.4   Determining the Balance Dimensions and Stock Removals (Step 3)

Schematic Diagram Development for One Cut (Refer to Figure 3.6). The development of stock removals depends on the development of balance dimensions and tolerances for each secondary cut. This is accomplished using schematics to create closed-loop circuit diagrams working from the desired cut *up* to all known previous cuts. For example, building a schematic for cut 9 in Figure 3.5 proceeds as follows:

1. Sketch the vertical planes: *A, B, C, D.* Label each plane at the top. Near the bottom of the schematic, draw in cut 9 (bullet on plane *C,* arrow touching plane *A*).
2. Starting from the arrow end first, trace up until the stock condition is encountered. Go *up* the chart until the first arrowhead is hit (line 4). Draw in line 4 (arrow at *A,* bullet at *D*). Proceed up the chart until the next arrowhead is hit—line 1, which is the stock condition.
3. Go *up* the chart from cut 9's bullet end looking for the first arrowhead. This is line 2. Draw in line 2 (arrow at *C,* bullet at *D*). This closes the diagram and one can determine balance line 5, the result of line 4 minus line 2 (tolerances added).

The repetitive pattern is to combine the *first two cuts* at the top of each schematic to form the first balance line. Then using the *first balance line and the next cut,* create the next balance line. This balance line and the next cut is used to generate the next balance line. This balance line and the next cut

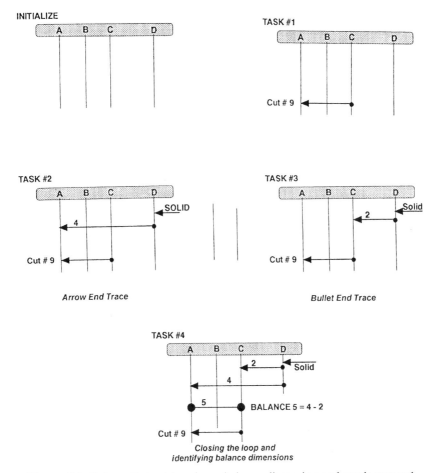

Figure 3.6. Schematic to determine a balance dimension and stock removal.

generate the next balance line, and so on. In short, the first balance line results from two cuts. All subsequent balance lines result from the just determined balance line and a single cut.

3.4.5 Determining the Balance Tolerances and Stock Removal Tolerances (Step 3 - continued)

Schematics to Develop All Tolerances for Balance Dimensions and Cuts. Using the initial tolerance chart (Figure 3.5) together with the schematics for each secondary cut (Figure 3.7), specific tolerances can now be calculated as described below. Note the difference between tolerances that result from balance dimensions and *stock removals* that result when "cuts" are involved (Figure 3.8):

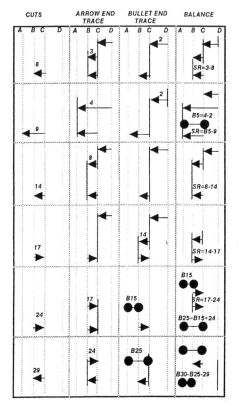

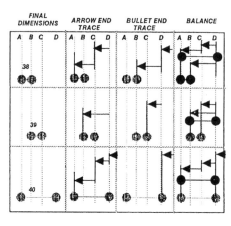

Figure 3.7. Schematics to determine balance dimensions and stock removals for *all* cuts in tolerance chart illustrated in Figure 3.4.

Primary Cuts

Cuts from solid stock: 1, 2, 3, 4.

Balance dimension 5 (*A–C*): equal to cut 4- cut 2.

Therefore, enter "Solid" in the *stock removal* column.
Tolerances 0.010 + 0.001 = 0.011

Secondary Cuts

Stock removal: line 8: cut 3- cut 8.

Stock removal: line 9 = balance 5 − cut 9.

Balance 10 (*A–D*): equal to cut 2 + cut 9.

Stock removal: cut 14 (*B–C*) = 8 − cut 14.

Balance 15 (*A–B*): equal to cut 9 − cut 14.

Tolerances: 0.010 + 0.003 = 0.013
Tolerances: 0.011 + 0.001 = 0.012

Tolerances: 0.001 + 0.001 = 0.002

Tolerances: 0.003 + 0.002 = 0.005

Tolerances 0.001 + 0.002 = 0.003

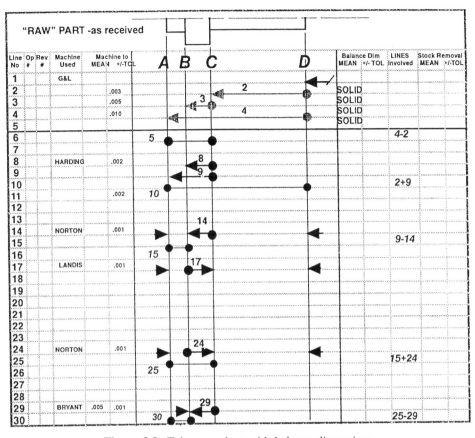

Figure 3.8. Tolerance chart with balance dimensions.

Stock removal: 17 = cut 14 − cut 17.

Tolerances 0.002 + 0.002 = 0.004

Stock removal: 24 (B–C) = cut 17 − cut 24.

Tolerances: 0.002 + 0.001 = 0.003

Balance 25 (A–E): equal to balance 15 + cut 24

Tolerances 0.003 + 0.001 = 0.004

Stock removal: Cut 29 (B–C) = cut 24 − cut 29.

Tolerances 0.001 + 0.001 = 0.002

Balance 30 (A–B): equal to balance 25 − cut 29.

Tolerances 0.004 + 0.001 = 0.005

Now read the final product requirement resultant lines below and post the corresponding values from the calculations above:

Resultant 38 (A–B) is equal to line 30 having a tolerance of 0.005.

Resultant 39 (*B–C*) is equal to line 29 having a tolerance of 0.002.

Resultant 40 (*A–D*) is equal to line 10 having a tolerance of 0.002.

3.4.6   Allocating the Tolerances (Step 4)

Arithmetic Allocation. The final step prior to completing a tolerance chart is to determine the mean working dimensions that will be broadcast to the shop at each manufacturing operation. Thus this step establishes the foundation for quality processing. This step works *up* from the bottom of the chart using the final product dimensions and tolerances. Referring to the schematics in Figure 3.9, each diagram begins at the bottom with the final product requirements.

- *Schematic for Final Dimension 38.* Final product requirement is ±0.005. This needs to be allocated above to cuts 24 and 25. An initial allocation might be ± 0.0025 to each.
- *Schematic for Final Dimension 39.* Final product requirement is ±0.001. This needs to be allocated above to cuts 24 and 29—perhaps as a first approximation using ± 0.0005 each.

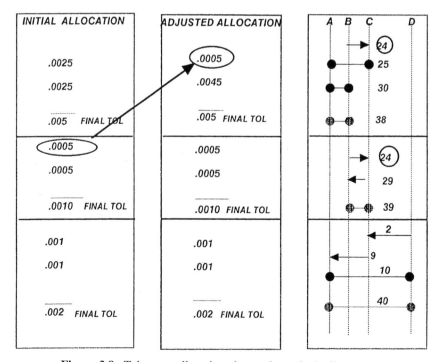

Figure 3.9. Tolerance allocations by mathematical adjustment.

• *Schematic for Final Dimension 40.* Final product requirement is ±0.002. This needs to be allocated above to cuts 2 and 9—perhaps as a first approximation using ± 0.001 each.

Now one looks for common dimensions across these schematics and notices that cut 24 occurs in schematics 38 and 39: Each common cut can only be allowed the size of the tolerance of the tightest condition. Therefore:

Cut 24 must all be ±0.0005.

Other cuts (i.e., 25) can be opened up to 0.0045.

Proportional Allocation. Recognizing that different manufacturing operations have greatly differing basic capabilities, the allocations above should be reallocated next, recognizing these different inherent levels of precision. Sta-

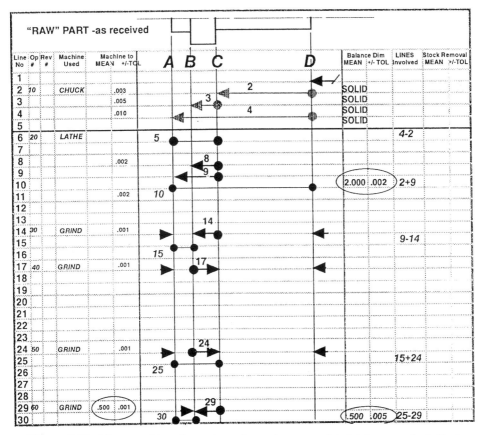

Figure 3.10. Mean working dimensions shown in a completed tolerance chart. Post final dimensions and tolerances within the chart: cuts on the left, balance dimensions on the right.

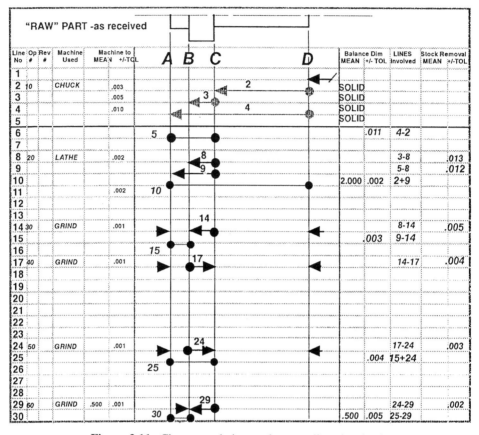

Figure 3.11. Chart completiont tasks enter lines involved.

tistical machine capability information would make this process even more accurate.

3.4.7   Completing the Chart

Figures 3.10 to 3.14 show the process leading up to the completed tolerance chart (Figure 3.15). It remains to post the final dimensions to the "resultant" area on the lower right corner and to compare these results with the final product requirements. In our example:

Line 38:   Results from balance line 30   Dimension 0.500   Tolerance 0.005
Line 39:   Results from cut 29   Dimension 0.500   Tolerance 0.001
Line 40:   Results from balance 10   Dimension 2.000   Tolerance 0.002

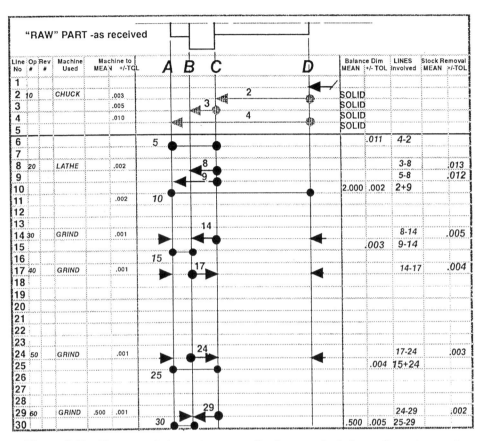

Figure 3.12. Chart completion tasks: enter all tolerances for balance dimensions and stock removals (calculated as outlined in Section 3.4.5).

Thus all the resultant dimensions and tolerances match the final product requirements. At this point, the process designer can assure the product designer that the part can be manufactured as intended and can also issue in-process shop instruction sheets and/or NC programs having the in-process dimensions and tolerances for each operation as indicated by the chart.

3.4.8   Tolerance Chart Retention

In the event that quality problems occur in the future, the tolerance charts should be retained as a key basis for analyzing problems and developing solutions. If group technology types of families of similar parts exists, the chart could be filed by the part family number and thus be readily available for future part planners.

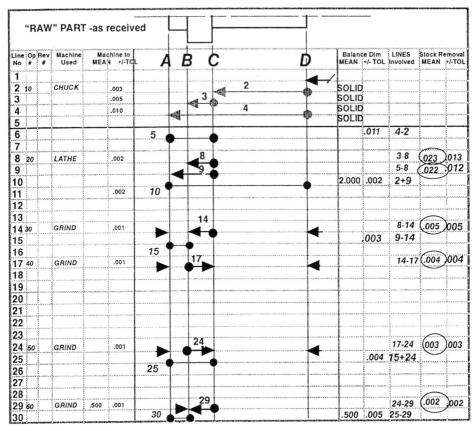

Figure 3.13. Chart completion tasks: calculate stock removal dimensions. Assume for grinding that the stock removals are equal to the stock removal tolerance; assume for machining that the stock removals are 0.010 plus the stock removal variation.

3.4.9 Advanced Tolerance Charting Techniques

The preceeding examples have illustrated a simple situation—no angles were involved. For processing tapers, critical corner radii, and free-form shapes, extensive added geometrical calculations are required to convert the basic ideas just presented into accurate predictors of part behavior. For these situations, the reader is again, referred to the two works by Wade (1967, 1983).

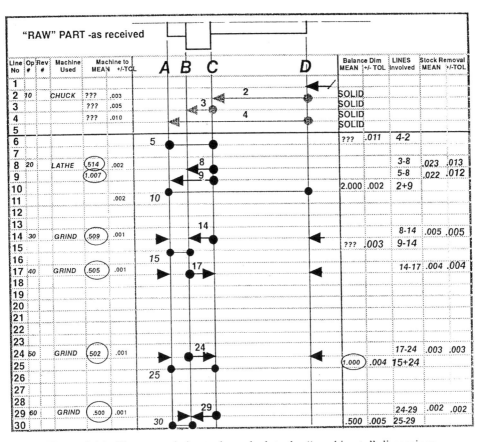

Figure 3.14. Chart completion tasks: calculate the "machine to" dimensions.

Line 29:   Final 39 = cut 29 =
                0.500

Line 24:   No. 24 − stock removal   No. 24 − 0.002 = 0.500   No. 24 = 0.502
                = No. 29

Line 17:   No. 17 − stock removal   No. 17 − 0.003 = 0.502   No. 17 = 0.505
                = No. 24

Line 14:   No. 14 − stock removal   No. 14 − 0.004 = 0.505   No. 14 = 0.509
                = No. 17

Line 8:   No. 8 − stock removal   No. 8 − 0.005 = 0.509   No. 8 = 0.514
                = No. 14

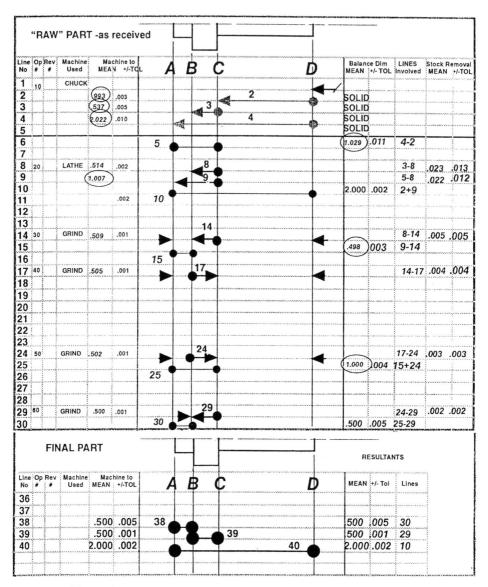

Figure 3.15. Chart completion tasks: calculate balance dimensions.

Bal. line 25:   No. 30 + No. 29   No. 25 = 0.500 + 0.500   No. 25 = 1.000
                    = No. 25

Bal. line 15:   No. 25 = No. 15   No. 15 = No. 25 − No.   No. 15 = 10.000 −
                    + No. 24   24   0.502 = 0.498

Bal. line 15:   No. 15 = No. 9   Cut No. 9 = No 15 +   No. 9 = 0.498 + 0.509
                    − No. 14   No. 14   = 1.007

Bal. line 5:   No. 5 − stock   No. 5 = 1.007 + 0.022
                    removal =   = 1.029
                    No. 9

REFERENCES

Curtis, M. A., 1988, Tolerance Charting, Chapter 6 in *Process Planning,* Wiley, New York.

Liggett, J. V., 1993, Process Charts, Chapter 2 in *Dimensional Variation Management Handbook: A Guide for Quality, Design, and Manufacturing Engineers,* Prentice Hall, Upper Saddle River, NJ.

Nicks, J. E., 1982, Tolerance Control, Chapter 11 in *Basic Programming Solutions for Manufacturing,* Society of Manufacturing Engineers, Dearborn, MI.

Tang, G.-R., Fuh, Y. M., and Kung, R., 1993, A List Approach to Tolerance Charting, *Comput. Ind.,* Vol. 22. No. 3 pp. 291–302.

Todd, R. H., Allen, D. K., and Alting, L., 1994, *Manufacturing Processes Reference Guid,* Industrial Press, New York.

Wade, O. R., 1967, *Tolerance Control in Design and Manufacturing,* Industrial Press, New York.

Wade, O. R., 1983, Tolerance Control, Chapter 2 in *Tool and Manufacturing Engineers Handbook,* Vol. 1, *Machining,* T. J. Drozda and C. Wick (eds.), Society of Manufacturing Engineers, Dearborn, MI, pp. 2-1 to 2-22.

4

TOLERANCE ANALYSIS: A TUTORIAL OF CURRENT PRACTICE

RICHARD J. GERTH

Ohio University
Athens, Ohio

4.1 INTRODUCTION

The purpose of most manufacturing systems is to fabricate a product of consistent quality at minimum cost. This requires the communication and coordination of all engineering and manufacturing activities, from conception and design, through production, to shipping and postproduction service. In the past 15 to 20 years much attention has been focused on the issue of quality and the methods by which it can be achieved. Over-the-wall engineering has been replaced by concurrent engineering and product design teams (PDTs). These teams consider the various interactions of their design decisions as they affect product function, manufacturability, service, reliability, and product quality.

At the core of statistical tolerancing is the concept of interchangeability of parts. If two parts can be exchanged in an assembly without altering the function or quality of the assembly, they are deemed to be interchangeable and can be considered to be the same part. The tolerance range sets the dimensional limits within which parts can vary and still be considered to be interchangeable. As a general rule, tolerances should be as close to zero as possible to ensure interchangeability and high product quality. But tighter tolerances generally require more precise manufacturing processes at in-

Advanced Tolerancing Techniques, Edited by Hong-Chao Zhang
ISBN 0-471-14594-7 © 1997 John Wiley & Sons, Inc.

creased costs. Thus one goal is to generate parts with as loose a tolerance as possible, to minimize the production cost and still guarantee part interchangeability.

A tolerance region defines the basis for determining whether a product is within specifications. Items that do not conform to specification must be either reworked or scraped, thus adding to product cost. Tolerances are related to product reliability and the customer's perception of product quality. In automobile body stampings and assembly, for example, the fit and finish of the "body-in-white" is extremely important to the customer's perception of vehicle quality and are thus held to very tight tolerances (Baron, 1992).

Product tolerances affect customer satisfaction, quality, inspection, manufacturing, and design. Yet proper tolerance assignment is one of the least well understood engineering tasks. For example, in a recent study by Gerth and Hancock (1995), improperly set tolerances were determined to be one of the greatest causes of scrap, rework, and warranty returns. This is, in part, because PDTs often pay a great deal of attention to "critical dimensions" but often pay little (or no) attention to tolerances of noncritical dimensions, so that their cumulative effect causes problems later in manufacturing.

Tolerance stackups, also known as *propagation of error,* is the term used to indicate how individual processes or component feature tolerances can combine to affect a final assembly dimension. Traditional tolerance assignment has been based on tabulated values, past designs, rules of thumb, blanket tolerances, and more recently, computer-aided design (CAD) system default settings. These methods are no longer blindly acceptable because of the impact that tolerances have on product cost.

Thus, the problem becomes one of selecting an acceptable set of tolerances that will satisfy functional product requirements, manufacturing requirements, and quality requirements. The tolerancing problem is difficult because the function that describes the relationship between component feature dimensions and the assembly dimension of interest is usually unknown. This function called the assembly function, tolerance stackup function, or tolerance chain, usually involves a large number of component feature dimensions, and a highly nonlinear geometry which is, at best, difficult for a human being to determine. In addition, with the adoption of ANSI Y14.5M (ASME, 1994), geometric dimensioning and tolerancing (GD&T), proper choice of part dimension reference frames (datums), and modeling of part feature dimensions can lead to vastly different assembly functions.

There are two basic approaches to the tolerancing problem. The first, termed *tolerance analysis,* addresses the question: Given a set of individual component tolerances, what is the resulting assembly tolerance? The second, termed *tolerance allocation* or *tolerance synthesis,* addresses the question: Given a required assembly tolerance, what should the resulting component tolerances be? This chapter focuses on the former, specifically how to conduct a one- and two-dimensional tolerance analysis through specific examples. Three-dimensional tolerance analysis will not be elaborated upon, since it is most commonly conducted with the aid of computer programs such as VSA (Applied Computer Solutions, 1990) and 3-DCS (Dimensional Control Systems, 1996) Finally, the author would like to offer some insight into the future of tolerancing research and identify critical issues yet to be addressed.

At this point it is appropriate to mention three areas that are related to but not a focus of this chapter: geometric design and tolerancing (GD&T), coordinate measuring machines (CMMs), and solid-model-based tolerancing research. GD&T is an ASME standard for geometric part representation, including nominal dimensions and tolerances (ASME Y14.5M). It is an important and related area because it provides a formal representation of tolerances and dimensional features. There are currently efforts directed at providing a more rigorous mathematical representation of feature dimensions and their tolerances. This is motivated by increased inspection and the introduction of CMMs. CMMs are high-precision, general-purpose, measuring instruments capable of measuring complex shapes. Because CMMs may be programmed to measure a variety of dimensions automatically, they are finding increased use as on-line inspection machines in flexible manufacturing environments. However, both of these areas tend to focus on the specification, representation, and measurement of *individual* component feature tolerances, not on the *relationships* among tolerances.

Finally, GD&T is a feature-based specification standard (i.e., it treats surfaces, volumes, and similar geometric entities as continuous entities). CMMs and B-Rep CAD systems represent features by parametric models (i.e., a series of points and lines). As the number of points used to model a feature increase, the closer the model approximates the feature. Solid model geometry is claimed to be the only method by which a proper tolerance analysis can be conducted, because it can accurately represent GD&T feature specifications. This area is still developing, and a major research issue is the mathematical modeling and representation of three-dimensional tolerances in different CAD environments. The interested reader is referred to Etesami (1988), Lu and Wilhelm (1991), Gupta and Turner (1993), and Roy and Liu (1993). This chapter focuses on the methods used to conduct a parametric tolerance analysis, because it is the most widely accepted and most developed method, the errors introduced by parametric models over solid models are practically negligible for many applications, and parametric models are amenable to validation by CMM measurements.

4.2   TOLERANCE ANALYSIS

The history of tolerance analysis has developed from mechanical systems; therefore, the terminology is related to physical objects and geometric relationships. However, the principles are applicable to nongeometric objects and relationships as long as the relationships between the dependent and the independent variables are known. For example, a tolerance analysis can be performed on the current and voltage (dependent variables) of an electric circuit as a function of the circuit (independent variables). Or it could be used to determine the variation in process yield of a chemical process as a function of raw material variation and process control procedures.

The first step in a tolerance analysis is to understand the system in question. In this chapter we assume that the system is based on geometric objects and relationships. An object consists of features that are defined by dimensions, x_i. When two objects are assembled together, they are called *components* of an assembly. The assembly has assembly dimensions, Y_S, which are a function of the component feature dimensions. The function between the assembly and component dimensions must be known or derived and is referred to as the *assembly function, tolerance chain,* or *stackup function.* Deriving the assembly function is usually the most difficult aspect of a tolerance analysis.

Tolerance analysis is concerned with determining the distribution of the assembly dimension Y_S, which is given by

$$Y_S = f(x_1, x_2, ..., x_n) \tag{4.1}$$

The variables, x_i, are dimensions of random length, and unless stated otherwise are assumed to be independent with distributions $w_i(x_i)$. The distributions, $w_i(x_i)$, are not necessarily normal, but have a mean μ_i and a finite variance σ_i^2.

The distribution of Y_S will be determined, in part, by the random variations in x_i or the component tolerances, tol_i. Component tolerances are often set at $\pm 3Cp_i\sigma_i$, although they could be tighter or looser. Cp_i is the process capability index associated with x_i given by

$$CP_i = \frac{USL_i - LSL_i}{6\sigma_i} \tag{4.2}$$

where USL_i and LSL_i denote the upper and lower specification limit of x_i, respectively. The process capability index is equivalent to a statement of satisfactory process yield. For example, a Cp value of 1 results in a process yield of 99.73%.

In general, the function f could be a linear or a nonlinear function. The tolerancing methods outlined below all assume a linear function. If the function is nonlinear, a Taylor series expansion is used as an approximation. The approximation requires that the analytical form of the assembly function be known and differentiable. The method is to expand the assembly function into a multivariable Taylor series by taking partial derivatives and retain only the lower-order terms (Evans, 1975a). If only first-order terms are retained, the procedure reduces a nonlinear function to a linear equation and results in the well-known linear propagation of errors. Linear approximation is only appropriate, however, if $f(x_1, x_2, ..., x_n)$ is approximately linear within the region about μ_i described by the $\pm 3Cp\sigma_i$ limits of the distribution, $w_i(x_i)$.

In practice, normal distributions are generally assumed for $w_i(x_i)$, and a first-order approximation is always used, assuming that f can be approximated by a straight line over the relatively narrow tolerance range of x_i. When the accuracy offered by the linear form is not adequate, a second-order approximation could be performed. The retention of higher-order terms improves the approximation, but also greatly increases the computational difficulty. The results of a rigorous second-order error propagation analysis was published in three technical reports by John Tukey in 1957 and is out of print (Evans, 1975a). In practice, only the linear form is used.

It should be pointed out that determining assembly and component tolerances is often an iterative process. Typically, the required assembly tolerance is known and an initial set of component tolerances are determined using traditional methods, such as handbooks or past designs. Then a tolerance analysis, using one of the methods below, is conducted to determine whether the initial component tolerance set results in an acceptable assembly tolerance. If this is not the case, it is possible to go back and change specific component tolerances and recalculate the assembly tolerance. This process is repeated until a satisfactory set of tolerances capable of producing the desired yield has been established.

The following sections are intended to illustrate a tolerance analysis process through two specific examples. First, the difficult process of determining an assembly function is illustrated for a one-dimensional application, followed by a variety of tolerance analysis methods. Finally, the proposed design is evaluated in terms of the analysis results, and design alternatives are discussed.

4.3 LINEAR ONE-DIMENSIONAL TOLERANCE ANALYSIS: DC MOTOR EXAMPLE

To illuminate the issues involved, consider the direct-current (dc) motor assembly depicted in Figure 4.1. The assembly consists of eight components: the armature (core and shaft), endplate, frame assembly (frame and magnets), housing, output shaft, and planetary gear system. The assembly dimension of interest is the airgap between the armature core and the magnets in the frame assembly. The airgap should be as small as possible to produce the maximum magnetic field strength and the strongest motor. Yet, if the airgap is zero anywhere, the armature core will rub against the magnets, creating excessive friction and resulting in product failure (a condition called *polerub*).

The question is: What is the airgap assembly function (i.e., which component feature dimensions determine the airgap, and what is their mathematical relationship to the airgap)? This is generally the most difficult aspect of tolerance analysis. In some cases the relationship is simple and obvious. For example, every unit increase in the armature core radius will result in a unit decrease in the airgap. For other dimensions, the relationship may be less obvious. For example, it is not clear how much the airgap will change for

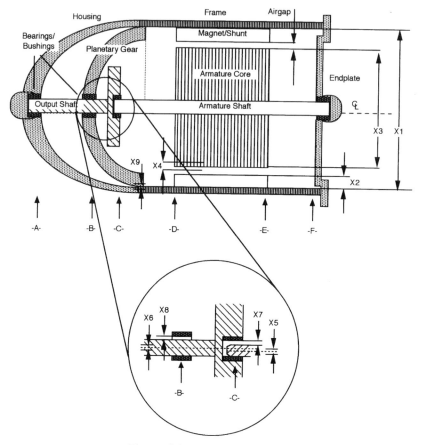

Figure 4.1. dc motor assembly.

every unit change in the clearance between the armature shaft and the endplate if the armature is allowed to wobble freely due to the clearance.

One may begin the analysis by considering how the airgap is defined and the underlying assumptions. The airgap is defined as the distance between the magnet surface and the armature core surface. If there is significant wobble of the armature in the bushings in planes C and F, the airgap will not be uniform across the length of the armature. However, this "wobble" is likely to be small, and thus one considers the airgap to be constant along the length of the armature. For this approach to be accurate, one must make several additional assumptions: The armature core is straight, flat, and parallel to the armature shaft; the magnets and shuts are straight, flat, and the sides are of uniform height; and all shafts move up and down in a parallel manner rather than in a skewed or wobbly manner.

The next step is to trace the path of dimensions that begins at one end of the assembly dimension of interest, say the armature surface, goes through the assembly, and ends at the other end of the assembly dimension, in this case the magnet surface. This path of dimensions is the assembly function. The component dimensions that may affect the assembly dimension, and therefore belong in the assembly function, are selected from the part drawings. It would be beyond the scope of this chapter to reproduce all the part drawings; however, the frame and magnets are shown in Figure 4.2. It is usually good practice to trace the path in the order of the processing steps used to assemble the product, because other dimensions, such as those of fixtures, may be the "true" aligning features, in which case their dimensions would also need to be included in the assembly function.

For example, consider the assembly of the frame to the rest of the motor in Figure 4.1. The assembly process consists of fitting the gear system onto the output shaft, placing the output shaft into the housing, and placing the frame onto the housing. If the housing is held by a fixture that has an externally defined reference frame, the motor frame will be located relative to that external coordinate system and will be independent of any variation in the housing dimension. However, if the fixture tightens around the housing and the frame is located relative to the moving portion of the fixture, the frame location will vary with the housing dimensions. Finally, if the frame is simply placed on the housing without a fixture and is allowed to "float" freely, the clearance between the gear system and the frame will define its location.

Returning to the airgap, one begins developing the assembly function with the armature surface, which is defined by the core outside diameter (OD) (x_3) and core runout (x_4). One must determine the relationship between the position of the armature shaft relative to the frame and magnets. There are

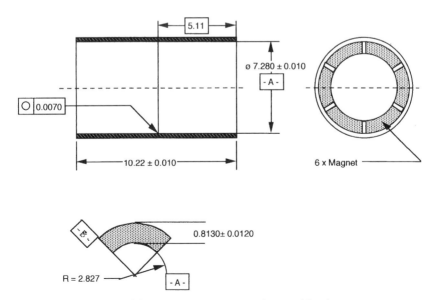

Figure 4.2. Frame and magnet print specifications.

three possible paths: (1) through the endplate, (2) through the output shaft and planetary gear system, and (3) through the output shaft and housing. Because of the assumption that the armature does not wobble but moves parallel to the frame (one-dimensional analysis), one need not consider the endplate path and the other paths simultaneously. If one were conducting a two-dimensional analysis (i.e., allowing the armature to wobble), all paths would need to be considered independently. Since the output-shaft planetary gear system has the greater clearances, that route will be followed.

Following the shaft, there is the clearance at plane C (x_7), the true position error of the machined bearing of the armature shaft (x_5), the clearance at plane B (x_8), the true position error of the machined bearing of the output shaft (x_6), the clearance at the gear–frame junction (x_9), and finally, the frame assembly inside diameter (ID), computed from the frame ID (x_1) and the magnet thickness (x_2). This results in nine dimensions, $x_1, x_2, ..., x_9$, which determine the airgap. Table 4.1 presents the variable index along with its description, and the nominal and tolerance value as it was stated in the GD&T part print.

The assembly function is given by

$$\text{airgap} = Y = f(x_1, ..., x_9) = \tfrac{1}{2}(x_1 - 2x_2 - x_3 - 2x_4) - \tfrac{1}{2}\sum_{i=5}^{9} x_i \quad (4.3)$$

The term in parentheses in equation (4.3) is the total gap on both sides of the armature and is divided in half to obtain the airgap on one side of the armature. The runout tolerance, x_4, is doubled because it applies to both sides of the armature surface. Thus in this example the runout acts like a plus/minus tolerance on diameter. Those familiar with GD&T will note that this

TABLE 4.1. Specification of DC Motor Component Features That Affect the Airgap Stackup Analysis. (All Dimensions in Inches)

i	Component Feature	Nominal	Tolerance
1	Frame ID	7.280	±0.0100
2	Magnet/shunt thickness	0.813	±0.0120
3	Armature core OD	5.550	±0.0120
4	Runout of armature core	—	0.0120
5	True position of output shaft bearing bore	—	0.0120
6	True position of planetary ring bearing ID	—	0.0113
7	Clearance at C: armature shaft to output shaft bearing ID clearance	—	0.0072
8	Clearance at B: output shaft journal and planetary ring bearing ID clearance	—	0.0105
9	Clearance: planetary ring OD to frame ID clearance	—	0.0370

use of runout is in violation of the ASME standard, which states that the diameter must be within its size limit after application of the runout tolerance and would therefore not normally be included in an assembly function. However, since the example is based on an actual case study where the engineers applied the tolerance incorrectly, the runout tolerance will remain in the assembly function. The remaining terms consist of true position and clearance tolerances. These tolerances make the model particularly difficult to understand and deserve elaboration.

These tolerances are all unilateral tolerances and have a nominal value of zero (i.e., they only can take on positive values). The true position tolerances are halved in equation (4.3) because the values stated in the part print are the diameter of a circular area around the nominal within which the true position of the features must lie (see Figure 4.3). Thus in a sense they look like bilateral tolerances since the true position can be off to the positive or negative side of nominal. But, in reality, they behave as unilateral tolerances because of the nature of the airgap. If the true position is off to one side, it will decrease the airgap on that side and increase it on the opposite side, and vice versa. But since polerub can occur anywhere the airgap is too small, any deviation from the nominal true position is "bad," and thus true position acts like a unilateral tolerance.

The clearance tolerance is more difficult to visualize. The maximum clearance between two features, such as a shaft journal outer diameter (OD) and a bearing inner diameter (ID), is the difference between the OD and the ID. If the features are concentric (nominal case), half the maximum clearance will be on either side of the journal. If the journal then moves to one side, the clearance is decreased on that side and increased to the maximum on the other side. Thus the airgap can only be reduced by the amount the journal can move in either direction: that is, half the clearance. Therefore, the clearance tolerances are also halved in equation (4.3) and they, too, behave as unilateral tolerances.

Substitution of the nominal values from Table 4.1 into equation (4.3) yields a nominal airgap value of 0.052 in. The question that tolerance analysis attempts to address is: What is the magnitude of variation around the nominal

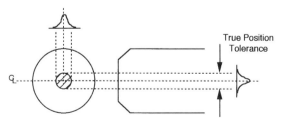

Figure 4.3. Relationship and distribution of true position tolerance relative to nominal centerline position for a shaft journal.

dimension? Once the assembly function has been determined, it is possible to conduct any or all of the following tolerance analyses.

4.3.1 Worst-Case Model

The worst-case model relies on the adage "the whole is equal to the sum of its parts" and assumes that each part has taken on its worst possible value. In this case all component dimensions are situated at one or the other extreme of the tolerance range. Assuming that the dimensions are independent and have bilateral tolerances, the resulting assembly tolerance T_S would be equal to the sum of the component tolerances tol_i. Using the Taylor series approximation yields for the general case

$$T_s = \sum_{i=1}^{n} \left| \frac{\partial f}{\partial x_i} \right| tol_i \tag{4.4}$$

where T_S is the assembly tolerance and tol_i is the component tolerance from Table 4.1.

Although the Taylor series approximation is not necessary because the one-dimensional analysis is by definition linear, the simple sum of the print tolerances is also not appropriate, because not all tolerances are bilateral and many GD&T tolerances require interpretation (see above). Thus the partial derivatives are a "reminder" of the coefficients in the assembly function.

For the airgap example above, equation (4.4) yields a value of 0.0740 in., which results in an upper and lower specification limit of 0.1260 and -0.0220 in., respectively. Unfortunately, 0.0740 in. is not the correct value in this case because (4.4) applies only when all x_i have bilateral tolerances, which in many instances is not the case. As discussed above, the true position and clearance tolerances in this example should be treated as unilateral tolerances even though they represent the diameter of a circle around the nominal indicating a bilateral tolerance.

A better approach is to examine the sign of the partial derivative for each x_i and substitute into the assembly function the appropriate USL or LSL value that results in the maximum and minimum assembly function value. This approach is demonstrated in Table 4.2 for the airgap. Examination of the partial derivatives indicates that the airgap is greatest when the frame ID is largest (USL), the magnets are thinnest (LSL), the core OD is smallest (LSL), the runout is smallest (LSL), and all true position tolerances are at their nominal dimension of 0.00 in. (LSL), and vice versa, to obtain the smallest airgap. Substituting these values into equation (4.3) yields a maximum airgap of 0.0750 in. and a minimum airgap of -0.0220 in. Note that the specification limits are not symmetric and that the limit farthest from the nominal, the LSL in this case, is equal to that computed from (4.4). This is because the airgap

TABLE 4.2. Worst-Case Airgap Stackup Analysis (All Dimensions in Inches)

i	Component Feature	Nominal	Tolerance	Partial	Large Gap	Small Gap
1	Frame ID	7.280	0.0100	0.5	7.2900	7.2700
2	Magnet/shunt thickness	0.813	0.0120	−1.0	0.8010	0.8250
3	Armature core OD	5.550	0.0120	−0.5	5.5380	5.5620
4	Runout of armature core	0.000	0.0120	−1.0	0.0000	0.0120
5	True position of output shaft bearing bore	0.000	0.0120	−0.5	0.0000	0.0060
6	True position of planetary ring bearing ID	0.000	0.0113	−0.5	0.0000	0.0057
7	True position at C: armature shaft to output shaft bearing ID clearance	0.000	0.0072	−0.5	0.0000	0.0036
8	True position at B: output shaft journal and planetary ring bearing ID clearance	0.000	0.0105	−0.5	0.0000	0.0053
9	True position: planetary ring OD to frame ID clearance	0.000	0.0370	−0.5	0.0000	0.0185
	Airgap	0.052			0.0750	−0.0220

LSL contains all tolerances, whereas the airgap USL does not because of the unilateral behavior of the clearances and true position. Finally, note that the LSL on the airgap indicates severe polerub and that the proposed component dimension tolerances would be unacceptable.

The major benefits of worst-case tolerancing are its computational simplicity and its intuitive appeal. If all components in the stack are manufactured within specification, the assembly is guaranteed to be within specification. The major drawback of the worst-case model is that the worst case can occur only when all components are produced at the positive or negative tolerance extreme. It should be evident that the probability of this event is very small and approaches zero as the number of components in the assembly increases. The result is an assembly tolerance that is too loose, or component tolerances that are too tight. For example, the airgap would be -0.0220 in. only if all component features were at the extreme of their distributions. The probability of such an occurrence is a function of the individual process capabilities. Assuming that each process had a Cp value of 1 (very conservative assumption given that most manufacturers have a goal of 1.67 or better), the probability of the airgap being -0.0220 is $0.0013^9 = 1.1 \times 10^{-26}$.

Since most assemblies have a large number of components, worst-cases tolerance analysis is not recommended. The method results in wide assembly tolerance specifications, overly conservative component feature specifications, and costly production processes. In the example above, the designer would have rejected the initial specifications and tightened component feature specifications to achieve the desired result of zero polerub. However, worst-case analysis is still quite common in practice because it protects the designer at the expense of manufacturing, since assemblies are guaranteed to be within specification if components are produced within specification.

4.3.2 Statistical Model

The *statistical model,* also known as the *root-mean-square (RMS) model,* makes use of the fact that the probability of all the components being at the extremes of their tolerance range is very low. Assuming that the component feature dimensions are independent and normally distributed, and again using the Taylor series expansion to approximate any nonlinear assembly function yields

$$\sigma_s^2 = \sum_{i=1}^{n} \left(\frac{\partial f}{\partial x_i}\right)^2 \sigma_i^2 \tag{4.5}$$

Substitution of equation (4.2) into (4.5) yields

$$T_S = \text{Cp}_S \cdot 3\sigma = \text{Cp}_S \cdot 3 \sqrt{\sum_{i=1}^{n} \left(\frac{\partial f}{\partial x_i}\right)^2 \left(\frac{\text{USL} - \text{LSL}}{6\text{Cp}_i}\right)^2} \qquad (4.6)$$

Note that if the process capabilities for the component feature and the assembly dimensions are the same, the CP_i values cancel.

Statistical tolerancing requires one to be concerned with the distribution of x_i, $w_i(x_i)$. In general, application of equation (4.5) requires that $w_i(x_i)$ be normally distributed, and in most cases this is assumed and validated with data. It is not clear how robust equation (4.5) is to the normality assumption, and the answer depends in part on the purpose of the tolerance analysis (see Section 4.5). With regards to the airgap example, one may safely assume that the x_1, x_2, and x_3 are normally distributed around their nominal dimensions with standard deviations given by equation (4.2). The true position dimensions are also commonly modeled as normal distributions in a one-dimensional model with mean 0.00 in. (nominal true position), and standard deviation given by (4.2), where the tolerance range is equal to the true position tolerance (see Figure 4.3). The runout will be assumed to be normally distributed but with a mean of 0.006 in., a USL value of 0.012 in., and an LSL value of 0.00 in.

The statistical tolerance analysis for the airgap example is presented in Table 4.3 assuming a process capability of 1.0 for all component and assembly dimensions. The resulting distribution on the airgap is 0.0460 ± 0.0263 in. The 0.006 in. smaller expected value is due to the assumption that the runout has a mean of 0.006 in. Thus the lower limit on the airgap would be 0.0197 in., indicating that the initial component tolerances are more than adequate to ensure the no-polerub condition. This is contrary to the worst-case analysis. The contradiction is explained by examining the probability of producing a nonconforming assembly. In the statistical case the probability is a function of the Cp; in this example the probability of a nonconforming assembly is approximately 0.0027 regardless of whether the individual components were conforming. In the worst-case analysis, the assembly is nonconforming only if all components are at the extreme of their tolerance range; an event that would occur on average only once in 1.1×10^{-26} times.

The statistical model is perhaps the most popular method currently used in practice. It is computationally more complex than the worst case but results in tighter assembly tolerances, looser component tolerances, and lower-cost products. It is also conceptually unappealing to many designers because a defective assembly can result even if all components are within specification. The probability of this occurring is, however, very slim (see Section 4.3.1). Statistical analysis also tends to predict higher yields than is normally encountered in practice (Chase and Greenwood, 1988). This is attributed primarily to the fact that the assumptions of normality and independence are not generally evident in reality. It also assumes that the distributions are always

TABLE 4.3. Statistical Stackup Analysis of the Airgap (All Dimensions in Inches)

i	Component Feature	E[airgap]	Tolerance	Partial	Sigma	$\left(\frac{\partial f}{\partial x_i}\right)^2 tol_i^2$	% Contrib.
1	Frame ID	7,280	0.0100	0.5	0.0033	2.78E-06	4
2	Magnet/shunt thickness (total of 2)	0.813	0.0120	−1.0	0.0040	1.60E-05	21
3	Armature core OD	5.550	0.0120	−0.5	0.0040	4.00E-06	5
4	Runout of armature core	0.006	0.0120	−1.0	0.0020	4.00E-06	5
5	Runout of output shaft bearing bore	0.000	0.0120	−1.0	0.0020	4.00E-06	5
6	Runout of planetary ring bearing ID	0.000	0.0113	−1.0	0.0019	3.55E-06	5
7	Maximum clearance at C: armature shaft to output shaft bearing ID	0.000	0.0072	−1.0	0.0012	1.44E-06	2
8	Maximum clearance at B output shaft journal and planetary ring bearing ID	0.000	0.0105	−1.0	0.0018	3.06E-06	4
9	Maximum clearance: planetary ring OD to frame ID	0.000	0.0370	−1.0	0.0062	3.80E-05	49
	Airgap	0.046	0.0263		0.0088	7.69E-05	100

centered around their nominal dimensions. This leads to the development of other models.

4.3.3 Constant Factor Models

The statistical model assumes that the component tolerances are independent and normally distributed, which most distributions encountered in practice generally are not. Shifts and drifts in the production process force shifts in the component dimensions and result in skewed, biased, double, and other "normal-like" distributions. A number of factors could force the average of a dimension to change: tool wear, change of materials, new suppliers, and so on. Since these factors often progress over time, the assumption of independence is also not completely justified.

A number of researchers have proposed models that contain a correction factor to the basic statistical model to compensate for the deviation in the component distributions from normal. The most common, *benderizing,* has the effect of spreading the tolerances out beyond the limits obtained from the pure statistical model. Bender's correction factor is based on probability, approximation, and experience involving studies of assemblies with many component parts (Bender, 1962). Application of the formula commonly referred to as benderizing, can be stated as

$$T_B = 1.5 T_S = \mathrm{Cp}_S \cdot 4.5 \sqrt{\sum_{i=1}^{n} \left(\frac{\partial f}{\partial x_i}\right)^2 \left(\frac{\mathrm{USL} - \mathrm{LSL}}{6 \mathrm{Cp}_i}\right)^2} \qquad (4.7)$$

For the airgap example, the benderized tolerance is simply $1.5 \times 0.0263 = 0.0395$ in., resulting in a LSL value of 0.0065 in., which is still adequate.

4.3.4 Estimated Mean Shift Model

The estimated mean shift model, developed by Chase and Greenwood (1988), incorporates both worst-case and statistical models by assigning weighting factors m_i for each assembly component. The factor m_i varies between 0 and 1 and is, in essence, the probability that the process will shift from its mean; the smaller the number, the less likely the process will shift from its known mean.

$$T_S = \sum_{i=1}^{n} \left| m_i \frac{\partial f}{\partial x_i} \mathrm{tol}_i \right| + \sqrt{\sum_{i=1}^{n} (1 - m_i)^2 \left(\frac{\partial f}{\partial x_i}\right)^2 \mathrm{tol}_i^2} \qquad (4.8)$$

The model allows a mix of shift factors in an assembly. Some parts may be worst case, while others may be statistical. For example, thermal expansion, which is often treated as a worst-case factor, can be included in a statistical assembly analysis by setting $m_i = 1$ for that element. It is no longer necessary to treat all components equally and thus penalize the entire assembly with a worst-case limit analysis because of a few poorly controlled components. The model no longer assumes normality for all components and accounts for bias in the individual component distributions via the shift factors. The shift factors, however, are not determined methodically but are at the designer's discretion, and the accuracy of the estimates depend on the designer's ability to pass judgment over the degree to which the process means will shift. Thus, in practice, the m_i are often either 0 or 1. In addition, the worst-case formula does not work with unilateral tolerances (see Section 4.3.1).

In the airgap example, the runouts and clearances were assumed to be a worst-case condition; the other distributions were assumed to suffer a mean shift of 0.25 of their tolerance range. Under those assumptions the mean shift model computations are shown in Table 4.4. The worst-case stackup contributions clearly overshadow the statistical component, resulting in an unacceptable lower specification limit of $0.052 - 0.0675 = -0.0155$ in.

4.3.5   Discussion and Redesign

Four tolerance analyses were conducted, and the question remains whether or not the design is acceptable. Of the four models presented above, the worst-case model clearly results in overly conservative specifications and therefore will not be considered further. Similarly, the mean shift model, which relies heavily on the values of m_i, will not be considered further, so as not to complicate the example unnecessarily. Thus, examining the results of the statistical and benderizing analysis both show that the design is more than adequate given a LSL value on the airgap of 0.001 in.

From an engineering point of view there are now two options: adjusting the nominal dimensions or loosening the tolerances. From a product performance point of view, enlarging the magnets by increasing their nominal thickness is the preferred choice. This will result in stronger magnets, which will increase field strength and hence improve motor performance. If this is not possible, the next choice would be to increase the armature core OD.

From a manufacturing cost point of view, the tolerances should be loosened: but which tolerances, and by how much? Chase and Greenwood present two methods of enlarging all tolerances: proportional scaling and the precision factor model (Chase and Greenwood, 1988). Both lead to a closed-form solution but are unappealing because they do not take into account that some tolerances have a greater impact than others on the assembly dimension or on production cost. Other methods based on minimum cost optimization have been developed but require knowing the relationship between manufacturing cost and tightening or loosening a tolerance, which is often unknown (Gerth, 1994; Spotts, 1973; Chase et al., 1990; Ostwald and Huang, 1977).

An alternative method to determine which tolerances to tighten and/or loosen is to examine the variance contribution of the different tolerances to

TABLE 4.4 Mean Shift Analysis of the Airgap (All Dimensions in Inches)

i	Component Feature	Nominal	Tolerance	m_i	Worst Case	Statistical
1	Frame ID	7.280	0.0100	0.25	0.0013	14.1E-06
2	Magnet/shunt thickness (total of 2)	0.813	0.0120	0.25	0.0030	81.0E-06
3	Armature core OD	5.550	0.0120	0.25	0.0015	20.3E-06
4	Runout of armature core	—	0.0120	1	0.0120	0.00E+00
5	Runout of output shaft bearing bore	—	0.0072	1	0.0060	0.00E+00
6	Runout of planetary ring bearing ID	—	0.0120	1	0.0057	0.00E+00
7	Maximum clearance at C: armature shaft to output shaft bearing ID	—	0.0105	1	0.0036	0.00E+00
8	Maximum clearance at B output shaft journal and planetary ring bearing ID	—	0.0113	1	0.0053	0.00E+00
9	Maximum clearance: planetary ring OD to frame ID	—	0.0370	1	0.0185	0.00E+00
	Airgap	0.052	0.0675		0.0568	115.4E-06

the overall assembly dimension variation. The variance contribution is the squared product of the partial derivatives and the tolerances [see equation (4.6)]. Examination of the variance contributions in Table 4.3 and the assembly function (4.3) leads to the following observations.

The clearance between the planetary gear system and the frame, and the magnet thickness, are the greatest contributors to the airgap variation, accounting for 70% of the variation. This is not particularly surprising since they have the greatest tolerances and the greatest partial derivatives. The variance contribution of the remaining dimensions seems to be more or less equally distributed. Thus the conclusion is to tighten the gap between the frame and the gear and possibly to tighten the tolerances on the magnets. The gap is a function of both the frame and gear nominal dimension and tolerances. The gap could be reduced by decreasing the nominal frame ID subject to the constraint that it still fits over the gear. This also has the benefit of reducing the average airgap, thereby increasing field strength and motor performance.

This simple one-dimensional example has hopefully introduced the reader to the basics of tolerance analysis and its impact on product design. A two-dimensional assembly with other design considerations, such as proper dimensioning, is presented in the next section.

4.4   NONLINEAR TWO-DIMENSIONAL TOLERANCE ANALYSIS: BRACKET EXAMPLE

Two-dimensional analysis is usually more complex because the assembly function involves more variables and is usually nonlinear. The analysis steps are essentially the same as in the one-dimensional case. However, this example will show the difficulty of deriving an assembly function for even a simple two-dimensional assembly and the importance of proper dimensioning and tolerance specification practices.

The analysis will be performed on the three-piece bracket in Figures 4.4 where the assembly dimension of interest, D, is the distance between the two end holes A1 and B1. The individual pieces are held together by pins. A full analysis would require a three-dimensional analysis; however, for simplicity the analysis will be restricted to two dimensions. This simplification implies certain assumptions, however. For example, if the holes A3 and B3 are not aligned along the z-axis (out of the page), B may appear rotated relative to part A, which may be undesirable. Furthermore, if the pin–hole fit is a clearance fit (which simplifies assembly), the bracket will be relatively loose and the exact shape of the bracket assembly will depend on the final forces acting on it when it is installed. Thus, whether the simplified two-dimensional analysis is justified or not depends on the impact on assembly and function. In this example it will be assumed that the holes are aligned along the z-axis, the pins are straight, and the hole–pin fit is an interference fit.

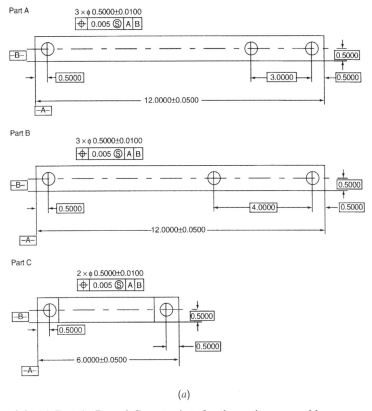

(a)

Figure 4.4. (a) Part A, B, and C part prints for three-piece assembly.

The assembly function will begin with hole A1. The position of A1 is important, but the diameter is not because D is measured from the center of A1. The next position of interest is A3. Part B will be affixed to part A by placing the center of B3 on the center of A3 using the interference fit assumption, which allows one to ignore the pin and hole diameters. If it were a clearance fit, the clearance between the pin and hole as determined from their outer and inner diameters, respectively, would be important. At this point, part B is free to rotate in the XY-plane about point B3. Its final position is determined by part C. Thus part C is affixed to part A by placing C1 on A2, and affixed to part B by placing C2 on B2. This defines the bracket completely and the distance between A1 and B1 can be computed.

To compute this distance it is necessary to generate the assembly function, which is usually the most difficult aspect of the tolerance analysis. Deriving the assembly function is when many of the system's relationships and much of its complexity become evident, and is thus often well worth the effort. There are few programs that do this automatically in two dimensions, such as Mechanical Advantage by Cognition (Cognition, 1991) and AutoCATS by

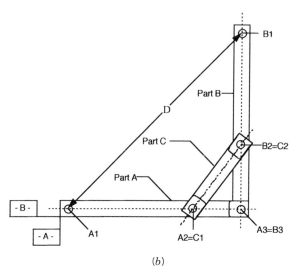

(b)

Figure 4.4. (b) Three-piece bracket assembly.

ADCATS (ADCATS, 1994), and none that can do it in three dimensions. Care should be taken, however, when using programs that generate the assembly function automatically, because many make assumptions about locating features, assembly methods, and so on. Thus the use of computer tolerancing programs, especially three-dimensional programs, require the user to create custom assembly modules that capture the variety and intricacy of many assembly operations.

The assembly function for the bracket is presented in its entirety to show the procedure and difficulty involved in determining the assembly function for even a simple assembly in only two dimensions, and to present the reader with sufficient detail that he or she may reproduce the analysis results. The following procedure is commonly used by commercial Monte Carlo simulation packages, such as VSA (Applied Computer Solutions, 1990) and 3-DCS (Dimensional Control Systems, 1996). They first create a set of component-part models in the part reference frame, containing all the part dimension specification information. The models are created directly in the program or from CAD files. The part models are then manipulated in the assembly reference frame through a set of "move" functions representing the assembly process. The moves may be preprogrammed or user defined. Although the bracket is sufficiently simple that the assembly function can be derived directly from the assembly print, the more structured procedure just described will be followed for illustration purposes.

The first step is to create the set of part feature models (see Figure 4.4a). Each part will be modeled as a set of points whose coordinates are random variables. The point variables that are of interest are the points that define the centers of the various holes. Their coordinates will be denoted by the hole name, an X or Y suffix to denote the X and Y coordinate as measured from their A and B datums, respectively, and a "part" subscript to denote the part versus assembly reference frame. The nominal and tolerance values of the part dimensions that are of interest in the bracket example are given in Table 4.5. The dimensions are assumed to be normally distributed about their nominal value with a standard deviation equal to one-third their tolerance, or a process capability of 1. For example, $B1X_{part}$ is the actual X coordinate of hole 1 in part B relative to its part datum A. It is equal to B1TPX, a normally distributed random variable with a mean of 0.5 and a standard deviation of 0.005/3. The other part variables are defined similarly. The complete part models are given below. Note the stackup of the part length and hole true position tolerances for the holes A2–A3, B2–B3, and C1–C2.

Part A:

$$A1X_{part} = A1TPX$$

$$A1Y_{part} = A1TPY$$

$$A3X_{part} = ALENGTH - A3TPX$$

$$A3Y_{part} = A3TPY$$

$$A2X_{part} = A3X_{part} - A2TPX$$

$$A2Y_{part} = A2TPY$$

Part B:

$$B1X_{part} = B1TPX$$

$$B1Y_{part} = B1TPY$$

$$B3X_{part} = BLENGTH - B3TPX$$

$$B3Y_{part} = B3TPY$$

$$B2X_{part} = B3X_{part} - B2TPX$$

$$B2Y_{part} = B2TPY$$

Part C:

TABLE 4.5 Nominal Values and Tolerances for Three-Piece Bracket Example (All Dimensions in Inches and Relative to the Part Reference Frame)

	X		Y	
Dimension	Nominal	Tolerance	Nominal	Tolerance
Part A	0.500	0.005	0.500	0.005
A1TP: hole A1 true position	3.000	0.005	0.500	0.005
A2TP: hole A2 true position	0.500	0.005	0.500	0.005
A3TP: hole A3 true position	12.000	0.050		
ALENGTH: bar length				
Part B	0.500	0.005	0.500	0.005
B1TP: hole B1 true position	4.000	0.005	0.500	0.005
B2TP: hole B2 true position	0.500	0.005	0.500	0.005
B3TP: hole B3 true position	12.000	0.050		
BLENGTH: bar length				
Part C	0.500	0.005	0.500	0.005
C1TP: hole C1 true position	0.500	0.005	0.500	0.005
C2TP: hole C2 true position	6.000	0.050		
CLENGTH: bar length				

$$C1X_{part} = C1TPX$$

$$C1Y_{part} = C1TPY$$

$$C2X_{part} = CLENGTH - C2TPX$$

$$C2Y_{part} = C2TPY$$

Now that the parts are defined, they must be assembled. The first step is to move part A from its part reference frame to the assembly reference frame. Since the two frames are identical, this step is simply the part model for part A and is not shown. The next is to assemble part B to part A, by locating B3 onto A3. since an interference fit is assumed, one can assume that the hole centers align exactly.

$$B3X = A3X$$

$$B3Y = A3Y$$

At this point part B is free to rotate around the A3–B3 point. Part C is then attached to A.

$$C1X = A2X$$

$$C1Y = A2Y$$

The exact position of parts B and C is defined by the coordinates where B2 and C2 meet. From Figure 4.4b one can see that the point is the intersection of two circles: one with center C1 and radius $R_C = \overline{C1,C2}$ (distance between C1 and C2); the other with center B2 and radius $R_B = \overline{B3,B2}$. The radii for the two circles are

$$R_C^2 = (C1X_{part} - C2X_{part})^2 + (C1Y_{part} - C2Y_{part})^2$$
$$R_B^2 = (B3X_{part} - B2X_{part})^2 + (B3Y_{part} - B2Y_{part})^2$$

The intersection points of the two circles is given by a similar equation in the assembly reference frame:

$$R_C^2 = (C1X - C2X)^2 + (C1Y - C2Y)^2 \tag{4.9}$$

$$R_B^2 = (B3X - B2X)^2 + (B3Y - B2Y)^2 \tag{4.10}$$

Note that the intersection implies that C2X = B2X and C2Y = B2Y. Expanding equations (4.9) and (4.10) and subtracting (4.10) from (4.9) yields the equation of a straight line:

$$C2Y = m \cdot C2X + b \tag{4.11}$$

where

$$m = \frac{C1X - B3X}{B3Y - C1Y} \tag{4.12}$$

$$b = \frac{R_C^2 - R_B^2 - C1X^2 + B3X^2 - C1Y^2 + B3Y^2}{1(B3Y - C1Y)} \tag{4.13}$$

Substituting equations (4.11) into (4.9) yields a quadratic equation of the form

$$C2X^2 + 2p \cdot C2X + q = 0$$

where p and q are given by

$$p = \frac{m(b - C1Y) - C1X}{1 + m^2}$$

$$q = \frac{C1X^2 + (b - C1Y)^2 - R_C^2}{1 + m^2}$$

The two solutions of the quadratic are given by

$$z_{1,2} = -p \pm \sqrt{p^2 - q} \qquad (4.14)$$

The correct solution, z^*, is the one that yields a positive C2Y coordinate value when equation (4.14) is substituted into (4.11):

$$C2Y = mz^* + b = B2Y$$

and

$$z^* = C2X = B2X$$

There is one special case that must be considered, namely when B3Y = C1Y, because the slope m is infinite [see equation (4.12)]. In this case the mathematics simplify subtracting equation (4.10) from (4.9), and C2X can be computed directly:

$$C2X = \frac{R_B^2 - R_C^2 + C1X^2 - B3X^2}{2(C1X - B3X)}$$

This value is then substituted into (4.9) to yield the value for C2Y:

$$C2Y = C1Y + \sqrt{R_C^2 - (C2X - C1X)^2}$$

The assembly is now fixed. The last step to determine the distance D is to compute the new coordinates for B1. Commercial Monte Carlo simulation programs use standard matrix translation and rotation equations to "move" the part models to the new assembly coordinates computed above. A more cumbersome approach is necessary if using spreadsheets to compute the new coordinates. Figure 4.5 shows the variables involved in computing the coordinates for B1.

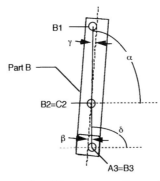

Figure 4.5. Diagram for computing B1 point coordinates (greatly exaggerated and not to scale).

$$\beta = \tan^{-1} \frac{B3Y_{part} - B2Y_{part}}{B3X_{part} - B2X_{part}}$$

$$\gamma = \tan^{-1} \frac{B1Y_{part} - B2Y_{part}}{B1X_{part} - B2X_{part}}$$

$$\delta = \tan^{-1} \frac{B2Y - B3Y}{B2X - B3X}$$

$$\alpha = \beta + \delta - \gamma$$

$$B1X = B2X + \sqrt{(B1Y_{part} - B2Y_{part})^2 + (B1X_{part} - B2X_{part})^2} \cos \alpha$$

$$B1Y = B2Y + \sqrt{(B1Y_{part} - B2Y_{part})^2 + (B1X_{part} - B2X_{part})^2} \cdot \sin \alpha$$

The assembly dimension D is thus given by

$$D = \sqrt{(B1X - A1X)^2 + (B1Y - A1Y)^2} \qquad (4.15)$$

This completes the derivation of the assembly of the assembly function.

A major task in any modeling is validation of the model. A first validation is to see whether the expected result is achieved under nominal conditions. For the bracket example above, substitution of the nominal dimensions results in a distance of 15.556 in., which is equal to the theoretical distance of $\sqrt{11^2 + 11^2} = 15.556$ in. Ideally, the model should be checked with production or prototype data if they are available.

4.4.1 Worst-Case Model

The worst-case condition can easily be determined by entering the extreme tolerances and noting the result. Unfortunately, for complex systems it is not always clear whether the positive tolerance of one variable, say the A1 true position, will combine in the same direction with another positive tolerance, say the A3 true position. Thus many computer programs will perform complete enumeration of all possible combinations and retain the maximum and minimum assembly dimension as the worst-case scenario. In some cases the sheer number of variables makes total enumeration infeasible, and alternative methods are required.

One alternative is to compute the partial derivatives of the assembly feature with respect to each of the component features. The partial derivatives represent the impact of the tolerance on the assembly dimension. Thus, by inspection of the sign of the partial derivative it can easily be determined which combination of tolerances will result in the worst-case conditions. The partial derivatives for simple nonlinear systems such as the three-member bracket may be determined using symbolic mathematics manipulation software such as Mathematica or Maple.

Another less accurate alternative is to estimate the partial derivatives from one-factor-at-a-time experiments. Each dimension is moved to its upper and/or lower specification limit while the other dimensions are held constant and the observed change in the assembly dimension is recorded. The ratio of this change to the change in component dimension can be interpreted as a partial derivative. The limitation is that interaction effects are not considered. However, in most cases interactions are negligible given the narrow tolerance range over which the dimensions are being varied.

The worst case analysis for the bracket example is shown in Table 4.6. The first three columns are the nominal and tolerance for every component dimension. The next column is the upper specification limit value for each dimension that was used to compute the distance D while holding the other dimensions at their nominal value. The partial is ratio of the difference of the distance from its nominal to the component dimension tolerance. The sign of the partial was used to determine the direction for each component dimension that would result in the upper and lower worst case airgap tolerance, which are presented in the last two columns. The resulting upper worst case limit is 15.556 + 0.348 = 15.905 in., and the lower worst case is 15.556 − 0.346 = 15.210 in. As one can see, the distances under the worst-case conditions can be quite large, and as discussed previously, unlikely to occur.

TABLE 4.6 Worst-Case Tolerance Analysis of Three-Piece Bracket (All Dimensions in Inches)

Variable	Nominal	Tolerance	Hi	Distance	Partial	USL	LSL
A1TPX	0.500	0.005	0.505	15.55281	−0.70703	0.495	0.505
A1TPY	0.500	0.005	0.505	15.55281	−0.70703	0.495	0.505
A2TPX	3.000	0.005	3.005	15.54663	−1.94353	2.995	3.005
A2TPY	0.500	0.005	0.505	15.5693	2.590023	0.505	0.495
A3TPX	0.500	0.005	0.505	15.55281	−0.70703	0.495	0.505
A3TPY	0.500	0.005	0.505	15.54691	−1.88765	0.495	0.505
ALENGTH	12.000	0.050	12.050	15.59174	0.707908	12.050	11.950
B1TPX	0.500	0.005	0.505	15.55281	−0.70703	0.495	0.505
B1TPY	0.500	0.005	0.505	15.55281	−0.70703	0.495	0.505
B2TPX	4.000	0.005	4.005	15.54339	−2.59219	3.995	4.005
B2TPY	0.500	0.005	0.505	15.56607	1.943345	0.505	0.495
B3TPX	0.500	0.005	0.505	15.55281	−0.70703	0.495	0.505
B3TPY	0.500	0.005	0.505	15.55016	−1.23811	0.495	0.505
BLENGTH	12.000	0.050	12.050	15.59174	0.707908	12.050	11.950
C1TPX	0.500	0.005	0.505	15.54014	−3.24097	0.495	0.505
C1TPY	0.500	0.005	0.505	15.55635	0.000589	0.505	0.495
C2TPX	0.500	0.005	0.505	15.54014	−3.24097	0.495	0.505
C2TPY	0.500	0.005	0.505	15.55635	0.000589	0.505	0.495
CLENGTH	6.000	0.050	6.050	15.71836	3.240238	6.050	5.950
D	15.556					15.905	15.210

4.4.2   Statistical Model

The statistical model is computed in much the same way as the worst-case model. Again the partial derivatives are required, and once they are known they can be substituted into equation (4.6), yielding the specification limits. The results of the statistical analysis are presented in Table 4.7. The specification on D is 15.556 ± 0.173 in., resulting in LSL and USL values of 15.383 and 15.729 in., respectively. This is significantly tighter that the specifications computed from the worst-case analysis. Note that the greatest contributor to the overall variation is the length of the C bracket. This is because of the strong leverage effect due to the bracket geometry, as evidenced by its large tolerance and large partial derivative. One conclusion would be that the C bracket length is a critical dimension that must be controlled very carefully. In the redesign section, however, it will be shown how the dimension can be removed from the tolerance analysis completely.

4.4.3   Constant Factor Model

The benderized assembly tolerance is simply 1.5 times the statistical tolerance and for the bracket is equal to 15.556 ± 0.260 in., which is still substantially smaller than the worst-case limits.

4.4.4   Discussion and Redesign

Assume that the resulting tolerance for the bracket is unacceptable regardless of the tolerancing method, and the design must be changed. There are three reasons for the large variation:

1. Improper datum selection
2. Excessive component tolerances
3. Large partial derivatives

Improper datum selection means that the features selected to define the part reference frame (datums) introduced additional and unnecessary tolerances into the assembly function. For example, consider the same bracket shown in Figure 4.6, where all component dimensions are referenced from the same datums as opposed to other features. This is called *base line dimensioning,* as opposed to *chain dimensioning,* which was used in Figure 4.4*a* (ASME, 1994). From inspection it is evident that the tolerances associated with the length of all three brackets no longer affects the assembly tolerance when baseline dimensioning is used.

The worst-case and statistical tolerance analysis for the redesigned bracket are shown in Table 4.8. As one can see from the partial derivatives and the variance contribution columns, the length tolerances no longer contribute to the overall assembly variation. Another interesting point is that the variance

TABLE 4.7 Statistical Tolerance Analysis of Three-Piece Bracket (All Dimensions in Inches)

Variable	Nominal	Tolerance	Partial	Cp	σ	Var. contrib.	% Contrib.
A1TPX	0.500	0.005	−0.70703	1	0.001667	1.39E-06	0
A1TPY	0.500	0.005	−0.70703	1	0.001667	1.39E-06	0
A2TPX	3.000	0.005	−0.70703	1	0.001667	1.39E-06	0
A2TPY	0.500	0.005	−1.94353	1	0.001667	1.05E-05	0
A3TPX	0.500	0.005	2.590023	1	0.001667	1.86E-05	1
A3TPY	0.500	0.005	−0.70703	1	0.001667	1.39E-06	0
ALENGTH	12.000	0.005	−1.88765	1	0.001667	9.9E-06	0
B1TPX	0.500	0.050	0.707908	1	0.001667	0.000139	4
B1TPY	0.500	0.005	−0.70703	1	0.001667	1.39E-06	0
B2TPX	4.000	0.005	−0.70703	1	0.001667	1.39E-06	0
B2TPY	0.500	0.005	−2.59219	1	0.001667	1.87E-05	1
B3TPX	0.500	0.005	1.943345	1	0.001667	1.05E-05	0
B3TPY	0.500	0.005	−0.70703	1	0.001667	1.39E-06	0
BLENGTH	12.000	0.005	−1.23811	1	0.001667	4.26E-06	0
C1TPX	0.500	0.050	0.707908	1	0.001667	0.000139	4
C1TPY	0.500	0.005	−3.24097	1	0.001667	2.92E-05	1
C2TPX	0.500	0.005	0.000589	1	0.001667	9.65E-13	0
C2TPY	0.500	0.005	−3.24097	1	0.001667	2.92E-05	1
CLENGTH	6.000	0.005	0.000589	1	0.001667	9.65E-13	0
		0.050	3.240238	1	0.001667	0.002916	87
D	15.55635	0.173221		1	0.05774	0.003334	100

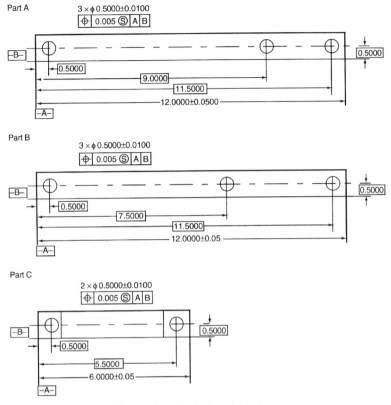

Figure 4.6. Redesigned brackets.

contribution is now more evenly distributed among the different component features, although generally speaking the true position of the holes that determine the position of part C, namely A2, B2, C1, and C2, have the greatest impact.

A comparison of the original and redesigned bracket is shown in Table 4.9. As one can see, the redesign has substantially reduced the assembly variance, without increasing cost. In general, one should attempt to assign datums to as many features in the assembly function as is reasonable. For example, a further reduction in assembly variation would be possible by using A1 and B1 as datums, since they are the two component features that define the assembly dimension. This would eliminate the true position tolerance of A1 and B1 from entering the tolerance chain, thereby further reducing the variation in D. Although it is desirable to use assembly features as component feature datums, this may not always be practical, due to other constraints. For example, it may be that the bracket surfaces must be flat to mate flush with the two surfaces to which the bracket will be connected, and therefore they

TABLE 4.8 Tolerance Analysis for Redesigned Bracked (All Dimensions in Inches)

Variable	Nominal	Tolerance	Hi	D-HI	Partial	Worst Case	σ	Var. Contrib.	% Contrib.
A1TPX	0.500	0.005	0.505	15.5528	-0.7070	0.495	0.001667	1.39E-06	1
A1TPY	0.500	0.005	0.505	15.5528	-0.7070	0.495	0.001667	1.39E-06	1
A2TPX	8.500	0.005	8.505	15.5661	1.9456	8.505	0.001667	1.05E-05	7
A2TPY	0.500	0.005	0.505	15.5693	2.5900	0.505	0.001667	1.86E-05	12
A3TPX	11.500	0.005	11.505	15.5502	-1.2368	11.495	0.001667	4.25E-06	3
A3TPY	0.500	0.005	0.505	15.5469	-1.8876	0.495	0.001667	9.9E-06	7
ALENGTH	12.000	0.050	12.050	15.5563	0.0000	11.950	0.001667	3.51E-31	0
B1TPX	0.500	0.005	0.505	15.5528	-0.7070	0.495	0.001667	1.39E-06	1
B1TPY	0.500	0.005	0.505	15.5528	-0.7070	0.495	0.001667	1.39E-06	1
B2TPX	7.500	0.005	7.505	15.5693	2.5933	7.505	0.001667	1.87E-05	12
B2TPY	0.500	0.005	0.505	15.5661	1.9433	0.505	0.001667	1.05E-05	7
B3TPX	11.500	0.005	11.505	15.5469	-1.8856	11.495	0.001667	9.88E-06	7
B3TPY	0.500	0.005	0.505	15.5502	-1.2381	0.495	0.001667	4.26E-06	3
BLENGTH	12.000	0.050	12.050	15.5563	0.0000	11.950	0.001667	3.51E-31	0
C1TPX	0.500	0.005	0.505	15.5401	-3.2410	0.495	0.001667	2.92E-05	19
C1TPY	0.500	0.005	0.505	15.5564	0.0006	0.505	0.001667	9.65E-13	0
C2TPX	5.500	0.005	5.505	15.5726	3.2408	5.505	0.001667	2.92E-05	19
C2TPY	0.500	0.005	0.505	15.5564	0.0006	0.505	0.001667	9.65E-13	0
CLENGTH	6.000	0.050	6.050	15.5563	0.0000	5.950	0.001667	3.51E-31	0
D	15.556					15.680	0.012268	0.000151	100

TABLE 4.9. Comparison of Tolerance Analysis Results for Original and Redesigned Brackets (All Dimensions in Inches)

Tolerance Method	Original Bracket	Redesigned Bracket
Worst case	15.556 ± 0.348	15.556 ± 0.123
Statistical	15.556 ± 0.173	15.556 ± 0.037
Benderized	15.556 ± 0.260	15.556 ± 0.055

are the preferred datums. Other considerations are inspection and manufacturing processes, which may not be able to key off certain component features.

As the example shows, proper selection and dimensioning of component features is the most effective method of improving product quality and cost. The manner in which features are dimensioned reflects how the feature will be produced and inspected. These considerations should enter into the analysis first, and tolerance analysis issues should be second.

The other two reasons for a large variation in D, the tolerances and the partial derivatives, are expressed in the variance contribution analysis. Consider the variance contribution matrix for the redesigned bracket shown in Table 4.8. Note that the contribution is a function of the squared product of the component dimension variance and the partial derivative. The dimension variance is a function of the manufacturing process capability, and the partial derivative is a function of the component features' nominal dimensions. In general, it is difficult to alter the nominal design because it is usually determined by functional considerations. Thus the most effective use of the variance contribution table is to determine where trade-off studies and continuous improvement efforts should be focused.

For example, the greatest contributors are the true position tolerances of C1 and C2. Thus these tolerances should be kept as tight as possible, and process improvement studies should be focused on reducing the process variation on true position. One the other hand, the true position tolerances of A1, A3, B1, and B3 are not as critical, primarily because of their low partial derivatives. This would indicate that tolerances may be loosened in these areas without a substantial impact on D. The tolerances of the remaining two holes, A2 and B2, have moderate impact and thus require a trade-off analysis to determine if any significant manufacturing cost gains can be realized.

4.5 SUMMARY AND CONCLUSIONS

The purpose of tolerance analysis is to determine the amount of variation in an assembly dimension that is produced by variation in the assembly's component features. The approaches outlined above do this mathematically by approximation techniques. The limitations of such methods are in their accuracy and generalizability. All techniques make certain assumptions regarding either the assembly functions or the component tolerance distributions.

Given the variety of methods as well as the availability of software solutions, it would seem appropriate at this point to reflect on the future of tolerancing. Tolerance analysis is in excellent shape. In general, the first-order Taylor series approximation to a non-linear assembly function is widely accepted, along with statistical tolerancing being preferred to worst-case tolerancing. In addition, the commercial availability of Monte Carlo simulators for tolerancing makes iterative tolerancing based on data (as opposed to heuristic methods such as proportional scaling) a reality. The major problem that still needs to be addressed is validation of the tolerance analysis model, in terms of both the assembly function and the component feature distributions. This is particularly a concern when simulation models are used to predict part-per-million defects.

Thus far it has been assumed that the feature distributions are normally distributed. Yet it is known that this is often not true. For example, the true position of a hole is really a bivariate normal distribution, which can be modeled in polar coordinates as a Rayleigh (radius) distribution convoluted with a uniform (angle) distribution. This is done because some manufacturers measure true position as a radial distance from the nominal, which follows a Rayleigh distribution and has a very different tail behavior than a normal distribution.

Figure 4.7 shows the effect of distribution type on the computation of tail area probabilities. For example, the normal distribution has approximately 0.0013 defects at the $+3$ sigma limit, whereas a chi-squared distribution with 18 degrees of freedom, $\chi^2(18)$, has the same level of defects at 4.5 sigma. Stated another way, at the 3σ limit the $\chi^2(18)$ has 0.01 defects, compared to the normal 0.0013, about 10 times more! Thus for very small defect levels, current statistical methods are inadequate to estimate the behaviors at the tails of distributions. It is also difficult to get data on this region because part-per-million defects are such rare events.

In practice, most designers do not worry about the "true" behavior and compensate for the lack of knowledge with large process capability indices. In addition, correction factor methods, such as benderizing, are often applied to the final assembly tolerance to protect against nonnormal distributions as well as shifts and drifts.

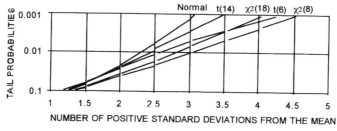

Figure 4.7. Tail area probabilities of the normal and some nonnormal distributions.

One of the main assumptions of tolerancing is that the assembly function is known or can be determined. However, this is often the most difficult and error-prone aspect of tolerance analysis. For dimensional tolerancing (i.e., tolerancing where the assembly function can be determined from geometry), it should be possible to automate the assembly function determination function from the CAD drawings. This requires that the tolerancing information be functionally linked to the part feature. This has been the focus of much research in solid model tolerancing (Etesami, 1988; Lu and Wilhelm, 1991; Gupta and Turner, 1993; Roy and Liu, 1993). For the two-dimensional assembly, CAD software, such as Mechanical Advantage (Cognition, 1991) and AutoCATS (ADCATS, 1994), exists which can determine whether the part drawing is complete and consistent (not over- or underconstrained), determine the assembly tolerance given the component feature tolerances, and identify the tolerance variation contributions. For tolerance analysis without specific process information, these are perhaps the most sophisticated tolerancing tools on the market to date. However in the three-dimensional arena, although there are powerful simulation packages available for tolerance analysis (Applied Computer Solutions, 1990; Dimensional Control Systems, 1996), they cannot automatically identify the assembly function and require significant model building effort on the part of the analyst.

Another difficulty is modeling certain aspects of the geometric design and tolerancing standards (ANSI Y 14.5M), e.g., hole and shaft true positions under maximum and least material conditions. Although models can be created that capture the mathematics of the standard, they are unsatisfying because they do not represent the behavior of the manufacturing process, which cares nothing of specifications or standards.

Finally, there is a great deal of novel research being conducted in the area of tolerancing, such as minimum cost tolerance allocation (Gerth, 1994; Spotts, 1973; Chase et al., 1990; Ostwald and Huang, 1977), Gerth's compensatory tolerancing (Gerth, 1992, 1996), and methods based on designed experiments, such as Taguchi's parameter design (Islam, 1995) and tolerance design (Phadke, 1989) and Bisgaard's pre- and postfractionated experiments for assemblies (Bisgaard, 1993). The designed experiments method has the advantage of being able to determine an empirical linear model of the assembly functions when they are unknown and/or unknowable. For example, in the design of a dc motor, the relationship between current draw and the tolerances on the various component dimensions (such as the commutator bar position relative to the brushes) and other design parameters (such as the field strength of the stationary poles) is not known to sufficient detail to set tolerances. Designed experiments could be used to determine an empirical relationship as well as the high-variance contributors so that designers would know which tolerances to tighten.

Statistical tolerancing can provide product designers with the information necessary to determine which design alternatives will result in the lowest-cost, highest-quality product. The tool that can provide the designer with the most pertinent and accurate information will find the greatest acceptance and yield the best product. The best tool, however, is no substitute for a good engineering design. If the designs under consideration are not "good" to begin with, statistical tolerancing will not make them good. Statistical tolerancing will, however, ensure that the final design can be produced within specifications at a given yield within a certain probability. Through proper engineering and manufacturing discipline it is possible both to increase quality and reduce cost. Statistical tolerancing is a tool that can improve communication between product design and process planning and contribute to the coordination of the entire manufacturing process.

REFERENCES

ADCATS, 1994, *Research Status Report,* Brigham Young University, Provo, UT.

Applied Computer Solutions, 1990, *VSA 2.2 Reference Manual,* Applied Computer Solutions, St. Clair Shores, MI.

ASME, 1994, *Dimensioning and Tolerancing,* ANSI Y14.5M-1994, American Society of Mechanical Engineers, New York.

Baron, J. S., 1992, Dimensional Analysis and Process Control of Body-in-White Processes, Ph. D. dissertation, University of Michigan, Ann Arbor, MI.

Bender, A., 1962, Benderizing Tolerances: A Simple Practical Probability Method of Handling Tolerances for Limit-Stack-Ups, *Graph. Sci.* December.

Bisgaard, S., 1993, *Designing Experiments for Tolerancing Assembled Products,* Technical Report 99, Center for Quality and Productivity, University of Wisconsin.

Chase, K. W., and Greenwood, W. H., 1988, Design Issues in Mechanical Tolerance Analysis, *Manuf. Rev.,* Vol. 1, No. 1, pp. 50–59.

Chase, K. W., Greenwood, W. H., Loosli, B. G., and Hauglund, L. F., 1990, Least Cost Tolerance Allocation for Mechanical Assemblies with Automated Process Selection, *Manuf. Rev.,* Vol. 3, No. 1, pp. 49–57.

Cognition, 1991, *Mechanical Advantage Ver. 3.1,* Cognition Corporation, Billerica, MA.

Dimensional Control Systems, 1996, *3-D CS User's Manual,* Dimensional Control Systems, Inc., Troy, MI.

Etesami, F., 1988, Tolerance Verification Through Manufactured Part Modeling, *J. Manuf. Syst.,* Vol. 7, No. 3, pp. 223–232.

Evans, D. H., 1975a, Statistical Tolerancing: The State of the Arft, part 2, *J. Quality Technol.,* Vol. 7, No. 1, pp. 1–12.

Gerth, R., 1992, Demonstration of a Process Control Methodology Using Multiple Regression and Tolerance Analysis, Ph.D. dissertation, University of Michigan, Ann Arbor, MI.

Gerth, R., 1994, A Spreadsheet Approach to Minimum Cost Tolerancing for Rocket Engines, *Comput. Ind. Eng.,* Vol. 27, No. 1–4, pp. 549–552.

Gerth, R., 1996, Compensatory Tolerancing, *Proceedings of the 1996 Industrial Engineering Research Conference,* Minneapolis, MN, May, 18–20, pp. 333–338.

Gerth, R., and Hancock, W. M., 1995, Reduction of the Output Variation of Production Systems Involving a Large Number of Processes, *Qual. Eng.,* Vol. 8, No. 1, pp. 145–163.

Gupta, S., and Turner, J. U., 1993, Variational Solid Modeling for Tolerance Analyis, *IEEE Comput. Graph. Appl.* May, pp. 64–74.

Islam, Z., 1995, A Design of Experiments Approach to Tolerance Allocation, M.S. thesis, Ohio University, Athens, OH.

Lu, S., and Wilhelm, R. G., 1991, Automating Tolerance Synthesis: A Framework and Tools, *J. Manuf. Syst.,* Vol. 10, No. 4, pp. 281–295,

Ostwald, P. F., and Huang, J., 1977, A Method for Optimal Tolerance Selection, *J. Eng. Ind.,* August, pp. 558–565.

Phadke, M. S., 1989, *Quality Engineering Using Robust Design,* Prentice Hall, Upper Saddle River, NJ.

Roy, U., and Liu, C. R., 1993, Integrated CAD Frameworks: Tolerance Representation Scheme in a Solid Model, *Comput. Ind. Eng.,* Vol. 24, No. 3, pp. 495–509.

Spotts, M. F., 1973, Allocation of Tolerances to Minimize Cost of Assembly, *J. Eng. Ind.,* August, pp. 762–764.

II
MATHEMATICAL DEFINITIONS AND THEORETICAL STUDIES

5

TOLERANCE ANALYSIS OF TWO- AND THREE-DIMENSIONAL MECHANICAL ASSEMBLIES WITH SMALL KINEMATIC ADJUSTMENTS

Kenneth W. Chase
Spencer P. Magleby

Brigham Young University
Provo, Utah

Jinsong Gao

Hewlett–Packard
San Diego, California

5.1 INTRODUCTION

Tolerance analysis and tolerance control have become the focus of increased activity as manufacturing industries strive to increase productivity and improve the quality of their products. The effects of tolerance specifications are far-reaching, as shown in Figures 5.1. Not only do the tolerances affect the ability to assemble the final product but also the production cost, process selection, tooling, setup cost, operator skills, inspection and gaging, and scrap and rework. Tolerances also directly affect engineering performance and robustness of a design. Products of lesser quality, excess cost, or poor performance will eventually lose out in the marketplace.

Advanced Tolerancing Techniques, Edited by Hong-Chao Zhang
ISBN 0-471-14594-7 © 1997 John Wiley & Sons, Inc.

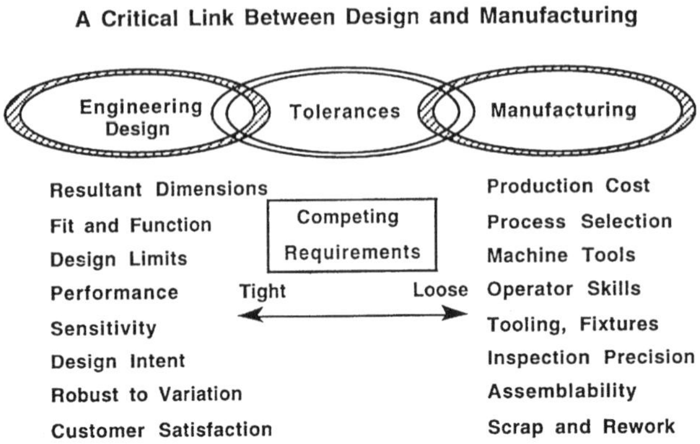

A Critical Link Between Design and Manufacturing

Figure 5.1. Effects of tolerance specifications are far-reaching.

Engineering design and manufacturing have competing tolerances requirements. Engineers want tight tolerances to assure proper performance; manufacturing prefers loose tolerances to reduce cost. There is a critical need for quantitative design tools for specifying tolerances. Tolerance analysis brings the engineering design requirements and manufacturing capabilities together in a common model, where the effects of tolerance specifications on both design and manufacturing requirements can be evaluated quantitatively.

Statistical tolerance analysis offers powerful analytical methods for predicting the effects of manufacturing variations on design performance and production cost. There are, however, many factors to be considered. Statistical tolerance analysis is a complex problem that must be formulated carefully to assure validity, and then interpreted carefully to determine the overall effect on the entire manufacturing enterprise.

New computer-aided design (CAD) tools for tolerance evaluation are being developed and integrated with commercial CAD systems so that assembly tolerance specifications may be created with a graphical preprocessor and evaluated statistically. Built-in modeling aids, statistical tools, and a manufacturing process database will allow the nonexpert to include manufacturing considerations in design decisions. The architecture for the Computer-Aided Tolerancing System (CATS) is shown in Figure 5.2. Use of these new tools will reduce the number of manufacturing design changes, reduce product development time, reduce cost, and increase quality. They will elevate tolerance analysis to the level of an accepted engineering design function, together with finite element analysis, dynamic analysis, and so on.

In this chapter we present a comprehensive system for analytical modeling of assembly variations. A versatile modeling procedure is described, which is adaptable to graphical modeling. With only a few basic elements, a designer can represent a wide range of assembly applications. An efficient solution procedure, based on linear algebra, is demonstrated, which requires analysis

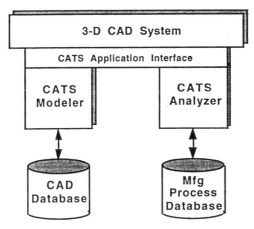

Figure 5.2. Computer-aided tolerance analysis system.

of only one assembly to estimate tolerance accumulation throughout the assembly. The system is ideally suited to CAD integration and iterative design. Three detailed examples illustrate the method.

5.2 THREE SOURCES OF VARIATION IN ASSEMBLIES

Manufactured parts are seldom used as single parts. They are used in *assemblies* of parts. The dimensional variations that occur in each component part of an assembly accumulate statistically and propagate kinematically, causing the overall assembly dimensions to vary according to the number of contributing sources of variation. The resultant critical clearances and fits that affect performance are thus subject to variation due to the stackup of the component variations. There are three main sources of variation that must accounted for in mechanical assemblies:

1. Dimensional variations (lengths and angles)
2. Form and feature variations (flatness, roundness, angularity, etc.)
3. Kinematic variations (small adjustments between mating parts)

Dimensional and form variations are the result of variations in the manufacturing processes or raw materials used in production. Kinematic variations occur at assembly time, whenever small adjustments between mating parts are required to accommodate dimensional or form variations. Tolerances are added to engineering drawings to limit variations. Dimensional tolerances limit component size variations. Geometric tolerances, defined by ANSI Y14.5M-1982 (ASME, 1982), are added to further limit the form, location, or orientation of individual part features. Assembly tolerance specifications are added to limit the accumulation of variation in assemblies of parts.

The two-component assembly shown in Figure 5.3 demonstrates the relationship between dimensional variations in an assembly and the small kinematic adjustments that occur at assembly time. The three component dimensions A, R, and θ vary as shown. The variations in the three dimensions have an effect on the distance U, locating the point of contact on the horizontal surface. U is important to the function of the assembly.

The parts are assembled by inserting the cylinder into the groove until it makes contact on the two sides of the groove. For each set of parts, the distance U will adjust to accommodate the current value of dimensions A, R, and θ. The assembly resultant U_1 represents the nominal position of the cylinder, while U_2 represents the position of the cylinder when the variations ΔA, ΔR, and $\Delta \theta$ are present. This adjustability of the assembly describes a kinematic constraint, or a closure constraint on the assembly.

It is important to distinguish between component and assembly dimensions in Figure 5.3. Whereas A, R, and θ are component dimensions, subject to random process variations, distance U is not a component dimension; it is a resultant assembly dimension. Variations in A, R, and θ occur during the manufacture of individual parts. U is not a manufacturing variable, it is a kinematic assembly variable. Variations in U can only be measured after the parts are assembled. A, R, and θ are the independent variables in this assembly. U is a dependent variable.

Figure 5.4 illustrates the same assembly with exaggerated geometric feature variations. For production parts, the contact surfaces are not really flat and the cylinder is not perfectly round. The pattern of surface waviness will differ from one part to the next. In this assembly the cylinder makes contact on a peak of the lower contact surface, while the next assembly may make contact in a valley. Similarly, the lower surface is in contact with a lobe of

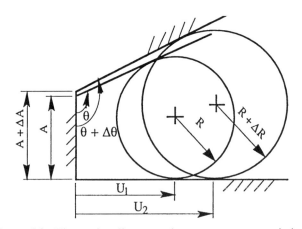

Figure 5.3. Kinematic adjustment due to component variations.

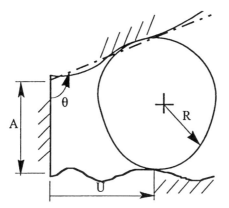

Figure 5.4. Adjustment due to geometric shape variations.

the cylinder, while the next assembly may make contact between lobes. Local surface variations such as these can propagate through an assembly and accumulate just as size variations do. Thus, in a complete assembly model, all three sources of variation must be accounted for to assure realistic and accurate results.

The objective of this chapter is to generalize the procedures for computer-aided tolerance modeling and analysis of two- and three-dimensional mechanical assemblies using vector-loop-based assembly models. In vector models, all three variation sources may be included. Of particular interest will be the assembly kinematics employed to set up the kinematic assembly constraints and their solution through a linearization procedure, by which assembly variations can be predicted and evaluated.

5.3 SAMPLE ASSEMBLY VARIATION PROBLEM

To illustrate the problems associated with two-dimensional tolerance analysis, consider the simple assembly shown in Figure 5.5, as described by Fortini (1967). This one-way mechanical clutch, is a common device used to transmit rotary motion in only one direction. When the outer ring of the clutch is rotated clockwise, the rollers wedge between the ring and hub, locking the two so they rotate together. In the reverse direction, the rollers just slip, so the hub does not turn. The pressure angle ϕ_1 between the two contact points is critical to the proper operation of the clutch. If ϕ_1 is too large, the clutch will not lock; if it is too small, the clutch will not unlock.

The primary objective of performing a tolerance analysis on the clutch is to determine how much the angle ϕ_1 is expected to vary due to manufacturing variations in the clutch component dimensions. The independent manufacturing variables are the hub dimension a, the cylinder radius c, and the ring

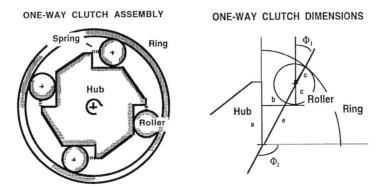

Figure 5.5. One-way clutch assembly and its relevant dimensions.

radius e. The distance b and angle ϕ_1 are not dimensioned. They are assembly resultants that are determined by the sizes of a, c, and e when the parts are assembled. Variations in b and ϕ_1 are examples of the small kinematic adjustments between the mating parts that occur at assembly time in response to the dimensional and geometric feature variations of the components in the assembly. For example, if the roller in the clutch assembly is produced undersized, as shown in Figure 5.6, the points of contact with the hub and ring will shift, causing kinematic variables b and ϕ_1 to increase.

Usually, limiting values of kinematic variations are not marked on a mechanical drawing, but tolerances on critical performance variables, such as a clearance or a location, may appear as assembly specifications. The task for the designer is to assign tolerances to each component in the assembly so that each assembly specification is met, in this case, the limits on pressure angle ϕ_1. Estimating the variation of an assembly parameter, such as ϕ_1, requires an assembly function which relates the assembly parameter to the relevant component dimensions that contribute to the assembly variation. The assembly function can take the form of explicit or implicit algebraic equations.

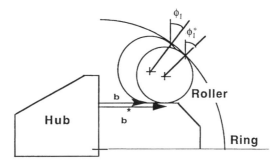

Figure 5.6. Kinematic or assembly variations due to a change in the roller size.

5.3.1 Explicit Assembly Equations

By trigonometry, the dependent assembly resultants, distance b and angle ϕ_1, can be expressed as explicit functions of a, c, and e.

$$\phi_1 = \cos^{-1} \frac{a + c}{e - c} \qquad b = \sqrt{(e - c)^2 - (a + c)^2} \qquad (5.1)$$

The expression for angle ϕ_1 may be analyzed statistically to estimate quantitatively the resulting variation in ϕ_1 in terms the specified tolerances for a, c, and e. If performance requirements are used to set engineering limits on the size of ϕ_1, the quality level and percent rejects may also be predicted.

5.3.2 Implicit Assembly Equations

Figure 5.7 shows a vector model overlaid on the clutch assembly. The vectors represent the part dimensions that contribute to the overall assembly dimensions. Kinematic joints are placed at the points of contact between mating parts. Assembly relationships are described by loops or chains of vectors from which a set of algebraic equations may be derived. Simultaneous solution of the algebraic kinematic equations permits the prediction of the resulting kinematic adjustments and assembly variations caused by small manufacturing variations. Form variations may be added to the model and their effects may be predicted as well. The resulting nominal assembly dimensions and variations may be analyzed statistically and compared to engineering design specs to predict the number of rejected assemblies to expect in production. Design iteration of the component tolerances may then be applied until the desired quality levels are achieved.

From the clutch assembly vector loop, three scalar loop equations may be derived:

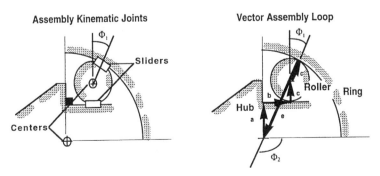

Figure 5.7. Kinematic joints and vector loop representing the one-way clutch assembly.

$$H_x = b + c \cos(90° + \phi_1) + e \cos(270° + \phi_1) = 0 \qquad (5.2)$$

$$H_y = a + c + c \sin(90° + \phi_1) + e \sin(270° + \phi_1) = 0 \qquad (5.3)$$

$$H_\theta = 90° - 90° + 90° - \phi_1 + 180° + \phi_2 = 0 \qquad (5.4)$$

The known independent variables in this set of equations are a, c, and e. The unknown dependent variables are b, ϕ_1, and ϕ_2. Examination of the system of equations reveals that they are nonlinear implicit functions, which must be solved simultaneously for all three dependent variables.

Establishing explicit assembly functions, such as equation (5.1), to describe assembly kinematic adjustments places a heavy burden on the designer. For most mechanical assemblies, this relationship may be difficult or impossible to obtain. It is very difficult to define such explicit assembly functions in a generalized manner for real-life mechanical assemblies. This difficulty makes the use of explicit functions impractical in a CAD-based system intended for use by mechanical designers.

A procedure for solving implicit systems of assembly equations, based on vector loop representations, is presented in section 5.6, but we must first present a general method for deriving the equations that describe assembly variations. It is the kinematic variations that result in implicit assembly functions. Current tolerance analysis practices fail to account adequately for this significant variation source. In a comprehensive assembly tolerance analysis model, all three variations should be included: dimensional, geometric, and kinematic. If any of the three is overlooked or ignored, it can result in significant error. Only when a complete model is constructed can the designer accurately estimate the resultant variations in an assembly.

5.4   METHODS AVAILABLE FOR TOLERANCE ANALYSIS

In this section we briefly review the methods available for nonlinear tolerance analysis when an explicit assembly function is provided which relates the resultant variables of interest to the contributing variables or dimensions in an assembly. The purpose of the review is to provide background for a discussion of a generalized method for treating implicit functions. A more comprehensive review may be found in a previous paper (Chase and Parkinson, 1991).

5.4.1   Linearized Method for Estimating Variation

The variation in b and ϕ_1 may be estimated by applying error analysis methods, in which the explicit function is linearized in terms of the independent variables a, c, and e, and the values of expected error in a, c, and e, that is, their tolerances. The linearization method is based on a first-order Taylor series expansion of the assembly function, equation (5.1). The partial derivatives of ϕ_1 are evaluated at the nominal values of a, c, and e. Then the variation $\Delta\phi_1$ may be estimated by a worst-case or statistical model for tolerance accumulation (Cox, 1986; Shapiro and Gross, 1981).

$$\Delta\phi_1 = \left|\frac{\partial\phi_1}{\partial a}\right| \text{tol}_a + \left|\frac{\partial\phi_1}{\partial c}\right| \text{tol}_c + \left|\frac{\partial\phi_1}{\partial e}\right| \text{tol}_e \qquad \text{(worst case)} \quad (5.5)$$

$$\Delta\phi_1 = \sqrt{\left(\frac{\partial\phi_1}{\partial a}\text{tol}_a\right)^2 + \left(\frac{\partial\phi_1}{\partial c}\text{tol}_c\right)^2 + \left(\frac{\partial\phi_1}{\partial e}\text{tol}_e\right)^2} \qquad \text{(statistical)} \quad (5.6)$$

The derivatives of ϕ_1 with respect to each of the independent variables a, c, and e are called the *tolerance sensitivities*. They are essential to the models for accumulation; hence the need for an explicit function is apparent.

5.4.2   System Moments

System moments is a statistical method for expressing assembly variation in terms of the moments of the statistical distributions of the components in the assembly. The first four moments describe the mean, variance, skewness, and kurtosis of the distribution, respectively. A common procedure is to determine the first four moments of the assembly variable and use these to match a distribution that can be used to describe system performance (Evans, 1975a, 1975b; Cox, 1979, 1986; Shapiro and Gross, 1981). Moments are obtained from a Taylor's series expansion of the assembly function $\phi_1(x_i)$ about the mean, retaining higher-order derivative terms:

$$\begin{aligned} E[m_k] = E\Bigg[&\sum_{i=1}^{n} \frac{\partial\phi_1}{\partial x_i}[\phi_1(x_i) - \phi_1(\mu_i)] + \sum_{i=1}^{n} \frac{\partial^2\phi_1}{\partial x_i^2} \\ &[\phi_1(x_i) - \phi_1(\mu_i)]^2 \\ &+ \sum_{i<j}^{n-1}\sum_{j=1}^{n} \frac{\partial^2\phi_1}{\partial x_i \partial x_j}[\phi_1(x_i) - \phi_1(\mu_i)] \\ &[\phi_1(x_j) - \phi_1(\mu_j)] + \cdots \Bigg]^k \end{aligned} \qquad (5.7)$$

where m_k is the kth moment, E is the expected value operator, x_i are the variables a, c, and e, and μ_i are their mean values. Expanding the truncated series to the third and fourth power yields extremely lengthy expressions for the third and fourth moments. Clearly, this method also relies on an explicit assembly function.

5.4.3 Quadrature

The basic idea of quadrature is to estimate the moments of the probability density function of the assembly variable by numerical integration of a moment generating function:

$$E[m_k] = \int_{-\infty}^{+\infty} \int_{-\infty}^{+\infty} \int_{-\infty}^{+\infty} [\phi_1(a,c,e) - \phi_1(u_a, u_c, u_e)]^k$$
$$w(a)w(c)w(e) \; da \; dc \; de$$

(5.8)

where m_k is the kth moment of the assembly distribution, $w(a)$, $w(c)$, and $w(e)$ are the probability density functions for the independent variables a, c, and e, and μ_a, μ_c, and μ_e are their mean values. Engineering limits are then applied to the resulting assembly distribution to estimate the statistical performance of the system (Evans, 1967, 1971, 1972).

5.4.4 Reliability Index

The Hasofer–Lind reliability index, also called the second moment reliability index, was originally developed for structural engineering applications (Hasofer and Lind, 1974; Ditlevsen, 1979a, b). This sophisticated method has been applied to mechanical tolerance analysis (Parkinson, 1978, 1982, 1983; Lee and Woo, 1990). The reliability index may be used to approximate the distance of each engineering limit from the mean of the assembly and estimate the percent rejects. It requires only the means and covariances of the independent variables, which assumes that all the independent variables are normally distributed and independent.

5.4.5 Taguchi Method

The general idea of the Taguchi method is to use fractional factorial or orthogonal array experiments to estimate the assembly variation due to component variations. It may further be applied to find the nominal dimensions and tolerances which minimize a specified *loss function*. The Taguchi method is applicable to both explicit and implicit assembly functions (Taguchi, 1978).

5.4.6 Monte Carlo Simulations

The Monte Carlo simulation method evaluates individual assemblies using a random number generator to select values for each manufactured dimension, based on the type of statistical distribution assigned by the designer or determined from production data. Each set of dimensions is combined through the assembly function to determine the value of the assembly variable for each simulated assembly. This set of computed assembly values is then used to calculate the first four moments of the assembly variable. Finally, the moments may be used to determine the system behavior of the assembly, such as the mean, standard deviation, and percentage of assemblies which fall outside the design specifications (Sitko, 1991; Fuscaldo, 1991; Craig, 1989). An explicit assembly function is required to permit substitution of random sets of component dimensions and compute the change in assembly variables for each assembly.

5.5   ASSEMBLY KINEMATICS

The kinematics present in a tolerance analysis model of an assembly is different from the traditional mechanism kinematics. The input and output of the traditional mechanism are large displacements of the corresponding components, such as the rotation of the input and output cranks of a four-bar linkage. The linkage is composed of rigid bodies, so all the component dimensions remain constant, or fixed at their nominal values. In contrast to this, the kinematic inputs of an assembly tolerance analysis model are small variations of the component dimensions around their nominal values, and the outputs are the variations of assembly features, including clearances and fits critical to performance as well as small kinematic adjustments between components.

The kinematic adjustments in an assembly have a meaning similar to kinematic degrees of freedom in a mechanism, but the input motions do not refer to displacements of a mechanism. They actually represent differences from the nominal dimension from one assembly to the next. It is the assembly-to-assembly variation of an assembly line, not the time variation of a single assembly, that is described by assembly kinematics. The kinematic assembly equations describe constraints on the interaction between mating component parts. These constraints also serve as functions by which assembly variations may be studied. Since the assembly model is similar to a classical kinematic mechanism model, the analysis methods developed for mechanism kinematics can be applied to assembly variation analysis.

5.6   VECTOR-LOOP-BASED ASSEMBLY MODELS

Using the concepts presented in Sections 3 and 4, vector-loop-based assembly models use vectors to represent the dimensions in an assembly. Each vector represents either a component dimension or kinematically variable dimension. The vectors are arranged in chains or loops representing those dimensions that "stack" together to determine the resulting assembly dimensions. The assembly tolerance specifications are the engineering design limits on those assembly feature variations that are critical to performance.

The other model elements include kinematic joints, datum reference frames, feature datums, assembly tolerance specifications, component tolerances, and geometric feature tolerances. Using relatively few model elements, vector models can describe a broad spectrum of assembly applications for tolerance analysis. Kinematic joints describe motion constraints at the points of contact between mating parts. Marler (1988) and Chun (1988) defined a set of kinematic joint types to accommodate the kinematic variations at the contact points in two-dimensional assemblies. Figure 5.8 shows the joints and datums for two-dimensional analysis. Corresponding modeling rules have been developed for correctly representing the kinematic degrees of freedom in an assembly. Larsen (1991) and Trego (1993) further developed Chun and Marler's work and automated the procedure of generating vector loop relationships for assemblies.

5.7 DLM: LINEARIZATION OF IMPLICIT ASSEMBLY FUNCTIONS

The direct linearization method (DLM) for assembly tolerance analysis is based on the first-order Taylor's series expansion of the assembly kinematic constraint equations with respect to both the assembly variables and the manufactured variables (component dimensions) in an assembly. Linear algebra is employed to solve the resulting linearized equations for the variations of the assembly variables in terms of the variations of the manufactured components. The resulting explicit expressions may be evaluated by either a worst-case or statistical tolerance accumulation model (Chase et al., 1995).

5.7.1 Kinematic Assembly Constraint

Figure 5.9 shows a vector loop model of an assembly in two dimensions. Each vector defines the relative rotation and translation from the previous vector. If a vector represents a component dimension, its variation is the specified component tolerance. If it is a kinematic variable, its variation must be determined by solving the vector equation. A similar interpretation holds for the relative angles. Whether a length or angle is a kinematic variable is

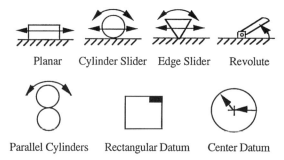

Planar   Cylinder Slider   Edge Slider   Revolute

Parallel Cylinders   Rectangular Datum   Center Datum

Figure 5.8. Two-dimensional kinematic joint and datum types.

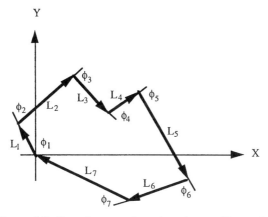

Figure 5.9. Sample vector-loop-based assembly model.

determined by the degrees of freedom of the corresponding kinematic joint defined at the points of contact between mating parts.

A closed vector loop, such as that shown in Figure 5.9, defines a kinematic closure constraint for the assembly. This means there is some adjustable element in the assembly which always permits closure. Closed-loop constraints can readily be expressed as implicit assembly functions. An open vector loop describes a gap or a stack dimension, corresponding to a critical assembly feature. The gap variation is the result of the accumulation of component tolerances.

Many assembly applications may be described by an implicit system of open and closed loops requiring simultaneous solution. When no adjustable elements are present, no closed loops are required, in which case open loops may be expressed as explicit assembly functions. Assembly tolerance limits are determined by performance requirements. Component tolerance limits are determined from process characterization studies but may have to be modified as a result of tolerance analysis, which reveals how each component variation contributes to the overall assembly variation. Engineering design limits may be placed on any kinematic variation in a closed loop or any assembly feature variation defined by an open loop. By comparing the computed variations to the specified limits, the percent rejects and assembly quality levels may be estimated.

By summing the vector components in the global x and y directions and summing the relative rotations, a vector loop produces three scalar equations, each summing to zero, as shown by the equations

$$H_x = \sum_{i=1}^{n} L_i \cos \left(\sum_{j=1}^{i} \phi_j \right) = 0 \tag{5.9}$$

$$H_y = \sum_{i=1}^{n} L_i \sin \left(\sum_{j=1}^{i} \phi_j \right) = 0 \qquad (5.10)$$

$$H_\phi \sum_{i=1}^{n} \phi_j = 0 \text{ or } 360° \qquad (5.11)$$

It is significant that each vector direction, represented by the arguments of the sine and cosine functions in the equations above, is expressed as the sum of the relative angles of all the vectors preceding it in the loop. Both the manufactured and kinematic angles are relative angles. This allows rotational variations to propagate realistically through an assembly, producing rigid-body rotations of stacked mating parts. This effect of individual angle variations could not be described if global angles were used in the equations.

5.7.2   Taylor's Expansion of Implicit Assembly Functions

The first-order Taylor's series expansion of the closed-loop assembly equations can be written in matrix form:

$$\{\Delta H\} = [A] \{\Delta X\} + [B] \{\Delta U\} = \{\Theta\} \qquad (5.12)$$

where $\{\Delta H\}$ = variations of the closure relation
    $\{\Delta X\}$ = variations of the manufactured variables
    $\{\Delta U\}$ = variations of the kinematic assembly variables
    $[A]$ = first-order partial derivatives of the closure functions with respect to the manufactured variables, $\partial H_i / \partial x_j$
    $[B]$ = first-order partial derivatives of the closure functions with respect to the kinematic assembly variables, $\partial H_i / \partial u_j$

Solving equation (5.12) for ΔU gives (assuming that $[B]$ is a full-ranked matrix)

$$\{\Delta U\} = -[B]^{-1} [A] \{\Delta X\} \qquad (5.13)$$

For an open-loop assembly constraint, there may also be one or more closed-loop assembly constraints that the assembly must satisfy. The strategy for such a system of assembly constraints is to solve the closed-loop constraints first, then substitute the solution in the open-loop assembly constraint. The variations in the open-loop variables may then be evaluated directly. This procedure may be expressed mathematically as follows:

$$\{\Delta V\} = [C] \{\Delta X\} + [D] \{\Delta U\} \qquad (5.14)$$

where ΔV represents the variations of the open-loop assembly variables, $[C]$

the first-order derivative matrix of the assembly variables with respect to the manufactured variables in the open loop, $\partial V_i / \partial x_j$, and $[D]$ the first-order derivative matrix of the assembly variables with respect to the kinematic assembly variables in the open loop, $\partial V_i / \partial u_j$. If $[B]$ is full-ranked, equation (5.12) may be written as

$$\{\Delta V\} = ([C] - [D][B]^{-1}[A])\{\Delta X\} \tag{5.15}$$

5.7.3 Estimation of Kinematic Variations and Assembly Rejects

The estimation of the kinematic variations can be obtained from equation (5.13) for the closed-loop constraint, or equation (5.15) for the open-loop constraint, by a worst-case or statistical tolerance accumulation model.

Worst case:

$$\Delta U_i = \sum_{j=1}^{n} S_{ij} |\text{tol}_j \leq T_{\text{ASM}} \tag{5.16}$$

Statistical model:

$$\Delta U_i = \sqrt{\sum_{j=1}^{n} (S_{ij}\text{tol}_j)^2} \leq T_{\text{ASM}} \tag{5.17}$$

where $i = 1,...,n$ assembly variables, tol_j the tolerance of the jth manufactured dimension, T_{ASM} the design specification for the ith assembly variable, and $[S]$ the tolerance sensitivity matrix of the assembly. For closed-loop constraints,

$$[S] = -[B]^{-1}[A] \tag{5.18}$$

For open-loop constraints,

$$[S] = [C] - [D][B]^{-1}[A] \tag{5.19}$$

The estimation of the assembly rejects is based on the assumption that the resulting sum of component distributions is normal or Gaussian, which is a reasonable estimate for assemblies of manufactured variables. If all the component tolerances are assumed to represent three standard deviations of the corresponding process, the estimate of the related assembly variation will be three standard deviations. Equation (5.17) may easily be modified to account for tolerance limits which represent a value other than three standard deviations. When tolerance limits have been specified for an assembly variation, the mean and standard deviation of the assembly variable can be used to calculate by either integration or table the assembly rejects for a given production quantity of assemblies.

5.8   EXAMPLES

As examples to demonstrate the procedure of applying the DLM assembly tolerance analysis method to real assemblies, the one-way clutch assembly and a geometric block assembly are examined in detail.

5.8.1   Example 1: One-Way Clutch

Figure 5.7 illustrated the vector-loop-based model of the one-way clutch assembly. Table 5.1 shows the detailed dimensions for the assembly. The question marks in the table indicate the kinematic variations which must be determined by tolerance analysis. From Figure 5.7 the loop equations of the assembly follow naturally as

$$H_x = b + c \cos(90° + \phi_1) + e \cos(270° + \phi_1) = 0 \qquad (5.20)$$

$$H_y = a + c + c \sin(90° + \phi_1) + e \sin(270° + \phi_1) = 0 \qquad (5.21)$$

$$H_\theta = 90° - 90° + 90° - \phi_1 + 180° + \phi_2 = 0 \qquad (5.22)$$

The known independent variables in this set of equations are a, c, and e. The

TABLE 5.1. Dimensions of One-Way Clutch Vector Loop

Part Name	Transformation	Nominal Dimension	Tolerance ($\pm$)	Variation
Height of hub	Rotation	90°		
	Translation a	27.645	0.0125	Independent
Position of roller	Rotaton	−90°		
	Translation b	4.81053	?	Kinematic
Radius of roller	Rotation	90°		
	Translation c	11.43	0.1	Independent
Radius of roller	Rotation ϕ_1	−7.01838°	?	Kinematic
	Translation c	11.43	0.01	Independent
Radius of ring	Rotation	180°		
	Translation e	50.8	0.05	Independent
Closing vector	Rotation ϕ_2	97.01838°	?	Kinematic
	Translation	0		

unknown dependent variables are b, ϕ_1, and ϕ_2. Note that dimension c appears twice in equation (5.21). Since both vectors are produced by the same process, they will both be oversized or undersized simultaneously.

Applying the DLM method, the first-order derivative matrices $[A]$ and $[B]$ and sensitivity matrix $[S]$ can be obtained.

$$
\begin{aligned}
[A] &= \begin{bmatrix}
\dfrac{\partial H_x}{\partial_a} & \dfrac{\partial H_x}{\partial c} & \dfrac{\partial H_x}{\partial e} \\[2ex]
\dfrac{\partial H_y}{\partial a} & \dfrac{\partial H_y}{\partial c} & \dfrac{\partial H_y}{\partial e} \\[2ex]
\dfrac{\partial H_\theta}{\partial a} & \dfrac{\partial H_\theta}{\partial c} & \dfrac{\partial H_\theta}{\partial e}
\end{bmatrix} \\[4ex]
&= \begin{bmatrix}
0 & \cos(90° + \phi_1) & \cos(270° + \phi_1) \\
1 & 1 + \sin(90° + \phi_1) & \sin(270° + \phi_1) \\
0 & 0 & 0
\end{bmatrix} \\[4ex]
&= \begin{bmatrix}
0 & 0.1222 & -0.1222 \\
1 & 1.9925 & -0.9925 \\
0 & 0 & 0
\end{bmatrix}
\end{aligned}
\tag{5.23}
$$

$$
\begin{aligned}
[B] &= \begin{bmatrix}
\dfrac{\partial H_x}{\partial b} & \dfrac{\partial H_x}{\partial \phi_1} & \dfrac{\partial H_x}{\partial \phi_2} \\[2ex]
\dfrac{\partial H_y}{\partial b} & \dfrac{\partial H_y}{\partial \phi_1} & \dfrac{\partial H_y}{\partial \phi_2} \\[2ex]
\dfrac{\partial H_\theta}{\partial b} & \dfrac{\partial H_\phi}{\partial \phi_1} & \dfrac{\partial H_\phi}{\partial \phi_2}
\end{bmatrix}
\end{aligned}
\tag{5.24}
$$

$$
\begin{aligned}
&= \begin{bmatrix}
1 & -c\,\sin(90° + \phi_1) - e\,\sin(270° + \phi_1) & 0 \\
0 & c\,\cos(90° + \phi_1) + e\,\cos(270° + \phi_1) & 0 \\
0 & 1 & 1
\end{bmatrix} \\[4ex]
&= \begin{bmatrix}
1 & 39.075 & 0 \\
0 & -4.811 & 0 \\
0 & 1 & 1
\end{bmatrix}
\end{aligned}
\tag{5.25}
$$

$$
\begin{aligned}
[S] &= [B]^{-1}[A] \\[2ex]
&= \begin{bmatrix}
-8.1220 & -16.305 & 8.1833 \\
0.2079 & 0.4142 & -0.2063 \\
-0.2079 & -0.4142 & 0.2063
\end{bmatrix}
\end{aligned}
$$

With the sensitivity matrix known, the variations of the kinematic or assembly variables can then be calculated by applying equation (5.16) or (5.17).

Worst case: Statistical model:

$$\begin{Bmatrix} \Delta b \\ \Delta\phi_1 \\ \Delta\phi_2 \end{Bmatrix} = \begin{Bmatrix} 0.6737 \\ 0.9772° \\ 0.9772° \end{Bmatrix} \qquad \begin{Bmatrix} \Delta b \\ \Delta\phi_1 \\ \Delta\phi_2 \end{Bmatrix} = \begin{Bmatrix} 0.4520 \\ 0.6540° \\ 0.6540° \end{Bmatrix} \tag{5.26}$$

In this assembly, dimension ϕ_1 is the one that has a specified design tolerance since its mean value and variation will affect the performance of the clutch. The design limits for ϕ_1 are set to be $T_{ASM} = \pm0.6°$, with a desired quality level of ±3.0 standard deviations. The number of standard deviations Z to which the design spec corresponds may be calculated from the relation

$$Z = \frac{T_{ASM}}{\sigma_1} = \frac{0.6}{0.6540} \times 3.0 = 2.7523 \tag{5.27}$$

The number of standard deviations Z can then be used to estimate the assembly reject rate η either from standard tables or by integration of the normal distribution between limits.

$$\text{reject rate} = \eta = 0.002959 \quad \text{defect per unit} \tag{5.28}$$

The assembly rejects for a production run of 1000 assemblies can be estimated by

$$\text{assembly rejects} = 2\eta \times \text{number of assemblies}$$
$$= 2(0.002959)(1000)$$
$$= 5.918 \tag{5.29}$$

So there would be about six that would function improperly (three at each design limit).

5.8.2   Example 2: Geometric Block Assembly

Figure 5.10 shows a geometric block assembly. The resultant assembly dimension U_1 is very difficult to express explicitly as a function of only the independent component dimensions a, b, c, d, e, and f. It is very difficult to define such explicit assembly functions in a generalized manner for real-life mechanical assemblies. This difficulty makes the use of explicit functions impractical in a CAD-based system intended for use by mechanical designers.

   The geometric block assembly requires three vector loops to describe the assembly relationship completely, even though it is only a simple three-component assembly. Figure 5.11 shows the vector loop assembly model. Table 5.2 gives all the dimensions for the three vector loops. For each vector loop, three equations having the same format as equation (5.20) to (5.22) can be obtained. Therefore, nine equations are required to describe the assembly.

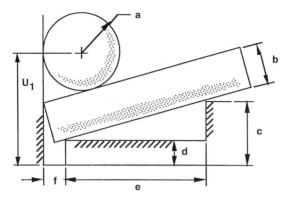

Figure 5.10. Simple geometric block assembly.

$$[A] = \left[\frac{\partial H^i_x}{\partial X_j}, \frac{\partial H^i_y}{\partial X_j}, \frac{\partial H^i_\theta}{\partial X_j}\right]^T$$

i = number of loops, j = number of independent variables

$$= \begin{array}{cccccc} a & b & c & d & e & f \end{array}$$
$$\begin{bmatrix}
-1.2635 & 0 & 0 & 0 & 0 & 0 \\
0.9647 & 0 & 0 & 0 & 0 & 0 \\
0 & 0 & 0 & 0 & 0 & 0 \\
0 & 0.2635 & 0 & 0 & 0 & -1 \\
0 & -0.9647 & 0 & -1 & 0 & 0 \\
0 & 0 & 0 & 0 & 0 & 0 \\
0 & 0.2635 & 0 & 0 & -1 & -1 \\
0 & -0.9647 & -1 & 0 & 0 & 0 \\
0 & 0 & 0 & 0 & 0 & 0
\end{bmatrix} \qquad (5.30)$$

$$[B] = \left[\frac{\partial H^i_x}{\partial U_j}, \frac{\partial H^i_y}{\partial U_j}, \frac{\partial H^i_\theta}{\partial U_j}\right]^T$$

i = loops; j = number of dependent variables

u_1	u_2	u_3	u_4	u_5	ϕ_1	ϕ_2	ϕ_3	ϕ_4
0	0.9647	0	0	0	18.7181	10.0477	0	0
−1	0.2635	1	0	0	−6.6200	0	0	0
0	0	0	0	0	1	1	0	0
0	0	0	0.9647	0	0	10.0477	4.0600	0
0	0	1	0.2635	0	0	0	−3.9050	0
0	0	0	0	0	0	1	1	0
0	0	0	0	0.9647	0	10.0477	0	10.6750
0	0	1	0	0.2635	0	0	0	−28.1250
0	0	0	0	0	0	1	0	1

$$(5.31)$$

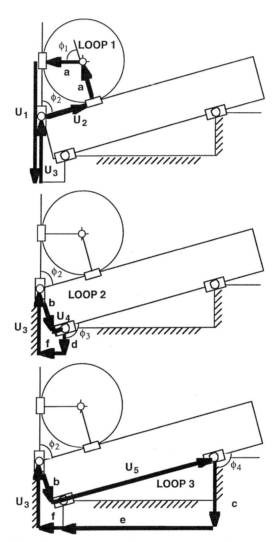

Figure 5.11. Vector loop model of the geometric block assembly.

TABLE 5.2. Dimensions of the Geometric Block Assembly

Loop Name	Part Name	Transformation	Nominal Dimensions	Tolerance ($\pm$)
Loop 1	Ground	Rotation	90°	
		Translation U_3	10.0477	?
	Block	Rotation ϕ_2	−74.7243°	?
		Translation U_2	8.6705	?
	Cylinder	Rotation	90°	
		Translation a	6.62	0.2
	Cylinder	Rotation ϕ_1	74.7243°	?
		Translation a	6.62	0.2
Closing angle 90°	Ground	Rotation	90°	
		Translation U_1	18.7181	?
Loop 2	Ground	Rotation	90°	
		Translation U_3	10.0477	?
	Block	Rotation $\phi_2 + 90°$	−164.7243°	?
		Translation b	6.805	0.075
	Block	Rotation	90°	
		Translation U_4	2.1894	?
	Ground	Rotation ϕ_3	−105.2761°	?
		Translation d	4.06	0.15
Closing angle 180°	Ground	Rotation	−90°	
		Translation f	3.905	0.125
Loop 3	Ground	Rotation	90°	
		Translation U_3	10.0477	?
	Block	Rotation $\phi_2 + 90°$	−164.7243°	?
		Translation b	6.805	0.075
	Block	Rotation	90°	
		Translation U_5	27.2965	?
	Ground	Rotation U_4	−105.2761°	?
		Translation c	10.675	0.125
	Ground	Rotation	−90°	
		Translatoin e	24.22	0.35
Closing angle 180°	Ground	Rotation	0°	
		Translation f	3.905	0.125

$$[S] = -[B]^{-1}[A]$$

$$
= \begin{bmatrix}
a & b & c & d & e & f \\
1.3098 & 1.0367 & 0.2581 & 0.7419 & -0.0705 & -0.2731 \\
1.3097 & 0 & 0.3453 & -0.3453 & -0.0943 & 0 \\
0 & 1.0367 & -0.0872 & 1.0872 & 0.0238 & -0.2731 \\
0 & -0.2731 & -0.2385 & 0.2385 & 0.0651 & 1.0366 \\
0 & -0.2731 & 0.0250 & -0.0250 & 1.0298 & 1.0366 \\
0 & 0 & -0.0384 & 0.0384 & 0.0105 & 0 \\
0 & 0 & 0.0384 & -0.0384 & -0.0105 & 0 \\
0 & 0 & -0.0384 & 0.0384 & 0.0105 & 0 \\
0 & 0 & -0.0384 & 0.0384 & 0.0105 & 0
\end{bmatrix} \quad (5.32)
$$

The variations of the kinematic or assembly variables can then be calculated by applying equation (5.16) or (5.17).

$$
\begin{array}{cc}
\text{Worst case:} & \text{Statistical model:} \\
\begin{Bmatrix} \Delta U_1 \\ \Delta U_2 \\ \Delta U_3 \\ \Delta U_4 \\ \Delta U_5 \\ \Delta \phi_1 \\ \Delta \phi_2 \\ \Delta \phi_3 \\ \Delta \phi_4 \end{Bmatrix} = \begin{Bmatrix} 0.5421 \\ 0.3899 \\ 0.2942 \\ 0.2384 \\ 0.5174 \\ 0.8156° \\ 0.8156° \\ 0.8156° \\ 0.8156° \end{Bmatrix} &
\begin{Bmatrix} \Delta U_1 \\ \Delta U_2 \\ \Delta U_3 \\ \Delta U_4 \\ \Delta U_5 \\ \Delta \phi_1 \\ \Delta \phi_2 \\ \Delta \phi_3 \\ \Delta \phi_4 \end{Bmatrix} = \begin{Bmatrix} 0.2998 \\ 0.2725 \\ 0.1844 \\ 0.1411 \\ 0.3836 \\ 0.4784° \\ 0.4784° \\ 0.4784° \\ 0.4784° \end{Bmatrix}
\end{array} \quad (5.33)
$$

In this assembly, dimension U_1 is the feature for which a design tolerance was specified, since its value and variation will affect the desired performance of the assembly. If the design limits for U_1 are set to be $T_{ASM} = \pm 0.28$ and the estimated variation ΔU_1 represents 3.0 standard deviations, the design spec corresponds to Z standard deviations, where

$$
Z = \frac{T_{ASM}}{\phi_1} = \frac{0.28}{0.2998} \times 3.0 = 2.8019 \quad (5.34)
$$

Then the predicted reject rate for each design limit is estimated from

$$
\text{reject rate} = \eta = 0.002540 \text{ defect per unit}
$$

$$
\text{assembly rejects} = 2\eta \times \text{number of assemblies}
$$

$$
- 2(0.002540)(1000)
$$

$$
= 5.08 \text{ per 1000 assemblies} \quad (5.35)
$$

5.9 TOLERANCE ANALYSIS OF MECHANICAL ASSEMBLIES IN THREE-DIMENSIONAL SPACE

Estimating variations in three-dimensional assemblies is considerably more difficult than in two-dimensional assemblies. The mathematical representation of the assembly functions requires concatenated transformation matrices. Global coordinate data for positioning and orienting important features and mating surfaces are difficult to calculate. The task is better suited for modeling the geometry on a three-dimensional CAD system and extracting the relevant geometry from the CAD model. Establishing a sound mathematical basis is a necessary first step. Engineering models of assembly variations must include all three contributing sources: (1) dimensional, (2) geometric, and (3) kinematic variations.

Gilbert (1992) and Whitney et al. (1994) proposed a system to estimate the propagation effect of the dimensional and geometric feature tolerances of components in an open-loop assembly using the linearized model of the 4 × 4 transformation matrix constraint developed by Veitschegger and Wu (1986). Several geometric feature tolerances were studied and geometric feature variations were characterized for certain mating conditions between the parts.

Lin and Chen (1994) developed a linearized scheme for tolerance analysis of closed-loop mechanisms in two-dimensional space using the 4 × 4 transformation matrix in the Denavit–Hartenberg (D-H) symbolic notation common to robotics. This method is significant since it has the capability of assessing the influence of each error component on a machine's accuracy, and it is also capable of including geometric feature variations. It is particularly useful to allocate a machine's tolerances at its design stage or to characterize the performance of an existing machine. Thus far the method has not been applied to static assemblies. It has only been applied to robotic mechanisms with revolute and prismatic joints (see the next section for joint classifications).

Robison (1989) and Gao et al. (to appear) developed the DLM for treating implicit assembly functions, such as kinematic constraints, for assembly tolerance analysis in three-dimensional space. It is a generalized approach, employing common engineering concepts of open or closed vector loops, kinematic joints, and geometric feature tolerances. The DLM is based on the first-order Taylor's series expansion of the assembly kinematic constraint equations with respect to both assembly variables and manufactured component dimensions in an assembly. The implicit equations describing the kinematic constraints are reduced to a set of linear equations for small changes about the nominal. Matrix algebra is employed to solve the linearized equations for the variations of each assembly variable in terms of the variations of the manufactured components by either worst-case or statistical models. Assembly geometry is reduced dramatically, so it is very efficient computationally. It is suited for integration with commercial CAD systems, by which the required geometry may be extracted directly from a solid model. The

DLM method has been extended to include all geometric feature variations in two- and three-dimensional assemblies. (Chase et al., 1996).

5.9.1   Kinematic Constraint Equations

The mating conditions for three-dimensional assemblies require a variety of kinematic joints to describe the possible relative motions between mating parts. Figure 5.12 shows a set that can handle a wide variety of assembly constraints. The kinematic degrees of freedom vary from zero for the rigid joint to five for the spherical slider and crossed cylinder joints.

The vector loop equations were derived for two-dimensional assemblies in Section 5.6. Each vector loop equation for two-dimensional assemblies resulted in three scalar equations, representing the sum of the x and y projections of the dimension vectors in global coordinates and the sum of the relative rotations. In three-dimensional space, the system of equations is more complicated. Each vector loop equation for three-dimensional assemblies results in six scalar equations, representing the sum of the x, y, and z projections of the dimension vectors in global coordinates and the sum of the relative rotations about the three global axes. The projection equations still hold in global x, y, and z, but the rotation constraints can only be expressed in matrix form.

A kinematic constraint equation for a three-dimensional mechanical assembly can be described by a closed vector loop. As the vector loop is traversed from the beginning point to the endpoint, the dimensions of each component

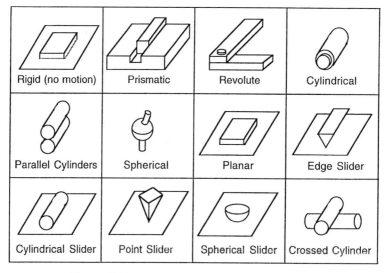

Figure 5.12. Three-dimensional kinematic joints.

and the relative rotations are summed. The translations must sum to zero and the rotations must be such that the coordinate system at the end is congruent with the one at the beginning. To express the loop equations in matrix form, each vector may be expressed as a product of transformation matrices. The transformation from joint $i - 1$ to i consists of a combination of three rotation and one translation matrix. We define the translation always to be along the local x-axis, following the next vector in the chain. The magnitude of the translation, L, is the length of the next vector. The rotation matrices describe the relative rotation from one vector to the next. For three-dimensional transformations we have

$$[R_x] = \begin{bmatrix} 1 & 0 & 0 & 0 \\ 0 & \cos \phi_z & -\sin \phi_z & 0 \\ 0 & \sin \phi_z & \cos \phi_z & 0 \\ 0 & 0 & 0 & 1 \end{bmatrix} \tag{5.36}$$

$$[R_y] = \begin{bmatrix} \cos \phi_y & 0 & \sin \phi_y & 0 \\ 0 & 1 & 0 & 0 \\ -\sin \phi_y & 0 & \cos \phi_y & 0 \\ 0 & 0 & 0 & 1 \end{bmatrix} \tag{5.37}$$

$$[R_z] = \begin{bmatrix} \cos \phi_z & -\sin \phi_z & 0 & 0 \\ \sin \phi_z & \cos \phi_z & 0 & 0 \\ 0 & 0 & 1 & 0 \\ 0 & 0 & 0 & 1 \end{bmatrix} \tag{5.38}$$

$$[T] = \begin{bmatrix} 1 & 0 & 0 & L \\ 0 & 1 & 0 & 0 \\ 0 & 0 & 1 & 0 \\ 0 & 0 & 0 & 1 \end{bmatrix} \tag{5.39}$$

With this convention, a closed-loop kinematic constraint can be written as equation (5.40), meaning that the product of all the transformation matrices is equal to the identity matrix:

$$[R_1][T_1][R_2][T_2] \cdots [R_i][T_i] \cdots [R_n][T_n][R_f] = [I] \tag{5.40}$$

where $[R_i]$ is the product of rotation matrices at joint i, $[T_i]$ the translation matrix at joint i, $[R_f]$ the final rotation required to bring the loop to closure, and $[I]$ the identity matrix.

From the theory of mechanisms, up to six independent equations can be drawn from each loop equation (5.40) (Sandor and Erdman, 1984). Therefore, each loop constraint may be solved for up to six unknowns or assembly variables. These are nonlinear equations and require a nonlinear solver to find solutions. However, for small variations about the nominal of each component in the assembly, the solutions can be approximated by the DLM. Therefore, only the derivatives are needed to formulate the linearized equations.

5.9.2 Derivative Evaluation

Robison (1989) used a small perturbation method for evaluating the derivatives on the assembly constraint equation (5.8). For a three-dimensional case, if the translation or rotation at joint i is the variable with respect to which the derivatives are desired, a small length perturbation dL or angle perturbation $d\phi$ is added to the original value, and the matrix multiplication of equation (5.40) is performed. For translation,

$$[R_1][T_1] \cdots [R_i][T_i(L + dL)] \cdots [R_n][T_n]\{0 \quad 0 \quad 0 \quad 1\}^t$$
$$= \{DX \quad DY \quad DZ \quad 1\}^t \quad (5.41)$$

For rotation,

$$[R_1][T_1] \cdots [R_i(\phi + d\phi)][T_i] \cdots [R_n][T_n]\{0 \quad 0 \quad 0 \quad 1\}^t$$
$$= \{DX \quad DY \quad DZ \quad 1\}^t \quad (5.42)$$

The derivatives can then be approximated numerically:

Translational variable: Rotational variable:

$$
\begin{aligned}
\frac{\partial H_x}{\partial L} &\approx \frac{DX}{dL} & \qquad \frac{\partial H_x}{\partial \phi} &\approx \frac{DX}{d\phi} \\
\frac{\partial H_y}{\partial L} &\approx \frac{DY}{dL} & \frac{\partial H_y}{\partial \phi} &\approx \frac{DY}{d\phi} \\
\frac{\partial H_z}{\partial L} &\approx \frac{DZ}{dL} & \frac{\partial H_z}{\partial \phi} &\approx \frac{DZ}{d\phi} \\
\frac{\partial H_{\theta x}}{\partial L} &= 0 & \frac{\partial H_{\theta x}}{\partial \phi} &= \cos \alpha \\
\frac{\partial H_{\theta y}}{\partial L} &= 0 & \frac{\partial H_{\theta y}}{\partial \phi} &= \cos \beta \\
\frac{\partial H_{\theta z}}{\partial L} &= 0 & \frac{\partial H_{\theta z}}{\partial \phi} &= \cos \gamma
\end{aligned}
\quad (5.43)
$$

where H_x, H_y, and H_z are the translational constraints and $H_{\theta x}$, $H_{\theta y}$, and $H_{\theta z}$ are the rotational constraints in the x, y, and z directions, respectively, and α, β, and γ are the global direction consine angles of the local axis around which the rotation is made (Gao, 1993).

5.9.3 Linearization of the Implicit Assembly Constraints

The first-order Taylor's series expansion of the *closed-loop* kinematic constraints, equation (5.38), can be written in matrix form as

$$\{\Delta H\} = [A]\{\Delta X\} + [B]\{\Delta U\} = \{\Theta\} \tag{5.44}$$

where $\{\Delta H\}$ = variations of the clearance
$\{\Delta X\}$ = variations of the manufactured variables
$\{\Delta U\}$ = variations of the assembly variables
$[A]$ = first-order partial derivatives of the manufactured variables
$[B]$ = first-order partial derivatives of the assembly variables

Matrixes $[A]$ and $[B]$ can be obtained by the method discussed above. Each column of the $[A]$ matrix takes the format

$$\{A_i\} = \left[\frac{\partial H_x}{\partial x_i}, \frac{\partial H_y}{\partial x_i}, \frac{\partial H_z}{\partial x_i}, \frac{\partial H_{\theta x}}{\partial x_i}, \frac{\partial H_{\theta y}}{\partial x_i}, \frac{\partial H_{\theta z}}{\partial x_i} \right]^t \tag{5.45}$$

where x_i is the ith manufactured dimension. Matrix $[B]$ has the same column notation except that the variable is u_i instead of x_i.

To map the derivatives into the A or B matrices correctly requires that each vector and rotation in the loop be identified as either a dependent or an independent variable or a constant. A set of modeling rules is required when creating the model which assure the proper relationships between the vectors passing through each joint and the joint axes. Consistent relationships permit construction of algorithms for making the determination and performing the mapping correctly (Chase and Trego, 1994).

Solving equation (5.44) for ΔU gives (assuming that $[B]$ is a full-ranked matrix)

$$\{\Delta U\} = [B]^{-1}[A]\{\Delta X\} \tag{5.46}$$

If the $[B]$ matrix is singular or nonsquare, that means the system is overconstrained or has too many equations, and a least square fit must be applied to solve equation (5.44):

$$\{\Delta U\} = -([B]^{\mathrm{T}}[B])^{-1}[B]^{\mathrm{T}}[A]\{\Delta X\} \tag{5.47}$$

For an *open-loop* kinematic constraint, there may also be one or more closed-loop kinematic constraints that the assembly must satisfy. The strategy for such a system of assembly constraints is to solve the closed-loop constraints first and then substitute the solution in the open-loop kinematic constraint. Finally, evaluate the variations of the open-loop variables:

$$\{\Delta V\} = [C]\{\Delta X\} + [D]\{\Delta U\} \tag{5.48}$$

where ΔV is the variations of the open-loop assembly variables, $[C]$ the first-order derivative matrix of the manufactured variables in the open loop, and $[D]$: the first-order derivative matrix of the assembly variables in the open loop. If $[B]$ is full-ranked, equation (5.48) can be written as

$$\{\Delta V\} = ([C] - [D][B]^{-1}[A])\{\Delta X\} \tag{5.49}$$

or when $[B]$ is not full-ranked:

$$\{\Delta V\} = ([C] - [D]([B]^{\mathrm{T}}[B])^{-1}[B]^{\mathrm{T}}[A])\{\Delta X\} \tag{5.50}$$

5.9.4   Estimation of Kinematic Variations and Assembly Rejects

The estimation of the kinematic variations can be obtained from equation (5.46) or (5.47) for the closed-loop constraint, or equation (5.49) or (5.50) for the open-loop constraint, by a worst-case or statistical accumulation model, as we obtained for two-dimensional assemblies.

Worst case:

$$\Delta U_i = \sum_{j=1}^{n} |S_{ij}| \, \mathrm{tol}_j \leq T_{\mathrm{ASM}} \tag{5.51}$$

Statistical model:

$$\Delta U_i = \sqrt{\sum_{j=1}^{n} (S_{ij}\mathrm{tol}_j)^2} \leq T_{\mathrm{ASM}} \tag{5.52}$$

where $i = 1,...,n$ assembly variables, tol_j is the tolerance of the jth manufactured dimension, T_{ASM} the design specification for the ith assembly variable, and $[S]$ the tolerance sensitivity matrix. For determined and over-determined systems, $[S]$ is given respectively by, for closed-loop constraints,

$$[S] = -[B]^{-1}[A] \qquad \text{(determined)} \tag{5.53}$$

$$[S] = -([B]^{\mathrm{T}}[B])^{-1}[B]^{\mathrm{T}}[A] \qquad \text{(overdetermined)} \tag{5.54}$$

and for open-loop constraints,

$$[S] = [C] - [D][B]^{-1}[A] \qquad \text{(determined)} \qquad (5.55)$$

$$[S] = [C] - [D]([B]^{\mathrm{T}}[B])^{-1}[B]^{\mathrm{T}}[A] \qquad \text{(over determined)} \qquad (5.56)$$

The estimation of the assembly rejects for specified tolerance limits may be obtained from standard tables or by integration of the resulting normal distribution between limits.

5.10 THREE-DIMENSIONAL EXAMPLE

As an example to demonstrate the procedure of applying DLM to a real assembly, the three-dimensional crank slider mechanism shown in Figure 5.13 is examined in detail. The assembly consists of a base, a crank, a link, and a slider. The crank rotates around its shaft. This rotation is transmitted by the crank lever to the connecting rod or link. The connecting rod pivots around the "ball joints" as the slider moves forward and backward. The dimensions that govern the operation of the crank slider assembly are shown in the figure. The assembly constraint equation for this mechanism can be obtained by rotating and translating the local joint coordinate system while traversing the assembly vector loop. When the vector loop has been traversed completely, an assembly constraint equation in the form of equation (5.40) results. The

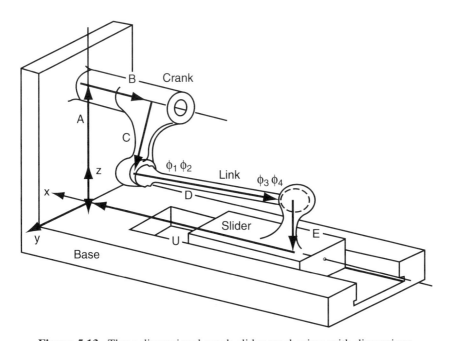

Figure 5.13. Three-dimensional crank slider mechanism with dimensions.

dependent kinematic variables must be identified before the system can be analyzed.

For analysis, the input crank is rotated to a predetermined position; therefore, it is not an unknown. The two spherical joints are defined with two rotational degrees of freedom (Sandor and Erdman, 1984). Together, they have four degrees of freedom. The slider can only move forward and backward, so it has 1 degree of freedom. The rotation of the connecting rod around its axis is arbitrary and therefore meaningless to the tolerance analysis. So the total kinematic or assembly variables in this system are five instead six, as expected for a three-dimensional assembly loop. The geometric data of the crank slider mechanism is given in Table 5.3, with the global reference frame defined in Figure 5.13.

The derivative matrix with respect to the assembly variables can be expressed as

$$
[B] = \begin{bmatrix}
\dfrac{\partial H_x}{\partial \phi_1} & \dfrac{\partial H_x}{\partial \phi_2} & \dfrac{\partial H_x}{\partial \phi_3} & \dfrac{\partial H_x}{\partial \phi_4} & \dfrac{\partial H_x}{\partial U} \\[2mm]
\dfrac{\partial H_y}{\partial \phi_1} & \dfrac{\partial H_y}{\partial \phi_2} & \dfrac{\partial H_y}{\partial \phi_3} & \dfrac{\partial H_y}{\partial \phi_4} & \dfrac{\partial H_y}{\partial U} \\[2mm]
\dfrac{\partial H_z}{\partial \phi_1} & \dfrac{\partial H_z}{\partial \phi_2} & \dfrac{\partial H_z}{\partial \phi_3} & \dfrac{\partial H_z}{\partial \phi_4} & \dfrac{\partial H_z}{\partial U} \\[2mm]
\dfrac{\partial H_{\theta x}}{\partial \phi_1} & \dfrac{\partial H_{\theta x}}{\partial \phi_2} & \dfrac{\partial H_{\theta x}}{\partial \phi_3} & \dfrac{\partial H_{\theta x}}{\partial \phi_4} & \dfrac{\partial Y_{\theta x}}{\partial U} \\[2mm]
\dfrac{\partial H_{\theta y}}{\partial \phi_1} & \dfrac{\partial H_{\theta y}}{\partial \phi_2} & \dfrac{\partial H_{\theta y}}{\partial \phi_3} & \dfrac{\partial H_{\theta y}}{\partial \phi_4} & \dfrac{\partial H_{\theta y}}{\partial U} \\[2mm]
\dfrac{\partial H_{\theta z}}{\partial \phi_1} & \dfrac{\partial H_{\theta z}}{\partial \phi_2} & \dfrac{\partial H_{\theta z}}{\partial \phi_3} & \dfrac{\partial H_{\theta z}}{\partial \phi_4} & \dfrac{\partial H_{\theta z}}{\partial U}
\end{bmatrix}
$$

$$
= \begin{bmatrix}
0.8579 & -13.967 & -3.1115 & -2.4467 & 1 \\
8.4857 & -9.8516 & 26.078 & 4.3604 & 0 \\
-8.4857 & -6.7201 & -24.716 & -19.436 & 0 \\
0 & -0.1566 & -0.3492 & 0.8721 & 0 \\
0.7071 & 0.6984 & 0.6223 & 0.4894 & 0 \\
0.7071 & -0.6984 & 0.7006 & 0 & 0
\end{bmatrix} \tag{5.57}
$$

The derivative matrix with respect to the manufactured variables can be obtained in the same way:

TABLE 5.3. Geometric Data of the Crank Slider Mechanism

Part Name	Transformation	Nominal Dimension	Tolerance ($\pm$)
Height of base	Rotation	$\psi = -90°, z = 0°$	
	Translation A	20	0.025
Position of crank	Rotation	$\psi = -90°, z = 0°$	
	Translation B	12	0.0125
Length of crank arm	Rotation	$\psi = -90°, z = 45°$	
	Translation C	15	0.0125
Length of link	Rotation $\phi_1 \ \phi_2$	$\psi = 99.007°, z =$	? ?
	Translation D	$-20.705°$	0.03
		30	
Height of slider	Rotation $\phi_3 \ \phi_4$	$\psi = -78.157°, z =$	? ?
	Translation E	$-44.473°$	0.0025
		5	
Position of slider	Rotation	$\psi = -90°, z = -29.298°$	
	Translation U	39.7164	?

$$[A] = \begin{bmatrix} \dfrac{\partial H_x}{\partial A} & \dfrac{\partial H_x}{\partial B} & \dfrac{\partial H_x}{\partial C} & \dfrac{\partial H_x}{\partial D} & \dfrac{\partial H_x}{\partial E} \\[2mm] \dfrac{\partial H_y}{\partial A} & \dfrac{\partial H_y}{\partial B} & \dfrac{\partial H_y}{\partial C} & \dfrac{\partial H_y}{\partial D} & \dfrac{\partial H_y}{\partial E} \\[2mm] \dfrac{\partial H_z}{\partial A} & \dfrac{\partial H_z}{\partial B} & \dfrac{\partial H_z}{\partial C} & \dfrac{\partial H_z}{\partial D} & \dfrac{\partial H_z}{\partial E} \\[2mm] \dfrac{\partial H_{\theta x}}{\partial A} & \dfrac{\partial H_{\theta x}}{\partial B} & \dfrac{\partial H_{\theta x}}{\partial C} & \dfrac{\partial H_{\theta x}}{\partial D} & \dfrac{\partial H_{\theta x}}{\partial E} \\[2mm] \dfrac{\partial H_{\theta y}}{\partial A} & \dfrac{\partial H_{\theta y}}{\partial B} & \dfrac{\partial H_{\theta y}}{\partial C} & \dfrac{\partial H_{\theta y}}{\partial D} & \dfrac{\partial H_{\theta y}}{\partial E} \\[2mm] \dfrac{\partial H_{\theta z}}{\partial A} & \dfrac{\partial H_{\theta z}}{\partial B} & \dfrac{\partial H_{\theta z}}{\partial C} & \dfrac{\partial H_{\theta z}}{\partial D} & \dfrac{\partial H_{\theta z}}{\partial E} \end{bmatrix}$$

$$= \begin{bmatrix} 0 & -1 & 0 & -0.9239 & 0 \\ 0 & 0 & 0.7071 & -0.3535 & 0 \\ 1 & 0 & -0.7071 & -0.1464 & -1 \\ 0 & 0 & 0 & 0 & 0 \\ 0 & 0 & 0 & 0 & 0 \\ 0 & 0 & 0 & 0 & 0 \end{bmatrix} \qquad (5.58)$$

Equation (5.53) cannot be used to find the sensitivity matrix since the [B] matrix is not a square matrix. Therefore, equation (5.54) must be used to find a least squares fit solution.

$$[S] = -([B]^T[B])^{-1}[B]^T[A]$$

$$= \begin{bmatrix} -0.0290 & 0 & 0.0355 & -0.0032 & 0.0290 \\ 0.0308 & 0 & 0.0064 & -0.0186 & -0.0308 \\ 0.0156 & 0 & -0.0345 & 0.0095 & -0.0156 \\ 0.0336 & 0 & -0.0102 & -0.0117 & -0.0336 \\ 0.5860 & 1 & -0.0735 & 0.6677 & -0.5860 \end{bmatrix} \qquad (5.59)$$

With the sensitivity matrix known, the variations of the kinematic or assembly variables can then be calculated by applying either equation (5.51) or (5.52).

<table>
<tr><td colspan="2" align="center">*Worst case:*</td><td colspan="2" align="center">*Statistical model:*</td></tr>
</table>

$$\begin{Bmatrix} \Delta\phi_1 \\ \Delta\phi_2 \\ \Delta\phi_3 \\ \Delta\phi_4 \\ \Delta U \end{Bmatrix} = \begin{Bmatrix} 0.0766° \\ 0.0851° \\ 0.0656° \\ 0.0804° \\ 0.0496 \end{Bmatrix} \qquad \begin{Bmatrix} \Delta\phi_1 \\ \Delta\phi_2 \\ \Delta\phi_3 \\ \Delta\phi_4 \\ \Delta U \end{Bmatrix} = \begin{Bmatrix} 0.0492° \\ 0.0549° \\ 0.0371° \\ 0.0528° \\ 0.0278 \end{Bmatrix} \qquad (5.60)$$

In this assembly, dimension U is the one that has the design specification, since its value and variation will affect the performance of the mechanism. If the design spec for U is set and the estimated variation ΔU equals 3.0 standard deviations, the assembly reject rate can be calculated from standard tables or by integration of the resulting normal distribution. This example is evaluated at the predetermined position of the crank arm. However, this procedure can be applied repeatedly to find the maximum variation of the position of the slider as the crank arm is incremented through 360°.

5.11   CONCLUSIONS

This chapter has presented a comprehensive method for modeling and ana-lyzing variation in two- and three-dimensional mechanical assemblies. It will make possible new CAD tools for engineering designers which integrate man-ufacturing considerations into the design process. Using this method, design-ers will be able to quantitatively predict the effects of variation on performance and producibility. After the product is in production, manufac-turing systems personnel can substitute more accurate data in the model. Tolerances may be reallocated among the components to reduce overall cost. "What-if" studies can be performed to determine the effect of vendor-supplied components that are out of spec.

If design and manufacturing personnel can adopt a common engineering model describing assembly variation, it can serve as a vehicle for resolving their often competing tolerance requirements. Tolerance analysis can become a common meeting ground where they can work together to pursue cost reduction and quality improvement.

The DLM approach for two- and three-dimensional tolerance analysis is compatible with product design and development processes. It is well suited to design iteration and optimization. Modeling elements are familiar to most engineers, making it suited to the technical skills of engineering designers. A high level of integration with commercial CAD systems should enhance its acceptance by the design community.

A preliminary study compared the DLM assembly tolerance analysis method with Monte Carlo simulation (Gao, 1993; Gao et al., 1995). The results show that the DLM produces accurate estimates of variation for a broad range of assembly applications.

The geometrical feature variations of the components in an assembly, mentioned in Section 5.1, were not discussed. A detailed treatment may be found in a separate paper by the authors (Chase et al., 1996).

ACKNOWLEDGMENTS

This work was sponsored by ADCATS, the Association for the Development of Computer-Aided Tolerancing Software, a consortium of 12 industrial sponsors and the Brigham Young University, including Allied Signal Aerospace, Boeing, Cummins, FMC, Ford, HP, Hughes, IBM, Motorola, Sandia Labs, Texas Instruments, and the U.S. Navy. Major portions of this research were performed by former graduate students Jaren Marler, Ki Soo Chun, Angela Trego, and Richard Robison, with helpful suggestions by Alan R. Parkinson of the Mechanical Engineering Department.

REFERENCES

ASME, 1982, *Dimensioning and Tolerancing,* ANSI Y14.5M-1982, American Society of Mechanical Engineers, New York.

Chase, K. W., and Parkinson, A. R., 1991, A Survey of Research in the Application of Tolerance Analysis to the Design of Mechanical Assemblies, *Res. Eng. Des.,* Vol. 3, pp. 23–37.

Chase, K. W., and Trego, A., 1994, *AutoCATS: Computer-Aided Tolerancing System, 2-D Modeling User Guide,* ADCATS Publication, Brigham Young University, Provo, UT.

Chase, K. W., Gao, J., and Magleby, S. P., 1995, General 2-D Assembly Tolerance Analysis of Mechanical Assemblies with Small Kinematic Adjustments, *J. Des. Manuf.,* 5, No. 4. pp 263–274.

Chase, K. W., Gao, J., Magleby, S. P., and Sorensen, C. D., 1996, Including Geometric Feature Variations in Tolerance Analysis of Mechanical Assemblies *Inst. Ind. Eng. Trans.,* IIE transactions, Vol. 28 No. 10 pp. 795–807.

Chun, K. S., 1988, Development of Two-Dimensional Tolerance Modeling Methods for CAD Systems, M.S. thesis, Mechanical Engineering Department, Brigham Young University, Provo, UT.

Cox, N. D., 1979, Tolerance Analysis by Computer, *J. Qual. Technol.*, Vol. 11, No. 2, April, pp. 80–87.

Cox, N. D., 1986, The ASQC Basic References in Quality Control: Statistical Techniques J. A. Cornell and S. S. Shapiro, series editors, Vol. 11. *How to Perform Statistical Tolerance Analysis,* American Society for Quality Control, Statistical Division, Milwaukee, WI.

Craig, M., 1989, Managing Variation Design using Simulation Methods, *Failure Prevention and Reliability—1989,* DE-Vol. 16, American Society of Mechanical Engineers, New York, pp. 153–163.

Ditlevsen, O., 1979a, Generalized Second Moment Reliability Index, *J. Struct. Mech.,* Vol. 7, No. 4, pp. 435–451.

Ditlevsen, O., 1979b, Narrow Reliability Bounds for Structural System, *J. Struct. Mech.,* Vol. 7, No. 4, pp. 453–472.

Evans, D. H., 1967, An Application of Numerical Integration Techniques to Statistical Tolerancing, *Technometric,* Vol. 9, No. 3, August. pp 441–456.

Evans, D. H.: 1971, An Application of Numerical Integration Techniques to Statistical Tolerancing, II: A Note on the Error, *Technometrics,* Vol. 13, No. 2, February. pp. 315–324.

Evans, D. H., 1972, An Application of Numerical Integration Techniques to Statistical Tolerancing, III: General Distributions, *Technometrics,* Vol. 14, No. 1, May. pp. 23–35.

Evans, D. H., 1975a, Statistical Tolerancing: State of the Art, Part 2: Methods of Estimating Moments, *J. Qual. Technol.,* Vol. 7, No. 1, January, pp. 1–12.

Evans, D. H., 1975b, Statistical Tolerancing: State of the Art, Part 3: Shift and Drifts, *J. Qual. Technol.,* Vol. 7, No. 2, April, pp. 71–76.

Fortini, E. T., 1967, *Dimensioning for Interchangeable Manufacturing,* Industrial Press, New York.

Fuscaldo, J. P., 1991, *Optimizing Collect Chuck Designs Using Variation Simulation Analysis,* SAE Technical Paper 911639, Society of Automative Engineers, Warrendale, PA, August.

Gao, J., 1993, Nonlinear Tolerance Analysis of Mechanical Assemblies, Dissertation, Mechanical Engineering Department, Brigham Young University, Provo, UT.

Gao, J., Chase, K. W., and Magleby, S. P., 1995, Comparison of Assembly Tolerance Analysis by the Direct Linearization and Modified Monte Carlo Simulation Methods, *Proceedings of the ASME Design Engineering Technical Conferences,* Boston, pp. 353–360.

Gao, J., Chase, K. W., and Magleby, S. P., General 3-D Tolerance Analysis of Mechanical Assemblies with Small Kinematic Adjustments, to appear in *Inst. Ind. Eng. Trans.*

Gilbert, O. L., 1992, Representation of Geometric Variations Using Matrix Transforms for Statistical Tolerance Analysis in Assemblies, M.S. thesis, Department of Mechanical Engineering, MIT, Cambridge, MA, June.

Hasofer, A. M., and Lind, N. C., 1974, Exact and Invariant Second Moment Code Format, *J. Eng. Mech. Div. ASME,* Vol. EM-1, February, pp. 111–121.

Larsen, G. C., 1991, A Generalized Approach to Kinematic Modeling for Tolerance Analysis of Mechanical Assemblies, M.S. thesis, Mechanical Engineering Department, Brigham Young University, Provo, UT.

Lee, W. J., and Woo, T. C., 1990, Tolerances: Their Analysis and Synthesis, *J. Eng. Ind.* Vol. 112, May, pp. 113–121.

Lin, P. D., and Chen, J. F., 1994, Analysis of Errors in Precision for Closed Loop Mechanisms, *J. Mech. Des.,* Vol. 116, pp. 197–203.

Marler, J. D., 1988, Nonlinear Tolerance Analysis Using the Direct Linearization Method, M.S. thesis, Mechanical Engineering Department, Brigham Young University, Provo, UT.

Parkinson, D. B., 1978, First-Order Reliability Analysis Employing Translation System, *Eng. Struct.,* Vol. 1, October, pp. 31–40.

Parkinson, D. B., 1982, The Application of Reliability Methods to Tolerancing, *J. Mech. Des.* Vol. 104, July, pp. 612–618.

Parkinson, D. B., 1983, Reliability Indices Employing Measures of Curvature, *Reliabil. Eng.,* June, pp. 153–179.

Robison, R. H., 1989. A Practical Method for Three-Dimensional Tolerance Analysis Using a Solid Modeler, M.S. thesis, Mechanical Engineering Department, Brigham Young University, Provo, UT.

Sandor, G. N., and Erdman, A. G., 1984, *Advanced Mechanism Design: Analysis and Synthesis,* Vol. 2, Prentice Hall, Upper Saddle River, NJ.

Shapiro, S. S., and Gross, A., 1981, *Statistical Modeling Techniques,* Marcel Dekker, New York.

Sitko, A. G., 1991, *Applying Variation Simulation Analysis to 2-D Problems,* SAE Technical Paper 910210, Society of Automotive Engineers, Warrendale, PA, February.

Taguchi, G., 1978, Performance Analysis Design, *Int. J. Prod. Des.,* Vol. 16, pp. 521–530.

Trego, A., 1993, A Comprehensive System for Modeling Variation in Mechanical Assemblies, M.S. thesis, Mechanical Engineering Department, Brigham Young University, Provo, UT.

Veitschegger, W. K., and Wu, C., 1986, Robot Accuracy Analysis Based on Kinematics, *IEEE J. Robot. Autom.,* Vol. RA-2, No. 3, September, pp. 171–179.

Whitney, D., O. L. Gilbert, and M. Jastrzebski, 1994, Representation of Geometric Variations Using Matrix Transforms for Statistical Tolerance Analysis of Assemblies, *Res. Eng. Des.* Vol. 6, pp. 191–210.

6

TECHNIQUES FOR COMPOSING A CLASS OF STATISTICAL TOLERANCE ZONES

VIJAY SRINIVASAN

IBM T. J. Watson Research Center
Columbia University
Yorktown Heights, New York
New York, New York

MICHAEL A. O'CONNOR

IBM T. J. Watson Research Center
Yorktown Heights, New York

FRITZ W. SCHOLZ

Boeing Information and Support Services
Seattle, Washington

6.1 INTRODUCTION

In 1994 the American Society of Mechanical Engineers issued its latest revision of the dimensioning and tolerancing standard (ASME, 1994). About three of a total of over 230 pages in this standard are devoted to statistical tolerancing. Previous revisions issued earlier contained no reference at all to statistical tolerancing. Although minimal, the latest attempt by ASME to address statistical tolerancing is a significant beginning. Widespread practice of statistical tolerancing within American companies, and those abroad, have

Advanced Tolerancing Techniques, Edited by Hong-Chao Zhang
ISBN 0-471-14594-7   © 1997 John Wiley & Sons, Inc.

forced ASME and other international bodies to take a serious look at possible codification of statistical tolerancing. Recently, ISO has set up a task group toward this purpose (Srinivasan and O'Connor, 1995). Indications thus far point to an increased level of activity in understanding and codification of statistical tolerancing.

An earlier paper (Srinivasan and O'Connor, 1994) proposed possible interpretations of statistical tolerancing in the context of geometric part specifications. That study was based on an understanding of the prevailing manufacturing practices in leading companies. An important class of interpretations depended on process capability indices: C_p, C_{pk}, and C_c. This led to statistical tolerance zones in the C_{pk}–C_p plane as well as in the μ–σ plane defined to specify what populations are statistically acceptable. The focus thus far has been strictly on part specifications and their interpretations.

Since most products are assemblies of parts, it is not sufficient to consider only part specifications. Ideally, a designer should start with an assembly "budget" for allowable variation, and distribute it to the constituent parts. In reality, this problem is attacked by several iterations of tolerance analysis where part variations are composed to determine the assembly variation. In statistical tolerance analysis, this means that techniques should be found to compose the part-level specifications that are in the form of statistical tolerance zones. In this chapter we provide such mathematical and computational techniques.

What is the use of composing part-level statistical tolerance zones? The composed zone gives us a compact geometric description of "all" possible statistical outcomes of a critical assembly-level characteristic. If we have the composed tolerance zone for an assembly-level characteristic, we can then reason more rationally about the risk that some instances of the product may not function properly. In the absence of the composed tolerance zone, people have often justified estimates of the assembly-level failure rates by heuristics that are hard to defend or explain, and even grossly erroneous. It is the combination of the attraction of a compact geometric representation of composed statistical tolerance zones and the possibility of subsequent risk analyses using these tolerance zones that makes our work relevant.

In Section 6.2 we describe the type of part-level statistical tolerance zones covered in this chapter. The actual task of composing the part specifications into assembly specification is addressed in Section 6.3. Some thoughts on the risk analysis of the assembled product based on the composition is the topic of Section 6.4. An example illustrating the use of the techniques is given in Section 6.5. The mathematical details involved in the composition is presented in the Appendix.

6.2 STATISTICAL TOLERANCE ZONES

Let x be a random variable and LSL and USL be the lower and upper specification limits. If μ and σ are the mean and standard deviation of x, then

$$C_p = \frac{\text{USL} - \text{LSL}}{6\sigma}$$

and

$$
\begin{aligned}
C_{pk} &= \min \left\{ \frac{\mu - \text{LSL}}{3\sigma}, \frac{\text{USL} - \mu}{3\sigma} \right\} \\
&= \frac{(\text{USL} - \text{LSL})/2 - |(\text{LSL} + \text{USL})/2 - \mu|}{3\sigma}
\end{aligned}
$$

are known as the process capability indices. In addition, it is useful to define

$$C_c = \frac{|\mu - (\text{LSL} + \text{USL})/2|}{(\text{USL} - \text{LSL})/2}$$

to quantify the mean shift from the target value of $(\text{LSL} + \text{USL})/2$. The variable x can be derived from a collection of actual parts (or features) by Gaussian or Chebyschev fitting. In this case x is called an *actual value.* A particular type of actual value is the *actual mating size,* which is found to be very useful for assembly analysis.

In many industrial practices a population of acceptable parts (or features) is statistically specified by

$$C_p \geq P, \quad C_{pk} \geq K, \quad \text{and} \quad C_c \leq F$$

for particular values P, K, and F, or some subset of these inequalities. This leads to a set of acceptable (μ, σ) pairs, that is, a statistical tolerance zone, hereafter referred to as an STZone. It can be shown that as a subset of the μ–σ plane it is a polygonal region between the μ-axis and the graph of a function defined piecewise by finitely many nonnegative linear functions with bounded domains (Srinivasan and O'Connor, 1994). Such an STZone will be said to be σ-polygonal. Table 6.1 summarizes the results of our survey of current part-level statistical tolerancing practices in five leading companies around the world. The names of the companies have been suppressed for reasons of confidentiality. All the STZones in the table are σ-polygonal.

One may construct a more complete statistical tolerance zone in the parametric space of the population of parts. The parameters can be, for example, all the (central) moments of the distribution. An STZone defined above may then be viewed as the intersection of the μ–σ plane with such a higher-dimensional statistical tolerance zone.

When two or more parts are assembled to form a product, we would like to know the STZone for a product characteristic. When some product characteristic is a linear combination of statistically independent primitive part-level characteristics, we can easily and explicitly provide this STZone. All

TABLE 6.1. Summary of Current Part-Level Statistical Tolerancing Practices in Companies Surveyed

Company	Practice	Shape of STZone
A	$C_p \geq 2.0,\ C_{pk} \geq 1.5,\ C_c \leq 0.25$	
B	$C_p \geq 2.0,\ C_{pk} \geq 0.75C_p$	
C	$C_{pk} \geq 1.0,\ C_c \leq 0.20$	
D	$C_p \geq 2.0,\ C_{pk} \geq 1.5$	
E	$C_{pk} \geq 1.33$	

primitive part-level STZones will be assumed to be σ-polygonal in what follows.

6.3 COMPOSING STZones

Let a linear "gap" function be

$$G = a_0 + a_1 X_1 + a_2 X_2 + \cdots + a_n X_n \qquad (6.1)$$

where X_1, X_2, ..., X_n are independent random variables with σ-polygonal STZones. Without loss of generality, equation (6.1) can be rewritten in terms of $g = G - a_0$ and $x_i = a_i X_i$ as

$$g = x_1 + x_2 + \cdots + x_n$$

where the x_i's are independent random variables. Hereafter, for simplicity of statement, we assume that all compositions are of this form, that is, linear with coefficients equal to unity involving only finitely many independent variables. Since x_i is a simple linear transformation of X_i, we see that the STZone for x_i is also σ-polygonal, because

$$\begin{Bmatrix} \mu_i \\ \sigma_i \end{Bmatrix} = \begin{bmatrix} a_i & 0 \\ 0 & |a_i| \end{bmatrix} \begin{Bmatrix} \mu_{X_i} \\ \sigma_{X_i} \end{Bmatrix} \qquad (6.2)$$

where μ_i and σ_i are the mean and standard deviation of x_i, and μ_{X_i} and σ_{X_i} those of X_i. Our task is to find the STZone for g; we call this the problem of composition. To accomplish this, we first move from the μ–σ plane to the μ–σ^2 plane, where the composition is reduced to simple Minkowski sums. To see this, we first define Minkowski sums.

The *Minkowski sum* of two sets A and B in $\mathbf{R}^n$ is defined as

$$A \quad B = \{a + b : a \in A, b \in B\}$$

Minkowski sums are also known as *vector sums,* for obvious reasons. If we are summing several independent random variables, from elementary probability theory the mean of the sum is the sum of the means, and the variance of the sum is the sum of the variances. Applying these simple rules, we can see that if we work in the mean–variance (i.e., μ–σ^2) plane, the composition problem becomes one of computing the vector sum (i.e., the Minkowski sum) of the STZones for the random variables in that plane. Minkowski sums are commutative and associative. They also distribute over unions, that is,

$$A \oplus (B \cup C) = (A \oplus B) \cup (A \oplus C)$$

For more details on Minkowski sums, see Matheron (1975) and Kaul (1993).

6.3.1   Composition in the μ–σ^2 Plane

Since σ is always nonnegative, each (μ, σ) pair is uniquely associated with a (μ, σ^2) pair by squaring σ or choosing the positive square root of σ^2. We can thus represent any STZone in either the μ–σ plane or the μ–σ^2 plane and pass trivially from one to the other. We claim that the STZone for x_i in the μ–σ^2 plane is bounded by line segments and parabolic arcs. Indeed, if we start with an inclined line in the μ–σ plane, given by

$$\sigma = m\mu + c \qquad m \neq 0, \infty$$

then when transformed to the μ–σ^2 plane, it becomes the parabola

$$\sigma^2 = (m\mu + c)^2 = m^2 \left(\mu + \frac{c}{m} \right)^2 \qquad (6.3)$$

A line segment in the line with nonnegative σ values is transformed to a parabolic arc, which can include the apex (the point closest to the directrix) of the parabola only as an endpoint, so that the arc is an increasing arc or a decreasing arc, that is, a monotonic arc. On the other hand, a horizontal line in the μ–σ plane with a constant σ value of $\sigma_0 > 0$ is transformed to a horizontal line in the μ–σ^2 plane with a constant σ^2 value of σ_0^2, and any line segment of the line in the μ–σ plane is transformed to a line segment of the line in the μ–σ^2 plane. It follows that the upper boundary of the STZone for x_i in the μ–σ^2 plane is composed of bounded nonnegative horizontal line segments and monotonic arcs of graphs of nonnegative parabolic functions. The STZone itself is the region between the μ-axis and this collection of arcs. For example, the house-shaped STZone in the μ–σ plane practiced by company C in Table 6.1 leads to the region in the μ–σ^2 plane of Figure 6.1.

In analogy to the property of being σ-polygonal, an STZone will be called σ^2-parabolic if it is represented in the μ–σ^2 plane by a region between the μ-axis and the graph of a function defined piecewise by finitely many bounded nonnegative horizontal line segments and bounded monotonic subarcs of nonnegative parabolic functions. In these terms we have shown that a σ-polygonal STZone will always be σ^2-parabolic.

Since the random variables are independent, we have

$$\left\{ \begin{matrix} \mu_g \\ \sigma_g^2 \end{matrix} \right\} = \left\{ \begin{matrix} \mu_1 \\ \sigma_1^2 \end{matrix} \right\} + \left\{ \begin{matrix} \mu_2 \\ \sigma_2^2 \end{matrix} \right\} + \cdots + \left\{ \begin{matrix} \mu_n \\ \sigma_n^2 \end{matrix} \right\}$$

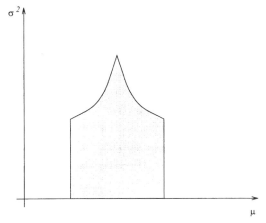

Figure 6.1. Statistical tolerance zone in the μ–σ^2 plane.

where μ_g and σ_g are the mean and standard deviation of g. Thus the rule of composition in the μ–σ^2 plane reduces to:

Lemma 1 *In the μ–σ^2 plane the STZone of g is equal to the Minkowski sum of the STZones of the x_i's.*

In general, finding the representation of the STZone of g in the μ–σ^2 plane by applying this last result requires computing the Minkowski sum of regions bounded by parabolic arcs. Computing the Minkowski sum of regions bounded by nonlinear arcs can be difficult at best. The algebraic complexity of the arcs bounding the sum may grow (Kaul, 1993), so that in a sum involving many summands, as is common in compositions of STZones, the complexity may become prohibitive. However, in the case of sums of σ^2-parabolic zones, these problems do not arise. The sums themselves are easy to obtain, and the algebraic complexity of the sum is unchanged. In particular, Corollary 1 of the appendix and the discussion following it imply the following two lemmas.

Lemma 2 *The composition of σ^2-parabolic STZones is σ^2-parabolic.*

Lemma 3 *Let A be a collection of finitely many σ^2-parabolic STZones. Let P be the collection of arcs formed by translating each defining arc in the upper boundary of each member of A by each endpoint of each defining arc in the upper boundary of every other member of A. The upper envelope of P is the upper boundary of the representation of the composition of the STZones of A in the μ–σ^2 plane.*

The first of the lemmas implies that σ^2-parabolic STZones are closed under the Minkowski sum of any finite number of them. The second implies that the arcs of the upper boundary of a Minkowski sum of σ^2-parabolic STZones merely come from translates of the arcs bounding the summands. These results permit the explicit calculation of the representation in the μ–σ^2 plane of the STZone of a composition of σ-polygonal STZones. An example that involves two STZones is shown in the top row of Figure 6.2. The upper boundary of the Minkowski sum consists of parabolic arcs that are just translates of those that form the upper boundary of the summands. The number of such arcs on the upper boundary of the sum may increase in some cases, as illustrated in Figure 6.12, but their algebraic complexity remains the same. To obtain the representation in the μ–σ plane, we continue the analysis in the next section.

6.3.2   Composition in the μ–σ Plane

Translation of the parabola of equation (6.3) by (X,Y) in the μ–σ^2 plane yields the parabola

$$\sigma^2 = m^2 \left(\mu - X + \frac{c}{m} \right)^2 + Y \tag{6.4}$$

In the μ–σ plane equation (6.4) is satisfied by a nondegenerate hyperbola, when $Y > 0$, and the degenerate hyperbola, a product of two lines,

$$0 = (m\mu - mX + c + \sigma)(m\mu - mX + c - \sigma) \tag{6.5}$$

when $Y = 0$.

First consider the case when $Y > 0$. If we transform the parabola of equation (6.4) in the μ–σ^2 plane to the μ–σ plane, it becomes the upper curve of the nondegenerate hyperbola defined by (6.4). In this case a monotonic arc of the parabola is transformed to a monotonic arc of the upper curve of the nondegenerate hyperbola.

On the other hand, when $Y = 0$, the parabola of (6.4) is mapped to the two half-lines with nonnegative σ-values defined by (6.5). If we call these two half-lines the upper curve of the degenerate hyperbola, a monotonic arc of the parabola becomes a line segment in the upper curve of the degenerate hyperbola.

Conversely, any monotonic arc of the upper curve of a hyperbola defined by an equation of the form of (6.4) with $Y \geq 0$ in the μ–σ plane transforms to a monotonic arc of a nonnegative parabola of the μ–σ^2 plane. See Figure 6.3 for an illustration. Proceeding as before, let a subset of the μ–σ plane be called σ-hyperbolic if it is a region between the μ-axis and the graph of a

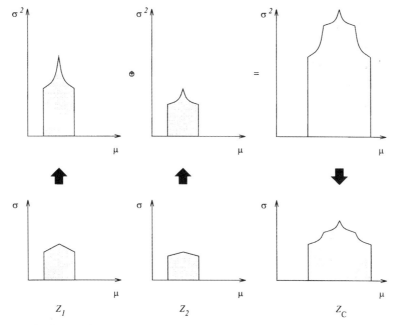

Figure 6.2. Composing two STZones. The bottom row shows the STZones Z_1 and Z_2 of x_1 and x_2, respectively, and their composition in the form of the STZone Z_C of $g = x_1 + x_2$, all in the μ–σ plane. In the process of composition, we transform the STZones of x_1 and x_2 from the μ–σ plane to the μ–σ^2 plane and obtain their Minkowski sum as shown in the top row. This sum is then transformed back to the μ–σ plane to get the desired composition.

function defined piecewise by finitely many bounded horizontal nonnegative line segments or bounded monotonic subarcs of the upper curves of a hyperbolas determined by equations of the form of (6.4) with $Y \geq 0$. In these terms we have shown

Lemma 4   *An STZone is σ^2-parabolic if and only if it is σ-hyperbolic.*

Since a σ-polygonal STZone is always σ^2-parabolic, Lemmas 2 and 4 yield part of the characterization of the STZone for g we have been pursuing.

Proposition 1   *The STZone of a composition of σ-polygonal STZones is σ-hyperbolic.*

In fact, Lemmas 2 and 4 say more.

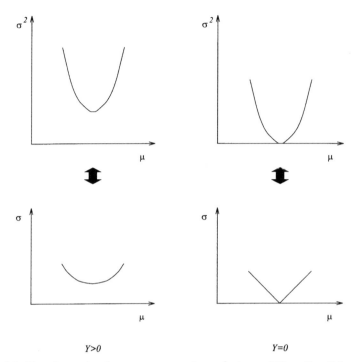

Figure 6.3. Transformation between μ–σ and μ–σ^2 planes. When $Y > 0$ in equation (6.4), a parabola in the μ–σ^2 plane goes to the upper curve of a hyperbola in the μ–σ plane (as σ is always nonnegative), and conversely. When $Y = 0$, the hyperbola degenerates into two half-lines.

Proposition 2   *The STZone of a composition of σ-hyperbolic STZones is σ-hyperbolic.*

This means, loosely put, that σ-hyperbolic STZones are closed under the operation of composition of STZones. This is of more than merely theoretical interest. Since STZones often represent parts in an assembly, which in turn are often subassemblies in a larger assembly, the last result assures that no new analysis is required to treat this more general application. To complete the characterization we need only to make more explicit the piecewise-hyperbolic function that defines the STZone of g. This is accomplished by the following

Proposition 3   *Let A be a collection of finitely many σ-hyperbolic STZones. If*

$$\mu \in D \longrightarrow \sqrt{(m\mu - c)^2 + d} \tag{6}$$

is a parameterization of one of the defining arcs of the upper boundary of one of the members of A, and (a,b) is an endpoint of a defining arc of the upper boundary of any other member of A, define the parameterization of a new arc by

$$\mu \in D + a \longrightarrow \sqrt{(m\mu - c - ma)^2 + d} + b \qquad (7)$$

Let H *be the collection of all possible such newly defined arcs. The upper envelope of* H *is the upper boundary of the representation of the composition of the STZones of A in the μ–σ plane.*

Proof. Passing from the μ–σ plane to the μ–σ^2 plane, applying Lemma 3, calculating the effect of the translations, and passing back from the μ–σ^2 plane to the μ–σ plane suffice to establish the claim. ∎

If one of the STZones is σ-polygonal with linear parameterizations, "squaring" each of the parameterizations yields ones compatible with the statement of the proposition. If $m = 0$, the hyperbolic arcs of (6.6) and (6.7) reduce to line segments. If $d = 0$, the hyperbolic arc of (6.6) reduces to a line segment, and if $b = d = 0$, the hyperbolic arc of (6.7) reduces to a line segment. All other arcs come from nondegenerate hyperbolas.

A calculation of the upper envelope called for in the last proposition requires the intersection of hyperbolic arcs. The intersection of two hyperbolas usually requires the solution of a quartic polynomial; however, the special form of (6.4) reduces the complexity to that of a simple quadratic. The bottom row in Figure 6.2 shows an example of the composed STZone Z_C for $g = x_1 + x_2$ in the μ–σ plane.

6.3.3 Special Case Simplifications

In this section to this point we have considered only the most general composition problem and only from a theoretical viewpoint. Even in the general case many simplifications are possible for algorithmic purposes. For example, judicious use of the observation following Lemma 5 of the appendix—that sums of increasing and decreasing functions need only use part of the vertical boundaries—can decrease the resultant number of potential upper boundary pieces by a factor of 4. Treating monotonic chains by a simple generalization of the lemma can add further reductions. Similar reductions are possible for pairs of increasing boundary elements.

In less general settings, prescribed design practices often allow obvious simplifications. For example, symmetric part-level STZones produce a symmetric assembly-level STZone, so that only half of the upper boundary of the STZone need be computed. More restrictive, though common practices can lead to very explicit compositions. For example, let us assume that all parts are specified by $C_{pk} \geq K$ and $C_c \leq F$, as practiced by company C of Table 6.1, with some globally fixed K and F. For ease of statement let us further assume that the nominal values of all part dimensions equal zero, so that the specification for any actual value, x_i, becomes

$$\frac{T_i - |\mu_i|}{3\sigma_i} = C_{pk} \geq K \quad \text{and} \quad \frac{|\mu_i|}{T_i} = C_c \leq F$$

where T_i is the USL of x_i. Finally, let us assume that $\{x_i\}_{i=1}^n$ is ordered such that $T_i \leq T_j$ for all $i \leq j$. If $g = \Sigma_{i=1}^n x_i$, let us say that g is a restricted type C composition defined by K, F, and $\{x_i, T_i\}_{i=1}^n$. In these terms we now have

Proposition 4 *Let g be a restricted type C composition defined by K, F, and $\{x_i, T_i\}_{i=1}^n$. Let*

$$f(0) = \frac{1}{3K} \left(\sum_{i=1}^n T_i^2 \right)^{1/2}$$

and

$$f(\mu) = \frac{1}{3K} \left[(1 - F)^2 \sum_{i=1}^{k-1} T_i^2 + T_k^2 (1 - \theta F)^2 + \sum_{i=k+1}^n T_i^2 \right]^{1/2}$$

for $|\mu| = F(\Sigma_{i=1}^{k-1} T_i + \theta T_k)$ with $0 < \theta \leq 1$. The graph of f is the upper boundary of the representation of the STZone of g in the μ–σ plane.

For the μ coordinate of any point in Z_C, the STZone of g in the μ–σ plane, there is a unique representation of $|\mu|$ as required in this proposition, so that the result completely and explicitly yields Z_C. We provide only a sketch of the proof of the proposition here. Let

$$D_\mu = \left\{ (\mu_1, ..., \mu_i, ..., \mu_n) : 0 \leq \mu_i \leq FT_i \text{ for all } 1 \leq i \leq n \text{ and } \sum_{i=1}^n \mu_i = |\mu| \right\}$$

The point, $(\mu, \tilde{\sigma})$, is in the upper boundary of Z_C if and only if $9K^2\tilde{\sigma}^2$ is the maximum of $h(\mu_1, ..., \mu_n) = \Sigma_{i=1}^n (T_i - \mu_i)^2$ over D_μ. This maximum occurs at an extreme point of D_μ, that is, a point in D_μ with $\mu_i = 0$ or $\mu_i = FT_i$ for all i except possibly one with $\mu_i = \theta FT_i$. Since

$$h(\mu_1, ..., \mu_i + \delta, ..., \mu_j - \delta, ..., \mu_n) - h(\mu_1, ..., \mu_i, ..., \mu_j, ..., \mu_n) = 2\delta h_1$$

for $h_1 = \mu_i - \mu_j + \delta + T_j - T_i$, a simple case analysis, comparing extreme points differing only on coordinate pairs, and considering the form of h_1

shows that h never decreases as we pass in proper direction from extreme point to extreme point to the claim of the proposition.

6.4 RISK ANALYSIS

As we pointed out in Section 6.1, a major motivation for computing the composed STZone is to perform assembly-level risk analysis more rationally. In this section we outline several ways to exploit the composed STZone to perform such analyses. The main scope of this chapter is not the risk analyses, but we include them here to show the usefulness of composed STZones.

Risk analysis entails evaluating or approximating the probability that the value of some random variable from some collection of random variables will fall outside an acceptable region. One-sided risk analysis considers probabilities of the form $\Pr(g \le k)$ [or the completely analogous $\Pr(g \ge k)$], and two-sided risk analysis the sum of probabilities $\Pr(g \le k) + \Pr(g \ge k')$, for random variables g and fixed real values k and k'. Of course, to provide such an analysis sufficient information about the distributions of the random variables must be available.

An STZone uses only the mean and standard deviation to define a class of random variables. This offers too little distributional information to say much about risk, in general. However, by imposing mild assumptions on the class of random variables, we can change the situation. For this purpose, if we assume that each random variable in the class of random variables defined by an STZone Z is completely determined by its mean and standard deviation, so that we can parameterize the random variables of the class by $(\mu,\sigma) \in Z \to g(\mu,\sigma)$, and that any variable of the class can be transformed to any other by shifting μ and scaling σ, so that for all (μ,σ) and (μ',σ') in Z,

$$\Pr(g(\mu,\sigma) \le k) = \Pr\left(\frac{\sigma}{\sigma'}(g(\mu',\sigma') - \mu') + \mu \le k\right) \qquad (6.8)$$

then much more can be said.

Let us first consider the one-sided risk $\Pr(g(\mu,\sigma) \le k)$ associated with $(\mu,\sigma) \in Z$. We begin by observing that (6.8) implies immediately that

$$\Pr(g(\mu,\sigma) \le k) = \Pr\left(g(\mu',\sigma') \le \frac{\sigma'}{\sigma}(k - \mu) + \mu'\right) \qquad (6.9)$$

It follows directly from this that for $\mu' \ne k$,

$$\Pr(g(\mu,\sigma) \le k) = \Pr(g(\mu',\sigma') \le k), \qquad \text{whenever} \qquad \sigma = \frac{\sigma'(k - \mu)}{k - \mu'}$$

so that each of the points in the intersection of a ray of the type

$$\left\{ \left(\mu, \frac{\sigma'(k - \mu)}{k - \mu'} \right) : \mu > k \text{ if } \mu' > k \right\}$$

or $\left\{ \left(\mu, \frac{\sigma'(k - \mu)}{k - \mu'} \right) : \mu < k \text{ if } \mu' < k \right\}$

with Z is associated with the same risk. Each of these rays is a nonvertical ray in the $\sigma > 0$ half-plane emanating from the point $(k,0)$. The vertical ray emanating from $(k,0)$ is the set of points $\{(k,\sigma) : \sigma > 0\}$. Since shifting the mean and scaling the standard deviation does not change the probability of the value of a random variable being less than the mean, (6.8) implies that there is a constant α such that $\Pr(g(\mu,\sigma) \le \mu) = \alpha$ for all μ and σ, and in particular $\Pr(g(k,\sigma) \le k) = \alpha$. Thus each of the rays in the half-plane emanating from $(k,0)$ is associated with a fixed risk. If μ increases without bound with σ and k fixed, (6.9) implies that the risk associated with (μ,σ) decreases to a limiting value of zero. If μ decreases without bound with σ and k fixed, (6.9) implies that the risk associated with (μ, σ) increases to a limiting value of 1. Hence as we rotate the rays through $(k,0)$ counterclockwise from the horizontal through the vertical to the horizontal again, sweeping out all the points in the half-plane in the process, the risk associated with the rays increases from zero at the horizontal through α at the vertical to 1 at the horizontal. We can also express this in terms of the slopes of the rays, which will be most convenient. As the slope of the ray increases from zero to the vertical and then through all negative slopes to zero again, the risk increases from zero to α to 1 (see Figure 6.4).

Determining the maximum one-sided risk, $\Pr(g(\mu,\sigma) \le k)$ over $(\mu,\sigma) \in Z$ is now a simple matter. If $k > \mu$ for some $(\mu,\sigma) \in Z$, rays emanating from $(k,0)$ with negative slopes arbitrarily close to zero must intersect Z, so the risk is 1. If k equals the minimum of all μ such that there exists some (μ,σ)

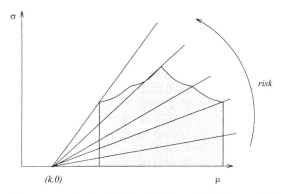

Figure 6.4. Isorisk contours in the STZone Z. The risk is the probability $\Pr(g \le k)$. The maximum risk is associated with a supporting line of Z passing through $(k,0)$.

$\in Z$, the risk is α. If $k < \mu$ for all $(\mu,\sigma) \in Z$, we need only determine the line through $(k,0)$ with maximum slope that intersects Z, that is, the line of support for Z through $(k,0)$, other than the μ-axis. This can be computed easily for σ-hyperbolic STZones. By definition, the hyperbolic arcs in the boundary of Z come from the upper curves of hyperbolas defined by equations of the form of (6.4). No interior point of such an arc can be on a line of support of Z unless the arc is linear. If the arc is linear and an interior point of it is on the line of support for Z, so are the endpoints. Thus only the finitely many endpoints of the arcs are needed to define the line of support through $(k,0)$. If Z comes from a composition of STZones, that is, $Z = Z_C$, an explicit evaluation of Z_C is not even required, as Proposition 3 enables us to obtain these endpoints directly. In any event, if (μ',σ') is a point defining this line of support, $\Pr(g(\mu',\sigma') \le k)$ is the maximum one-sided risk.

We wish to stress again that the assumptions made in this development are mild. They are met by many common families of random variables. For example, assuming the random variables to be normal more than suffices. In particular, for sum aggregates, such as $g = x_1 + \cdots + x_n$, which form Z_C, normality is a reasonable and common assumption. In the case of normality and many others, then, a final evaluation of the maximum one-sided risk can easily be found by appeal to widely available tables.

Finding the analogous maximum two-sided risk, $\Pr(g(\mu,\sigma) \le k) + \Pr(g(\mu,\sigma) \ge k')$, over $(\mu,\sigma) \in Z$, is more difficult. A simple extension of the analysis of the one-sided case shows that the two-sided risk is maximum for some (μ,σ) in the upper boundary of Z; however, the risk associated with $(\mu,\sigma) \in Z$ need no longer be monotonic in a simple parameter such as the slope of a line, and the simple extension does not allow any claim that the risk must be maximized by an endpoint of an arc. Thus rather than the simple calculation of the maximum slope over finite set in the one-sided case, we are left with a one-dimensional nonlinear optimization problem in the two-sided case. Approximation techniques exist and may suffice in many cases, but more work on this remains.

In some situations, maximum risk analysis may be overly pessimistic. This may be particularly so if Z is an assembly-level STZone. If individual and joint distributions can reasonably be assumed on the part-level means, a distribution of the assembly-level means is implied. Often, this assembly-level distribution can be calculated or approximated. For example, if the part-level means are independent and normal, the assembly-level means are normal. Or if part-level means have independent, uniform distributions (or more generally, piecewise-polynomial densities), the assembly-level distribution has a piecewise-polynomial density which can be explicitly calculated. It may even be justified to directly assume that the assembly-level means have some distribution, for example, a normal distribution by an application of the central limit theorem. In any case, if we assume that the distribution of the means in Z is known, this distribution can be used to provide a less pessimistic risk analysis. For this purpose let $\Pr(\mu \in I) = p$ for some interval I, and let $Z_I =$

$\{(\mu,\sigma) \in Z : \mu \in I\}$. If we apply the previous techniques to Z_I to obtain a maximum risk, r, over Z_I, we can claim that with probability p the risk is no worse than r. This reduced maximum risk analysis would be useful, when very unlikely μ values far from the mean of the means determine a highly pessimistic risk assessment.

It is worthwhile to point out the distinction between the two types of probabilities p and r. The risk r concerns the fall out rate for fixed levels of (μ,σ). On the other hand, p concerns the chance that μ falls within certain limits. For one particular process this chance is taken just once. A long-run frequency interpretation is only possible over many such processes.

Even the reduced maximum risk analysis in certain cases may be less useful than information about an average risk R, that is,

$$R = \int_Z R(\mu,\sigma) \, dF(\mu,\sigma)$$

where $R(\mu,\sigma)$ is the risk associated with (μ,σ), and dF is the joint probability measure for μ and σ over Z. It is quite possibly too much to assume full knowledge of the joint probability, but since either the one- or two-sided risk is increased by increasing σ, we can bound R by

$$R \leq \int_\Delta R(\phi(\mu)) \, dH(\mu)$$

where Δ is the intersection of Z with the μ-axis, ϕ is a parameterization of the upper boundary of Z, and dH is the probability measure associated with μ. The simple explicit form of ϕ that can be obtained from Proposition 3 implies that at least for the one-sided risk analysis the integral can easily be approximated by standard techniques.

6.5   EXAMPLE

To illustrate the techniques described thus far, let us consider as a concrete example the statistical specifications of a magnetic storage product. Figure 6.5 shows a vertical section view of a disk-drive assembly. Four critical parts in this assembly are identified for our consideration: a magnetic disk that stores the data, an arm that swings in an angular movement to enable a magnetic head to read and write data onto the disk, and two bearings that support the disk and the arm. This figure is a grossly simplified version of an actual disk-drive cross-sectional view but is sufficient for our purpose of illustration.

A critical assembly characteristic in such a disk-drive is the arm-to-disk clearance, shown in Figure 6.6 as the gap g. For proper functioning of the disk-drive, this clearance g should be neither too large nor too small. It is, of

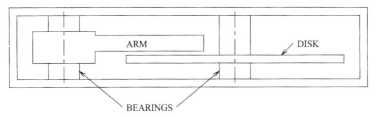

Figure 6.5. Disk-drive assembly indicating critical parts.

course, controlled by the critical dimensions of the four parts we identified earlier. More precisely, we have an instance of the linear "gap" function (6.1) in the form

$$g = l_1 + l_2 - l_3 - l_4 \tag{6.10}$$

Each of the critical dimensions belongs to a different part in the disk-drive assembly and can be justifiably assumed to be mutually independent random variables.

Part-level statistical specifications of these critical dimensions are summarized in Table 6.2. We have chosen a practice similar to that of company C described in Table 6.1. For ease of explanation, the lower bound K for C_{pk} and the upper bound F for C_c have been chosen to be the same, 1.5 and 0.25, respectively, for all the four critical dimensions. STZones that correspond to these statistical specifications are plotted in Figure 6.7. The STZone for l_1 in Figure 6.7(1) is the region under the piecewise linear function of $\mu \in [1.7375, 1.7625]$. These limits for μ are obtained from the constraint that $C_c = |\mu - 1.75|/0.05 \leq 0.25$. The other constraint that $C_{pk} = (0.05 - |1.75 - \mu|)/3\sigma \geq 1.5$ leads explicitly to

$$\sigma(\mu) = \begin{cases} \dfrac{\mu - 1.70}{4.5} & \text{for } \mu \in [1.7375, 1.7500] \\[2ex] \dfrac{1.80 - \mu}{4.5} & \text{for } \mu \in (1.7500, 1.7625] \end{cases}$$

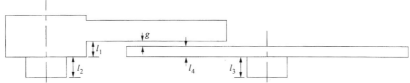

Figure 6.6. Critical part dimensions and an assembly characteristic in a disk-drive assembly. The critical assembly characteristic is the arm-to-disk spacing g, which is related to the critical part dimensions l_1, l_2, l_3, and l_4 by the linear "gap" function $g = l_1 + l_2 - l_3 - l_4$.

TABLE 6.2. Statistical Specifications on Critical Dimensions of the Disk-Drive Example[a]

Part	C.D.	N.D.	T.I.	USL	LSL	K	F
Arm	l_1	1.75	0.10	1.80	1.70	1.5	0.25
Arm_Bearing	l_2	2.00	0.14	2.07	1.93	1.5	0.25
Disk_Bearing	l_3	2.00	0.14	2.07	1.93	1.5	0.25
Disk	l_4	1.00	0.06	1.03	0.97	1.5	0.25

[a]C.D., critical dimension; N.D., nominal dimension; T.I., tolerance interval.

as the piecewise linear function. Similarly, Figure 6.7(2) and (3) involve the piecewise linear function

$$\sigma(\mu) = \begin{cases} \dfrac{\mu - 1.93}{4.5} & \text{for } \mu \in [1.9825, 2.0000] \\ \dfrac{2.07 - \mu}{4.5} & \text{for } \mu \in (2.0000, 2.0175] \end{cases}$$

and Figure 6.7(4) has

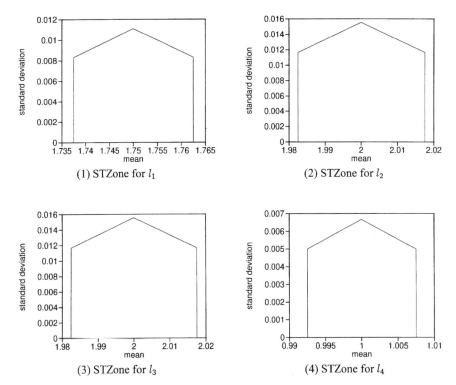

(1) STZone for l_1

(2) STZone for l_2

(3) STZone for l_3

(4) STZone for l_4

Figure 6.7. Plots of the STZones for the four critical dimensions.

$$\sigma(\mu) = \begin{cases} \dfrac{\mu - 0.97}{4.5} & \text{for } \mu \in [0.9925, 1.0000] \\[2mm] \dfrac{1.03 - \mu}{4.5} & \text{for } \mu \in (1.0000, 1.0075] \end{cases}$$

as its piecewise linear function. Notice the change in scales in these plots, from the mean to the standard deviation axes as well as from dimension to dimension, adopted for clarity.

The next task is to compute the composed STZone for g. Since we have set the same bounds $K = 1.5$ and $F = 0.25$ for all four variables, after rearranging the critical dimensions in the increasing order of tolerance intervals, the conditions of Proposition 4 are met. Therefore, the piecewise hyperbolic function, which is the upper boundary of the STZone of g, is given by Proposition 4 explicitly as

$\sigma(\mu) =$

$$\begin{cases} \dfrac{1}{4.5}\sqrt{0.00466875 + 0.0049\left(1 - 0.25\,\dfrac{0.7125 - \mu}{0.0175}\right)^2} & \text{for } 0.6950 \le \\ & \mu < 0.7125 \\[3mm] \dfrac{1}{4.5}\sqrt{0.00681250 + 0.0049\left(1 - 0.25\,\dfrac{0.7300 - \mu}{0.0175}\right)^2} & \text{for } 0.7125 \le \\ & \mu < 0.7300 \\[3mm] \dfrac{1}{4.5}\sqrt{0.01030625 + 0.0025\left(1 - 0.25\,\dfrac{0.7425 - \mu}{0.0125}\right)^2} & \text{for } 0.7300 \le \\ & \mu < 0.7425 \\[3mm] \dfrac{1}{4.5}\sqrt{0.01230000 + 0.0009\left(1 - 0.25\,\dfrac{0.7500 - \mu}{0.0075}\right)^2} & \text{for } 0.7425 \le \\ & \mu < 0.7500 \end{cases}$$

and

$\sigma(\mu) =$

$$\begin{cases} \dfrac{1}{4.5}\sqrt{0.01230000 + 0.0009\left(1 - 0.25\,\dfrac{\mu - 0.7500}{0.0075}\right)^2} & \text{for } 0.7500 \le \\ & \mu < 0.7575 \\[3mm] \dfrac{1}{4.5}\sqrt{0.01030625 + 0.0025\left(1 - 0.25\,\dfrac{\mu - 0.7575}{0.0125}\right)^2} & \text{for } 0.7575 \le \\ & \mu < 0.7700 \\[3mm] \dfrac{1}{4.5}\sqrt{0.00681250 + 0.0049\left(1 - 0.25\,\dfrac{\mu - 0.7700}{0.0175}\right)^2} & \text{for } 0.7700 \le \\ & \mu < 0.7875 \\[3mm] \dfrac{1}{4.5}\sqrt{0.00466875 + 0.0049\left(1 - 0.25\,\dfrac{\mu - 0.7875}{0.0175}\right)^2} & \text{for } 0.7875 \le \\ & \mu < 0.8050 \end{cases}$$

Figure 6.8 shows a plot of the composed STZone for g. At the resolution

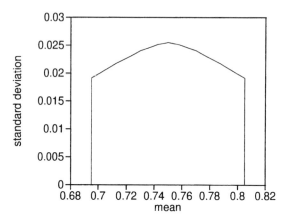

Figure 6.8. Composed STZone for the arm-to-disk clearance g.

in which it is plotted, the curvatures of the hyperbolic arcs in the upper envelope of the STZone are barely visible.

Once we have the composed STZone, risk analyses can be performed in many ways, as pointed out earlier. First we need to make an assumption about the distribution of the arm-to-disk clearance g, and we will assume that it is normal. We may appeal to the central limit theorem to justify that normality is a good approximation. If we are concerned about this clearance reaching below a critical value of, say, 0.65 units, we can estimate the one-sided risk as the probability $\Pr(g(\mu,\sigma) \leq 0.65)$. This risk will vary as the (μ,σ) point varies within the STZone for g. However, following discussions of Section 6.4 we know that the maximum risk is attained at the tangent point (μ,σ) on the supporting line of the STZone of g passing through $(0.65,0)$. This tangency occurs at one of the end points of the hyperbolic arcs of the upper boundary of the STZone. Thus, since $\sigma(\mu)$ is increasing for $\mu \leq 0.75$ and decreasing thereafter, only the five endpoints $\{(0.6950,0.01915),$ $(0.7125,0.02174),$ $(0.7300,0.02405),(0.7425,0.02515),(0.7500,0.02553)\}$ are possible points of tangency. Visual inspection (see Figures 6.9(1) and 6.4 for comparison) indicates that and simple calculations prove that tangency occurs at $(0.6950,0.01915)$, so that maximum risk occurs there. Using easily available approximations for the error function, we find that this maximum risk is approximately 9.4 parts per thousand. If we want to estimate the maximum risk for a different critical value k of the lower limit for the arm-to-disk clearance g, we first find the supporting line that passes through $(k,0)$. A simple calculation shows that for $0.566 < k < 0.695$ the supporting lines pass through the same tangent point $(0.695,0.01915)$. Then the associated risk is given by $\Pr(z \leq (k - 0.695)/0.01915)$ of a unit normal variate z. Figure 6.9(2) illustrates this for several values of k. Figure 6.10 shows a plot of the risk for a small range of k. The designer can then decide if such risks are acceptable and, if not, start the redesign process. If other types of risk analyses are needed, the STZone for g provides a basis for them.

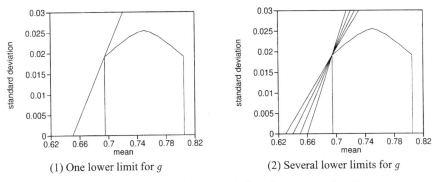

(1) One lower limit for g (2) Several lower limits for g

Figure 6.9. Maximum risk analysis based on supporting lines.

6.6   SUMMARY AND CONCLUSIONS

National and international standards committees are currently investigating how to codify the statistical specifications of part tolerances. Because of the widespread use of process capability indices (C_p, C_{pk}, and C_c) in industry, a case can be made for using them in standardizing the statistical tolerancing of parts. However, it will not be a strong case in the absence of techniques that the designers can use to infer assembly-level variation from such part specifications. In this chapter we provide the needed mathematical and computational techniques.

In developing these techniques we depended only on two major assumptions: that the assembly characteristic is a linear function of the part characteristics and that the part variations are mutually independent. With these we showed that the STZones in the μ–σ plane can be composed by first transforming them to the μ–σ^2 plane, where the composition is reduced to Min-

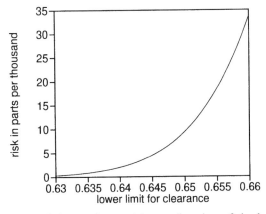

Figure 6.10. Variation of the maximum risk as a function of the lower limit for the arm-to-disk clearance g.

kowski sums and then transforming the sum back to the $\mu-\sigma$ plane. We proved that the algebraic complexity of the Minkowski sum remains low and unchanged irrespective of the number of parts that participate in the assembly. More precisely, the STZone for the assembly-level characteristic is bounded only by line segments and hyperbolic arcs. This enabled us to present an exact and explicit representation for the composed STZone. In a somewhat restricted but useful special case, an explicit expression for the piecewise hyperbolic curve that forms the upper boundary of the composed STZone was given. Under some mild assumptions, the one-sided isorisk contours become line segments in the composed STZones so that designers can reason about the chances they take with such statistical designs. We also outlined several avenues for exploring the risk analysis further. Thus a critical assembly analysis gap has been, at least partially, filled.

As we pointed out in the body of the chapter, the risk analysis may be refined further. Other advances may be attempted, if deemed important, in at least three directions: the geometric form of primitive part-level STZones may be enriched, and the assumptions of linearity and independence in the parts-to-product characteristics relation may be relaxed. The first is perhaps the easiest. If the geometric shape of the primitive part-level STZones differs from those in Table 6.1 but is still σ-polygonal (or, more generally, σ-hyperbolic), all our results apply and are sufficient to achieve the composition. STZones in the form of semicircular disks, proposed in some literature, can be composed. For more difficult cases, linear approximations may be used to reduce them to σ-polygonal cases, which we know how to handle.

ACKNOWLEDGMENT AND DISCLAIMER

We gratefully acknowledge numerous colleagues in industry and in the ASME and ISO standards committees for valuable information and support. The opinions expressed in this chapter are, however, our own and not those of ASME, ISO, or any of their member bodies.

REFERENCES

ASME, 1994, *Dimensioning and Tolerancing,* ANSI Y14.5M-1994, American Society of Mechanical Engineers, New York.

Kaul, A., 1993, Computing Minkowski Sums, Ph.D. thesis, Department of Mechanical Engineering, Columbia University, New York.

Matheron, G., 1975, *Random Sets and Integral Geometry,* Wiley, New York.

Srinivasan, V., and O'Connor, M. A., 1994, On Interpreting Statistical Tolerancing, *Manuf. Rev.,* Vol. 7, No. 4, pp. 304–311.

Srinivasan, V., and O'Connor, M. A., 1995, Towards an ISO Standard for Statistical Tolerancing, *Proceedings of the 4th CIRP Seminar on Computer Aided Tolerancing,* University of Tokyo, Tokyo.

APPENDIX

In this appendix we present and prove a collection of mathematical properties of certain Minkowski sums that support the chapter. Results presented here are more general than needed in the body of the chapter. However, we found that the proofs are no more difficult in the general case and hope that they may lead to new applications in the future. We begin with some terminology.

Recall [Theorem 4.1, part IV of Valentine (1964)] that a closed set $S \subseteq \mathbf{R}^2$ with nonempty interior is convex if and only if for each x in ∂S, the boundary of S, there is a line of support of S at x, that is, a line through x which defines an open half-plane that is disjoint from S.

For a piecewise-continuous function $f : \text{Dom}(f) \to \mathbf{R}^2$ with $\text{Dom}(f) \subset \mathbf{R}$ let

$$U(f) = \{(x,y) \in \mathbf{R}^2 : x \in \text{Dom}(f) \text{ and } y \geq f(x)\}$$

Note that the graph of f is a subset of the boundary of $U(f)$ and that the piecewise continuity of f assures that $U(f)$ will have interior points. If f is also nonnegative, define $L(f)$, the region below f, as

$$L(f) = \{(x,y) \in \mathbf{R}^2 : x \in \text{Dom}(f) \text{ and } 0 \leq y \leq f(x)\}$$

Call a continuous function f convex if $\overline{U(f)}$ is convex (where the bar indicates closure). Recall that a function is monotonic either if it is an increasing function on its entire domain or it is a decreasing function on its entire domain. Let a simple function be a nonnegative continuous monotonic function defined on a closed bounded interval with at most one root. For a simple function f with domain, $[a,b]$, define $V(f)$, the vertical part of the boundary of $L(f)$, as

$$V(f) = \{(a,y) \in \mathbf{R}^2 : 0 \leq y \leq f(a)\} \cup \{(b,y) \in \mathbf{R}^2 : 0 \leq y \leq f(b)\}$$

$V(f)$ is composed of two closed vertical line segments or a point on the x-axis and a closed vertical line segment. The continuity of f implies that the interior of $L(f)$ is a nonempty open set bounded by $V(f)$, the graph of f, and those points in $L(f)$ with y-coordinate equal to 0. In these terms we have the following principal lemmas.

Lemma 5 *If f is a simple increasing function and g is a simple decreasing function, then*

$$L(f) \oplus L(g) = \{L(f) \oplus V(g)\} \cup \{V(f) \oplus L(g)\}$$

Proof.   Let $[a,b]$ be the domain of f and $[c,d]$ that of g. If $p \in L(f) \oplus L(g)$, there exist $(x,y) \in L(f)$ and $(u,v) \in L(g)$ with $p = (x,y) + (u,v)$. Trivially, $p = (x + t,y) + (u - t,v)$ for all $t \in \mathbf{R}$, and in particular, for $t \geq 0$. If

$t \geq 0$, then since f is increasing, $f(x + t) \geq f(x) \geq y$, whenever $x + t \in [a,b]$, so that $(x + t, y) \in L(f)$. Similarly, $(u - t, v) \in L(g)$, whenever $u - t \in [c,d]$. Thus if we let t_0 be the minimum of $b - x$ and $u - c$, then either $(x + t_0, y) \in V(f)$ and $(u - t_0, v) \in L(g)$ or $(x + t_0, y) \in L(f)$ and $(u - t_0, v) \in V(g)$. In either case we have shown that $p \in \{L(f) \oplus V(g)\} \cup \{V(f) \oplus L(g)\}$, so that $L(f) \oplus L(g) \subset \{L(f) \oplus V(g)\} \cup \{V(f) \oplus L(g)\}$. The opposite inclusion in trivial. ∎

First, note that since $V(f)$ consists of points and vertical line segments, the computation of each Minkowski sum in the union is trivially obtained by a translation of the defining functions. Next note that the proof in fact shows that not all of $V(f)$ and $V(g)$ are required. The right part of $V(f)$ and the left part of $V(g)$ would suffice. This is illustrated in Figure 6.11.

Lemma 6 *If f and g are simple convex functions, then*

$$L(f) \oplus L(g) = \{L(f) \oplus V(g)\} \cup \{V(f) \oplus L(g)\}$$

Proof. If for $S \subset \mathbf{R}^2$, we let $S_- = \{(x,y) : (-x,y) \in S\}$, then for subsets $D, E \subset \mathbf{R}^2$, $D \oplus E = (D_- \oplus E_-)_-$. Thus we may assume that f is increasing, or otherwise we may reflect the problem around the y-axis. If g is decreasing, the claim follows from the preceding lemma, so we may also assume that g is increasing.

Let $[a,b]$ be the domain of f and $[c,d]$ that of g. If $p \in L(f) \oplus L(g)$, there exist $(x,y) \in L(f)$ and $(u,v) \in L(g)$ with $p = (x,y) + (u,v)$. Trivially, $p = (x, y + t) + (u, v - t)$ for all $t \in R$. If τ_1 is the minimum of $f(x) - y$ and v, then either $y + \tau_1 = f(x)$ and $0 \leq v - \tau_1 \leq g(u)$, or $0 \leq y + \tau_1 \leq f(x)$ and

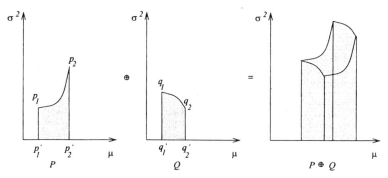

Figure 6.11. Minkowski sum of a region below a simple increasing function and one below a simple decreasing function. If $l(p,q)$ denotes the line segment between points p and q, Lemma 5 states that $P \oplus Q$ can be obtained as the union of $P \oplus l(q_1, q_1')$, $P \oplus l(q_2, q_2')$, $Q \oplus l(p_1, p_1')$, and $Q \oplus l(p_2, p_2')$. However, unioning only the first and the last of the summands will suffice in this case.

$v - \tau_1 = 0$, so that either p is the sum of $(x, f(x))$ and $(u, v + y - f(x)) \in L(g)$, or p is the sum of $(x, y + v) \in L(f)$ and $(u, 0)$. We now discuss these two cases separately.

First consider the case where p is the sum of $(x, y + v) \in L(f)$ and $(u, 0)$. If we let τ_2 be the minimum of $b - x$ and $u - c$, then either $x + \tau_2 = b$ and $c \leq u - \tau_2 \leq d$, or $a \leq x + \tau_2 \leq b$ and $u - \tau_2 = c$. Whether $x + \tau_2 = b$ or $u - \tau_2 = c$, $y + v \leq f(x) \leq f(x + \tau_2)$, since f is increasing. Thus either p is the sum of $(b, y + v)$ in $V(f)$ and $(u + x - b, 0)$ in $L(g)$, or p is the sum of $(x + u - c, y + v)$ in $L(f)$ and $(c, 0)$ in $V(g)$, establishing the claim in this case.

Now consider the case where p is the sum of $(x, f(x))$ and $(u, v + y - f(x)) \in L(g)$. For simplicity of notation let $w = v + y - f(x)$. We may assume that neither $f(x) = 0$ nor $w = 0$, since these conditions would be covered by the previous case by relabeling f and g, if necessary. We may also assume that $x \neq a, b$ and that $u \neq c, d$, or nothing remains to be demonstrated. The convexity of f implies the existence of a line of support M for $U(f)$ at $(x, f(x))$. Since M is a line of support, it could be vertical only at a or b, so that M is not vertical. We can thus parameterize M as $\{(x, \gamma(x)) : x \in \mathbf{R}\}$ for some linear function γ with $\gamma(x) \leq f(x)$ for all x in the domain of f. It follows that the intersection of M and $L(f)$ is a closed bounded line segment containing $(x, f(x))$ in its interior with endpoints in $V(f)$ or having a y-coordinate of 0.

Let N be the line through (u, w) parallel to M. We wish to show the existence of a half-line N^+ of N emanating from (u, w), such that N^+ intersected with $L(g)$ is a closed interval with (u, w) as one endpoint and a point in $V(g)$ or one with a y-coordinate of 0, as the other. If $w = g(u)$ and N is a line of support for $U(g)$ at (u, w), an analysis identical to that just completed for $(x, f(x))$ and M verifies the existence of N^+. If $w = g(u)$ but N is not a line of support for $U(g)$ at (u, w), at least one of the half-lines of N emanating from (u, w) is strictly below any line of support for $U(g)$ at (u, w), and hence below the graph of g. This half-line serves as N^+. Finally, if (u, w) is not in the graph of g, it is in the interior of $L(g)$. The convexity of $U(g)$ implies that at least one of the half-lines of N emanating from (u, w) is disjoint from $U(g)$, and this half-line fulfills the requirements on N^+.

Parameterize N^+ as $N^+ : t \geq 0 \rightarrow N^+(t) = (u, w) + tv$ for some nonzero vector v, and a half-line of M as $M^- : t \geq 0 \rightarrow M^-(t) = (x, f(x)) - tv$. Note that $p = N^+(t) + M^-(t)$. Let τ_3 be the minimum t such that $N^+(\tau_3)$ is in $V(g)$ or has a y-coordinate of 0 or $M^-(\tau_3)$ is in $V(f)$ or has a y-coordinate of 0. If $N^+(\tau_3)$ is in $V(g)$, then $M^-(\tau_3)$ is in $L(f)$, or if $M^-(\tau_3)$ is in $V(f)$, then $N^+(\tau_3)$ is in $L(g)$, as claimed in the lemma. If $N^+(\tau_3)$ has a y-coordinate of 0, then $M^-(\tau_3)$ is in $L(f)$, or if $M^-(\tau_3)$ has a y-coordinate of 0, then $N^+(\tau_3)$ is in $L(g)$, which are situations we have already considered. The opposite inclusion is trivial.   ∎

The parabolic arcs used to define σ^2-parabolic STZones are simple convex functions. If a σ^2-parabolic STZone can be represented in the μ–σ^2 plane by a single defining parabolic arc, call it simple. Lemmas 5 and 6 are then applicable to the Minkowski sum of simple σ^2-parabolic STZones. Figure 6.12 partially illustrates this. Most often, σ^2-parabolic STZones are not simple. To treat Minkowski sums of σ^2-parabolic STZones in general, however, we need only a little more.

If for a set D in $\mathbf{R}^2$ there is a set of nonnegative piecewise continuous functions, $\mathfrak{F}$, such that $D = \cup_{f \in \mathfrak{F}} L(f)$, say that D is generated by $\mathfrak{F}$. If each of the functions in $\mathfrak{F}$ is simple, define $V(\mathfrak{F})$ to be $\cup_{f \in \mathfrak{F}} V(f)$.

Corollary 1   *If F is generated by a finite set of simple convex functions, $\mathfrak{F}$, and G is generated by a finite set of simple convex functions, $\mathfrak{G}$, then*

$$F \oplus G = \{F \oplus V(\mathfrak{G})\} \cup \{V(\mathfrak{F}) \oplus G\}$$

Proof.   Since $\{A \cup B\} \oplus C = \{A \oplus C\} \cup \{B \oplus C\}$ for any subsets A, B, and C, the claim follows directly from Lemma 6.   ■

Since $V(\mathfrak{F})$ is a union of points and vertical line segments and G is a union of regions below simple convex functions, $V(\mathfrak{F}) \oplus G$ reduces to the union of the Minkowski sums of points or vertical line segments with regions below simple convex functions. Each of these Minkowski sums is easily obtained by translating a simple convex function to a new simple convex function and defining a new region below this new function. The union of this myriad of new regions that yields $F \oplus G$ is itself easy to obtain, at least conceptually, since it reduces to the region below the upper envelope of the graphs of the functions defining the regions in the union.

We close by considering the implications of this corollary on the Minkowski sum of σ^2-parabolic STZones. If we denote the set of the defining

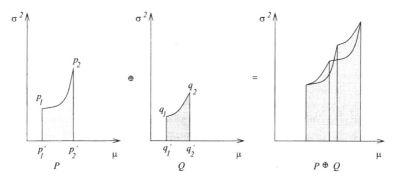

Figure 6.12. Minkowski sum of two simple σ^2-parabolic STZones P and Q. If $l(p,q)$ denotes the line segment between points p and q, Lemma 6 states that $P \oplus Q$ can be obtained as the union of $P \oplus l(q_1,q_1')$, $P \oplus l(q_2,q_2')$, $Q \oplus l(p_1,p_1')$, and $Q \oplus l(p_2,p_2')$. In some cases, one of the two vertices $p_1 + q_2$ and $p_2 + q_1$ may lie in the interior of the sum, as shown in Figure 6.2.

arcs of a σ^2-parabolic STZone by $\mathfrak{F}$, the STZone is generated by $\mathfrak{F}$, a finite set of simple convex functions. Corollary 1 thus applies to the Minkowski sums of σ^2-parabolic STZones in the μ–σ^2 plane. The translations of the defining monotonic parabolic arcs drawn from nonnegative parabolas yield arcs of the same type. The upper envelope is then defined piecewise again by arcs of the same type, determined by intersections of the arcs, yielding another σ^2-parabolic region.

REFERENCE

Valentine, F. A., 1964, *Convex Sets,* McGraw-Hill, New York.

7

CONCURRENT TOLERANCING FOR ACCURACY AND COST[1]

M. M. SFANTSIKOPOULOS

National Technical University of Athens
Athens, Greece

7.1 INTRODUCTION

Tolerance assignment in mechanical engineering product design and manufacturing is critical for product quality and performance and for manufacturing cost. On the other hand, the rising importance of precision products contributes to a strong demand for finer tolerances and clearances at a competitive cost. Dimensional tolerances are in general described as design/functional and manufacturing tolerances. There are also permissible machining variations in dimensions without tolerance indication in the ISO and other standards. Design/functional tolerances are related to the operational requirements of a mechanical assembly (i.e., critical clearances or interferences) or of a component and are determined through analytical and/or experimental methods. Manufacturing tolerances are connected primarily with processes that are considered for component manufacturing. In any case, manufacturing

---

[1]The contents of this chapter are based primarily on previous publications of the author—alone or with coauthors—first published in the *International Journal of Advanced Manufacturing Technology* (Sfantsikopoulos, 1990, 1993; Sfantsikopoulos et al., 1995), the *Interantional Journal of Robotics and Computer-Integrated Manufacturing* (Sfantsikopoulos and Diplaris, 1991) and the *International Journal of Machine Tools and Manufacture* (Sfantsikopoulos et al., 1994). They are included here by permission of the publishers.

Advanced Tolerancing Techniques, Edited by Hong-Chao Zhang
ISBN 0-471-14594-7 © 1997 John Wiley & Sons, Inc.

tolerances must obviously respect and satisfy the functional tolerances. Nevertheless, manufacturing tolerances play a major role in the development of the manufacturing cost, and therefore every effort for their optimum designation is justified. Numerous parameters affect the manufacturing accuracy of a dimension and the manufacturing cost related to this accuracy. They have primarily to do with the process plan, machine tool capabilities, tooling, inspection equipment, operator skill, lot size, and certainly, with the workpiece material properties and the workpiece shape, size, and machining allowances. The consequent derivation of any detailed analytical expression relating the manufacturing cost of a dimension with its specified tolerance zone has proved extremely difficult, if not impossible. Therefore, simple design rules of the type "the lower the tolerance, the higher the cost of manufacturing" and "do not specify higher accuracy than is really needed," industry-available comparative experimental data, tables, and charts, and personal expertise are commonly used for cost optimum tolerancing. However, such a methodology is not very practical, requires considerable time and effort, and is not always suitable for a CAD/CAM environment.

In this chapter an alternative approach to concurrent tolerancing for accuracy and optimum cost of accuracy is presented. It is based on a new cost–tolerance function and does not require detailed parameter analysis for its use in cost–tolerance optimization problems. For the solution of this kind of problem the method offers definite advantages compared with those mentioned earlier and the limited number of other approaches that have appeared, however systematic. It can lead to efficient and, with computer aid, timesaving tolerancing within the concurrent engineering concept. Typical problems thus treated in the following sections are those of tolerance transfer, coordinate tolerancing, tolerance compatibility, workpiece location for accurate machining, and peg-and-hole assembly accuracy. Computer programs for applications have been developed on the basis of Box's "complex" optimization algorithm, although other optimization algorithms may also be used.

7.2 COST–TOLERANCE FUNCTION

A cost–tolerance function typically relates a specified tolerance with the imposed by the tolerance cost of the manufacturing process involved. This manufacturing cost takes into account all related cost components, such as machine-tool time cost; tool, jig, and fixture expenditures; labor costs; cost of inspection equipment; cost of coolants and lubricants; and so on. The manufacturing cost is usually expressed as a relative ratio of the cost required for a specific tolerance value against the cost of producing a dimension with a very broad tolerance zone. A broad tolerance zone is considered one that presents no special or expensive accuracy requirements regarding skilled labor, high-accuracy inspection equipment and tooling, and special machining operations and/or conditions. Within the ISO tolerance quality grading it might be represented for turning, for example, by IT 11 or IT 12.

Different cost–tolerance functions have been proposed in the literature, all of them more or less exponential in type, following research and industry-wide data. These data generally agree with a trend such as that shown in Figure 7.1. From the total production cost $C(D,t)$ of a dimension D with an assigned tolerance $\pm t$, the cost component $C'(D)$ apparently corresponds to the upper threshold of the usual manufacturing accuracy, which does not have any specific quality requirements (i.e., relative cost equal to 1). It is the second component, $C''(D,t)$, of the total cost,

$$C(D,t) = C'(D) + C''(D,t) \tag{7.1}$$

which actually has to do with the "accuracy cost," which is specified by the tolerance zone $2t$ ($\pm$ t). For this additional accuracy cost, manufacturing practice suggests that for a relative tolerance variation dt/t, a multiple relative variation of the manufacturing cost is caused, which can be formulated as

$$\frac{dC''(D,t)}{C''(D,t)} = -r\frac{dt}{t} \qquad r > 0 \tag{7.2}$$

where r is a cost sensitivity to tolerancing exponent characteristic of the manufacturing process considered. Taking into account that apparently $dC'(D)/dt = 0$, from equations (7.1) and (7.2) the total cost $C(D,t)$ [equation (7.1)] is obtained as

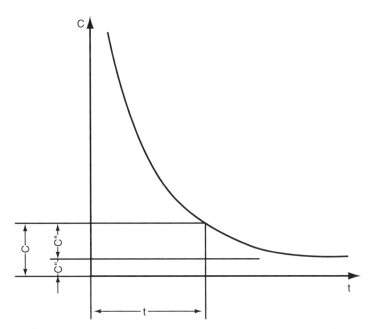

Figure 7.1. Typical manufacturing cost–tolerance relationship.

$$C(D,t) = C'(D) + \frac{C_0''(D)}{t^r} \tag{7.3}$$

with $C_0''(D)$ a constant that depends on the size of the dimension D.

Equation (7.3) can now be used to produce a cost–tolerance function that lies closer—as a general analytical expression—to manufacturing practice. The accuracy cost $C''(D,t)$ can thus be analyzed further if two dimensions D_1 and D_2 of the same workpiece are considered, which are produced by the same manufacturing setup with tolerances t_1 and t_2 of the same ISO standard tolerance grade. Because of this, and taking into account the industrial evidence reflected in the ISO standard tolerance grade methodology, the ratio of the accuracy costs $C''(D_1,t_1)$ and $C''(D_2,t_2)$ may be taken:

$$\frac{C''(D_1,t_1)}{C''(D_2,t_2)} = \frac{i(D_1)}{i(D_2)} \tag{7.4}$$

where $i(D_1)$ and $i(D_2)$ are the ISO standard tolerance factors corresponding to D_1 and D_2.[2] From equations (7.3) and (7.4),

$$\frac{C''(D_1,t_1)}{C''(D_2,t_2)} = \frac{C_0''(D_1)}{C_0''(D_2)} \left(\frac{t_2}{t_1}\right)^r = \frac{C_0''(D_1)}{C_0''(D_2)} \left[\frac{i(D_2)}{i(D_1)}\right]^r = \frac{i(D_1)}{i(D_2)}$$

or

$$C_0''(D_1) = C_0''(D_2) \left[\frac{i(D_1)}{i(D_2)}\right]^{r+1} \tag{7.5}$$

given that for the same tolerance grade (IT5-IT18), $t_2/t_1 = i(D_2)/i(D_1)$. Combining equations (7.3) and (7.5), a final model of the cost–tolerance function is obtained as

$$C(D,t) = C'(D) + C_0'' \frac{i(D)^{r+1}}{t^r} \tag{7.6}$$

where the constant C_0'' in this case depends on other than the size of the dimension D cost parameters of the specific application.

Equation (7.6) can certainly be used as is for concurrent manufacturing accuracy and cost optimization. It permits optimum tolerancing to be based on an explicit relationship between the size of the particular dimension, its specified tolerance zone, and the related manufacturing cost. In this way the

---

[2]For IT5-IT18 and $D \leq 500$ mm, $i(D) = 0.45 \sqrt[3]{D} + 0.001D$, D in millimeters and i in micrometers (ISO, 1988).

process planning decision making is supported and an optimum determination of the part's locating surfaces and working dimensions and tolerances can be produced. The cost sensitivity to tolerance r, an exponent characteristic of the manufacturing process, can be evaluated as lying, for different applications, inside a range of $\frac{1}{2} < r < 2$ (e.g., Figure 7.2). Nevertheless, lower r values ($\frac{1}{2} < r \leq 1$) appear to be more reasonable. The conclusion is supported by the relatively less expensive accuracy that modern machine tools and inspection equipment offer. In Figure 7.2 these lower r values demonstrate the actual industrial fact that lower ISO standard tolerance grades are now more readily accessible and less costly.

When for a toleranced dimension more than one stage of machining is used for which alternative processes (machines) may exist, the corresponding cost–tolerance function spectrum will finally appear as it is shown in the example of Figure 7.3 (Zhang et al., 1992). For design applications a cost–tolerance function may then represent the total cost of the processes employed.

7.3   TOLERANCE TRANSFER

Tolerance transfer of linear dimensions is of particular importance in process planning because it is associated directly with establishment of the locating surfaces for machining and inspection. Practice has nevertheless shown that functional/design datums and manufacturing datums of a part do not usually

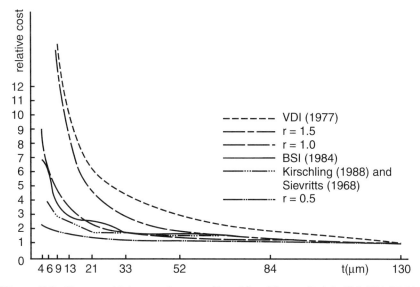

Figure 7.2. Cost sensitivity to tolerance, $D = 18,...,30$ mm (hole). ISO IT4-IT 11.

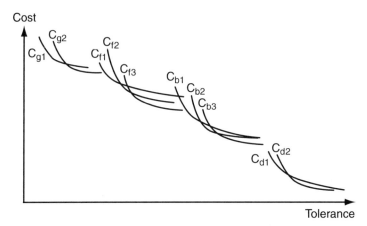

Figure 7.3. Manufacturing cost–tolerance curves for producing a cylinder bore. Cd, Cb, Cf, Cg for drill, bore, finishing bore, and grind operations, respectively. (From Zhang et al., 1992.)

coincide. As a consequence, new manufacturing tolerances have to be determined in the process plan operations. For these cases the single condition that the transferred new manufacturing tolerances must observe the initial functional/design tolerances is not sufficient for their direct and precise evaluation. Additional constraints, related to the manufacturing cost and the capabilities of the available machine tools, are required for an optimum and systematic assignment of the new tolerances.

To be machined accurately a mechanical component must, depending on its shape and details, possess one or more locating surfaces. Which of these surfaces will be used and in what succession is determined when planning the number and the sequence of the processing operations. To restrict machining errors as much as possible and to keep machine-tool idle time as low as possible, the fewer the location surfaces used in the process plan, the better. For every one of them, however, and for most applications, new working dimensions have to be calculated and new working tolerances have to be assigned on the basis of the initial design/functional dimensions and tolerances (i.e., the latter have to transferred to a new reference datum).

The general layout of the task described above corresponds to a linear dimensional chain with n dimensions D_j ($j = 1,2,...,n$) and design/functional tolerances $\pm t_j$, which are transferred for manufacturing reasons to a new reference or locating surface LS (Figure 7.4). The tolerances for convenience are taken here symmetrically relative to the nominal size of the dimensions. However, alternative positioning of the tolerance zone ($2t$) is possible through an arithmetical adjustment of the nominal size of the dimension. For the new manufacturing tolerances $\pm t_j'$ of the new dimensions D_j', the *tolerance transfer principle,* necessary for maintenance of the specified part's functional quality/accuracy, is expressed through the relationship.

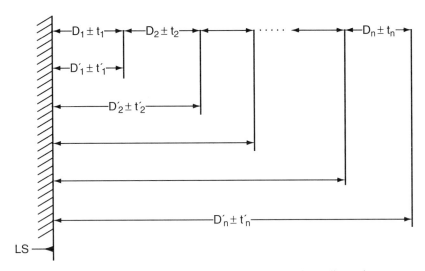

Figure 7.4. Transfer of tolerances. Datum and chained dimensions.

$$t'_{m-1} + t'_m \leq t_m, \qquad t_m, t'_m > 0 \qquad 1 < m \leq n \qquad (7.7)$$

Inequalities (7.7) alone obviously allow for an infinite number of correct (from the tolerancing point of view) solutions for the new tolerances. The appropriate solution has nevertheless to be established by the manufacturing engineer or process planner using methods such as those mentioned previously. In any case, the manufacturing costs induced by the new tolerances will also have to be considered in some way, on the basis of the existing data and rules. If the cost–tolerance function [equation (7.6)] is now introduced for linear dimensions and the resulting total manufacturing cost of the dimensions D'_j is required to be minimal, this additional condition may be used for evaluation of the new tolerances. For the same manufacturing process and setup,

$$\sum_{j=1}^{n} C(D'_j, t'_j) = \sum_{j=1}^{n} C'(D'_j) + C''_0 \sum_{j=1}^{n} \frac{i(D'_j)^{r+1}}{t'^r_j} \qquad (7.8)$$

from which, by taking into consideration that the first sum on the right-hand side of equation (7.8) is independent from the values of the new tolerances t'_j to be calculated, the condition for a cost optimum tolerance transfer is obtained as

$$\sum_{j=1}^{n} \frac{i(D'_j)^{r+1}}{t'^r_j} \longrightarrow \min \qquad (7.9)$$

The new tolerances t'_j can, in this way, finally be generated through the minimization of (7.9) with the tolerance transfer principle (7.7) acting as constraint. It is clear that this approach observes a full (100%) interchangeability (worst case). An example of a geared shaft for which Box's complex optimization algorithm was used is shown in Figure 7.5. For this case the cost sensitivity to tolerance exponent was taken $r = 0.8$.

7.4   COORDINATE TOLERANCING

7.4.1   Three-Dimensional Tolerance Model

Design and/or functional dimensions and tolerances do not always coincide with the computer numerical control (CNC) coordinate machining or coordinate measuring machine (CMM) axes. For reasons of manufacturing and inspection, however, these dimensions and tolerances must often be transferred to the orthogonal two- or three-dimensional system of coordinates that fits the application datums. The mathematical derivation of the new dimensions or coordinates usually does not present major problems, and they are therefore evaluated immediately or obtained by CAD. On the other hand, this is not the case with evaluation of the new coordinate tolerances. For them it is not possible to establish strict analytical relationships. They must simply observe, according to the tolerance transfer principle, the initial design and functional tolerances. This single condition is not sufficient for a systematic computational approach. Additional conditions, which will refer to the man-

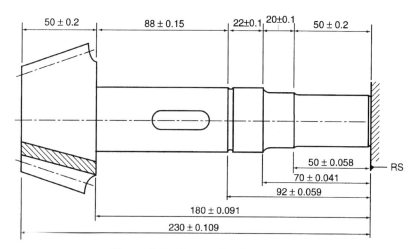

Figure 7.5. First application example.

ufacturing cost and to the capabilities of the processes and machine tools, must therefore be taken into account for the final designation of the new coordinate tolerances.

The boundary of a mechanical component consists of faces, edges, and vertices. The latter are at coordinate points. Coordinate dimensions are related to the design dimensions of component edges and the positioning dimensions of axes of symmetry, holes, slots, fillets, grooves, and so on. Location coordinate tolerances do not, however, refer to size tolerances such as for hole or cylinder diameters or fillet radii. For a design dimension D_{01} with an assigned symmetrical tolerance $\pm t_{01}$, not necessarily parallel to any of the three coordinate axes X, Y, Z of the given datum system, the dimensional relationship is (Figure 7.6)

$$D_{01} = \sqrt{(X_0 - X_1)^2 + (Y_0 - Y_1)^2 + (Z_0 - Z_1)^2} \qquad (7.10)$$

In equation (7.10) X_j, Y_j, and Z_j are the coordinates of the component points A_j, $j = 0,1$, of the dimension D_{01} and all are independent variables. For small variations ΔX_j, ΔY_j, and ΔZ_j, the corresponding variation ΔD_{01} of the dimension D_{01} can be approximated satisfactorily through the Taylor expansion of equation (7.10) by neglecting the second- and higher-order terms,

$$\Delta D_{01} \simeq \sum_{j=0}^{1} \left(\frac{\partial D_{01}}{\partial X_j} \Delta X_j + \frac{\partial D_{01}}{\partial Y_j} \Delta Y_j + \frac{\partial D_{01}}{\partial Z_j} \Delta Z_j \right) \qquad (7.11)$$

From equation (7.11) the symmetrical tolerance $\pm t_{01}$ is obtained for

$$|\Delta X_j| = t_{xj}$$
$$|\Delta Y_j| = t_{yj} \qquad (7.12)$$
$$|\Delta Z_j| = t_{zj}$$

as

$$t_{01} = \sum_{j=0}^{1} \left(\left| \frac{\partial D_{01}}{\partial X_j} \right| t_{xj} + \left| \frac{\partial D_{01}}{\partial Y_j} \right| t_{yj} + \left| \frac{\partial D_{01}}{\partial Z_j} \right| t_{zj} \right) \qquad (7.13)$$

where t_{xj}, t_{yj}, and t_{zj} are coordinate tolerances. Substitution in equation (7.13) of the partial derivatives calculated from (7.10) produces the final tolerance relationship,

$$t_{01} = \frac{1}{D_{01}} \sum_{j=0}^{1} (|X_0 - X_1| t_{xj} + |Y_0 - Y_1| t_{yj} + |Z_0 - Z_1| t_{zj}) \qquad (7.14)$$

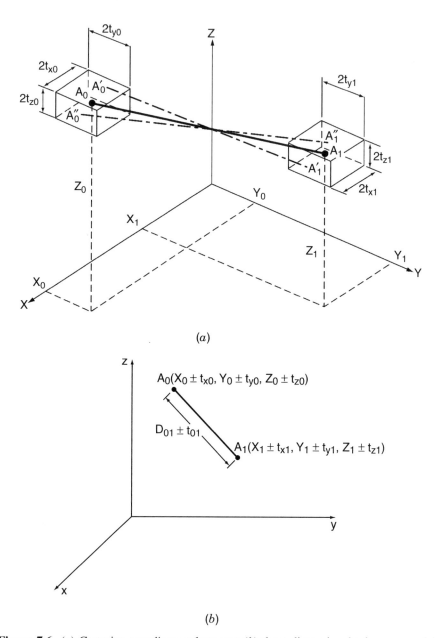

(a)

(b)

Figure 7.6. (a) Cartesian coordinate tolerances; (b) three-dimensional tolerance model.

As $\pm t_{01}$ is here considered a given design/functional tolerance that the new coordinate tolerances t_{xj}, t_{yj}, and t_{zj} must observe, the tolerance transfer principle can now be expressed, because of equation (7.14), through the conditions

$$\frac{1}{D_{01}} \sum_{j=0}^{1} (|X_0 - X_1|t_{xj} + |Y_0 - Y_1|t_{yj} + |Z_0 - Z_1|t_{zj}) \le t_{01} \qquad (7.15)$$

$$t_{xp} = \bar{t}_{xp} \qquad j = 0,1, \quad p,q,s, = 0,1$$

$$t_{yq} = \bar{t}_{yq}$$

$$t_{zs} = \bar{t}_{zs}$$

where $\bar{t}_{xp}$, $\bar{t}_{yq}$, and $\bar{t}_{zs}$ are eventually known or given coordinate tolerances. Machine-tool accuracy certification standards now provide for machining tolerances of ISO tolerance grades IT5, IT6, and IT7, whereas for less demanding operations, the target values may be considerably higher, up to 50 to 150%. If t is the required dimensional tolerance, it generally holds that

$$t \ge kT \qquad (7.16)$$

where T is the available machining accuracy and k is a safety factor taken approximately as ≤ 6. In addition to complying with the tolerance transfer principle (7.15), the coordinate tolerances will thus also have to comply with the process and machine tool accuracy capabilities, that is,

$$t_{xj} \ge kT_x$$

$$t_{yj} \ge kT_y \qquad (7.17)$$

$$t_{zj} \ge kT_z$$

where T_x, T_y, and T_z are the available accuracies along the X, Y, and Z machining axes for the operation planned.

7.4.2   Evaluation of the Coordinate Tolerances

Apparently, conditions (7.15) and constraints (7.17) alone do not suffice for evaluation of the coordinate tolerances t_{xj}, t_{yj}, and t_{zj}. In principle, an infinite number of solutions is always possible (even if just one of the six coordinate tolerances is unknown). It may also be possible that the problem has no solution for the process accuracies foreseen. On the other hand, every one of these solutions will be different and taking the accuracy-related manufacturing cost into account, they may or may not be worthy of consideration. An additional condition is necessary and the cost–tolerance relationship is used for this reason. The manufacturing cost for the coordinate tolerances of dimension D_{01} is, by equation (7.1),

$$C(D_{01}, t_{01}) = \sum_{j=0}^{1} [C(X_j, t_{xj}) + C(Y_j, t_{yj}) + C(Z_j, t_{zj})] \qquad (7.18)$$

and because of equation (7.6),

$$C(D_{01}, t_{01}) = \sum_{j=0}^{1} [C'(X_j) + C'(Y_j) + C'(Z_j)]$$

$$+ C_0'' \sum_{j=0}^{1} \left[\frac{i(X_j)^{r+1}}{t_{xj}^r} + \frac{i(Y_j)^{r+1}}{t_{yj}^r} + \frac{i(Z_j)^{r+1}}{t_{zj}^r} \right] \qquad (7.19)$$

where for the same machining process and setup the constant C_0'' is the same for all three axes. From equation (7.19), the cost-optimum coordinate tolerance transfer can be obtained by minimizing its accuracy-related part, with the tolerance transfer conditions (7.15) and (7.17) acting as constraints:

$$F(D, t, k) = \sum_{j=0}^{1} \left[\frac{i(X_j)^{r+1}}{t_{xj}^r} + \frac{i(Y_j)^{r+1}}{t_{yj}^r} + \frac{i(Z_j)^{r+1}}{t_{zj}^r} \right] - fk \rightarrow \min \qquad (7.20)$$

where f is a weighting factor aiming to optimize the accuracy safety factor k toward its maximum attainable value.

For a three-dimensional chain $D_{01}, D_{12}, \dots$ of $1, 2, \dots, n$ members (Figure 7.7), the optimization function $F(D, k, t)$ and the corresponding constraints that will lead to evaluation of the coordinate tolerances $t_{xm}, t_{ym}, t_{zm}, m = 0, 1, \dots, n$, are

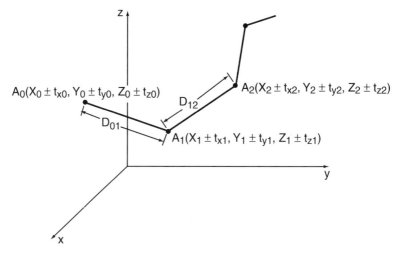

Figure 7.7. Three-dimensional chain.

derived on the basis of the preceding analysis. This time the constraints imposed by the tolerances of the intermediate coordinate points (component vertices, hole centers, etc.) must also be taken into consideration. Constraints (7.15) and (7.17) apply to every dimension $D_{m,m+1} \pm t_{m,m+1}$ of the chain,

$$\frac{1}{D_{m,m+1}} \sum_{j=m}^{m+1} (|X_m - X_{m+1}|t_{xj} + |Y_m - Y_{m+1}|t_{yj} + |Z_m - Z_{m+1}|t_{zj}) \le t_{m,m+1}$$

$$
\begin{aligned}
t_{xp} &= \bar{t}_{xp} & m &= 0,1,2,...,n \\
t_{yq} &= \bar{t}_{yq} & p,q,s &= 0,1,2...,n & (7.21) \\
t_{zs} &= \bar{t}_{zs}
\end{aligned}
$$

and

$$
\begin{aligned}
t_{xm} &\ge kT_x \\
t_{ym} &\ge kT_y & & (7.22) \\
t_{zm} &\ge kT_z & m &= 0,1,2,...,n
\end{aligned}
$$

For the total number of chain members the optimization function (7.20) is then formulated as

$$F(D,t,k) = \sum_{j=0}^{n} \left[\frac{i(X_j)^{r+1}}{t_{xj}^r} + \frac{i(Y_j)^{r+1}}{t_{yj}^r} + \frac{i(Z_j)^{r+1}}{t_{zj}^r} \right] - fk \rightarrow \min \quad (7.23)$$

As pointed out earlier, the problem (7.21)–(7.23) may or may not have a solution. In the second case, another datum system (i.e., location surface) for the component must be tried. If this next step is not possible or fails again, a modification of the machining process plan—through the employment of intermediate stages—and/or replacement of the equipment will have to be effected. It is noteworthy that the linear tolerance transfer problem and two-dimensional case represent special cases of the three-dimensional model by taking as the one-dimensional case

$$Y_m = \text{const}, Z_m = \text{const} \quad (7.24)$$

and as the two-dimensional case

$$Z_m = \text{const} \quad m = 0,1,2,...,n \quad (7.25)$$

The coordinate tolerances of the two-dimensional application example of Figure 7.8 were obtained using the same optimization algorithm as in Section 7.3, taking into account the relationships (7.21)–(7.23) and (7.25). Figure

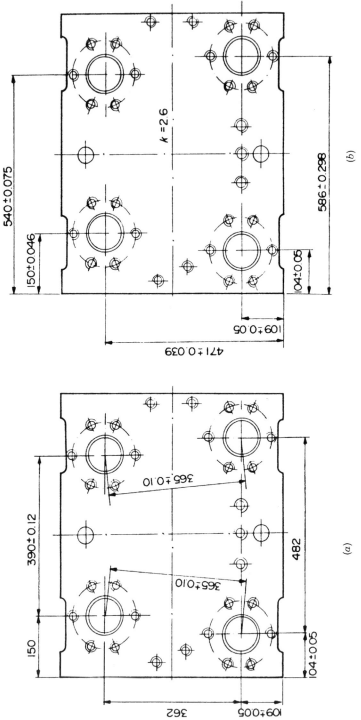

Figure 7.8. Second application example.

7.8(*a*) represents design/functional dimensioning and tolerancing. The coordinate dimensions and tolerances computed are given in Figure 7.8(*b*).

7.5 TOLERANCE COMPATIBILITY

7.5.1 Toleranced and Untoleranced Dimensions

Dimensions that define the shape, size, and details of an engineering product or an engineering component are inevitably, and for practical reasons, assigned with tolerances. Tolerances are applied to all three kinds of dimensions: functional, manufacturing, and inspection. Specific tolerance values known as toleranced dimensions (TDs) are used for the critical dimensions (i.e., those that are directly credited with the product or part functionality, manufacturing, or inspection). For dimensions that do not have particular accuracy requirements [usually described as untoleranced dimensions (UDs)] the permissible variations are specified either through the general categories—fine, medium, coarse—of the international standards or on the basis of the company's own rules, the designer's experience, and so on. Toleranced or untoleranced dimensions may also be designated as implicit dimensions, those not quoted in a two-dimensional engineering drawing or three-dimensional model because of conventional and/or currently applicable engineering dimensioning rules. It should be noted here, however, that two kinds of implicit dimensions exist: direct implicit dimensions (e.g., 90° angles, dimensions defined by symmetry, radii) and indirect implicit dimensions. The latter are always determined indirectly via one conventional dimension. Geometrical tolerances (straightness, parallelism, perpendicularity, flatness, etc.) are also translated to implicit toleranced dimensions (ITDs). All these types of dimensions—conventional, implicit, toleranced, untoleranced—are shown in Figure 7.9. If M_{TD} is the number of toleranced dimensions of a part, M_{UD} the number of its untoleranced conventional dimensions, M_{ITD} the number of implicit toleranced dimensions, and M_{IUD} the number of implicit untoleranced dimensions, it generally holds that

$$M_{TD} + M_{UD} + M_{ITD} + M_{IUD} = \mu N \qquad (7.26)$$

where N is the number of the characteristic coordinate points that define the shape in space (vertices, points of section lines, centers of circular arcs, etc.) and μ is a coefficient that accounts for the degrees of freedom of the shape. For one-, two-, or three-dimensional environments, $\mu = 1$, 2, or 3, respectively. Equation (7.26) also includes the implicit dimensions necessary for the shape location.

In principle, three different tolerance situations may be present in a dimensional accuracy specification: (1) all conventional dimensions are toleranced (i.e., $M_{UD} = 0$), (2) all conventional dimensions are untoleranced (i.e., $M_{TD} = 0$), or (3) toleranced and untoleranced dimensions coexist. The third

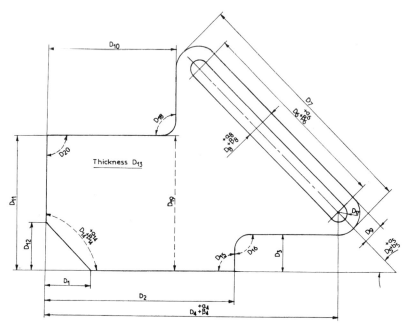

Figure 7.9. Conventional $(D_1,...,D_{13})$ and implicit $(D_{14},...,D_{20})$ dimensions; toleranced (D_4,D_5,D_6,D_8,D_{14}) and untoleranced dimensions.

case is the most representative of engineering practice. Similar situations apply for the implicit dimensions of the part. Their tolerances may, however, have to be adjusted by taking into consideration the presence of geometrical tolerances.

When specific, general, or other types of tolerances are assigned to conventional and/or implicit dimensions, a problem of compatibility of all four groups of tolerances generally arises for the entire *tolerance model. Tolerance compatibility* means that full use can be made of the tolerance zone of every dimension without affecting the tolerance zones of the other dimensions and that all tolerances can be produced by the machinery available. If tolerance compatibility does not occur, tolerance adjustments should then take place, mainly through suitable modification of the tolerances of the UD and/or IUD dimensions. This is a difficult and complicated task which has, in any case, to be performed and verified in an early design phase.

7.5.2   Tolerance Analysis

From N coordinate points $(X_1,Y_1,Z_1),...,(X_N,Y_N,Z_N)$ that define a mechanical component in a Cartesian coordinate system $(X,Y,Z,)$, only some of them are used in conventional engineering dimensioning. The use of coordinate points for dimensioning can therefore be direct and/or indirect. TD, UD, ITD, and

IUD dimensions, including those attributed to geometrical tolerances, may be regarded as consisting of constraints for the relative positions between them. Consequently, dimensional constraint equations have been used in the literature for study of the validity of the dimensional schemes. These equations may be linear or nonlinear and may be written as

$$D_i = D_i(\mathbf{X}) \qquad i = 1,...,n, \quad n = \mu N \tag{7.27}$$

where

$$\mathbf{X} = [X_1,...,X_n]^{\mathrm{T}} \tag{7.28}$$

is the geometry vector of the point coordinates. For small dimensional variations the Taylor series expansion of equations (7.27) yields

$$\Delta D_i = d_i = \sum_{j=1}^{n} \frac{\partial D_i}{\partial X_j} \Delta X_j = \sum_{j=1}^{n} \frac{\partial D_i}{\partial X_j} x_j \tag{7.29}$$

with the second- and higher-order terms of the series omitted. In matrix form, equations (7.29) are summarized as

$$\mathbf{d} = \mathbf{Jx} \tag{7.30}$$

where $\mathbf{J}$ is the Jacobian matrix of $D_i(\mathbf{X})$. For a proper component dimensioning, implicit dimensions included, this matrix can be inverted, and from equation (7.30) the coordinate variations will be

$$\mathbf{x} = \mathbf{J}^{-1}\mathbf{d} \tag{7.31}$$

If symmetrical dimensional and coordinate tolerances around the nominal size are now used,

$$d_i = \pm t_{D_i} \qquad x_j = \pm t_{xj} \qquad t_{D_i}, t_{x_j} \ge 0 \qquad i,j = 1,2,...,n \tag{7.32}$$

from equations (7.31), using the max-min tolerance evaluation method (worst case), the coordinate tolerances are obtained:

$$\mathbf{t}_x = \mathbf{J}_a^{-1} \cdot \mathbf{t}_D \tag{7.33}$$

The notation $\mathbf{J}_a$ in matrix equation (7.33) means that only the absolute values of the partial derivatives $|\partial D_i / \partial X_j|$ are taken into account in the Jacobian matrix of $D_i(\mathbf{X})$. Nonsymmetrical tolerances are also accounted for in equation (7.33) through arithmetical adjustment of the nominal size of the dimension.

For a compatible tolerancing scheme and given TD tolerances and UD permissible variations, the system of equations (7.33) should obviously provide positive signs for the coordinate tolerances t_{x_j}. On the other hand, solution of the system also has to satisfy the machining axes accuracy capability requirement, relationship (7.16),

$$t_{x_j} \geq k(T_x, T_y, T_z) \qquad 1 \leq k \leq 6 \qquad (7.34)$$

where T_x, T_y, and T_z are the available accuracies along the X, Y, and Z machining axes. This second constraint covers the positive sign condition. Conditions (7.34) are therefore necessary and sufficient for the compatibility of the dimensional and geometrical tolerances. When this is not true, the assigned tolerances are then not compatible and one, at least, or more of them have to be modified and/or new dimensioning and tolerancing schemes tried. Given that TD tolerances are related to the overall product/component functionality and must, in principle at least, remain unchanged, any tolerance adjustment will have to be attempted mainly through the UD and/or IUD dimensions. Because of their definition, these UD and IUD dimensions have, in any case, considerably larger zones of allowable variations and can consequently sustain the modifications that will eventually be needed.

7.5.3 Assignment of Compatible Tolerances

In a noncompatible tolerancing scheme with only TD tolerances valid, equations (7.33) consist of a system with $n + M_{UD} + M_{IUD} > n$ unknown variables (i.e., the new coordinate UD and IUD tolerances). Any new tolerancing scheme can therefore be evaluated only through an optimization procedure based on an appropriate function. The accuracy cost for producing a dimension D with a tolerance t results through the cost–tolerance relationship [equation (7.6)] and for dimensions that are not very large ($D < 500$ mm) proportional to the ratio $D^{2/3}/t$ with a satisfactory approximation. By taking $D^{2/3}$ as a weighting factor for a balanced tolerance distribution with reference to the cost of accuracy, the new tolerances may then be obtained by making the sum

$$\text{TF} = \sum_{i=1}^{n} X_i^{2/3} \, t_{X_i} \qquad (7.35)$$

attain its maximum possible value. This should, of course, be done in conjunction with constraints (7.34) and

$$\mathbf{J}_a \cdot \mathbf{t}_x \leq \mathbf{t}'_D \qquad (7.36)$$

where

$$\mathbf{t}'_D = [\mathbf{t}^{\mathrm{T}}_{\mathrm{TD}}, \mathbf{t}^{\mathrm{T}}_{\mathrm{ITD}}]^{\mathrm{T}} \tag{7.37}$$

is the vector of the given tolerances. The new UD and IUD allowances will then be equal to

$$\mathbf{t}''_D = \mathbf{J}_a \cdot \mathbf{t}_x \tag{7.38}$$

where $\mathbf{t}_x$ is the coordinate tolerance vector of the cost-optimization approach, [(7.34)–(7.36)]. These values will replace the initial general or other tolerances.

For a component dimensioning with $M_{\mathrm{UD}} = 0$ or $M_{UD} + M_{\mathrm{IUD}} = 0$, one or more of the TD tolerances have apparently to be modified if a nontolerance compatibility is verified. The TD dimensions will then have to be classified in groups M_{TD1}, $M_{\mathrm{TD2}},...,M_{\mathit{ITD1}}$, $M_{\mathrm{ITD2}},...(M_{\mathrm{TD}} = M_{\mathrm{TD1}} + M_{\mathrm{TD2}} + \cdots$, $M_{\mathrm{ITD}} = M_{\mathrm{ITD1}} + M_{\mathrm{ITD2}} + \cdots)$ of decreasing technical importance for the functionality or manufacturability of the part, and one may then proceed as with the UD tolerances. An important role for the analysis of the general tolerance compatibility problem relates to the number and type of implicit dimensions—direct or indirect of singular or multiple origin. Particular attention should be given where indirect multiple implicit dimensions exist, as their presence then has to be treated with modified Jacobian matrices. In Figure 7.10 an application example is shown that is based on the preceding tolerance compatibility analysis and use of an appropriate computer program.

7.6 OPTIMUM TOLERANCING FOR MACHINING

7.6.1 Machining Accuracy Requirements

Shape, size, and details (holes, threads, grooves, filets, etc.) of a mechanical part are determined through its dimensions and dimensional and geometrical tolerances. Part dimensioning rules and relative standards require manufacturing process considerations to be taken into account as far as possible.

Manufacturing dimensions, both conventional and implicit, should refer to one or more datum surfaces (i.e., locating surfaces). This is the "classical approach," which does not necessarily exploit all the capabilities of modern machine tools and the way in which the accuracy requirements for the part will best be met. The latter is particularly true for nonrotational parts. For this type of part, if suitable fixturing is provided for a given machine tool, there are, in fact, an infinite number of possible part positions inside the workspace (Figure 7.11).

Only a few of these positions can produce the dimensional and geometrical tolerances specified, and still fewer can produce this accuracy with minimum manufacturing cost. Tolerance assignment and/or synthesis under the concurrent engineering requirement for optimization of the manufacturing impli-

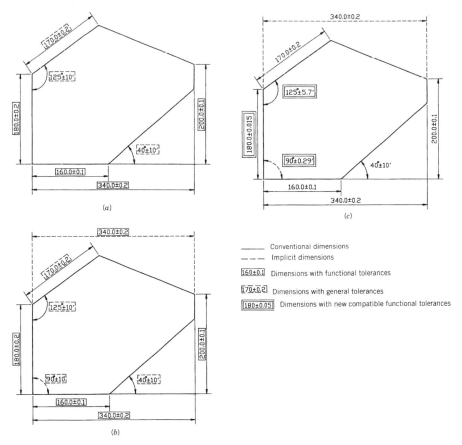

Figure 7.10. Third application example: (a) functional and general tolerances/conventional dimensions; (b) functional and general tolerances/conventional and implicit dimensions; (c) compatible tolerances.

cations, caused by the accuracy specified, has recently been given particular consideration. A common characteristic of most of these methodologies is that they are applied in a given part-dimensioning scheme (i.e., datum system). Coordinate dimensioning and tolerancing, studied in Section 7.4, allows for direct reference to the machine tool axes. Among all possible answers to the problem, it may therefore also be used to find an optimum part location for manufacturing. In terms of the Cartesian coordinate system to be selected, this means that part dimensioning will have to be accomplished for the highest possible accuracy and the lowest possible cost. This type of coordinate dimensioning is thus seen here from a point of view different from that in standards. It must guarantee not only the shape and size of the part but also conformance to dimensional and geometrical functional tolerances, within the available accuracies of the machining axes and with a minimum accuracy

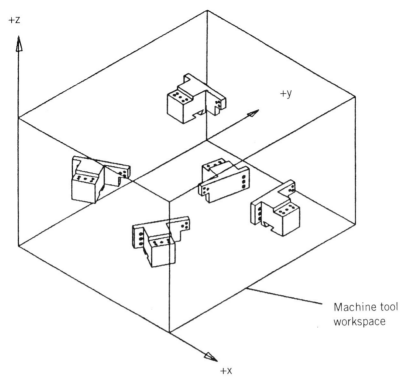

Figure 7.11. Alternative part positioning for machining.

cost. Industrial practice has shown that for nontrivial and/or simple cases, this task constitutes an extremely difficult job demanding expert knowledge, experience, and considerable effort and time.

7.6.2 Workpiece Cartesian Location System

As has been discussed, a part is geometrically and dimensionally described by a number of explicit and implicit dimensions equal to the product μN, where N is the number of characteristic coordinate points that define the shape of the part in space and μ is a coefficient that accounts for the degrees of freedom of the shape. This product value also includes the implicit dimensions that are necessary for the part location. An additional number of ν dimensions is used for the allocation and sizing of part details. For the number ν a theoretical limit does not apparently exist. All $\mu N + \nu$ dimensions are directly or indirectly assigned specific, general, or other types of tolerances, and included in this figure are the implicit toleranced dimensions that are generated through translation of the geometrical tolerances. The infinite number of possible part positions within the workspace of the machine tool, shown in Figure

7.11, may now be considered to represent an equally infinite number of Cartesian coordinate systems, which can, in principle, be used for part coordinate dimensioning (Figure 7.12). From this series of possible Cartesian systems $^0R(^0X,^0Y,^0Z)$, $^1R(^1X,^1Y,^1Z)$, $^2R(^2X,^2Y,^2Z)$,...,solution of the problem described in Section 7.6.1 evolves through the application of:

· Fundamental guidelines used to establish part locating surfaces
· The machining cost–tolerance relationship
· Constraints imposed by the accuracy capabilities of the machine tool axes

In the present approach the following six rules are used, first, to filter out datums inconsistent with engineering practice and to generate all others possible:

1. All external flat surfaces constitute possible (X,Y), (Y,Z), (Z,X) datums.

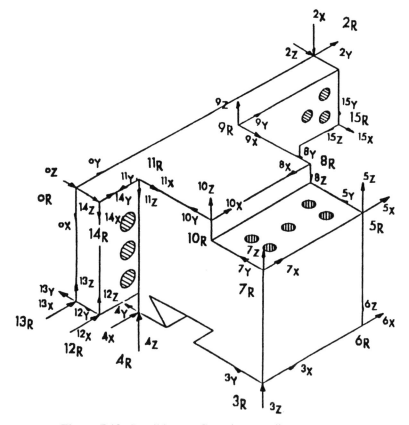

Figure 7.12. Possible part Cartesian coordinate systems.

2. An external flat surface is not taken as a possible datum if its size is smaller than one-fifth the size of the largest flat surface of the part.
3. If there is no flat surface, artificial datums are created that house external surfaces of the part. Rule 2 is also applicable.
4. Natural and artificial datums may coexist.
5. Every part straight edge in a datum plane may be taken as an X-axis. If there is no straight edge, any straight line preferably tangent to at least two part points may be taken as an X-axis.
6. For artificial datums the X-axis may be taken as any straight edge (if there is one) or tangent determined as in rule 5.

Application of rules 1 to 6 will generally still create a large number of possible datums. Any point on their X-axes may then be taken as a zero workpiece point (ZWP) [i.e., the origin of the Cartesian coordinate system $^lR(^lX,^lY,^lZ)$, $l = 0, 1, 2,...$]. For such a system a cost-optimum coordinate tolerancing is formulated through the system of relationships (7.21)–(7.23), which for the present application are written

$$\frac{1}{D_{m,m+1}} \sum_{j=m}^{m+1} (|^lX_m - ^lX_{m+1}| \; ^lt_{xj} + |^lY_m - ^lY_{m+1}| \; ^lt_{yj} + |^lZ_m - ^lZ_{m+1}| \; ^lt_{zj}) \leq t_{m,m+1}$$

$$(7.39)$$

$$^lt_{xp} = \bar{t}_{xp} \qquad ^lt_{yq} = \bar{t}_{yq} \qquad ^lt_{zs} = \bar{t}_{zs} \qquad m,p,q,s,j = 0,1,2,...,n \quad (7.40)$$

$$^lt_{xm} \geq kT_x \qquad ^lt_{ym} \geq kT_y \qquad ^lt_{zm} \geq kT_z \qquad (7.41)$$

$$F(D,t,k) = \sum_{j=0}^{n} \left[\frac{(i(^lX_j))^{r+1}}{(^lt_{xj})^r} + \frac{(i(^lY_j))^{r+1}}{(^lt_{yj})^r} + \frac{(i(^lZ_j))^{r+1}}{(^lt_{zj})^r} \right] - fk \rightarrow \min \quad (7.42)$$

In the relationships above, in addition to the notation already used:

- $D_{m,m+1}$ ($m = 0,1,2,...,n$) is a dimension of a dimensional chain of n members with point coordinates $(^lX_m,^lY_m,^lZ_m)$, $(^lX_{m+1},^lY_{m+1},^lZ_{m+1})$ referenced to the Cartesian coordinate system ^lR.
- $t_{m,m+1}$ is the tolerance of the dimension $D_{m,m+1}$.
- $^lt_{xm}, \; ^lt_{ym}, \; ^lt_{zm}, \; ^lt_{xm+1},$ and $^lt_{ym+1} ^lt_{zm+1}$ are the corresponding coordinate tolerances.
- $i(^lX_j)$ is the ISO standard tolerance factor for the coordinate ^lX_j.

Solution of the system of relationships (7.39)–(7.42) will produce, for the particular coordinate system, coordinate tolerances that will be consistent with the dimensional tolerances of the part, lie within the accuracy capabilities of the machining axes, and provide for minimum accuracy cost. The part will thus have to be located such that its Cartesian coordinate system ^lR coincides with the Cartesian coordinate system of the machine tool. However, such a part location in the machine tool workspace does not necessarily imply that the origins of both coordinate systems must coincide. Indeed, the required optimum results can be obtained if the part coordinate system ^lR is allowed to make a translational movement parallel to its axes (Figure 7.13), in accordance with the equations

$$^lX_m = {}^0X_m - a \qquad {}^lY_m = {}^0Y_m - b \qquad {}^lZ_m = {}^0Z_m - c \qquad (7.43)$$

where a, b, and c are the coordinate-system translation parameters. The relationships (7.39) and (7.42) can now be written

$$\frac{1}{D_{m,m+1}} \sum_{j=m}^{m+1} (|{}^0X_m - {}^0X_{m+1}| \, {}^lt_{xj} + |{}^0Y_m - {}^0Y_{m+1}| \, {}^lt_{yj} + |{}^0Z_m - {}^0Z_{m+1}| \, {}^lt_{zj})$$

$$\leq t_{m,m+1} \qquad (7.44)$$

$$F_i(D,t,k) = \sum_{j=0}^{n} \left[\frac{(i({}^0X_j - a))^{r+1}}{({}^lt_{xj})^r} + \frac{(i({}^0Y_j - b))^{r+1}}{({}^lt_{yj})^r} + \frac{(i({}^0Z_j - c))^{r+1}}{({}^lt_{zj})^r} \right]$$

$$- fk \rightarrow \min \qquad (7.45)$$

with $i = 1,2,3,...$ representing the number of coordinate systems considered.

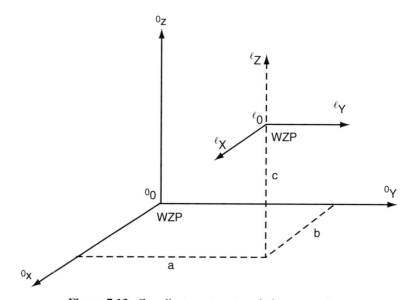

Figure 7.13. Coordinate system translation parameters.

For a solution of the system that comprises the relationships (7.40), (7.41), and (7.44) and the requirement (7.45), the Box optimization algorithm is used. Values for $^l t_{xj}$, $^l t_{yj}$, $^l t_{zj}$, a, b, c, and min F_i are obtained in this way. The same procedure has to be repeated for the other possible part datum schemes generated by application of the rules 1 to 6. The optimum part dimensioning and tolerancing coordinate system $^{opt}R(^{opt}X, ^{opt}Y, ^{opt}Z)$ is finally the one generated with the minimum-cost function value, that is,

$$F_{opt}(D,t,k) = \min \; [\min \; F_i(D,t,k)] \qquad i = 1,2,... \qquad (7.46)$$

The flowchart of the respective algorithm is shown in Figure 7.14.

Example. In Figure 7.15 a 2.5D oil pump body is shown with its three functional dimensions: $D_{01} = 110 \pm 0.1$ mm, $D_{12} = 54 \pm 0.05$ mm, and $D_{23} = 140 \pm 0.12$ mm. For simplicity all other part dimensions and details are omitted. The algorithm aims to define a ZWP and a relevant Cartesian coordinate dimensioning scheme that will provide for manufacturing tolerances compatible both with those specified (± 0.1, ± 0.05, ± 0.12 mm) and with the machining accuracies/safety margins available, $kT_x = kT_y = kT_z = 15$ μm. In addition, a minimum machining accuracy cost has to be attained.

Rules 1, 2, and 4 determine three flat surfaces A, B, C, all three with straight edges, for the part coordinate system to be referenced. For every X-axis edge a theoretically infinite number of possible ZWPs may be considered since X and Y axes may be allowed to float parallel to themselves, from the top to the bottom and from the left-hand to right-hand boundaries of the part (Figures 7.15 to 7.17).

For every datum surface A, B, C, part coordinate systems $^\circ R_A, ^\circ R_B, ^\circ R_C$ are determined whose X-axes coincide with the part straight edges as shown in the figures. The ZWP is taken arbitrarily to be a point on the X-axis. The algorithm then calculates the corresponding part Cartesian coordinate dimensions from the drawing input data. Based on random units, several acceptable schemes are thus created and the relevant tolerance and cost function values are computed. The convergence to an optimum solution after several iterations is controlled by the accepted arithmetic accuracy of the numerical process. The procedure is then repeated for the other two cases. The ZWP is fixed each time through the values of a and b (i.e., the distances from the origin of the initial system). Comparison of the values of the accuracy cost function F_1, F_1, F_3 of these three cases leads directly to the optimum part dimensioning and tolerancing scheme, which is case 3 in Figure 7.17. The results for the machining of the oil pump body are summarized in Table 7.1.

7.7 PEG-AND-HOLE ASSEMBLY PROBLEM

7.7.1 Problem Description

A peg-and-hole assembly may adopt clearance, transition, or interference fits, depending on the functional requirements of the particular design. For each

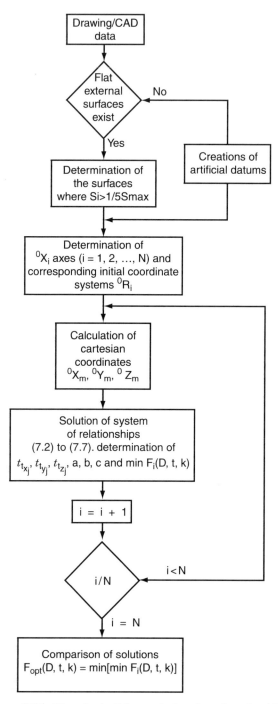

Figure 7.14. Flowchart of the workpiece location algorithm.

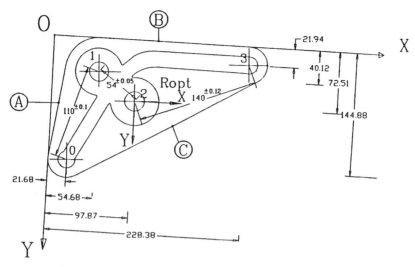

Figure 7.15. Fourth application example: determination of coordinate system for minimum cost, case 1.

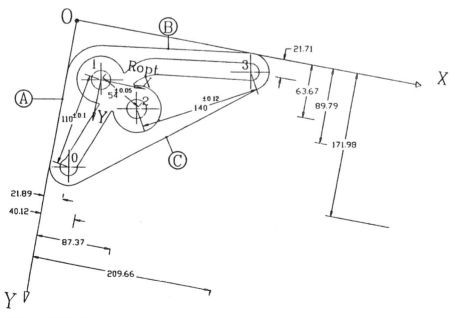

Figure 7.16. Fourth application example: determination of coordinate system for minimum cost, case 2.

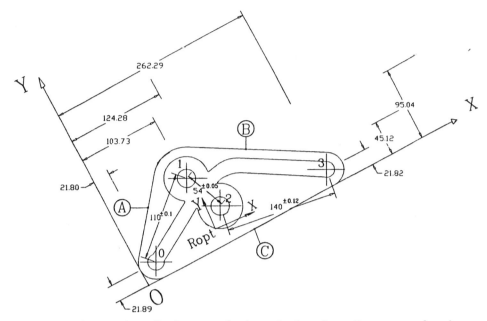

Figure 7.17. Fourth application example: determination of coordinate system for minimum cost, case 3 (optimum solution).

fit a maximum and a minimum functional clearance or interference are specified that are characteristic of the assembly. The international standards used to specify the type of fit refer only to the nominal dimensional size of the assembly; geometrical deviations and surface roughness are not considered.

TABLE 7.1. Optimum Tolerancing (μm) for Machining

Cartesian Coordinate Dimensions	Case		
	1	2	3
X_0	14	17	15
Y_0	15	13	14
X_1	18	19	19
Y_1	19	19	19
X_2	14	16	18
Y_2	21	20	19
X_3	30	32	31
Y_3	24	30	27
ZWP position	$a = 97.87$ mm	$a = 40.12$	$a = 103.73$
	$b = 72.5$ mm	$b = 63.67$	$b = 21.89$
Cost function value	$F_1 = 1.756$	$F_2 = 1.636$	$F_3 = 1.395$

Thus the clearances (or interferences) calculated are accounted for in the ideal component geometries. In practice, however, geometrically ideal components are not met. The actual clearances are a function not only of the intended dimensional deviations from the nominal size of the assembly but are also a function of the unavoidable geometrical deviations and, to a certain extent, of the surface roughness of the assembly components. Geometrical tolerances are actually assigned in addition to the dimensional tolerances, whereas for certain applications the latter may also include some provision for them (e.g., the envelope requirement). This is especially true for the tolerance data obtained through experimental work and/or quoted in most of the relative design guidelines. Regarding the surface roughness, the Centre Line Average (CLA) value is usually 5 to 20% of the respective dimensional tolerance (approximately 5 to 7% for clearance, 8 to 10% for transition, and 10 to 15% for interference fits).

Maximum and minimum functional clearances and dimensional and geometrical tolerances may, in general, be considered as consisting of an *assembly system,* where the values of the first two variables are known or given and the tolerance values have to be suitably allocated (Figure 7.18). This tolerance allocation should, of course, observe, on the other hand, the mini-

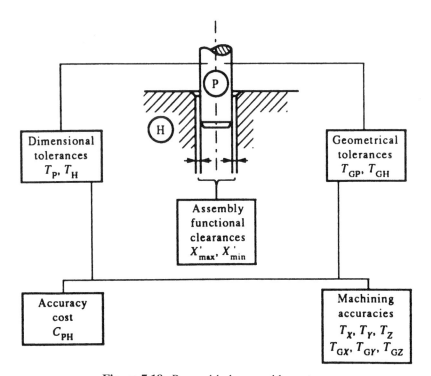

Figure 7.18. Peg-and-hole assembly system.

mum manufacturing cost requirement and conform with the accuracy capabilities of the production machinery available.

7.7.2   Functional Requirements

In a round peg-and-hole assembly (i.e., clearance fit) with ideal component geometries, the functional requirements for minimum clearance, X_{min}, and maximum clearance, X_{max}, lead directly to the ISO dimensional tolerances required. The tolerance evaluation is based on the industrial practice that provides, for economical manufacture, at least equal and usually coarser tolerance grades for the hole than for the peg (e.g., one tolerance grade higher) (Figure 7.19)

$$X_{max} = a_P + a_H + T_P + T_H \tag{7.47}$$

$$X_{min} = a_P + a_H = a \tag{7.48}$$

$$\nu = \frac{T_H}{T_P} = 1 - 1.6 - 1.6^2 - \cdots \tag{7.49}$$

$$T_P = \frac{1}{\nu + 1} (X_{max} - X_{min}) \tag{7.50}$$

$$T_H = \frac{\nu}{\nu + 1} (X_{max} - X_{min}) \tag{7.51}$$

When actual parts are considered, however, their cylindricity deviations act on the dimensional clearance of the tolerance model of Figure 7.19. The new model is shown in Figure 7.20, where T_{GP} and T_{GH} are now the geometrical

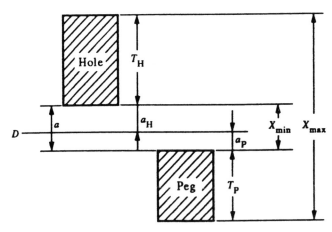

Figure 7.19. Dimensional tolerances for geometrically ideal assembly components.

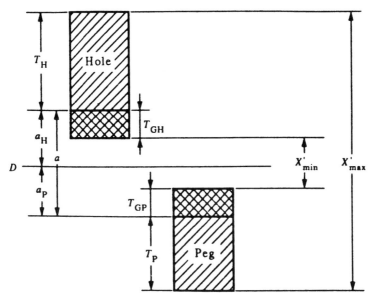

Figure 7.20. Dimensional and geometrical tolerances for actual assembly components.

tolerance zones for the peg and hole, respectively. They may, for convenience, be expressed in terms of their dimensional counterparts as

$$T_{GP} = \lambda_P T_P \qquad T_{GH} = \lambda_H T_H \tag{7.52}$$

with the geometric tolerance coefficients λ_P and λ_H varying in general between

$$0 \le \lambda_P \quad \text{and} \quad \lambda_H < 1 \tag{7.53}$$

The worst-case functional clearance is thus obtained as

$$X'_{min} = a - T_{GP} - T_{GH} < X_{min} \tag{7.54}$$

whereas the maximum clearance, X'_{max}, remains the same as that of Figure 7.19 ($X'_{max} = X_{max}$). Taking into consideration the relationships (7.52), equations (7.47) and (7.54) may also be written

$$X'_{max} = a + T_P(1 + \nu) \tag{7.55}$$

$$X'_{min} = a - T_P(\lambda_P + \nu\lambda_H) \tag{7.56}$$

The tolerance grade ratio, ν, of the assembly, defined by equation (7.49), now takes the discrete values ..., $1/1.6^2$, $1/1.6$, 1, 1.6, 1.6^2,... in accordance with the R5 series scaling of the ISO tolerance grades, as has been already noted.

Provided that accuracy and cost requirements are satisfied, an ideal value for ν is 1. Dimensional tolerance zones, geometrical tolerances, and basic deviations should thus satisfy, for the case of the clearance fit of a peg-and-hole assembly, the two conditions

$$a + T_P(1 + \nu) \leq X'_{\max} \tag{7.57}$$

$$a - T_P(\lambda_P + \nu\lambda_H) \geq X'_{\min} \tag{7.58}$$

with $X'_{\min}$ and $X'_{\max}$ now representing design/function fixed values.

7.7.3 Accuracy Specification

Conditions (7.57) and (7.58) do not apparently suffice alone for a systematic and technologically sound derivation of the unknown tolerances T_P, T_H, T_{GP}, and T_{GH}. Additional considerations and constraints related to the manufacturing cost and accuracy capabilities of the production machinery for this case also have to be taken into account. The accuracy cost of the peg-and-hole assembly components is evaluated through the cost–tolerance function described by equation (7.6) and is considered here to represent the total accuracy cost of all the machining stages eventually required. For the tolerance zones $T_P + T_{GP}$ and $T_H + T_{GH}$,

$$C''_{PH} = C''_P + C''_H = \frac{C''_{OP}i(D)^{r+1}}{(T_P + T_{GP})^r} + \frac{C''_{OH}i(D)^{r+1}}{(T_H + T_{GH})^r}$$

or from equations (7.49) and (7.52),

$$C''_{PH} = \frac{C''_{OP}i(D)^{r+1}}{T_P^r}\left[\frac{1}{(1 + \lambda_P)^r} + \frac{c}{\nu^r(1 + \lambda_H)^r}\right] \tag{7.59}$$

where

$$c = \frac{C''_{OH}}{C''_{OP}} \tag{7.60}$$

C''_{OH} and C''_{OP} are constants depending on the machining setup for the peg and the hole, respectively, and the cost sensitivity to the tolerance exponent is for simplicity taken to be approximately the same for both assembly components.

The design requirement for functional clearances $(X'_{\min}, X'_{\max})$ on the one hand and minimum cost C''_{PH} on the other can therefore be pursued simultaneously by applying the Lagrange multipliers method for constraints (7.57) and (7.58) and equation (7.59), through minimization of the *assembly function* AF, that is,

$$AF = \frac{C''_{OP} i(D)^{r+1}}{T_P^r} \left[\frac{1}{(1 + \lambda_P)^r} + \frac{c}{\nu^r (1 + \lambda_H)^r} \right]$$

$$+ K_1 [a + T_P(1 + \nu) - X'_{max}]$$

$$+ K_2 [-a + T_P(\lambda_P + \nu \lambda_H) + X'_{min}] \tag{7.61}$$

where K_1, and K_2 are the Lagrange multipliers. For $\partial AF / \partial T_P = \partial AF / \partial \lambda_P = \partial AF / \partial \lambda_H = \partial AF / \partial K_1 = \partial AF / \partial K_2 = 0$, the following three relationships are obtained:

$$T_P = \frac{X'_{max} - a}{1 + \nu} \tag{7.62}$$

$$\lambda_P = \left(\frac{a - X'_{min}}{X'_{max} - a} \frac{1 + \nu}{1 + c^{1/(r+1)}} \right) - \frac{c^{1/(r+1)} - \nu}{1 + c^{1/(r+1)}} \tag{7.63}$$

$$\lambda_H = \frac{c^{1/(r+1)}}{\nu} (1 + \lambda_P) - 1 \tag{7.64}$$

It may be noted that in equation (7.63) the quantity

$$A = \frac{a - X'_{min}}{X'_{max} - a} = \frac{T_{GP} + T_{GH}}{T_P + T_H} \tag{7.65}$$

reflects the contribution of the geometrical tolerances to the character of the peg-and-hole fit. By considering that $\lambda_P \geq 0$ from equations (7.63) and (7.65), a minimum value for the tolerance grade ratio, ν is derived:

$$\nu \geq \frac{c^{1/(r+1)} - A}{1 + A} \tag{7.66}$$

Equations (7.62)–(7.64) and constraint (7.66), in conjunction with the fact that all four tolerance zones T_P, T_H, T_{GP}, and T_{GH} should also comply with the available machining accuracies (T_{PO}, T_{GPO}), (T_{HO}, T_{GHO}) plus an appropriate safety margin,

$$T_P, T_{GP} \geq k T_{PO}, k T_{GPO} \qquad T_H, T_{GH} \geq k T_{HO}, k T_{GHO} \tag{7.67}$$

have been taken into account for the formulation of the algorithm shown in Figure 7.21. All ISO standard clearance fits for both basic hole and basic

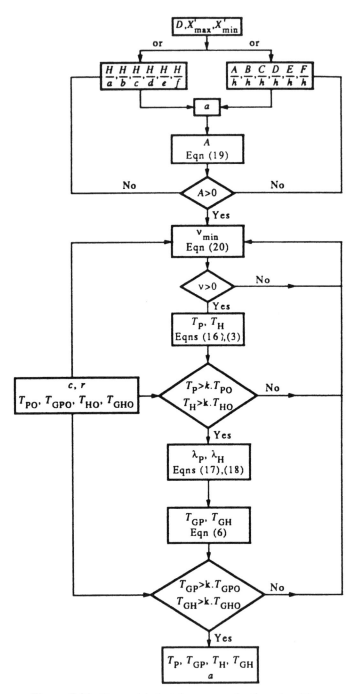

Figure 7.21. Peg-and-hole tolerance evaluation algorithm.

shaft systems $(H/a,...,H/f - A/h,...,F/h)$ are taken into account by the algorithm. The program for a given set of functional clearances (X'_{max}, X'_{min}) for a round peg-and-hole assembly of nominal size D and a given machining setup $(c, r, T_{PO}, T_{GPO}, T_{HO}, T_{GHO})$ produces a set of optimum cost, clearance, and machining compatible dimensional and geometrical tolerances.

Example. For a round peg-and-hole assembly of nominal size $D = 30$ mm, two sets of functional clearances are considered: No 1 ($X'_{max} = 350$ μm, $X'_{min} = 100$ μm) and No. 2 ($X'_{max} = 160$ μm, $X'_{min} = 40$ μm). For a maximum assembly length of $3D = 90$ mm, the machining setup ratio, c, is equal to 1.25 for the first set and 1.75 for the second, whereas for both cases the machining capabilities are specified as $k(T_{PO}, T_{HO}) = 15$ μm and $k(T_{GPO}, T_{GHO}) = 10$ μm. For a cost sensitivity to tolerance exponent $r = 0.8$, six basic hole clearance fits, H/a, H/b, H/c, H/d, H/e, and H/f, have been taken into account. For clearance set 1, two possible values for the minimum tolerance grade ratio are derived, $\nu = 1$ and $\nu = 1.6$, from which $\nu = 1.6$ leads immediately to a negative value of λ_H. Thereafter the algorithm produces the optimum problem solution for $\nu = 1$. Of the six possible clearance fits of set 2, there is only one minimum value for the tolerance grade ratio, $\nu = 1$. Six algorithm iterations were needed for the final tolerance assignment. For both sets, Table 7.2 summarizes the results obtained. ISO standardization has been used for presentation of the dimensional tolerances.

7.8 CONCLUSIONS

In most applications, the majority of the functional tolerances cannot readily be translated into manufacturing tolerances. This tolerance translation prob-

TABLE 7.2. Tolerances (μm) for Peg-and-Hole Assembly

	Clearance Set				
	1		2		
X'_{max}	350		160		
X'_{min}	100		40		
ν	1		1.6		
T_P	95		36.5		
T_H	95	$\varnothing 30 \dfrac{H10}{b10}$	58.5	$\varnothing 30 \dfrac{H9}{d8}$	
a	160		65		
T_{GP}	22		14		
T_{GH}	38		11		
λ_P	0.234		0.388		
λ_H	0.397		0.184		

lem becomes more difficult for solution when manufacturing costs for the specified accuracy are considered and there exist limits for the machining accuracies available. All these limitations have to be faced and optimized concurrently in both the design stage and, especially, in process planning.

Tolerance analysis, which refers to the part coordinates of a given Cartesian coordinate system (e.g., that of the machine tool), in conjunction with observation of the tolerance transfer principle and the tolerance compatibility requirement, constitutes a technological approach that can effectively support the assignment of accuracy consistent with cost-optimum and process-feasible manufacturing tolerances. The cost–tolerance function introduced makes it possible to pursue an optimum solution that is based on an explicit relationship between the size of a dimension, the tolerance assigned to it, and the related manufacturing cost. Tolerance evaluation computer programs based on known optimization algorithms are simple and quick and can be interfaced with a CAD/CAM environment. In a related consideration, the cost–tolerance function is effective for the analysis and accuracy requirements of the assembly components of a peg-and-hole clearance fit.

ACKNOWLEDGMENTS

Acknowledgements are due Springer-Verlag Ltd., Pergamon Press Ltd., and Hallwag Publishing Ltd. for granting permission to include material by the author first published in their journals.

NOMENCLATURE

a	fundamental dimensional tolerance deviation
a,b,c	coordinate system translation parameters
$C(D,t)$ or C	manufacturing cost of dimension D with tolerance $\pm t$
$C'(D)$ or C'	manufacturing cost of dimension D without specific accuracy requirement
$C''(D,t)$ or C''	additional manufacturing cost for producing dimension D with tolerance $\pm t$
C_0''	accuracy cost constant
D	linear dimension or diameter
$F(D,t,k)$	tolerance optimization function
f	weighting factor
$i(D)$	ISO standard tolerance factor for nominal size D
K_1, K_2	Lagrange multipliers
k	accuracy safety factor
M	number of dimensions
N	number of coordinate points
r	cost sensitivity to tolerance exponent

T	dimensional tolerance zone
T_G	geometrical tolerance
T_X, T_y, T_z	accuracies of the machining axes
t	tolerance
$t_{m,m+1}$	tolerance of the dimension $D_{m,m+1}$
t_{xm}, t_{ym}, t_{zm}	coordinate tolerances
t_{XP}, t_{YP}, t_{ZP}	design/function restricted coordinate tolerances
X_{max}, X_{min}	clearances
X, Y, Z	point coordinates
λ	geometric tolerance coefficient
μ	degrees of freedom coefficient
ν	tolerance grade ratio

Indices

H	hole
ITD	implicit toleranced dimension
IUD	implicit untoleranced dimension
P	peg
PH	peg-and-hole assembly
TD	toleranced dimension
UD	untoleranced dimension

REFERENCES

Beck, C., 1984, Einbeziehung von Form und Lageabweichungen in die Berechnung linearer Massketten, *Feingeratetechnik*, Vol. 33, pp. 6–9.

Box, M. J., 1965, A New Method for Constrained Optimization and Comparison with Other Methods, *Comput. J.*, Vol. 8, No. 6, pp. 42–52.

BSI, 1990, *Working Limits on Untoleranced Dimensions*, BS 4500, Part 3, British Standards Institution, London.

BSI, 1984, *Manual of British Standards in Engineering Drawing and Design*, British Standards Institution in association with Hutchinson, London.

Cagan, J., and Kurfess, T. R., 1991, *Optimal Design for Tolerance and Manufacturing Allocation*, EDRC, 24-67-91. Engineering Design Research Center, Carnegie Mellon University, Pittsburgh, PA.

Cogun, C., 1990, A Correlation Between Deviations in Circularity, Cylindricity, Roughness and Size Tolerances, *Int. J. Mach. Tools Manuf.*, Vol. 30, pp. 561–567.

Deutsches Institut für die Normung e.V., Berlin, 1975, *Dimensioning in Drawings: Kinds, Rules*, DIN 406, Parts 1 and 2, DIN,

DIN, 1986, *General Tolerances of Linear and Angular Dimensions*, DIN 7168, Part 1, DIN,

Greenwood, W. H., and Chase, K. W., 1988, Worst Case Tolerance Analysis with Nonlinear Problems, *J. Eng. Ind.*, Vol. 110, pp. 232–235.

He, J. R., and Gibson, P. R., 1992, Computer Aided Geometrical Dimensioning and Tolerancing for Process-Operation Planning and Quality Control, *Int. J. Adv. Manuf. Technol.*, Vol. 7, pp. 11–20.

He, J. R., and Lin, G. C. I., 1992, Computerized Trace Method for Establishing Equations for Dimensions and Tolerances in Design and Manufacture, *Int. J. Adv. Manuf. Technol.*, Vol. 7, pp. 210–217.

Hillyard, R. C., and Braid, I. C., 1978, Analysis of Dimensions and Tolerances in Computer-Aided Mechanical Design, *Comput.-Aid. Des.*, Vol. 10, No. 3, pp. 161–166.

ISO, 1983, *Technical Drawings—Geometrical Tolerancing, Tolerancing of Form, Orientation, Location and Run-out: Generalities, Definitions, Symbols, Indications on Drawings*, ISO 1101, International Standardization Organization, Paris.

ISO, 1985, *Technical Drawings—Dimensioning: General Principles, Definitions, Methods of Execution and Special Indications*, ISO 129, International Standardization Organization, Paris.

ISO, 1985, *Technical Drawings—Fundamentals of Tolerancing Principle*, ISO 8615, International Standardization Organization, Paris.

ISO, 1988, *System of Limits and Fits*, ISO 286, Parts 1 and 2, International Standardization Organization, Paris.

ISO, 1989, *Permissible Machining Variations in Dimensions Without Tolerances Indication*, ISO 2768, International Standardization Organization, Paris.

Iwata, K., and Sugimura, N., 1987, An Integrated CAD/CAPP System with "Know Hows" on Machining Accuracies of the Parts, *J. Eng.*, Vol. 109, pp. 108–133.

Kirschling, G., 1988, *Qualitätssicherung und Toleranzen*, Springer-Verlag, Berlin.

Koenigsberger, F., and Burdekin, M., 1982, *Testing Machine Tools*, Pergamon Press, Elmsford, NY.

Korsakov, V. S., 1979, *Fundamentals of Manufacturing Engineering*, Mir, Moscow.

Light, R., and Gossard, D., 1982, Modification of Geometric Models Through Variational Geometry, *Comput.-Aid. Des.*, Vol. 14, No. 4, pp. 209–214.

Ngoi, B. K. A., and Kai, C. C., 1993, A Matrix Approach to Tolerance Charting, *Int. J. Manuf. Technol.*, Vol. 8, pp. 175–181.

Ngoi, B. K. A., 1992, Applying Linear Programming to Tolerance Chart Balancing, *Int. J. Adv. Manuf. Technol.*, Vol. 7, pp. 187–182.

Osanna, P. H., 1979, Surface Roughness and Size Tolerance, *Wear*, Vol. 57, pp. 222–236.

Ostwald, P. F., and Huang, J., 1977, A Method for Optimal Tolerance Selection, *J. Eng. Ind.*, Vol. 99, pp. 558–565.

Peters, J., 1970, Tolerancing the Components of an Assembly for Minimum Cost, *J. Eng. Ind.*, Vol. 92, pp. 677–682.

Sfantsikopoulos, M. M., 1990, A Cost-Tolerance Analytical Approach for Design and Manufacturing, *Int. J. Adv. Manuf. Technol.*, Vol. 5, No. 2, pp. 126–134.

Sfantsikopoulos, M. M., and Diplaris, S. C., 1991, Coordinate Tolerancing in Design and Manufacturing, *Int. J. Robot. Comput.-Integrat. Manuf.*, Vol. 8, No. 4, pp. 219–222.

Sfantsikopoulos, M. M., 1993, Compatibility of Tolerancing, *Int. J. Adv. Manuf. Technol.*, Vol. 8, pp. 25–28.

Sfantsikopoulos, M. M., Diplaris, S. C., and Papazoglou, P. N., 1995, *Concurrent Dimensioning for Accuracy and Cost*, *Int. J. Adv. Manuf. Technol.*, Vol. 10, pp. 263–268.

Sfantsikopoulos, M. M., Diplaris, S. C., and Papazoglou, P. N., 1994, An Accuracy Analysis of the Peg-and-Hole Assembly Problem, *Int. J. Mach. Tools Manuf.*, Vol. 34, No. 5, pp. 617–623.

Sievritts, A., 1968, *Toleranzen und Passungen fur Längenmasse*, Beuth-Vertrieb, Berlin.

Speckhart, F. H., 1972, Calculation of Tolerance Based on a Minimum Cost Approach, *J. Eng. Ind.*, Vol. 94, pp. 447–453.

Spotts, M. F., 1973, Allocation of Tolerances to Minimize Cost of Assembly, *J. Eng. Ind.*, Vol. 95, pp. 762–764.

VDI, 1977, *Technisch-Wirtschsftliches Konstruieren*, VDI-Richtlinie 2225, VDI-Verlag, Dusseldorf.

Warnecke, H. J., and Dutschke, W., 1984, *Fertigungsmesstechnik*, Springer-Verlag, Berlin.

Weil, R., 1988, Integrating Dimensioning and Tolerancing in Computer-Aided Process Planning, *Robot. Comput.-Integrat. Manuf.*, Vol. 4, No. 1–2, pp. 41–48.

Willhelm, R. G., and Lu, S. C. Y., 1992, Tolerance Synthesis to Support Concurrent Engineering, *Ann. CIRP*, Vol. 41, No. 1, pp. 197–200.

Zhang, C., and Wang, H. P., 1993, Integrated Tolerance Optimization with Simulated Annealing, *Int. J. Adv. Manuf. Technol.*, Vol. 8, pp. 167–174.

Zhang, C., Wang, H. P., and Li, J. K., 1992, Simultaneous Optimization of Design and Manufacturing-Tolerances with Process (Machine) Selection, *Ann. CIRP*, Vol. 41, No. 1, pp. 569–572.

Zhang, H. C., Mei, J., and Dudek, R. A., 1991, Operational Dimensioning and Tolerancing in CAPP, *Ann. CIRP*, Vol. 40, No. 1, pp. 419–422.

8

SIMULTANEOUS TOLERANCING FOR DESIGN AND MANUFACTURING

G. ZHANG

Chongqing University,
Chongqing, China

8.1 INTRODUCTION

Tolerance design is a critical issue in design and manufacturing. It affects both product and process design, because the tolerance is the bridge between product requirements and manufacturing cost. Currently, tolerance design is being carried out in the two separate domains: tolerance design in computer-aided design (CAD) and tolerance design in computer-aided process planning (CAPP).

In product design, as input data, we know Y, the tolerance of product specification (or the performance parameter tolerance of a product), and assembly drawings, hence the mechanical construction of the product. From the assembly drawings, the functional equation that represents the relationship between the deviation of product specification y and part functional tolerance x_i can be derived by

$$y = f_y(x_1, x_2, ..., x_i, ..., x_n) \tag{8.1}$$

The task of tolerance analysis is to calculate the deviation y using formula (8.1) and to compare with tolerance Y, to determine whether or not $y \leq Y$,

Advanced Tolerancing Techniques, Edited by Hong-Chao Zhang
ISBN 0-471-14594-7   © 1997 John Wiley & Sons, Inc.

while the objective of tolerance synthesis is to distribute tolerance Y over part tolerances x_i with the minimized manufacturing cost.

In manufacturing, as input data, we know x_i, the part functional tolerance, and the process selected to produce the part. From the process selected, the machining equations that represent the relationships between each machining deviation Δ_i corresponding to functional tolerance X_i and its machining tolerance δ_{ij} can be formed by

$$\Delta_i = f_x(\delta_{i1}, \delta_{i2}, ..., \delta_{ij}, ..., \delta_{im_i}) \qquad i = 1,2...,n \qquad (8.2)$$

The task of tolerance analysis is to calculate deviation Δ_i using formula (8.2) and to compare with functional tolerance x_i, in order to know whether or not $\Delta_i \leq x_i$, while the objective of tolerance synthesis is to distribute tolerance x_i over the machining tolerances δ_{ij} with a minimized manufacturing cost.

Such a procedure of tolerance design is illustrated in Figure 8.1. This development in tolerance design has the following shortcomings:

- As a bridge between design and manufacturing, functional tolerances are absolutely necessary. However, it is impossible to obtain optimal functional tolerances in tolerance synthesis in CAD because we cannot determine the manufacturing cost precisely without the manufacturing information. Quite often, the predicted manufacturing cost (or functional tolerances) in CAD is not the same as that of manufacturing (or machining tolerances) (see Walker, 1991), because the manufacturing cost depends on the process selected to produce the workpiece. That is why the

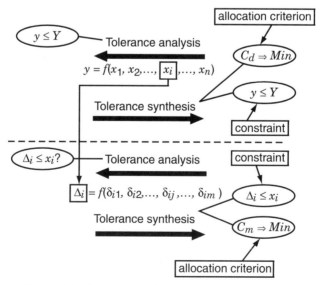

Figure 8.1. Tolerance design in a traditional meaning.

manufacturing engineer must communicate with the designer in order to modify functional tolerances and obtain an appropriate machining process; however, it is a time-consuming procedure.

- Functional tolerances need to be redistributed over machining tolerances in CAPP. Normally, machining tolerances cannot fully occupy the space supplied by functional tolerances. As a result, the manufacturing cost will be higher than the necessary cost.
- There is no direct relation between the deviation of product specification and the machinining tolerances. The manufacturing engineer is limited to the part level. He or she does not have an overview of the entire product, and his or her subjective initiative is constrained.
- No direct relation between the deviation of product specification and the machining tolerances obstructs the application of dynamic tolerance control in manufacturing.
- It requires more memory space, because we need to do the same things twice: to form a cost model and to solve a constrained optimization problem.
- It needs more lead time.

To overcome those shortcomings, we need to abandon functional tolerances, establish a direct relation between the deviation of product specification and the machining tolerances, and directly determine the optimal machining tolerances (also machining process) in product design. The introduction of concurrent engineering into product design and manufacturing offers us such an opportunity. That is, we need to develop a simultaneous tolerancing theory in a concurrent engineering context. In a report on an international conference on concurrent engineering (Weill, 1992), Weill wrote: Concurrent engineering should focus on the connection between CAD and CAM. For example, in dimensioning and tolerancing, the connection is very important because design dimensions and production dimensions are not identical. This subject has been addressed by some published works (Zhang et al., 1992; Dong, 1992; Willhelm and Lu, 1992; Bourne et al., 1989; Cagan and Kurfess, 1992; Srinivasan and Wood, 1992), but none have given an overview of the problem. Furthermore, a general mathematical model of simultaneous tolerancing has not been developed. In the following sections we describe the elements of simultaneous tolerancing, develop two commonly used mathematical models, and propose a method to determine the machining process without using functional tolerances.

8.2 ELEMENTS OF SIMULTANEOUS TOLERANCING

First, a definition: *Simultaneous tolerancing* directly determines machining tolerances, which include simultaneous tolerance analysis and simultaneous tolerance synthesis, in product design without the bridge of functional tolerances. With simultaneous tolerancing theory, the relationship between the deviation of product specification y and the machining tolerance δ_{ij} is direct (see Figure 8.2). In simultaneous tolerance analysis, the deviation of product specification y can be calculated directly from machining tolerance δ_{ij}. In simultaneous tolerance synthesis, the tolerance of product specification Y can be distributed directly over the machining tolerance δ_{ij} with minimal manufacturing cost. The elements of simultaneous tolerancing are shown in Figure 8.3.

Although it is a new concept, simultaneous tolerancing has elements similar to those of conventional tolerancing techniques. The function of each element is described below.

E1—Input Data. As input data, we need Y, the tolerance of product specification, assembly drawings, and a part description. The tolerance of product specification can be determined by the Taguchi method (Taguchi et al., 1989). Assembly drawings and description of parts can be retrieved from a CAD database.

E2—Identification of Functional Components and Tolerance Types. To form the functional equation, we need to know which component is functional and which tolerance should be included in the functional equation. Solid modeling or variational geometry techniques (Turner, 1990; SDRC, 1991), dimensional chains (Wang and Ozasoy, 1990; Dong and Soom, 1990; Bjorke, 1989; He and Lin, 1992; Soderberg, 1992), and functional relationship graphs (Eyada and Ong, 1991; Juster et al., 1992; Shah and Zhang, 1992; Zhang and Porchet, 1993a) are techniques suitable for this task. In our experience, an oriented functional relationship graph (OFRG) is especially useful for this task.

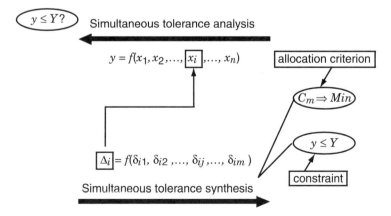

Figure 8.2. Tolerance design in concurrent engineering.

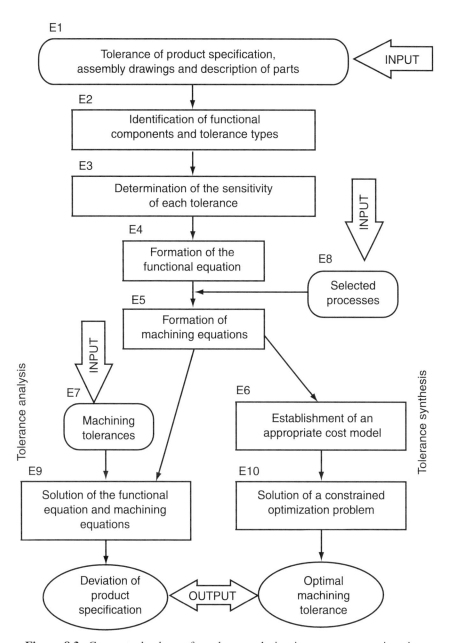

Figure 8.3. Conceptual schema for tolerance design in concurrent engineering.

E3—Determination of the Sensitivity of Each Functional Tolerance.
Variational geometry (SDRC, 1991), dimensional chains (Bjork, 1989), simulation techniques (Zhang and Porchet, 1993b), and algebraic reasoning methods (Inui and Kimura, 1991) can be used to determine sensitivities.

E4—Formulation of the Functional Equation. Two models, worst-case (Greenwood and Chase, 1988) and statistical (Lehtihet and Dindelli, 1989), are commonly used. Worst-case model:

$$y = \sum_{i=1}^{n} |\xi 1_i| x_i \tag{8.3}$$

Statistical model:

$$y = \sqrt{\sum_{i=1}^{n} \xi 1_i^2 \sigma_{xi}^2} \tag{8.4}$$

E5—Formation of Machining Equations. After a process is selected, we can form machining equations. Several techniques are suitable for this task, such as tolerance charts (Irani et al., 1989), dimensional chains (Zhang et al., 1991), Bourdet's strategy (Weill, 1988), and the matrix method (Zhang and Porchet, 1992). Like the functional equation, a machining equation can formed by two models. Worst-case model:

$$\Delta_i = \sum_{j=1}^{m_i} |\xi 2_{ij}| \delta_{ij} \tag{8.5}$$

In general, $\xi 2_{ij} = 1$ or -1. Statistical model:

$$\Delta_i = \sqrt{\sum_{j=1}^{m_i} \xi 2_{ij}^2 \sigma_{\delta ij}^2} \tag{8.6}$$

E6—Establishment of an Appropriate Cost Model. It is well known that the manufacturing cost is the best criterion for tolerance synthesis; however, to develop a practical cost model is not an easy task. The existing cost models are based on two principles: empirical data (Wu et al., 1988; Cheikh and McGoldrick, 1988) and process data (Zhang, 1992). The former is based on the fact that the smaller the tolerance, the higher the manufacturing cost. This observation results in simple cost models such as $c = k/x^a$ and $c = ke^{-bx}$. The coefficients in these cost models can be determined from the empirical tolerance–cost data by means of a curve-fitting technique, while process-data-based cost models can be determined from process data, which depend strictly on the machining process selected. For simultaneous tolerancing, process-data-based cost models are more useful, because the manufacturing cost can be calculated precisely from the machining process selected.

E7—Machining (Operational) Tolerances. These are the input data in tolerance analysis. Machining tolerances can be determined from past experience.

E8—Selection of Appropriate Process. This element is dealt with in Section 8.4.

E9—Calculation of the Deviation of Product Specification. This element gives the deviation of product specification y, to compare with the tolerance of product specification Y. If $y > Y$, we need to reselect machining tolerances or/and processes.

E10—Formation and Solution of a Constrained Optimization Problem. Mathematical models are discussed in Section 8.3. After a mathematical model is set up, we can choose an appropriate non linear programming technique to solve the model. The result will be optimal machining tolerances. If the mathematical model can be linearized, linear programming will be suitable.

8.3 MATHEMATICAL MODELS FOR OPTIMAL TOLERANCING

In this section we give mathematical models for optimal tolerancing (tolerance synthesis).

Objective Function. We take the manufacturing cost as the objective function. Assume that the manufacturing cost of ith machining tolerance is c_{ij}:

$$c_{ij} = f_c(\delta_{ij}) \qquad i = 1,\ldots,n, \quad j = 1,\ldots,m_i \tag{8.7}$$

The total manufacturing cost of a product will be C, where

$$C = \sum_{i=1}^{n} \sum_{j=1}^{m_i} c_{ij} \tag{8.8}$$

Constraint of the Tolerance of Product Specification. The deviation of product specification y must be smaller than or equal to the tolerance of product specification Y.

$$y = f_y(x_1,...,x_n) \leq Y \qquad\qquad (8.9)$$

Intermediate Constraints of Machining Equations. Machining equations need to be respected.

$$x_i = f_x(\delta_{i1},...,\delta_{im_i}) \qquad i = 1,...,n \qquad\qquad (8.10)$$

Allowance Constraints. The resultant value of an allowance chain must be smaller than its maximum and minimum limits.

$$a_{k,min} \leq a_k = f_a(\delta_{k1},...,\delta_{kq_k}) \leq a_{k,max} \qquad k = 1,...,p \qquad (8.11)$$

In general, the relationship between a_k and δ_{ki} is linear. Hence a_k is the sum of δ_{ki}:

$$a_k = \sum_{i=1}^{q_k} \xi 2_{ki}\delta_{ki} \qquad\qquad (8.12)$$

Process Bounds

$$\delta_{ij,min} \leq \delta_{ij} \leq \delta_{ij,max} \qquad\qquad (8.13)$$

The mathematical model for tolerance synthesis is as follows:

$$\text{minimize } C = \sum_{i=1}^{n} \sum_{j=1}^{m_i} c_{ij}$$

$$c_{ij} = f_c(\delta_{ij}) \qquad i = 1,...,n, \quad j = 1,...,m_i$$

$$\text{subject to } y = f_y(x_1,...,x_n) \leq Y$$

$$x_i = f_x(\delta_{i1},...,\delta_{im_i})$$

$$a_{k,min} \leq a_k = f_a(\delta_{k1},...,\delta_{kq_k}) \leq a_{k,max} \qquad k = 1,...,p$$

$$\delta_{ij,min} \leq \delta_{ij} \leq \delta_{ij,max}$$

Functional tolerances, used only for inspection, can be calculated using formula (8.10). Taking notice of the difference between machined deviation and machining tolerances, the proposed mathematical model opens up possibilities for dynamic tolerance control to achieve more economical machining; that is, we measure and register the deviations achieved and replace the tolerances achieved by these deviations. To redistribute the nonachieved machining tolerances produces larger tolerances. For the two commonly used tolerancing models, the mathematical model proposed can be further simplified.

Worst-Case Model. Let $\Delta_i = x_i$; then formulas (8.3) and (8.5) give the direct relation between δ_{ij} and y.

$$y = \sum_{i=1}^{n} \sum_{j=1}^{m_i} |\xi 1_i||\xi 2_{ij}|\delta_{ij} \tag{8.14}$$

With formulas (8.12) and (8.13), the mathematical model becomes

$$\text{minimize } C = \sum_{i=1}^{n} \sum_{j=1}^{m_i} c_{ij}$$

$$c_{ij} = f_c(\delta_{ij}) \qquad i = 1,...,n, \quad j = 1,...,m_i$$

$$\text{subject to } y = \sum_{i=1}^{n} \sum_{j=i}^{m_i} |\xi 1_i||\xi 2_{ij}|\delta_{ij} \leq Y$$

$$a_{k,\min} \leq a_k = \sum_{i=1}^{q_k} \xi 2_{ki}\delta_{ki}^2 \leq a_{k,\max} \qquad k = 1,...,p$$

$$\delta_{ij,\min} \leq \delta_{ij} \leq \delta_{ij,\max}$$

Statistical Model: Let $\Delta_i = \sigma_{xi}$; then formulas (8.4) and (8.6) give the direct relation between $\sigma_{\delta ij}$ and y:

$$y = \sqrt{\sum_{i=1}^{n} \sum_{j=1}^{m_i} \xi 1_i^2 \, \xi 2_{ij}^2 \sigma_{\delta ij}^2} \tag{8.15}$$

With the statistical model, formula (8.12) becomes

$$a_k = \sqrt{\sum_{i=1}^{q_k} \xi 2_{ki}^2 \sigma_{\delta ki}^2} \tag{8.16}$$

With formulas (8.15) and (8.16), the mathematical model becomes

$$\text{minimize } C = \sum_{i=1}^{n} \sum_{j=1}^{m_i} c_{ij}$$

$$c_{ij} = f_c(\delta_{ij}) \qquad i = 1,...,n \quad j = 1,...,m_i$$

$$\text{subject to } y = \sqrt{\sum_{i=1}^{n} \sum_{j=1}^{m_i} \xi 1_i^2 \xi 2_{ij}^2 \sigma_{\delta ij}^2}$$

$$a_{k,\min} \le a_k = \sqrt{\sum_{i=1}^{q_k} \xi 2_{ki}^2 \sigma_{\delta ki}^2} \le a_{k,\max} \qquad k = 1,...,p$$

$$\delta_{ij,\min} \le \delta_{ij} \le \delta_{ij,\max}$$

8.4 PROCESS SELECTION

The selection of a machining process, including equipment accuracy, setup mode, machining sequence, and cutting parameters, is strongly affected by the tolerance (tolerance magnitude and tolerance types) of the part to be machined. In a conventional tolerancing system, functional tolerance magnitudes determined, often badly, in CAD are used as a criterion for selecting the machining process. In turn, the process selected needs to ensure the functional tolerances. In a simultaneous tolerancing system, the relation between the deviation of product specification and machining tolerances is direct, without the bridge of functional tolerances. However, a criterion is necessary in selecting an appropriate machining process. Such a criterion cannot be created without foundation. We propose to use the information on the assembly construction and the features of the parts to be machined. The information on assembly construction is represented by sensitivities in a functional equation, while the information on the features of the parts reflects the degree of machining difficulty. Using this information, we can assign an *interim tolerance* to each tolerance item in the functional equation. The interim tolerances can be used as the criterion for selecting an appropriate machining process. It should be noted that once a machining process is selected, interim tolerances lose their significance. It is not necessary to guarantee the interim tolerances by the process selected, while in a conventional tolerancing system, functional tolerances are served not only as the criterion for selecting a process, but also as the objective that the process selected must guarantee.

Interim tolerances can be determined by the following procedure.

1. Use formula (8.17), to determine preliminarily a set of tolerances:

$$X_i^{(a)} = \frac{Y}{n\,\xi 1_i} \tag{8.17}$$

$X_i^{(a)}$ reflects the impact of the sensitivity of tolerances X_i.

2. According to the geometric feature of tolerance X_i and the tolerance determined in step 1, from Table 8.1 determine the machining process and ISO tolerance grade or IT number (Farmer and Harris, 1984).

TABLE 8.1. Machining Process and Its Tolerance Grade Under Average Conditions

Machining Process	Tolerance Grade
Drilling	12
Coarse turning and boring	11
Milling, slotting, and planing	10
Horizontal of vertical boring	9
Reaming	8
Turning and boring	8
Coarse grinding	7
Fine turning and boring	7
Broaching	7
Honing	7
Grinding	6
Fine honing	6
Fine boring (jig borer)	5
Fine grinding	5
Machining lapping	5

3. According to the IT number and feature size (distance between two faces), using formula (8.18), determine the tolerance reflecting the feature size.

$$X_i^{(f)} = (0.45\sqrt[3]{L} + 0.001L)10^{(\text{IT}-16)/5} \qquad (8.18)$$

4. Using formula (8.19), $X_i^{(f)}$ is made to conform to the deviation of product specification.

$$X_i^{(p)} = \frac{YX_i^{(f)}}{\xi1_i \sum_{j=1}^{n} X_j^{(f)}} \qquad (8.19)$$

5. $X_I^{(a)}$ and $X_i^{(p)}$ are combined as the formula

$$X_i^{(c)} = \sqrt{[X_i^{(a)}]^2 + [X_i^{(p)}]^2} \qquad (8.20)$$

6. Finally, use formula (8.21) to calculate interim tolerances.

$$X_i = \frac{YX_i^{(c)}}{\xi1_i \sum_{k=1}^{n} X_k^{(c)}} \qquad (8.21)$$

In fact, such a procedure can be used to calculate functional tolerances in a traditional tolerance distribution process in CAD.

8.5 EXAMPLE

To show the feasibility of simultaneous tolerancing theory, we take the spindle shown in Figure 8.4 as an example. For reasons of simplicity, only a part of the spindle is taken into account (shown in Figure 8.5) and the components are renumbered.

8.5.1 Functional Analysis

To guarantee good functioning (sealing, reliability, etc.) of the spindle, the gaps δ_1, δ_2, δ_3, and δ_4 should be controlled. Suppose that the values of four gaps are equal; that is,

$$\delta_1 = \delta_2 = \delta_3 = \delta_4 = 0.5 \text{ mm}$$

and the tolerances are

$$T_{\delta 1} = T_{\delta 2} = T_{\delta 3} = 0.2 \text{ mm} \quad \text{and} \quad T_{\delta 4} = 0.3 \text{ mm}$$

Component 6 is a standardized part, its tolerance given as

$$t_{61-62} = 0.03 \text{ mm}$$

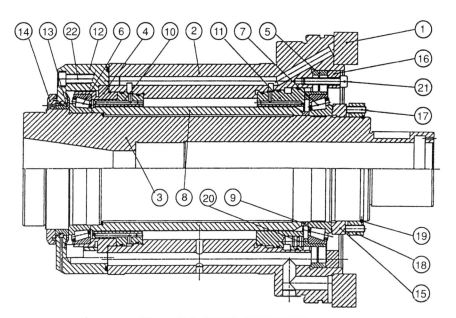

Figure 8.4. Spindle EPFL/LMO.

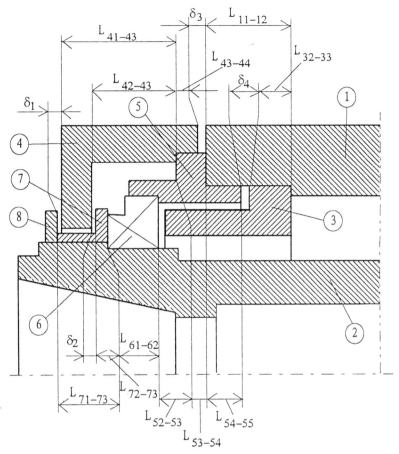

Figure 8.5. Part of the spindle.

8.5.2 Dimensional Chains (Functional Equations)

Using any of the methods proposed (Bjorke, 1989; Wang and Ozasoy, 1990; Zhang and Porchet, 1993a), we can find four dimensional chains. In the equations the subscript 41–43 means the dimension between two surfaces, 41 and 43.

- Dimensional chains δ_1 (see Figure 8.6)
- Dimensional chains δ_2 (see Figure 8.7)
- Dimensional chains δ_3 (see Figure 8.8)
- Dimensional chains δ_4 (see Figure 8.9)

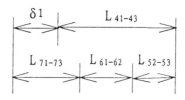

$$\delta_1 = L_{71-73} + L_{61-62} + L_{52-53} - L_{41-43}$$

Figure 8.6. Dimensional chains δ_1.

8.5.3   Dimensions of Components

From the four dimensional chains we can note that only six components are concerned in dimensional chains. The dimensions of these components and the numbers of surfaces are illustrated in Figure 8.10, the functional dimensions are framed.

8.5.4   Calculation of Interim Tolerances

The tolerance of component 6 is known:

$$t_{61-62} = 0.03 \text{ mm}$$

The following equations are tolerance chains corresponding to δ_1, δ_2, δ_3, and δ_4:

$$T_{\delta 1} = t_{71-73} + t_{61-62} + t_{52-53} + t_{41-43} \qquad T_{\delta 3} = t_{53-54} + t_{43-44}$$
$$T_{\delta 2} = t_{42-43} + t_{72-73} + t_{61-62} + t_{52-53} \qquad T_{\delta 4} = t_{11-12} + t_{54-55} + t_{32-33}$$

Substituting the known values into the equations above, the tolerance chains become

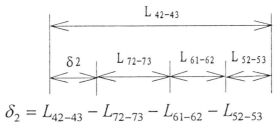

$$\delta_2 = L_{42-43} - L_{72-73} - L_{61-62} - L_{52-53}$$

Figure 8.7. Dimensional chains δ_2.

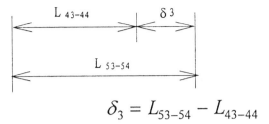

$$\delta_3 = L_{53-54} - L_{43-44}$$

Figure 8.8. Dimensional chains δ_3.

$$0.17 = t_{71-73} + t_{52-53} + t_{41-43} \quad 0.2 = t_{53-54} + t_{43-44}$$
$$0.17 = t_{42-43} + t_{72-73} + t_{52-53} \quad 0.3 = t_{11-12} + t_{54-55} + t_{32-33}$$

The calculation of interim tolerances is shown in Table 8.2. According to the interim tolerance values, all the interim tolerances can be ensured by the finishing cut.

8.5.5 Machining Process

The tolerances t_{32-33}, $t_{/1-73}$, and t_{72-72} can be ensured directly by the machining processes; the tolerance t_{11-12} is the sum of the errors produced by the process machining of two surfaces, 11 and 12. The machining processes of components 4 and 5 are quite complicated; we can use a tolerance chart to establish the relationship between the functional tolerances and the machining tolerances. The tolerance charts for components 4 and 5 are shown in Figure 8.11 and Figure 8.12. It should be indicated that the tolerance chart is used only to set up the machining equations, so columns that are not needed are not included in the tolerance charts.

8.5.6 Machining Equations

From the tolerance chart shown in Figure 8.11, to ensure three functional tolerance, three machining equations can be written:

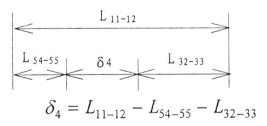

$$\delta_4 = L_{11-12} - L_{54-55} - L_{32-33}$$

Figure 8.9. Dimensional chains δ_4.

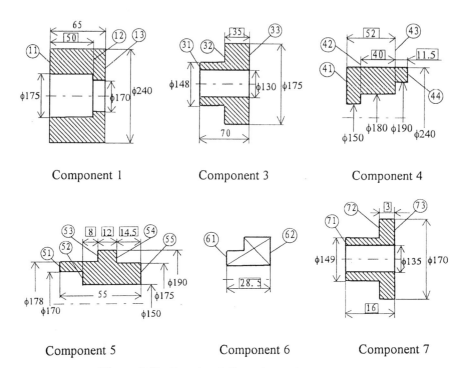

Figure 8.10. Functional dimensions of components.

$$t_{43-44} = t_{45} + t_{46} \qquad t_{42-43} = t_{46} + t_{47} \qquad t_{41-43} = t_{45} + t_{46} + t_{48}$$

From the tolerance chart shown in Figure 8.12, three machining equations can be written:

TABLE 8.2. Calculation of Interim Tolerances

		Formula (8.17)	Formula (8.18)	Formula (8.19)	Formula (8.20)	Formula (8.21)
$T_{\delta 1}$	$t_{41-43} = 0.057$	0.027	0.078	0.097	0.068
	$t_{71-73} = 0.057$	0.018	0.052	0.077	0.054
	$t_{52-53} = 0.057$	0.014	0.04	0.07	0.049
	$t_{61-62} = 0.03$	0.03	0.03	0.03	0.03
$T_{\delta 2}$	$t_{42-43} = 0.057$	0.025	0.093	0.109	0.075
	$t_{72-73} = 0.057$	0.01	0.037	0.068	0.046
	$t_{52-53} = 0.057$	0.014	0.04	0.07	0.049
	$t_{61-62} = 0.03$	0.03	0.03	0.03	0.03
$T_{\delta 3}$	$t_{43-44} = 0.01$	0.017	0.103	0.144	0.102
	$t_{53-54} = 0.01$	0.016	0.097	0.139	0.098
$T_{\delta 4}$	$t_{11-12} = 0.01$	0.027	0.117	0.154	0.109
	$t_{32-33} = 0.01$	0.024	0.104	0.144	0.102
	$t_{54-55} = 0.01$	0.018	0.078	0.127	0.09

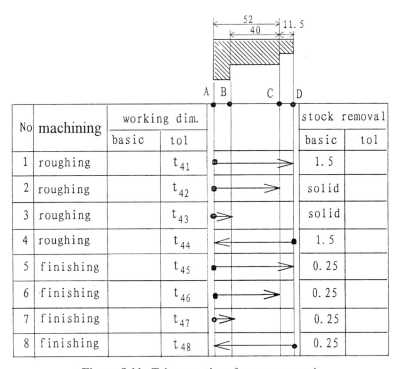

Figure 8.11. Tolerance chart for component 4.

No	machining	working dim.			stock removal	
		basic	tol		basic	tol
1	roughing		t_{41}		1.5	
2	roughing		t_{42}		solid	
3	roughing		t_{43}		solid	
4	roughing		t_{44}		1.5	
5	finishing		t_{45}		0.25	
6	finishing		t_{46}		0.25	
7	finishing		t_{47}		0.25	
8	finishing		t_{48}		0.25	

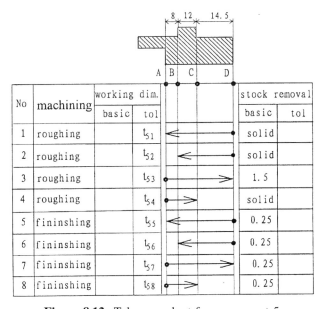

Figure 8.12. Tolerance chart for component 5.

No	machining	working dim.			stock removal	
		basic	tol		basic	tol
1	roughing		t_{51}		solid	
2	roughing		t_{52}		solid	
3	roughing		t_{53}		1.5	
4	roughing		t_{54}		solid	
5	fininshing		t_{55}		0.25	
6	fininshing		t_{56}		0.25	
7	fininshing		t_{57}		0.25	
8	fininshing		t_{58}		0.25	

$$t_{54-55} = t_{57} + t_{58} \qquad t_{52-53} = t_{55} + t_{56} \qquad t_{53-54} = t_{55} + t_{56} + t_{58}$$

We can also write the other machining equations:

$$t_{11-12} = t_{11} + t_{12} \qquad t_{32-33} = t_{31} \qquad t_{71-73} = t_{71} \ t_{72-73} = t_{72}$$

In total, there are 10 machining equations.

8.5.7   Allowance Equations

From the tolerance chart shown in Figure 8.11, four allowance equations can be written:

$$A_{A1A2} = 0.25 + t_{48} + t_{45} \qquad\qquad A_{C1C2} = 0.25 + t_{41} + t_{42} + t_{44} + t_{46}$$

$$A_{B1B2} = 0.25 + t_{41} + t_{43} + t_{44} + t_{47} \qquad A_{D1D2} = 0.25 + t_{41} + t_{44} + t_{45}$$

Since t_{41}, t_{42}, t_{43}, and t_{44} do not affect the function tolerances (they are not included in the machining equations), they can be replaced by the constant; let

$$t_{41} = t_{42} = t_{43} = t_{44} = 0.03 \text{ mm}$$

The allowance equations become

$$A_{A1A2} = 0.25 + t_{48} + t_{45} \qquad A_{C1C2} = 0.34 + t_{46}$$

$$A_{B1B2} = 0.34 + t_{47} \qquad A_{D1D2} = 0.31 + t_{45}$$

The allowance equations for component 5 (see Figure 8.12) can be written as follows:

$$B_{A1A2} = 0.25 + t_{55} + t_{53} + t_{51} \qquad B_{C1C2} = 0.25 + t_{58} + t_{55}$$

$$B_{B1B2} = 0.25 + t_{56} + t_{53} + t_{52} + t_{51} \qquad B_{D1D2} = 0.25 + t_{57} + t_{55}$$

With $t_{51} = t_{52} = t_{53} = t_{54} = 0.03 \text{ mm}$, the allowance equations can be rewritten as

$$B_{A1A2} = 0.31 + t_{55} \qquad B_{C1C2} = 0.25 + t_{58} + t_{55}$$

$$B_{B1B2} = 0.34 + t_{56} \qquad B_{D1D2} = 0.25 + t_{57} + t_{55}$$

8.5.8   Cost Model

The relationship between tolerance and machining cost can be expressed as follows:

$$C = A_0 + \frac{A_1}{t^{A_2}} \tag{8.22}$$

The coefficients A_0, A_1, and A_2 can be determined by using a regression technique. Normally, there are two kinds of surfaces in face turning operation. The internal face is more difficult to be machined than the external face; therefore, the machining cost of internal face is higher. Suppose that the relationship between tolerance and machining cost could be expressed by the curves in Figure 8.13. Given $t_1 = t_2 = 0.066$ mm, from formula (8.22) we have

$$8.8 = 7.2 + \frac{A_{11}}{0.066^{A_{21}}} \qquad 4.4 = 3.5 + \frac{A_{12}}{0.066^{A_{22}}} \tag{8.23}$$

Given $t_1 = t_2 = 0.04$, we have

$$11 = 7.2 + \frac{A_{11}}{0.04^{A_{21}}} \qquad 5.8 = 3.5 + \frac{A_{12}}{0.04^{A_{22}}} \tag{8.24}$$

Formulas (8.23) and (8.24) give four coefficients:

$$A_{11} = 0.015 \qquad A_{21} = 1.727 \qquad A_{12} = 0.006 \qquad A_{22} = 1.87$$

Finally, we get two cost models:

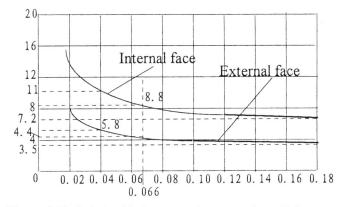

Figure 8.13. Relationship between tolerance and machining cost.

$$C_1 = 7.2 + \frac{0.015}{t^{1.727}} \qquad C_2 = 3.5 + \frac{0.006}{t^{1.87}}$$

8.5.9 Optimization Model

From the analysis above, we can set up the optimization model. Objective function:

$$C = C_{71} + C_{55} + C_{56} + C_{45} + C_{46} + C_{48} + C_{46} + C_{47} + C_{72} + C_{55}$$

$$+ C_{56} + C_{45} + C_{46} + C_{55} + C_{56} + C_{58} + C_{11} + C_{12} + C_{31} + C_{57}$$

$$+ C_{58}$$

$$= 13 \times 3.5 + 8 \times 7.2 + \frac{0.012}{t_{71}^{1.87}} + \frac{0.018}{t_{56}^{1.87}} + \frac{0.012}{t_{45}^{1.87}} + \frac{0.006}{t_{48}^{1.87}} + \frac{0.012}{t_{72}^{1.87}}$$

$$+ \frac{0.006}{t_{11}^{1.87}} + \frac{0.012}{t_{31}^{1.87}} + \frac{0.006}{t_{57}^{1.87}} + \frac{0.012}{t_{58}^{1.87}} + \frac{0.045}{t_{55}^{1.727}} + \frac{0.045}{t_{46}^{1.727}} + \frac{0.015}{t_{47}^{1.727}} + \frac{0.015}{t_{12}^{1.727}}$$

Functional equation constraints:

$$t_{71-73} + t_{52-53} + t_{41-43} \le 0.17 \qquad t_{53-54} + t_{43-44} \le \qquad 0.2$$

$$t_{42-43} + t_{72-73} + t_{52-53} \le 0.17 \qquad t_{11-12} + t_{54-55} + t_{32-33} \le 0.3$$

Machining equation constraints:

$$t_{71-73} = t_{71} \qquad\qquad t_{53-54} = t_{55} + t_{56} + t_{58}$$
$$t_{52-53} = t_{55} + t_{56} \qquad\qquad t_{43-44} = t_{45} + t_{46}$$
$$t_{41-43} = t_{45} + t_{46} + t_{48} \qquad t_{11-12} = t_{11} + t_{12}$$
$$t_{42-43} = t_{46} + t_{47} \qquad\qquad t_{54-55} = t_{57} + t_{58}$$
$$t_{72-73} = t_{72} \qquad\qquad t_{32-33} = t_{31}$$

Allowance constraints:

$$0.24 + t_{45} + t_{48} \le 0.6 \qquad 0.25 - t_{45} - t_{48} \ge 0.1$$
$$0.34 + t_{47} \le 0.6 \qquad\qquad 0.16 - t_{47} \ge 0.1$$
$$0.34 + t_{46} \le 0.6 \qquad\qquad 0.16 - t_{46} \ge 0.1$$
$$0.31 + t_{45} \le 0.6 \qquad\qquad 0.19 - t_{45} \ge 0.1$$
$$0.25 + t_{58} + t_{55} \le 0.6 \qquad 0.25 - t_{58} - t_{55} \ge 0.1$$
$$0.25 + t_{57} + t_{55} \le 0.6 \qquad 0.25 - t_{57} - t_{55} \ge 0.1$$
$$0.34 + t_{56} \le 0.6 \qquad\qquad 0.16 - t_{56} \ge 0.1$$
$$0.31 + t_{55} \le 0.6 \qquad\qquad 0.19 - t_{55} \ge 0.1$$

Process bounds:

$$0 \le t_{71} \le 0.06$$
$$0 \le t_{56} \le 0.06$$
$$0 \le t_{45} \le 0.06$$
$$0 \le t_{48} \le 0.06$$
$$0 \le t_{72} \le 0.06$$
$$0 \le t_{11} \le 0.06$$
$$0 \le t_{31} \le 0.06$$

$$0 \le t_{57} \le 0.06$$
$$0 \le t_{58} \le 0.06$$
$$0 \le t_{55} \le 0.06$$
$$0 \le t_{46} \le 0.06$$
$$0 \le t_{47} \le 0.06$$
$$0 \le t_{12} \le 0.06$$

8.5.10   Final Result

The optimization model is solved using a nonlinear programming technique (mixed penalty function). The optimal machining cost is

$$C = \$219.17$$

The optimal operational tolerances are

$$t_{71} = 0.026 \text{ mm}$$
$$t_{56} = 0.026 \text{ mm}$$
$$t_{45} = 0.026 \text{ mm}$$
$$t_{48} = 0.02 \text{ mm}$$
$$t_{72} = 0.034 \text{ mm}$$
$$t_{11} = 0.06 \text{ mm}$$
$$t_{31} = 0.06 \text{ mm}$$

$$t_{57} = 0.06 \text{ mm}$$
$$t_{58} = 0.06 \text{ mm}$$
$$t_{55} = 0.036 \text{ mm}$$
$$t_{46} = 0.036 \text{ mm}$$
$$t_{47} = 0.037 \text{ mm}$$
$$t_{12} = 0.06 \text{ mm}$$

The optimal functional tolerances are (if necessary)

$$T_{71-73} = 0.026 \text{ mm}$$
$$T_{41-43} = 0.082 \text{ mm}$$
$$t_{42-43} = 0.073 \text{ mm}$$
$$t_{53-54} = 0.122 \text{ mm}$$
$$t_{72-73} = 0.034 \text{ mm}$$

$$t_{43-44} = 0.062 \text{ mm}$$
$$t_{52-53} = 0.062 \text{ mm}$$
$$t_{11-12} = 0.12 \text{ mm}$$
$$t_{54-55} = 0.12 \text{ mm}$$
$$t_{32-33} = 0.06 \text{ mm}$$

Now let's verify the tolerances of four gaps:

$$t_{d1} = t_{71-73} + t_{61-62} + t_{52-53} + t_{41-43}$$

$$= t_{71} + 0.03 + t_{55} + t_{56} + t_{45} + t_{46} + t_{48}$$

$$= 0.026 + 0.03 + 0.036 + 0.026 + 0.026 + 0.036 + 0.02$$

$$= 0.2 \text{ mm}$$

$$t_{d2} = t_{42-43} + t_{61-62} + t_{72-73} + t_{52-53}$$

$$= t_{46} + 0.03 + t_{47} + t_{72} + t_{55} + t_{56}$$

$$= 0.036 + 0.03 + 0.037 + 0.034 + 0.036 + 0.026$$

$$= 0.199 \text{ mm}$$

$$t_{d3} = t_{53-54} + t_{43-44}$$

$$= t_{55} + t_{56} + t_{58} + t_{45} + t_{46}$$

$$= 0.036 + 0.026 + 0.06 + 0.026 + 0.036$$

$$= 0.184 \text{ mm}$$

$$t_{d4} = t_{11-12} + t_{54-55} + t_{32-33}$$

$$= t_{11} + t_{12} + t_{57} + t_{58} + t_{31}$$

$$= 0.06 + 0.06 + 0.06 + 0.06 + 0.06$$

$$= 0.3 \text{ mm}$$

8.6   CONCLUSIONS

Tolerance design in a concurrent engineering context is addressed in this chapter. The following result is obtained.

- The shortcomings of conventional tolerancing systems are described. It is noticed that conventional tolerancing systems are not suitable in a concurrent engineering context; therefore, a simultaneous tolerancing theory is necessary.
- A complete simultaneous tolerancing theory is developed and its elements discussed.
- A general mathematical model for simultaneous tolerancing has been established. Two commonly used models, worst-case and statistical, are developed in detail.
- A method for setting up a set of interim tolerances is developed. The interim tolerances can be used to select appropriate machining processes. Furthermore, the interim tolerances are an excellent replacement for functional tolerances in a conventional tolerancing system because they are more reasonable and simpler.
- With a simultaneous tolerancing system, dynamic tolerance control is possible.

· The example shows that the simultaneous tolerancing theory is feasible and can be used in practice.

ACKNOWLEDGMENTS

This research has been supported by the Doctoral Research Fund of the State Education Commission of China and the National Key Laboratory for Mechanical Manufacturing System Engineering.

NOMENCLATURE

a_k calculated allowance
C_d predetermined manufacturing cost in design
C_{ij} manufacturing cost of ith machining tolerance
C_m calculated manufacturing cost in process planning
L distance between two faces
m_i number of machining tolerances in a machining equation
n number of functional tolerances in the functional equation
p number of allowance chains
q_k number of machining tolerances in an allowance chain
X_i part functional tolerance
Y tolerance of product specification
y deviation of product specification
Δ_i machined deviation corresponding to X_i
δ_{ij} machining tolerance
$\xi1_i$ sensitivity of ith functional tolerance
$\xi2_{ij}$ sensitivity of ith machining tolerance
$\sigma_{\delta ij}$ standard deviation of ith machining tolerance
σ_{xi} standard deviation of ith functional tolerance

REFERENCES

Bjorke, O., 1989, *Computer-Aided Tolerancing,* 2nd ed., ASME Press, New York.

Bourne, D., Navinchandra, D., and Ramaswamy, R., 1989, Tolerance-Free Designs for Manufacturing, *Proceedings of the Concurrent Product and Process Design, The Winter Annual Meeting of ASME,* San Francisco. December 10–15, 1989.

Cagan, J., and Kurfess, T. R., 1992, Optimal Tolerance Allocation over Multiple Manufacturing Alternatives, *Proceedings of the ASME 18th Design Automation Conference,* Scottsdale, AZ.

Cheikh, A., and McGoldrick, P. F., 1988, The Influence of Cost Function and Process Capability on Tolerance, *Int. J. Qual. Reliabil. Manage.*, Vol. 5, No. 3, pp. 15–28.

Dong, Z., 1992, Design for Automated Manufacturing, in *Concurrent Engineering: Automation, Tools and Techniques,* A. Kusiak (ed.), Wiley, New York, pp. 207–234, 1992.

Dong, Z., and Soom, A., 1990, Automatic Optimal Tolerance Design for Related Dimension Chains, *Manuf. Rev.,* Vol. 3, No. 4, pp. 262–271.

Eyada, O. K., and Ong, J. B., 1991, An Assembly Recognition Algorithm for Automatic Tolerancing, *Transactions of NAMRI/SME, NAMRC XIX,* May 22–24, University of Missouri–Rolla, Rolla, MO.

Farmer, L. E., and Harris, A. G., 1984, Change of Datum of the Dimensions on Engineering Design Drawings, *Int. J. Mach. Tool Des. Res.,* Vol. 24, pp. 267–275.

Greenwood, W. H., and Chase, K. W., 1988, Worst Case Tolerance Analysis with Nonlinear Problems, *J. Eng. Ind.,* Vol. 110, No. 3, pp. 232–235.

He, J. R., and Lin, G. C. I., 1992, Computeraized Trace Method for Establishing Equations for Dimensions and Tolerances in Design and Manufacturing, *Int. J. Manuf. Technol.,* Vol. 7, No. 4, pp. 210–217.

Inui, M., and Kimura, F., 1991, Algebraic Reasoning of Position Uncertainties of Parts in an Assembly, Proceedings of the Symposium on Solid Modeling Fundations and CAD/CAM Applications. Texas, June 5–7, pp. 419–428.

Irani, S. A., Mittal, R. O., and Lehtihet, E. A., 1989, Tolerance Chart Optimization, *Int. J. Prod. Res.,* Vol. 27, No. 9, pp. 1531–1552.

Juster, N. P., Dew, P. M., and de Pennington, A., 1992, Automating Linear Tolerance Analysis Across Assemblies, *J. Mech. Des.,* Vol. 114, pp. 174–179.

Lehtihet, E. A., and Dindelli, B. A., 1989, Tolcon: Microcomputer-Based Module for Simulation of Tolerances, *Manuf. Rev.,* Vol. 2, No. 3, pp. 179–188.

Shah, J. J., and Bing-Chun Zhang, B. C., 1992, Attributed Graph Model for Geometric Tolerancing, *Proceedings of the ASME 18th Design Automation Conference,* Scottsdale, AZ. Sept. 13–16, 1992.

Soderberg, R., 1992, CATI: A Computer Aided Tolerancing Interface, *Proceedings of the ASME 18th Design Automation Conference,* Scottsdale, AZ. September 13–16, 1992.

Srinivasan, R. S., and Wood, K. L., 1992, "Fractal-Based Geometric Tolerancing for Mechanical Design," *Proceedings of the ASME 4th International Conference on Design Theory and Methodology,* Scottsdale, AZ. September 13–16, 1992.

SDRC, 1991, *I-DEAS User's Guide: Tolerance Analysis,* Structural Dynamics Research Corporation, Milford, Ohio.

Taguchi, G., et al., 1989, *Quality Engineering in Production Systems,* McGraw-Hill, New York.

Tang, X., and Davies, B. J., 1988, Computer Aided Dimensional Planning, *Int. J. Prod. Res.,* Vol. 26, No 2, pp. 283–297.

Turner, J. U., 1990, Exploiting Solid Models for Tolerance Computation, in *Geometric Modeling for Product Engineering,* M. J. Wozny, J. U. Turncr, and K. Preiss (eds.), Elsevier, Amsterdam.

Walker, R., 1991, The Effects of the Mathematical Definition of Tolerances, *Proceedings of the 1991 ASME International Computers in Engineering Conference and Exposition,* Santa Clara, CA. August 18–22, 1991.

Wang, N., and Ozasoy, T. M., 1990, Automatic Generation of Tolerance Chains from Mating Relations Represented in Assembly Models, *Proceedings of the 1990 ASME 16th Design Automation Conference,* Chicago. Sept. 16–19, 1990.

Weill, R., 1988, Integrating Dimensioning and Tolerancing in Computer-Aided Process Planning, *Robot. Comput. Integrat. Manuf.,* Vol. 4, No. 1–2, pp. 41–48.

Weill, R., 1992, Round Table Discussion on Concurrent Engineering Requirements and Perspectives, IFIP Tc5/WG 5.3/Wg 5.2, Working Conference on Manufacturing in the Era of Concurrent Engineering, Herzlya, Israel.

Willhelm, R. G., and Lu, S. C.-Y., 1992, Tolerance Synthesis to Support Concurrent Engineering, *Ann. CIRP,* Vol. 41, No. 1, pp. 197–200.

Wu, Z., Elmaraghy, W. H., and Elmaraghy, H. A., 1988, Evaluation of Cost-Tolerance Algorithms for Design Tolerance Analysis and Synthesis, *Manuf. Rev.,* Vol. 1, No. 3, pp. 168–179.

Zhang, G., 1992, *An Analytical Cost Model for Tolerancing in a CAD/CAM Environments,* Internal Report of CAD Laboratory of EPFL—Swiss Federal Institute of Technology in Lausanne, Switzerland.

Zhang, G., and Porchet, M., 1992, Automatic Calculation of Dimensions and Tolerances in Manufacturing, *Proceedings of the IMACS/SICE International Symposium on Robotics, Mechatronics and Manufacturing Systems,* Kobe, Japan. Sept. 16–20, 1992.

Zhang, G., and Porchet, M., 1993a, Assembly Modeling for Tolerancing in CAD: An Approach of Oriented Functional Relationship Graph, *Proceedings of the 9th International Conference on Engineering Design ICED'93,* The Netherlands. August 17–19, 1993, Hague, Netherlands.

Zhang, G., and Porchet, M., 1993b, Some New Progresses in Tolerance Design in CAD, *Proceedings of the 19th ASME Annual International Design Automation Conference and Exposition,* Albuquerque, NM. Sept. 19–22, 1993.

Zhang, H. C., Mei, J., and Dudek, R. A., 1991, Operational Dimensioning and Tolerancing in CAPP, *Ann. CIRP,* Vol. 40, No. 1. p. 419–p. 422.

Zhang, C., Wang, H. P., and Li, J. K., 1992, Simultaneous Optimization of Design and Manufacturing Tolerances with Process (Machine) Selection, *Ann. CIRP,* Vol. 41, No. 1., pp. 569–572.

9

TOLERANCE SYNTHESIS BY MANUFACTURING COST MODELING AND DESIGN OPTIMIZATION

ZUOMIN DONG

University of Victoria
Victoria, British Columbia, Canada

9.1 INTRODUCTION

As an integral part of mechanical design, tolerances have a profound influence on the functional performance and manufacturing costs of the designed product. Tolerance specification is a complex and demanding task and is carried out traditionally on a trial-and-error basis. Research on tolerance synthesis presents a systematic approach through quantitative modeling and design optimization but imposes an even greater challenge to tolerance specification in mechanical design.

In this chapter the major steps for carrying out tolerance synthesis are reviewed, and several key issues on the implementation of tolerance synthesis using real manufacturing data are addressed. The chapter begins with a general review of the tolerance synthesis research. Methods of building manufacturing cost models used in tolerance optimization are then explained. Tolerance synthesis methods under two different design criteria are discussed: the conventional minimum manufacturing cost design and the newly introduced balanced performance and cost design. Using a typical design example, a method is presented for jointly evaluating and optimizing the functional

Advanced Tolerancing Techniques, Edited by Hong-Chao Zhang
ISBN 0-471-14594-7   © 1997 John Wiley & Sons, Inc.

performance and manufacturing costs of a mechanical design in terms of tolerance. The result of the tolerance optimization is compared with the manually specified tolerance values according to design codes to illustrate the advantages of quantitative modeling and optimization.

In addition, the author uses tolerance synthesis as an example to illustrate the method for jointly modeling, evaluating, and optimizing the functional performance and manufacturing costs of a mechanical product. The ultimate goal of this research is to develop the methodology for carrying out concurrent engineering design to address far more complex and demanding real-world design problems.

9.1.1   Background

In mechanical design, tolerances are specified to control the undesirable variations of part geometry from the perfect geometric form and dimensions specified in the design. These unavoidable variations or manufacturing inaccuracies are due to the physical limitations of the manufacturing process and material property. The tolerances as specified by a designer, ensure the expected functional performance, and provide guidelines for manufacture of the parts.

The specification of appropriate design tolerance, however, is a nontrivial task. At present, the assignment of design tolerances is performed largely on a trial-and-error basis using the *tolerance analysis* method. Several key tolerances are specified based on the given design requirements. Other tolerances are determined based on one's design experience and manufacturing knowledge, or assigned with default values. The consistency requirements, imposed by the stackup constraints of the related tolerances, are verified using the tolerance analysis method, and uncritical tolerances are modified to satisfy these constraints. Although the approach produces consistent tolerances and assures the fit of an assembly, it will not lead to the optimized tolerances, due to the lack of quantitative cost and performance measures, as well as systematic evaluation and optimization.

As an alternative to this undesirable approach, *tolerance synthesis* (or tolerance optimization, optimal tolerance design) actually takes into account the influence of tolerance values to the functional performance and/or manufacturing costs of the design. The method seeks the best combination of all related tolerance values under the consistency constraint and certain design criteria. The design criterion employed in tolerance synthesis covers one or multiple aspects of product performance. Among the many life-cycle performance aspects of a mechanical product, functional performance and manufacturing cost are considered as the two most influential ones with contradictory design directions. A tighter tolerance leads to better functional performance and ease of assembly but also requires extensive manufacturing efforts, which in turn translate into higher manufacturing costs. To solve this controversy, two different design criteria were developed in tolerance synthesis research: *minimum manufacturing cost design* and *balanced performance and cost design.*

In the minimum manufacturing cost–based tolerance synthesis, given design functional performance and stackup consistency requirements are considered as design constraints, the influence of tolerance values to the manufacturing costs of a design is modeled and used as the objective function of the optimization. The tolerances values that satisfy given functional requirements and require minimum manufacturing costs are identified through a constrained optimization. Some earlier research on tolerance synthesis introduced this concept without considering the design functional performance constraints (Speckhart, 1972; Spotts, 1973; Sutherland and Roth, 1975). Most recent research on tolerance synthesis developed this scheme further over the last decade (Roy et al., 1991; Zhang and Huo, 1992; Kumor and Raman, 1992).

The balanced performance- and cost-based tolerance synthesis method is also based on a formulation of constrained, nonlinear optimization. However, the functional performances of the design are now embedded in the objective function rather than treated as the constraints of the optimization (Xue and Dong, 1994, 1996; Xue et al., 1995, 1996). The tolerance values leading to the best overall performance of the designed product are identified. In other words, the influence of design tolerances to the functional performance and manufacturing costs of the product are jointly modeled, evaluated, and optimized. This effort is aimed at the ultimate goal of concurrent engineering design, to achieve the optimal life-cycle performance of a mechanical product.

In either case, tolerance synthesis is carried out through detailed modeling, analysis, and optimization at different levels of sophistication. The tolerance synthesis methods impose an even greater challenge to the specification of design tolerances. Each of the tasks involved needs to be studied further to make tolerance synthesis a convenient design tool.

9.1.2 Related Work

Tolerance synthesis involves many subtasks to be completed before or during the execution of the tolerance assignment. These include the modeling of cost-tolerance and performance-tolerance relations for various type of tolerances, identification of related tolerances (or dimension chain) from a design, formulation of the tolerance synthesis optimization problem, and solution of the optimization problem. In practice it is unrealistic to ask a designer to carry out these tasks manually in design. Computer automation of all of these tasks becomes critical.

Earlier research on tolerance synthesis focused on the formulations of a tolerance assignment as a nonlinear optimization problem (Speckhart, 1972; Spotts, 1973; Sutherland and Roth, 1975). Based on the general characteristics of a manufacturing cost–tolerance data curve, several general cost–tolerance relation models, including the exponential model (Speckhart, 1972), the reciprocal squared model (Spotts, 1973), and the reciprocal powers model (Sutherland and Roth, 1975), were introduced. The tolerance optimization was formulated using these cost–tolerance models, and closed-form solutions to the optimization were given. These formulations pioneered the area of tolerance synthesis. The closed-form solutions are easy to calculate, and no numerical solution is needed. The approach also suffers from several drawbacks. First, relatively large model-fitting errors were introduced due to the simple forms of these mathematical models (Wu et al., 1988; Dong et al., 1994). Second, it is difficult to consider the valid range of a cost–tolerance curve using the closed-form solution method. Because a production process always has an upper bound and a lower bound for its accuracy-improving capability (Trucks, 1976), each tolerance variable should be bounded in the optimization to avoid solutions that are unfeasible. Third, the optimization problem has to be formulated manually, thus inhibiting the convenient use in day-to-day mechanical design. Following these earlier efforts, the method for tolerance specification has been improved significantly by many researchers in several related area. Most of the progress was made on the modeling of cost–tolerance relations and formulation of the optimization problem. Michael and Siddall solved optimal design problems with both design parameters and their tolerances as design variables and introduced the powers and exponential hybrid model (Michael and Siddall, 1981, 1982).

Parkinson further investigated the optimal design of mechanical tolerances in statistical tolerance assignment (Parkinson, 1985). Chase and Greenwood introduced the reciprocal model with better empirical data-fitting capability (Chase et al., 1990). Lee and Woo presented a discrete cost–tolerance model and associated tolerance optimization method, using reliability index and integer programming to eliminate modeling errors (Lee and Woo, 1989). Zhang and Wang introduced simulated annealing to discrete tolerance optimization as a better solution method (Zhang and Wang, 1993). Cagan and Kurfess studied tolerance optimization over multiple manufacturing considerations (Cagan and Kurfess, 1992). Turner and Wozny focused on the automated tolerance analysis in a solid modeling system and developed a method for representing and analyzing tolerance using model variations in a normed vector space (Turner and Wonzy, 1987, 1990). Wu et al. studied various existing continuous cost–tolerance models and compared their modeling errors based on a general empirical cost–tolerance curve (Wu et al., 1988). Dong and Soom have extended the tolerance optimization formulation to include multiple-dimension chains sharing common design tolerances and incorporated the valid tolerance range into the formulation (Dong and Soom, 1990, 1991). Dong et al. carried out an in-depth study on the empirical cost–tolerance data from typical production processes and introduced several new cost–tolerance models and a hybrid-model tolerance optimization formulation (Dong et al., 1994). These introduced models and formulation better represent empirical production data and provide more reliable results for tolerance synthesis. Lately, a method that combines a nontraditional optimization method and the Monte Carlo–based tolerance analysis was introduced with improved results on classical examples (Iannuzzi and Sandgren, 1994).

Another emerging research area in tolerance analysis and synthesis is computer automation and interface to CAD systems (Bjorke, 1989; Roy et al., 1991; Zhang and Huo, 1992; Shah, 1991). Dong and Soom first developed a method for automated tolerance analysis and synthesis in conventional CAD environments and automated formulation of tolerance optimization using an intelligent system (Dong and Soom, 1986, 1990, 1991). The method was later extended to a feature-based CAD environment (Dong, 1992, 1993; Xue and Dong, 1993) as well as integrated concurrent engineering design (Xue and Dong, 1994, 1996; Xue et al., 1995, 1996). Martino and Gabriele developed a method for analyzing conventional, statistical, and some geometric tolerances of a part using solid models and variational geometry (Martino and Gabriele, 1989). Software tools for automated tolerance analysis were also made available (VSA, 1993).

Robust design, aimed at making a product less sensitive to design variations and thereby improving the quality of product, is another area that has emerged from tolerance optimization research. The Taguchi method (Taguchi, 1986) and measures of quality loss (Kapur et al., 1990) are frequently used in this type of work to achieve the objective of robust design.

9.2   MODELING OF TOLERANCE-RELATED MANUFACTURING PROCESS COSTS

9.2.1   Tolerance and Manufacturing Costs

In practice, design tolerances are often interrelated and contribute to a given assembly (or resultant) tolerance of the design. These design tolerances specify various mechanical features, and the features are manufactured using different production processes. These production processes, however, have different manufacturing cost–tolerance (or cost-to-accuracy) relations. The sensitivity of total manufacturing cost with respect to each tolerance depends on the feature specified by the tolerance and the production process used for forming the feature. To identify the best combination of the interrelated design tolerances, which satisfies the stackup constraint of the given assembly tolerance and leads to the least total manufacturing costs, the influence of tolerance value to the manufacturing costs of a design needs to be modeled.

In the author's previous work, a method of accurately estimating the manufacturing costs of a design through the identification of its minimum-cost production process was introduced (Dong and Hu, 1991; Dong, 1994). The approach examines all feasible production processes for manufacturing the given design and calculates the manufacturing costs of each process by adding the costs of all manufacturing operations of the process. In a related work, several new cost–tolerance models for all commonly used machining operations were introduced to provide a better means for calculating the tolerance-related manufacturing costs (Dong et al., 1994). These exercises allow us to predicate accurately the manufacturing costs of each design tolerance. Adding the manufacturing costs of all related design tolerances, one can thus form the objective function of the optimization problem in tolerance synthesis.

9.2.2   Tolerance-Related Cost for a Production Operation

The tolerance-related cost for a production operation, or the cost–tolerance relation model, is at the lowest level of the cost modeling. The model is based on the empirical data acquired from various machining operations. Despite the variations of design features and their tolerances, most mechanical parts are produced using the commonly used machining operations. The cost–tolerance relation models for various machining operations thus serve as the building blocks for constructing the tolerance-related production process cost model and the tolerance synthesis model. The cost–tolerance relation is modeled as a function, $c(\delta)$, where c is the relative manufacturing cost and δ is the size of the tolerance. The valid tolerance range of a production operation, $[\delta_{min}, \delta_{max}]$, represents the accuracy-improving capability of a production operation. Within this valid range, a tighter tolerance or higher accuracy demand leads to higher manufacturing costs, and a looser tolerance or lower accuracy demand leads to lower manufacturing costs.

Empirical manufacturing cost–tolerance data curves for various production operations and processes were obtained from an earlier publication (Trucks, 1976). One of these curves, the empirical relation of manufacturing cost versus the true position tolerance in hole making, is illustrated in Figure 9.1, where δ denotes a tolerance that describes the hole location accuracy and $c(\delta)$ represents the manufacturing costs for producing a hole to the location accuracy of δ. The curve covers production operations of drilling, jig boring, and machining using special equipment. The empirical curve represents the cost–tolerance relation of a hole-making production sequence that consists of production operations of rough machining, semifinish machining, finish machining, and accurate forming using special equipment.

To incorporate the empirical cost–tolerance data collected for various production operations into the calculation of production costs of a production sequence, the appropriate mathematical model for these data curve segments needs to be introduced. Many researchers worked on this subject and introduced close to a dozen cost–tolerance models. These models are listed in Table 9.1.

Among the models given in the table, the first six introduced by earlier researchers have no valid bounds. In addition, these unbounded models were built using one specific set of cost–tolerance data. They also have relatively large modeling errors due to their simple form (Wu et al., 1988; Dong et al., 1994). The last six models listed in the table are given with a valid tolerance range. These models were introduced in the author's previous work (Dong

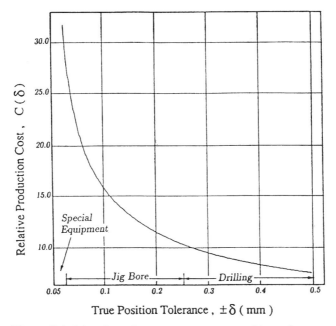

Figure 9.1. Manufacturing cost versus true position tolerance.

and Soom, 1991; Dong et al., 1994). In these studies, various published empirical manufacturing cost–tolerance data curves were collected and used for testing the existing and new models. The collected cost–tolerance data cover a wide variety of production processes, including hole producing (true position), face milling, turning, rotational surface grinding, internal grinding, die casting, and investment casting, compiled in (Trucks, 1976), and the general cost–tolerance relation published by Dieter (Dieter, 1983). These empirical manufacturing cost–tolerance data curves are plotted in Figure 9.2. Since each production operation has only limited accuracy-improving capability, curve fitting of the new cost–tolerance models (the last six models listed in Table 9.1) was conducted over the valid tolerance range of the operation, as shown in Figure 9.2. The new bounded models showed significant modeling accuracy improvements over the existing unbounded models (Dong et al., 1994). This can be illustrated using Dieter's empirical data as a test curve, as shown in Figure 9.3, where the curves of the new models and the empirical data curve overlaps and the old models present considerable errors.

9.2.3 Tolerance-Related Cost for a Production Process

In manufacturing each mechanical feature is produced through a sequence of production or machining operations, called a *production process*. Different features with different tolerances require different production processes. The

TABLE 9.1 Models for Empirical Cost–Tolerance Relation Data

No Bounds for δ	
Exponential (Speckhart, 1972)	$c(\delta) = a_0 e^{-a_1 \delta}$
Reciprocal squared (Spotts, 1973)	$c(\delta) = \dfrac{a_0}{\delta^2}$
Reciprocal powers (Sutherland and Roth, 1975)	$c(\delta) = a_0 \delta^{-a_1}$
Reciprocal powers/Exponential (Michael and Siddall, 1981)	$c(\delta) = a_0 \delta^{-a_1} e^{-a_2 \delta}$
Reciprocal (Greenwood and Chase, 1988)	$c(\delta) = \dfrac{a_0}{\delta}$
Discrete (Lee and Woo, 1989)	$\delta_i, c(\delta_i) \ (1 \le i \le N)$
With a Valid Range $\delta \in [\delta_{\min}, \delta_{\max}]$	
Modified exponential (Dong and Soom, 19990, 1991)	$c(\delta) = a_0 e^{-a_1(\delta - a_2)} + a_3$
Reciprocal powers/exponential II (Dong et al., 1994)	$c(\delta) = a_0 + a_1 \delta^{-a_2} + a_3 e^{-a_4 \delta}$
Combined linear and exponential (Dong et al., 1994)	$c(\delta) = a_0 + a_1 \delta + a_2 e^{-a_3 \delta}$
Cubic polynomial (Dong et al., 1994)	$c(\delta) = a_0 + a_1 \delta + a_2 \delta^2 + a_3 \delta^3$
Fourth polynomial (Dong et al., 1994)	$c(\delta) = a_0 + a_1 \delta + a_2 \delta^2 + a_3 \delta^3 + a_4 \delta^4$
Fifth polynomial (Dong et al., 1994)	$c(\delta) = a_0 + a_1 \delta + a_2 \delta^2 + a_3 \delta^3 + a_4 \delta^4 + a_5 \delta^5$

tolerance-related costs of these processes cannot be modeled easily, due to the infinite variations of these processes with any small change of the design.

On the other hand, each mechanical feature of a designed part is modified from its raw material form to the designed shape and accuracy through a production process. The mechanical tolerance and surface finish of a part are continually improved by the selected production operations applied in tandem. Based on this fact, a method for calculating the tolerance-related production process cost, by assembling the cost–tolerance models of all involved machining operations, was developed (Dong and Hu, 1991; Dong, 1994). The manufacturing cost for each given design tolerance can thus be modeled and calculated using the low-level, cost-tolerance relation models for commonly used machining operations.

In manufacturing a reasonably tight tolerance can be obtained with high-quality producing and inefficient cutting parameter settings, such as small depth of cut, slow feed rate, and fine turning of the machine tool. The infinite initial tolerance can be considered to be corresponding to zero cost. A mechanical feature with small original errors (tight blank part tolerance) requires

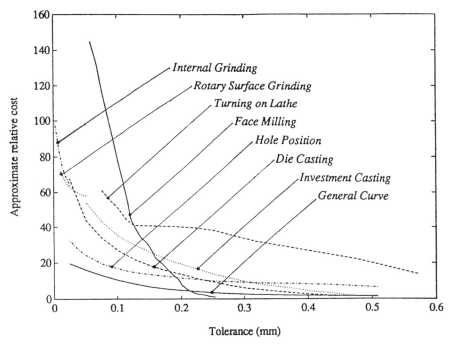

Figure 9.2. Empirical manufacturing cost–tolerance curves.

less manufacturing effort in machining and leads to low manufacturing costs. The relative manufacturing cost for improving a feature from the original accuracy (or tolerance), δ_1, to the improved accuracy (or tolerance), δ_2, can thus be calculated by

$$\Delta C_{\delta_1-\delta_2} = c(\delta_2) - c(\delta_1) \tag{9.1}$$

where $c(\delta_1)$ and $c(\delta_2)$ are the corresponding relative manufacturing costs to change the tolerance of a feature from an infinite tolerance (very large error) to the tolerances of δ_1 and δ_2. The result, $\Delta C_{\delta_1-\delta_2}$, is the relative manufacturing cost for improving the tolerance of a feature from δ_1 to δ_2.

The process for changing a mechanical feature from its raw material state with considerably larger error to the finished-part state with the designed tolerance can be modeled as shown in Figure 9.4, provided that the production process consists of three production operations: rough machining (r), semi-finish machining (sf), and finish machining (f). The manufacturing cost of this process sequence can be calculated by

$$C_{process} = \Delta C_r + \Delta C_{sf} + \Delta C_f$$
$$= [c_r(\delta_1) - c_r(\delta_0)] + [c_{sf}(\delta_2) - c_{sf}(\delta_1)] + [c_f(\delta_3) - c_f(\delta_2)] \tag{9.2}$$

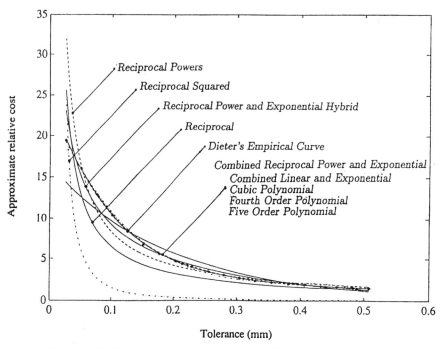

Figure 9.3. Model comparisons using Dieter's cost–tolerance curve.

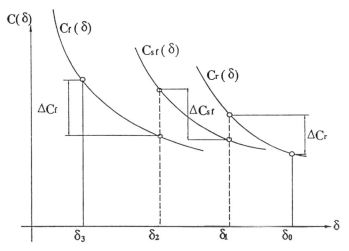

Figure 9.4. Cost–tolerance model for a production process.

where δ_0 is the tolerance of the blank part. δ_3 is the tolerance accomplished by the production process; $c_r(\delta)$, $c_{sf}(\delta)$, and $c_f(\delta)$ are the production cost–tolerance models for the rough machining, semifinish machining, and finish machining operations; and ΔC_i; ob$i \in r$, sf, f] represents the cost for improving the tolerance using a specific production operation.

In practice, a part is often unloaded from a machine tool after one production operation and loaded onto another machine tool to proceed to the next production operation. The setup error for the ith production operation, $e_{\text{setup},i}$, will be introduced. When a part is reloaded, the errors that the production operation has to handle include the error of the part, which has been inherited from the previous production operation, and the error that has been introduced in the reloading. In other words, the existing tolerance for the ith production operation is $\delta'_{i-1} = \delta_{i-1} + e_{\text{setup},i}$. A general formulation of the manufacturing costs for a production process with p production operations thus becomes

$$C(\delta_1,...,\delta_{p-1})_{\text{process}} = \sum_{i=1}^{p} \Delta C'_i = \sum_{i=1}^{p} [c_i(\delta_i) - c_i(\delta'_{i-1})]$$

$$\delta'_{i-1} = \delta_{i-1} + e_{\text{setup},i}$$

(9.3)

where δ_0 is the tolerance of the blank part and δ_p is the tolerance accomplished by the production process. The value of δ_p is equal to (or slightly smaller than) the designed tolerance δ of the feature.

These cost models allow the tolerance-related manufacturing costs of a specific production process to be calculated. While a given design might be manufactured through many different processes, the production process that has the lowest manufacturing costs is considered to be the process representing the true manufacturing costs of the design (Dong and Hu, 1991; Dong, 1994). This process and its manufacturing cost can be identified by minimizing the cost given in Eq. (9.3) for all feasible production processes and by comparing their minimized manufacturing costs. Further discussions on the identification of the true manufacturing costs of a design and the automated generation of all feasible production processes for a given design feature and tolerance are given in several publications (Dong and Hu, 1991; Dong, 1994; Xue and Dong, 1993).

The minimum-cost production process for the designed feature and tolerance, once identified, leads to the desired cost–tolerance curve for the design tolerance. One will find that this cost–tolerance curve is actually the embracing curve for the cost–tolerance curves of its production operations. Curve parameters can be determined using the same method. This relation is illustrated in Figure 9.5.

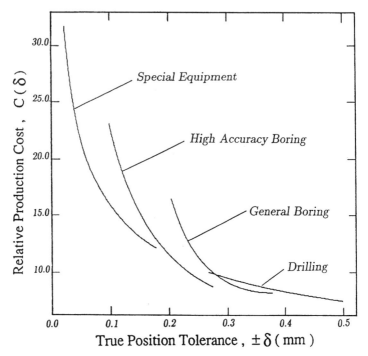

Figure 9.5. Manufacturing cost–tolerance curves for hole machining operations and process.

9.3 TOLERANCE SYNTHESIS BY DESIGN OPTIMIZATION

Traditionally, design tolerances was assigned manually by a designer largely on a trial-and-error basis. Computer automation of manual tolerance analysis that proportionally distributed the tolerance values among related tolerances and satisfied the stackup consistency requirement was reported by Dong and Soom (1986) and included later in a textbook by Zeid (1993). Detailed discussion of this issue also appears in a book by Bjorke (1989).

More recent tolerance synthesis work tends to follow the theme of design optimization, as discussed earlier. In optimization-based tolerance synthesis, the related tolerances in a dimension chain first have to be identified from the drawings. This can be carried out automatically with the help of an intelligent system (Dong and Soom, 1990; Dong, 1992). The tolerances identified are then treated as design variables, and their values are determined through the optimization. The majority of the tolerance synthesis research seeks a minimum-cost design solution. More recently, tolerance design under the concurrent engineering principle was introduced.

9.3.1 Minimum Cost-Tolerance Design

Minimum cost-tolerance design is carried out by minimizing the tolerance-related manufacturing costs, subject to the stackup consistency constraint and the functional performance constraints. The approach is aimed at the lowest possible manufacturing costs that can just meet the functional performance required.

Based on the manufacturing cost–tolerance models discussed previously, a quantitative measure of manufacturing costs for producing the related dimensions of the dimension chain to the specified tolerances can be formulated as

$$\min_{\text{w.r.t.}\delta_i} \ C(\delta_1,...,\delta_p) = \sum_{i=1}^{p} c_i(\delta_i) \tag{9.4}$$

subject to

$$\delta_1 + \delta_2 + \cdots + \delta_n = \delta_R \tag{9.5}$$

$$\delta_{0i} \le \delta_{\min,i} < \delta_i < \delta_{\max,i} \qquad i = 1,...,p, \quad p \le n \tag{9.6}$$

and

$$F(\mathbf{d}) \ge F_0 \tag{9.7}$$

where $\delta_1,...,\delta_n$ are n component tolerances of a dimension chain, with $\delta_{p+1},...,\delta_n$ prespecified to meet design specifications and $\delta_1,...,\delta_p$ to be optimized. δ_R represents the given design requirement or resultant tolerance. $C(\delta_1,...,\delta_p)$ is the manufacturing cost for producing the mechanical features associated with the adjustable design tolerances in the dimension chain. The stackup consistency constraint of equation (9.5) is based on stackup under a worst-case consideration, and a statistical approach could be introduced with minor modifications of this equation. Equation (9.6) specifies the valid bounds of the design tolerance, and (9.7) represents one or more design functional performance constraints. The stackup consistency and performance constraints may or may not appear, depending on the design. In general, the stackup consistency constraint shows up when dealing with multiple tolerances in a dimension chain, while the performance constraints show up in a performance-oriented design such as the one given in (Michael and Siddall, 1981, 1982). With this formulation, the best combination of design tolerances $\delta_1,...,\delta_p$ with the least manufacturing costs, as the solution to the multivariate, nonlinear constrained optimization problem, can be determined. More complicated cases with several nested dimension chains can be handled using the method introduced by Dong and Soom (1991).

Minimum-cost tolerance design, similar to minimum-cost design, is widely used in industry for commercial products. In developing a complex mechanical product with many components, the overall design functions are often decomposed into many subfunctions. These subfunctions are the minimum design requirements to be satisfied by different components. Since manufacturing cost is a critical measure of product competibility, reduction of manufacturing costs, satisfying the minimum functional performance, is often the goal for some uncritical components.

9.3.2 Balanced Performance and Cost–Tolerance Design

More recently, concurrent engineering has become the new underlying principle for carrying out mechanical design and product development (Kusiak, 1993; O'Grady and Young, 1991; Rosenblatt and Watson, 1991). Instead of a sequentially arranged product development process, considerations from all important product development aspects are incorporated simultaneously into the early design phase. Products with better balanced life-cycle performance can be formed in the original design, thus considerably reducing the number of redesigns and product development lead times. Among all contributing life-cycle performance aspects, functional performance and manufacturing cost are the two most important areas.

Balanced performance and cost–tolerance design, introduced in the author's recent work (Xue et al., 1995, 1996; Xue and Dong, 1996), is aimed at identifying the best trade-off between functional performance and manufacturing costs, subject to all functional and cost constraints. To merge the two "performance" measures of distinct nature, the functional performance and manufacturing costs are transformed into a comparable form—functional performance index, $I^{(F)}(\mathbf{d})$, and manufacturing cost index, $I^{(C)}(\mathbf{d})$, respectively.

A design with balanced functional performance and production costs can be accomplished by

$$\min_{\mathbf{d}} - I(\mathbf{d}) = \lambda_C I^{(C)}(\mathbf{d}) - \lambda_F I^{(F)}(\mathbf{d}) \qquad \lambda_F + \lambda_C = 1 \qquad (9.8)$$

where $\mathbf{d}$ represents the design variables, including all design tolerances; $I(\mathbf{d})$ is the overall design performance; λ_F and λ_C are application-dependent weighting factors; and $I^{(C)}(\mathbf{d})$ and $I^{(F)}(\mathbf{d})$ are the overall manufacturing cost and functional performance increase in the design, respectively.

The overall manufacturing cost increase can be calculated by

$$I^{(C)}(\mathbf{d}) = \frac{C(\mathbf{d}) - C(\mathbf{d}_0)}{C(\mathbf{d}_0)} \qquad (9.9)$$

where $C(\mathbf{d}_0)$ is the manufacturing cost of a reference design point.

The functional performance increase in the ith functional performance aspect, $I_i^{(F)}(\mathbf{d})$, and the overall functional performance increase of the design are defined and measured by

$$I_i^{(F)}(\mathbf{d}) = \frac{F_i(\mathbf{d}) - F_i(\mathbf{d}_0)}{|F_i(\mathbf{d}_0)|} \qquad i = 1,2,...,p \tag{9.10}$$

$$I^{(F)}(\mathbf{d}) = \frac{\sum_{i=1}^{p} w_i I_i^{(F)}(\mathbf{d})}{\sum_{i=1}^{p} w_i} \qquad w_1 + w_2 + \cdots + w_p = 1 \tag{9.11}$$

where $F_i(\mathbf{d}_0)$ is the functional performance of a reference design point, $F_i(\mathbf{d})$ and $F_i(\mathbf{d}_0)$ are calculated using the ith functional performance aspect model, and $w_1,...,w_p$ are coefficients for weighting the inputs from all related functional performance aspects.

This formulation is a good representation of the concurrent engineering principle. It can be extended by adding additional terms to include more life-cycle aspects. The model puts more control in the hand of the designers. An ideal design can be accomplished by improving the design with more functional performance gain and less manufacturing cost increase rather than just going to one extreme—the minimum manufacturing cost design.

In this formulation the tolerance-related manufacturing costs are modeled using the methods discussed earlier. The only difference is now a dimensionless cost reading: The relative cost increase (or decrease) is used to make the cost reading comparable to the performance reading.

The modeling of the functional performance is, however, case dependent. The critical factors that determine the functional performance of a design normally include the design geometry, the accuracy of designed geometry (or tolerance), the material of the part, and its manufacturing methods. The method of functional performance modeling and functional performance index calculation is discussed below using a design example. Many related issues, including modeling of mechanical features, automated generation of candidate designs, and optimization of design geometry, need to be addressed. These issues are beyond the scope of this chapter. Interested readers may refer to work of Xue and Dong (1993, 1994, 1996).

9.4 EXAMPLE OF BALANCED PERFORMANCE AND COST–TOLERANCE DESIGN

9.4.1 Tolerance Specification in Drill Head Spindle Hole Design

To demonstrate how an optimal design is identified using the methods introduced earlier, a design example is given. In this example we design a multiple-spindle drill head. For ease of illustration we focus on only two design variables of the drill head: the size tolerance, δ_D, and location tolerance, δ_{xy}, of the hole on the drill head case for the spindle, as illustrated in Figure 9.6. The design objective is to identify the optimal values of these two tolerances, which lead to the best life-cycle performance of the drill head.

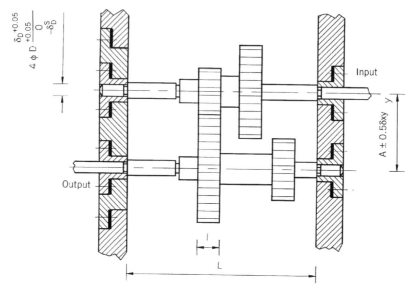

Figure 9.6. Spindle hole of a multispindle drill head.

The considered life-cycle performance of the spindle hole is limited to the design and manufacturing aspects. The hole can be modeled as a design feature "spindle/shaft hole" for supporting spindle rotation and maintaining accurate spindle location. The spindle hole is also a manufacturing feature "hole" to be produced by drilling and reaming. The two spindle holes are machined after installation of the journal bearings.

The size tolerance of the spindle hole, δ_D, influences the clearance between the spindle and the journal bearings. Two functional performance measures, power-loss variation, ΔPL, and spindle case working temperature variation, ΔT, are related to the clearance change. The location tolerance of the spindle hole, δ_{xy}, and the size tolerance of the spindle hole, δ_D, both influence the alignment of spindle and other shafts. Misalignments between two shafts with mating gears will change the distribution of the contact stress on the tooth of the gear. Stress concentration will reduce the designed lifetime of the gears.

Manufacturing costs are sensitive to both the hole location tolerance, δ_{xy}, and the hole dimension tolerance, δ_D. The manufacturing cost for machining the hole to an accurate size, C_D, and the manufacturing cost for producing the hole at an accurate location, C_{xy}, are modeled using the methods discussed earlier. Relations between the two design variables and the functional performance and manufacturing costs are illustrated in Figure 9.7.

9.4.2 Functional Performance and Manufacturing Cost Indices

Modeling Functional Performance. The functional performance indices, including power-loss variation, temperature variation, and gear life, are modeled based on the data curves from the mechanical design handbooks (Faires,

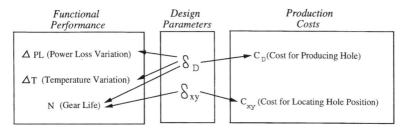

Figure 9.7. Design parameters and their influences.

1965; Shigley and Mischke, 1989). The original curves of (a) clearance versus power loss, (b) clearance versus temperature variation, and (c) gear life versus maximum stress are illustrated in Figure 9.8.

The clearance between the spindle and the bearing depends on the tolerances of the hole, δ_D, and the tolerance of the shaft, δ_D^S, as illustrated in Figure 9.6. The minimum clearance, Δ_{cr}, is selected as 0.05 mm. In practice, because it is much easier to achieve a higher accuracy in shaft machining than in hole machining, the tolerance of the shaft is often half that of the hole. The maximum and minimum clearances can thus be calculated by

$$\Delta_{min} = D_{min}^H - D_{max}^S = \Delta_{cr} \tag{9.12}$$

$$\Delta_{max} = D_{max}^H - D_{min}^S = \delta_D + \delta_D^S + \Delta_{cr} = \tfrac{3}{2}\delta_D + \Delta_{cr} \tag{9.13}$$

A larger tolerance δ_D can introduce a larger variation of clearance Δ, thereby increasing the variations of power loss and temperature, and leading to poor product quality in mass production. The functional performance measures are defined as follows:

Power-loss variation:

$$\Delta PL(\delta_D) = -|PL(\Delta_{max}) - PL(\Delta_{min})| \tag{9.14}$$

Temperature variation:

$$\Delta T(\delta_D) = -|T(\Delta_{max}) - T(\Delta_{min})| \tag{9.15}$$

The direct relations between the hole diameter tolerance, δ_D, and the power-loss variation, ΔPL, as well as the temperature variation, ΔT, are plotted in Figure 9.9a and b.

The size and location tolerances, δ_D and δ_{xy}, both contribute to the alignment of the shafts and influence the lifetime of the gear. Their influence can be counted through calculation of the alignment angle, θ:

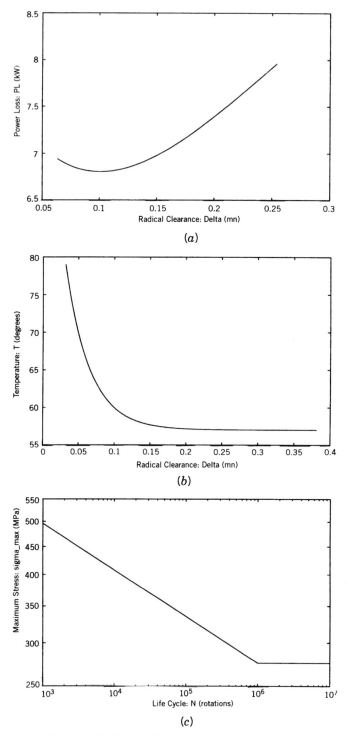

Figure 9.8. Original functional performance data.

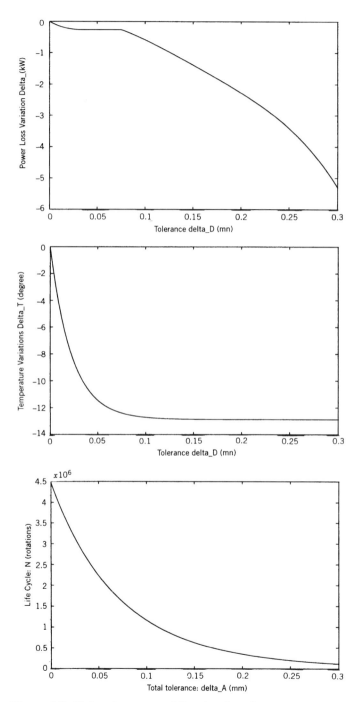

Figure 9.9. Hole tolerances and functional performance measures.

$$\theta = 2\frac{\delta_A}{L} \tag{9.16}$$

where δ_A is the worst-case variation caused by the inaccuracy allowed by the two tolerances, as illustrated in Figure 9.10a, and

$$\delta_A = 0.5\delta_{xy} + 0.5\Delta_{max} = 0.5\delta_{xy} + 0.5(1.5\delta_D + \Delta_{cr}) \tag{9.17}$$

where Δ_{max} is the maximum clearance equation (9.13).

The misalignment of shafts (illustrated in Figure 9.10a) changes the equal distribution of contact stress on the surface of the gear tooth across the width of the gear. Higher contact stresses will be imposed on certain areas of the tooth (at the edges), as shown in Figure 9.10b. This part of the tooth will then experience earlier failure and a shorter lifetime than expected. The average stress σ_0, or the contact stress with no alignment, can be calculated according to design handbooks (Shigley and Mischke, 1989):

$$\sigma_0 = \sqrt{\frac{2W_t}{\pi l \cos \phi} \frac{1/d_{p1} \sin \phi + 1/d_{p2} \sin \phi}{(1 - v_1^2)/E_1 + (1 - v_2^2)/E_1}} \tag{9.18}$$

where W_t = tangential load applied to the gear surface

l = width of the gear

ϕ = pressure angle

d_{p1} and d_{p2} = two pitch diameters of the gears

v_1 and v_2 = two Poisson's ratio parameters of the two gears

E_1 and E_2 = two elasticity module parameters of the two gears

Poisson's ratio and elasticity module are related only to the material.
Using the stress σ and unit strain ϵ relation

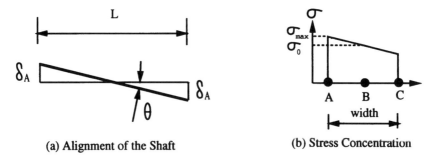

(a) Alignment of the Shaft   (b) Stress Concentration

Figure 9.10. Shaft alignment and gear tooth contact stress concentration.

$$\sigma = E\epsilon \tag{9.19}$$

the small changes of these two variables can be derived as

$$\Delta\sigma = E\,\Delta\epsilon \tag{9.20}$$

The strain change can be represented by

$$\Delta\epsilon = \frac{\Delta z}{t/2} \tag{9.21}$$

where Δz is the small strain at the left side of the gear surface due to compression and t is the pitch of the gear. Since the pitch can be represented by the module, m, using

$$t = \pi m \tag{9.22}$$

Δz can be represented as

$$\Delta z = \frac{\pi m}{2E}\,\Delta\sigma \tag{9.23}$$

The two gears share the same module, m, and stress change, $\Delta\sigma$. The sum of the strains at the left side of the gear tooth is calculated using

$$\Delta z_1 + \Delta z_2 = \frac{\pi m}{2}\,\Delta\sigma\left(\frac{1}{E_1} + \frac{1}{E_2}\right) \tag{9.24}$$

Because this sum can also be represented as

$$\Delta z_1 + \Delta z_2 = \frac{l}{2}\,\theta \tag{9.25}$$

the stress change is then represented as

$$\Delta\sigma = \frac{2l}{\pi m L}\frac{E_1 E_2}{E_1 + E_2}\,\delta_A \tag{9.26}$$

The maximum stress, as shown in Figure 9.10b, is calculated as

$$\sigma_{\max} = \sigma_0 + \Delta\sigma \tag{9.27}$$

The misalignment, δ_A, can be represented by tolerances δ_D and δ_{xy}, as described in equation (9.17), and the life cycle regarding the maximum stress can be found in Figure 9.8c. Based on these relations, mapping from the size and location tolerances to the gear lifetime functional performance can be accomplished:

$$N(\delta_D, \delta_{xy}) = N_0(\sigma_{max}) \tag{9.28}$$

If the following structure and material parameters are chosen: $W_t = 1.2 \times 10^3$ N, $\phi = 20°$, $l = 25.4 \times 10^{-3}$ m, $d_{p1} = 101.6 \times 10^{-3}$ m, $d_{p2} = 371.5 \times 10^{-3}$ m, $E_1 = 207.0 \times 10^9$ Pa, $E_2 = 100.0 \times 10^9$ Pa, $\nu_1 = 0.292$, $\nu_2 = 0.211$, $m = 4 \times 10^{-3}$ m, and $L = 762 \times 10^{-3}$ m, the gear lifetime and size/location tolerance relation areas shown in Figure 9.9c. Because the three functional performance measures discussed have different units, these measures must be transformed into a comparable measure. The method introduced in equations (9.9)–(9.11) can be used for this transformation. In this example the design with the minimum manufacturing costs is selected as the reference design, $\mathbf{d}_0 = (\delta_{D0}, \delta_{xy0})^T$. The three functional performance measures selected—power-loss variation, temperature variation, and gear lifetime—are represented with respect to the reference design as ΔPL_0, ΔT_0, and N_0, respectively. The three functional performance indices are obtained by

$$I_1^{(F)}(\mathbf{d}) = I_{\Delta PL}^{(F)}(\delta_D) = \frac{\Delta PL(\delta_D) - \Delta PL_0}{|\Delta PL_0|} \tag{9.29}$$

$$I_2^{(F)}(\mathbf{d}) = I_{\Delta T}^{(F)}(\delta_D) = \frac{\Delta T(\delta_D) - \Delta T_0}{|\Delta T_0|} \tag{9.30}$$

$$I_3^{(F)}(\mathbf{d}) = I_N^{(F)}(\delta_D, \delta_{xy}) = \frac{N(\delta_D, \delta_{xy}) - N_0}{|N_0|} \tag{9.31}$$

The overall functional performance index is calculated by

$$I^{(F)}(\mathbf{d}) = I^{(F)}(\delta_D, \delta_{xy}) = \frac{I_{\Delta PL}^{(F)}(\delta_D) + I_{\Delta T}^{(F)}(\delta_D) + I_N^{(F)}(\delta_D, \delta_{xy})}{3} \tag{9.32}$$

Modeling Manufacturing Costs. The manufacturing costs for machining the spindle hole will vary according to the size and location tolerances specified in the design. High accuracy and small tolerance require more manufacturing effort and thus higher costs. The manufacturing cost–tolerance relations were obtained from the machine shop and experiments (Trucks, 1976) and were modeled using the methods discussed previously. The manufacturing cost–tolerance models have the following forms. For the size tolerance-related cost,

$$C(\delta_D) = a_0 + a_1\delta_D + a_2\delta_D^2 + a_3\delta_D^3 + a_4\delta_D^4 + a_5\delta_D^5 \qquad (9.33)$$

where the parameters a_0, a_1, a_2, a_3, a_4, and a_5 are 112.3, -1.061×10^3, 5.833×10^3, -1.534×10^4, 1.845×10^4, and -8.269×10^3, respectively (Dong et al., 1994a). For the location tolerance-related cost,

$$C(\delta_{xy}) = b_0 + b_1\delta_{xy}^{-b2} + b_3e^{-b4\delta_{xy}} \qquad (9.34)$$

where the parameters b_0, b_1, b_2, b_3, and b_4 are 1.000×10^{-5}, 5.326, 0.4475, 6.652, and 11.72, respectively (Dong et al., 1994). The manufacturing costs for producing the holes are calculated using

$$C(\delta_D,\delta_{xy}) = 4[C(\delta_D) + C(\delta_{xy})] \qquad (9.35)$$

The overall manufacturing cost increase can be calculated by

$$I_{(C)}(\mathbf{d}) = I^{(C)}(\delta_D,\delta_{xy}) = \frac{C(\mathbf{d}) - C(\mathbf{d}_0)}{C(\mathbf{d}_0)} \qquad (9.36)$$

Feasible Design Space. The feasible design space of this simplified problem is defined by the minimum functional performance and maximum manufacturing costs imposed on the design. These constraints include $\Delta PL \leq 4$ kW, $\Delta T \leq 13°$, $N \geq 5^4$ rotations, $\delta_D \in [0.02$ mm, 0.5 mm], and $\delta_{xy} \in [0.02$ mm, 0.5 mm]. If plotted with respect to the two selected design variables, δ_D and δ_{xy}, the feasible design space has a closed area, as illustrated in Figure 9.11.

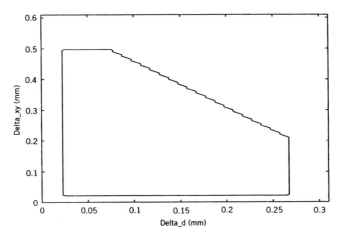

Figure 9.11. Feasible design space.

9.4.3   Formulation of Optimization and Solution

The optimal life-cycle performance design, or the design with the best balanced functional performance and manufacturing costs, is achieved by joint optimization of functional performance and manufacturing costs. As discussed previously, the design with the minimum manufacturing costs is used as the reference for design improvement. This reference is at the point $(\delta_{D0}, \delta_{xy0})^T$ in the feasible design space.

Assuming that $\lambda_F = \lambda_C$, a design with balanced functional performance and manufacturing costs can be accomplished by

$$\min_{\delta_D, \delta_{xy}} -I(\delta_D, \delta_{xy}) = 0.5 \times I^{(C)}(\delta_D, \delta_{xy}) - 0.5 \times I^{(F)}(\delta_D, \delta_{xy})$$

The identified optimal design is at $(\delta_D, \delta_{xy})^T = (0.105$ mm, 0.100 mm$)^T$. The contour map of the product life-cycle performance is illustrated in Figure 9.12.

Present design practice follows a quite different approach from the discussed life-cycle performance optimization approach. In general, a designer first specifies a rough design target in terms of functional performance. Based on the design objective determined, the values of design parameters are determined according to the recommendations of design handbooks and/or experience. The design is then evaluated and fine-tuned to meet the specified target and achieve minimum manufacturing costs.

Three major problems lie within this design practice. First, the values of design parameters are determined following the "one size fits many" design handbook. These values will seldom be optimal. Second, the actual functional performance and manufacturing costs of a design can hardly be known before

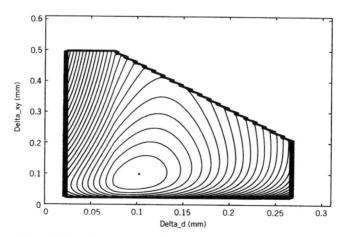

Figure 9.12. Contour map of product life-cycle performance.

the design is manufactured and tested, and the appropriate balance of these two product life-cycle aspects is very difficult to accomplish. Third, the design and redesign processes are carried out in darkness. Many trials and a great deal of experience are required to upgrade a design to a satisfactory form.

For the multiple-spindle-drill head design, the values of the two design variables considered, the spindle size tolerance and spindle hole location tolerance, are chosen as $(\delta_D, \delta_{xy})^{\mathrm{T}} = (0.0084 \text{ mm}, 0.115 \text{ mm})^{\mathrm{T}}$, according to the recommendations of ANSI tolerance grades IT10 and IT9, respectively.

The functional performance, manufacturing cost, and life-cycle performance readings of the manual design and from the design optimization are listed in Table 9.2. The results from the tolerance optimization present a higher life-cycle performance reading. This is achieved by a relatively small increase in manufacturing costs (8%) and a bigger gain in functional performance (12.3%). The manual design is by no mean targeted to either the peak functional performance or the minimum manufacturing cost and is intended to achieve a balanced design. However, without quantitative modeling and optimization, the manual approach is unable to reach the best balanced design. Other design methods, including the peak functional performance-oriented design and the minimum manufacturing cost–oriented design, if compared, produce even stronger bias with poor life-cycle performance readings, although optimization is used (Xue et al., 1995, 1996).

9.5 SUMMARY

Design tolerance has a profound influence on the functional performance and production costs of a mechanical product. However, tolerance specification is a complex and demanding task and is carried out traditionally on a trial-and-error basis. Research on tolerance synthesis presents a systematic approach through quantitative modeling and design optimization but imposes an even greater challenge to its use in mechanical design. In this chapter the major steps for carrying out tolerance synthesis are reviewed and several key issues on the implementation of tolerance synthesis, using real manufacturing data, are addressed. The chapter begins with a general review of the tolerance synthesis research. The methods for building manufacturing cost models used in tolerance optimization are then explained. Tolerance synthesis methods

TABLE 9.2 Comparison of Design Results

Performance Measures	Manual Design	Balanced Performance and Cost Design
$I^{(F)}(\delta_D, \delta_{xy})$	1.14	1.28
$I^{(C)}(\delta_D, \delta_{xy})$	0.62	0.67
$I(\delta_D, \delta_{xy})$	0.52	0.61

under two different design criteria, the conventional minimum manufacturing cost design and the newly introduced balanced performance and cost design, are discussed. Using a typical design example, a method for jointly evaluating and optimizing the functional performance and manufacturing costs of a mechanical design, in terms of tolerance, is presented. The result from the tolerance optimization is compared with the tolerance values from manual guided design to illustrate the advantages of quantitative modeling and optimization. Due to space limitation, only modest discussion and references to many related side issues are given throughout this chapter. The presented work contributes to automated tolerance synthesis, integrated concurrent design, and design optimization. The joint optimization can produce a better design with improved life-cycle performance and shorter product development lead times.

ACKNOWLEDGMENTS

Several graduate students and research associates have contributed to the research leading to this chapter. In particular, contributions from Weiping Hu, Joe H. Rousseau, and Deyi Xue (now a professor at the University of Calgary) are gratefully acknowledged. The author is also thankful to continuous financial support from the Natural Science and Engineering Research Council (NSERC) of Canada.

REFERENCES

Bjorke, O., 1989, *Computer-Aided Tolerancing,* 2nd ed., ASME Press, New York.

Cagan, J., and Kurfess, J. R., 1992, Optimal Tolerance Design over Multiple Manufacturing Choices, *Proceedings of the Advances in Design Automation Conference,* Scottsdale, AZ, Sept. 13–16, 1992, pp. 165–172.

Chase, K., Greenwood, W., Loosli, B., and Hauglund, L., 1990, Least Cost Tolerance Allocation for Mechanical Assemblies with Automated Process Selection, *Manuf. Rev.,* Vol. 3, No. 1, pp. 49–59.

Dieter, G. E., 1983, *Engineering Design: A Materials and Processing Approach,* McGraw-Hill, New York.

Dong, Z., 1992, Automation of Tolerance Analysis and Synthesis in Conventional and Feature-Based CAD Environments, *Int. J. Syst. Autom. Res. Appl.,* Vol. 2, pp. 151–166.

Dong, Z., 1993, Design for Automated Manufacturing, in *Concurrent Engineering: Automation, Tools, and Techniques,* A. Kusiak (ed.), Wiley, New York, pp. 207–234.

Dong, Z., 1994, Automated Generation of Minimum Cost Production Sequence, in *Artificial Intelligence in Optimal Design and Manufacturing,* Z. Dong (ed.), Prentice Hall, Upper Saddle River, NJ, pp. 153–172.

Dong, Z., and Hu, W., 1991, Optimal Process Sequence Identification and Optimal Process Tolerance Assignment in Computer-Aided Process Planning, *Comput. Ind.,* Vol. 17, pp. 19–32.

Dong, Z., and Soom, A., 1986, *Automatic Tolerance Analysis from a CAD Database,* ASME Technical Paper 86-DET-36, American Society of Mechanical Engineers, New York.

Dong, Z., and Soom, A., 1990, Automatic Optimal Tolerance Design for Related Dimension Chains, *Manuf. Rev.,* Vol. 3, No. 4, pp. 262–271.

Dong, Z., and Soom, A., 1991, Some Applications of Artificial Intelligence Techniques to Automatic Tolerance Analysis and Synthesis, in *Artificial Intelligence in Design,* D. T. Pham (ed.), Springer-Verlag, New York, pp. 101–124.

Faires, V., 1965, *Design of Machine Elements,* Macmillan, New York.

Greenwood, W., and Chase, K., 1988, Worst Case Tolerance Analysis with Nonlinear Problem, *J. Eng. Ind.,* Vol. 110, No. 3, pp. 232–235.

Iannuzzi, M., and Sandgren, E., 1994, Optimal Tolerancing: The Link Between Design and Manufacturing Productivity, in *Design Theory and Methodology, DTM'94,* DE-Vol. 68, American Society of Mechanical Engineers, New York, pp. 29–42.

Kapur, K. C., Raman, S., and Pulet, S., 1990, Methodology for Tolerancing Design Using Quality Loss Function, *Comput. Ind. Eng.,* Vol. 19, pp. 254–257.

Kumor, S., and Roman, S., 1992, Computer Aided Tolerancing Past, Present, and Future, *J. Des. Manuf.,* Vol. 2, pp. 29–41.

Kusiak, A. (ed.), 1993, *Concurrent Engineering: Automation, Tools, and Techniques,* Wiley, New York.

Lee, W.-J., and Woo, T. C., 1989, Optimum Selection of Discrete Tolerances, *J. Mech. Transm. Autom. Des.,* Vol. 111, No. 2, pp. 243–251.

Martino, P. M., and Gabriele, G. A., 1989, Application of Variational Geometry to the Analysis of Mechanical Tolerances, in *Advances in Design Automation—1989,* DE-Vol. 19-1, American Society of Mechanical Engineers, New York, pp. 19–28.

Michael, W., and Siddall, J. N., 1981, The Optimization Problem with Optimal Tolerance Assignment and Full Acceptance, *J. Mech. Des.,* Vol. 103, pp. 842–848.

Michael, W., and Siddall, J. N., 1982, The Optimal Tolerance Assignment with Less Than Full Acceptance, *J. Mech. Des.,* Vol. 104, pp. 855–860.

O'Grady, P., and Young, R. E., 1991, Issues in Concurrent Engineering Systems, *J. Des. Manuf.,* Vol. 1, pp. 27–34.

Parkinson, D. B., 1985, Assessment and Optimization of Dimensional Tolerances, *Comput.-Aid. Des.,* Vol. 17, No. 4, pp. 191–199.

Rosenblatt, A., and Watson, G. F., 1991, Concurrent Engineering, *IEEE Spectrum,* Vol. 28, No. 7, pp. 22–37.

Roy, U., Liu, C. R., and Woo, T. C., 1991, Review of Dimensioning and Tolerancing: Representation and Processing, *Comput.-Aid. Des.,* Vol. 23, No. 7, pp. 466–483.

Shah, J. J., 1991, Assessment of Features Technology, *Comput-Aid. Des.,* Vol. 23, No. 5, pp. 331–343.

Shigley, J. E., and Mischke, C. R., 1989, *Mechanical Engineering Design,* McGraw-Hill, New York.

Speckhart, F. H., 1972, Calculation of Tolerance Based on a Minimum Cost Approach, *J. Eng. Ind.,* Vol. 94, No. 2, pp. 447–453.

Spotts, M. F., 1973, Allocation of Tolerances to Minimize Cost of Assembly, *J. Eng. Ind.,* Vol. 95, pp. 762–764.

Sutherland, G. H., and Roth, B., 1975, Mechanism Design: Accounting for Manufacturing Tolerances and Costs in Function Generating Problems, *J. Eng. Ind.,* Vol. 98, pp. 283–286.

Taguchi, G., 1986, *Introduction to Quality Engineering,* Asian Productivity Organization, UNIPUB, White Plains, NY.

Trucks, H. E., 1976, in *Designing for Economical Production,* H. B. Smith (ed.), Society of Manufacturing Engineers, Dearborn, MI.

Turner, J. U., and Wonzy, M. J., 1987, Tolerance in Computer-Aided Geometric Design, *Visual Comput.,* Vol. 3, pp. 214–226.

Turner, J. U., and Wozny, M. J., 1990, The M-Space Theory of Tolerances, in *Advances in Design Automation—1990,* DE-Vol. 23-1, American Society of Mechanical Engineers, New York, pp. 217–226.

VSA 1993, *VSA-3D,* Variation Systems Analysis, Inc., St. Clair Shores, MI.

Wu, Z., Elmaraghy, W. H., and Elmaraghy, H. A., 1988, Evaluation of Cost-Tolerance Algorithms for Design Tolerance Analysis and Synthesis, *Manuf. Rev.,* Vol. 1, No. 3, pp. 168–179.

Xue, D., and Dong, Z., 1993, Feature Modeling Incorporating Tolerance and Production Process for Concurrent Design, *Concurrent Eng. Res. Appl.,* Vol. 1, pp. 107–116.

Xue, D., and Dong, Z., 1994, Developing a Quantitative Intelligent System for Implementing Concurrent Engineering Design, *J. Intell. Manuf.,* Vol. 5, pp. 251–267.

Xue, D., and Dong, Z., 1996, Integrated Concurrent Design Using a Quantitative Intelligent System, in *Integrated Product, Process and Enterprise Design,* B. Wang (ed.), Chapman & Hall, London.

Xue, D., Rousseau, J. H., and Dong, Z., 1995, Joint Optimization of Functional Performance and Production Costs Based upon Mechanical Features and Tolerances, *Proceedings of the 1995 ASME Design Engineering Technical Conferences,* DE-Vol. 83, Vol. 2, American Society of Mechanical Engineers, New York, pp. 1013–1028.

Xue, D., Rousseau, J. H., and Dong, Z., 1996, Joint Optimization of Performance and Costs in Integrated Concurrent Design: The Tolerance Synthesis Part, *J. Eng. Des. Autom.,* Vol. 2, No. 1, pp. 73–90; Special Issue on Tolerancing and Metrology for Precision Manufacturing, C. Zhang and B. Wang (eds.).

Zeid, I., 1993, *CAD/CAM Theory and Practice,* McGraw-Hill, New York.

Zhang, H. C., and Huo, M. E., 1992, Tolerance Techniques: The State-of-the-Art, *Int. J. Prod. Res.,* Vol. 30, No. 9, pp. 2111–2135.

Zhang, C., and Wang, H. P., 1993, The Discrete Tolerance Optimization Problem, *Manuf. Rev.,* Vol. 6, No. 1, pp. 60–71.

10

OPTIMAL TOLERANCE DESIGN FOR INTEGRATED DESIGN, MANUFACTURING, AND INSPECTION WITH GENETIC ALGORITHMS

SHUI-SHUN LIN

National Chinyi Institute of Technology
Taichung, Taiwan

HSU-PIN (BEN) WANG and CHUN (CHUCK) ZHANG

Florida A&M University/Florida State University
Tallahassee, Florida

10.1 INTRODUCTION

A substantial amount of research has been carried out regarding optimal tolerance allocation using cost–tolerance functions. Since design tolerancing has a great impact on product cost, much attention has been given to design tolerancing research. Various functions have been proposed to describe the cost–tolerance relationship, and various optimization methods have been applied.

Speckhart (1972) presented an exponential cost–tolerance model, and design tolerancing was formulated as an optimization problem. The same solution approach was used by Spotts (1973), in which a squared reciprocal cost–tolerance model was established and was solved by Lagrange multipliers. Both worst-case and simple statistical approaches were used in their work.

Advanced Tolerancing Techniques, Edited by Hong-Chao Zhang
ISBN 0-471-14594-7   © 1997 John Wiley & Sons, Inc.

Ostwald and Huang, (1977) introduced a model for optimal selection of tolerances for functional dimensions assuming discrete production costs. This model was solved by using linear programming with 0–1 variables.

Hauglund (1987) formulated tolerance design as an optimization problem by combining manufacturing process selection based on a least-manufacturing-cost criterion. The emphasis was on solution techniques of optimization and no general tolerance calculation methods were developed.

A probabilistic optimization approach similar to that used by Parkinson (1982) was introduced by Lee and Woo (1990). In their work, the tolerancing problem was formulated as a probabilistic optimization method and was further simplified into a deterministic nonlinear programming problem. An algorithm was developed and was proven to converge to the global optimum through an investigation of the monotonic behavior among tolerance, the reliability index, and cost. Chase et al. (1990) introduced a procedure for tolerance specification based on quantitative estimates of the cost from tolerance information, which permitted the selection of processes and component tolerances in mechanical assemblies for minimum production cost.

Most of the recent methods for optimal tolerance design have been developed for a single-dimension chain. Multiple-dimension chains sharing common design dimensions and tolerances cannot be handled readily. Applying contemporary optimal tolerancing methods to related dimension chains may lead to design conflicts. To solve the optimal tolerance design problems efficiently with multiple related dimension chains, Dong and Soom (1990) introduced an optimization model coupled with a solution procedure. In their work, a modified exponential cost–tolerance function was proposed to provide more flexibility in modeling. A method for generating the objective function of the optimal tolerance design automatically using the information provided by a computer-aided design (CAD) system was also developed.

Zhang and Wang (1993) introduced a solution procedure to solve the mixed discrete nonlinear optimization problems. A simulated annealing algorithm was modified and applied in their work. Zhang (1993) applied the simulated annealing algorithm to solve the cost–tolerance model for concurrent optimization of design and manufacturing tolerances.

Current analytical tolerance synthesis methods provide industry with data adequate enough for experienced engineers to plan a product's design. Tolerance design and optimization efficiency are two key issues. In this paper, two major aspects of the optimal tolerance design are presented: the optimization of integrated tolerance synthesis and the solution procedures.

10.2   PROBLEM STATEMENT

A product is composed of a number of component parts. The manufacturing cost of a product consists of the costs of fabricating those parts and others. Tolerance, including design tolerance (the final component's tolerance), manufacturing tolerance (intermediate component tolerance during fabrication), and measurement tolerance (inspection tolerance), is one of the most important parameters that affect product quality and cost. Rather than the current methods of tolerance allocation, a reliable analytical model for concurrent tolerance optimization will greatly improve and expand the foundations of design and manufacturing for computer-integrated manufacturing.

In general design practice, final assembly specifications are usually derived from customer requirements. These specifications are translated to product design parameters such as the mean and standard deviation of an assembly dimension, the defective rate (the percentage of product out of specification), and so on. From a design viewpoint, an assembly dimension is determined by a set of component dimensions and its tolerance is stacked up by the tolerances on those components. From a manufacturing viewpoint, a component dimension may require several stages of machining. At each stage, one process will be chosen for the component fabrication from a number of alternative machines that have different characteristics (accuracy and capability). For a specific process, the relationship among cost, tolerance, and process capability for part fabrication is shown in Figure 10.1. In the figure, *Cpl* and *Cpu* are the lower and upper bounds of process capability, respectively; *Tl* and *Tu* are the lower and upper bounds of tolerance, respectively; and *C* is manufacturing cost. The tighter the tolerance and the better the process capability, the higher the manufacturing cost. In addition, the capability of a process varies depending on machine accuracy, setups, and related factors.

The problem is to determine the optimal combination of processes and their parameters, which include process capability and processing (manufacturing) tolerance. By solving this problem, design and manufacturing toler-

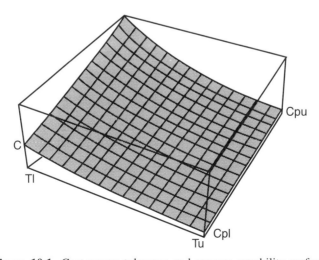

Figure 10.1. Cost versus tolerance and process capability surface.

ances are optimized and inspection tolerances are considered. Optimality is measured by the total manufacturing cost of the product.

10.3   TOLERANCE–COST RELATIONSHIP

10.3.1   Process Capability

Manufacturing processes undergo several stages of development prior to actual production. The stages of prove-out testing at the machine supplier, initial testing at the manufacturing facility, and preproduction testing all seek to determine whether machinery can produce, on a continuous basis, production units that meet the engineering specifications required. A widely used criterion to assess process capability is the process capability index (C_p):

$$C_p = \frac{\text{allowable process spread}}{\text{actual process spread}} = \frac{\text{USL} - \text{LSL}}{\text{NT}} = \frac{\delta}{3\sigma} \qquad (10.1)$$

where USL = upper specification limit
LSL = lower specification limit
σ = standard deviation of the process
NT = natural tolerance (6σ)
δ = dimensional semitolerance

This formulation implies that the confidence level is $\pm 3\sigma$ and is 99.73% by default. The capability of a process depends on a number of factors, such as machine capability, jigs and fixtures, skill level of the workers, maintenance, and so on. A high level of process capability, in conjunction with a stringent quality assurance system, will result in consistent production and quality products, thereby reducing the costs of scrap and rework of substandard products, wasted materials, and labor hours.

Another process capability index is C_{pk}, which is defined by

$$C_{pk} = \min \left(\frac{\text{USL} - \mu}{3\sigma}, \frac{\mu - \text{LSL}}{3\sigma} \right) \qquad (10.2)$$

where μ is the process mean value. C_{pk} is a tool to detect process mean shift. If the process mean is well controlled (i.e., without any mean shift), $C_p = C_{pk}$. In this research the process mean shift is assured to be zero, and therefore $C_p = C_{pk}$.

Traditionally, a C_p of 1.0 indicates that a process is "capable." However, as time changes, the customers demand higher levels of quality at a lower cost. Thus a larger C_p value is required as acceptable. For example, Motorola's 6σ philosophy (Harry and Stewart, 1988) enforces process with a C_p of 2.0. Figure 10.2 shows how different C_p values relate to the spread of a process relative to the specification width.

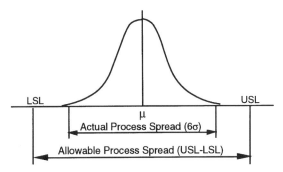

Figure 10.2. Relationship among process capability parameters.

10.3.2   Cost Versus Tolerance and Process Capability Model

Manufacturing cost, tolerance, and process capability are interrelated. As for the relationship between cost and tolerance, a decreasing monotonic curve is well recognized (i.e., the tighter the tolerance, the higher the manufacturing cost). It should be noted that this relationship exists under a certain level of process capability. As for the relationship between cost and process capability, a similar increasing monotonic relationship exists (i.e., the larger the C_p, the higher the manufacturing cost). This is because better process capability can be achieved only through substantial investment in plant, machinery, equipment, and employee training, which will increase the production overhead and result in higher unit costs of production.

Considering the effects of both tolerance and process capability on manufacturing cost, a cost versus tolerance and process capability model is formulated as follows:

$$g(\delta, C_p) = (ae^{-b(\delta - \delta_0)} + c)C_p^d \tag{10.3}$$

In the model, a, b, c, d, and δ_0 are nonnegative constants associated with a specific process. These parameters can be obtained through experimental or empirical data.

To check the feasibility of the model, let us consider the following two cases. In the model, when the process capability is fixed (i.e., $C_p = p_1$) the following decreasing monotonic relationship exists between cost and tolerance:

$$g_1(\delta) = ap_1^d e^{-b(\delta - \delta_0)} + cp_1^d \tag{10.4}$$
$$= a_1 e^{-b(\delta - \sigma_0)} + c_1$$

where $a_1 = ap_1^d$ and $c_1 = cp_1^d$ are constants. Similarly, when the tolerance is set to a certain value p_2, an increasing monotonic relationship exists between cost and process capability as follows:

$$g_2(C_p) = (ae^{-b(p_2-\delta_0)} + c)C_p^d$$
$$- a_2 C_p^d \tag{10.5}$$

where $a_2 = ae^{-b(p_2-\delta_0)} + c$ is a constant. The analyses above show that equation (10.3) is consistent with an understanding of the relationships among cost, tolerance, and process capability.

10.4 OPTIMIZATION MODEL

10.4.1 Constraints

Constraints imposed on the concurrent tolerance optimization model come from design requirements of critical elements, stock removal limitation, and process selection. Measurement errors and the attendant risks of incorrect decisions are also taken into account. They are described in the following sections.

Design Tolerance Constraint. Tolerances on some critical elements directly affect the functionality of a product. Thus tolerances on these design dimensions are important parameters in a product's design. For a product or an assembly, a design dimension may be affected by a set of interrelated dimensions. These dimensions form a closed loop, often called a *dimension chain*. The dimension that is obtained indirectly is called the *resultant dimension* and others are called *constituting* (or *component*) *dimensions* in a dimension chain. One example of resultant design elements is the clearance between the piston and the cylinder bore in an engine assembly.

Considering the worst case of tolerance stackup, regardless of the final measurement error, the tolerance on the resultant dimension is the summation of those on constituting dimensions. It can be described as follows:

$$\sum_{i=1}^{n} \delta_i = \delta_\Sigma \tag{10.6}$$

where δ_i's are constituting tolerances in the dimension chain of a resultant tolerance δ_Σ.

In the worst-case method, tolerance calculations are conducted based on the extreme conditions of a production error distribution. This approach is mathematically simple but may lead to unreasonably tight tolerances, resulting in unnecessarily high-manufacturing costs.

Since the dimensional variations of manufactured mechanical parts (other than single piece production) naturally follow statistical distributions, the chance of a part being manufactured at the extreme dimensions is usually small. Statistical tolerancing methods make more sense as far as economics is concerned.

In this study an assumption is made that each tolerance is tied to a normal random variable and each is independent of the others.[1] Based on this assumption, the following relationship exists:

$$\sigma_\Sigma^2 = \sigma_1^2 + \sigma_2^2 + \cdots + \sigma_n^2 \qquad (10.7)$$

where σ_Σ is the standard deviation of the resultant dimension and σ_i is the standard deviation of the ith component dimension. From equation (10.1),

$$\sigma = \frac{\delta}{3C_p} \qquad (10.8)$$

Substituting equation (10.4), the variance of the resultant tolerance can be expressed as follows:

$$\sigma_\Sigma^2 = \left(\frac{\delta_1}{3C_{p1}}\right)^2 + \left(\frac{\delta_2}{3C_{p2}}\right)^2 + \cdots + \left(\frac{\delta_n}{3C_{pn}}\right)^2 \qquad (10.9)$$

where δ_i is the design tolerance for the ith component and C_{pi} is the capability of the process used to fabricate this component.

Considering the process selection effects, the design tolerance constraint can be represented as

$$\sum_{i=1}^{n} \sum_{k=1}^{p_i} X_{ik} \left(\frac{\delta_{ik}}{3C_{pik}}\right)^2 \le \left(\frac{\delta_R}{3C_{pR}}\right)^2 \qquad (10.10)$$

where p_i = number of alternative machines for fabricating the ith component
X_{ik} = process selection coefficient
1, if process k is selected to produce the ith component;
0, otherwise
δ_{ik} = tolerance of the ith component machined by process k
δ_R = required resultant tolerance
C_{pR} = pseudo process capability for the resultant tolerance [it can be calculated from the required acceptance rate (a known parameter) of the resultant tolerance]

Stock Removal Constraint. In manufacturing tolerance allocation, consideration should be given not only to process accuracy bounds, but also to the amount of stock removal. The stock removal is the layer of material that is to be removed from the surface of a workpiece in manufacturing in order to

---

[1]In some cases other types of assumptions are made: uniform, triangular, beta, etc.; however, the most commonly applied model is the normal distribution. Because of this, the normal distribution is used in this study.

obtain the required dimension, accuracy, and surface quality. The determination of stock removal greatly influences the quality and the production efficiency of a machined part. Excessive stock removal will increase the consumption of material, machining time, tool, and power, and subquentially increase the manufacturing cost. On the other hand, if there is insufficient stock removal, the surface roughness and defective surface layer caused by the preceding process cannot be removed completely from the workpiece surface. Thus it will influence the surface quality of the part (Wang and Li, 1991).

In part machining, appropriate stock removal should be provided for each process operation. To make manufacturing processes more effective and efficient, the standard deviation of the stock removal (σ_z) for each process should be controlled under a certain value. In a manufacturing environment implementing statistical process control, the data for σ_z are available for each process. The amount of stock removal is the difference between the machining dimension obtained in the preceding operation and that in the current operation. Since the machining dimensions were not fixed and each of them had a tolerance range, the actual stock removals taken from workpiece surfaces varied. The stock removal was its nominal value. The variation of stock removal is summed by the dimensional variations of the current operation and the preceding operation. From a statistical viewpoint, the following relationship exists:

$$\sigma_{z_i}^2 = \sigma_{\delta_i}^2 + \sigma_{\delta_{i-1}}^2 \tag{10.11}$$

where σ_{δ_i} is the standard deviation of the manufacturing dimension in the ith operation.

From equations (10.8) and (10.11),

$$\sigma_{z_i}^2 = \left(\frac{\delta_i}{3C_{p_i}}\right)^2 + \left(\frac{\delta_{i-1}}{3C_{p_{i-1}}}\right)^2 \tag{10.12}$$

By giving the standard deviation of the stock removal for a process, the stock removal constraint can be expressed as follows:

$$\left(\frac{\delta_i}{3C_{p_i}}\right)^2 + \left(\frac{\delta_{i-1}}{3C_{p_{i-1}}}\right)^2 \leq \sigma_{z_i}^2 \tag{10.13}$$

where σ_{z_i} is the standard deviation of the stock removal for process i.

In addition to design tolerance and stock removal, the process selection and the limitations on tolerance and capability for a process form the necessary constraints for the optimization model. They are expressed as follows. For process selection:

$$\sum_{k=1}^{p_{ij}} X_{ijk} = 1 \qquad i = 1,...,n, \quad j = 1,...,m \tag{10.14}$$

where $X_{ijk} = 1$ if process k is chosen to produce the jth manufacturing tolerance on component i; 0, otherwise. This constraint set ensures that one and only one process is selected for each manufacturing tolerance. Process bounds are

$$\delta_{ijk}^l \le \delta_{ijk} \le \delta_{ijk}^u \tag{10.15}$$

$$\sigma_{ijk}^l \le \frac{\delta_{ijk}}{3C_{P_{ijk}}} \le \sigma_{ijk}^u \tag{10.16}$$

10.4.2   Measurement Tolerance Consideration

All measurements are subject to uncertainty. Uncertainty of measurement can be defined as the interval where the unknown difference between the true value of the feature measured and the measured value can be found. In dimensional metrology, different standards recommend ratios of 4:1 to 10:1 between the tolerance to be measured and the uncertainty of the measurement (Nielsen, 1992). The actual variation on a finished dimension includes the variation incurred by the manufacturing process used to produce the dimension and the measurement error. To give the customer the confidence of receiving only functionally acceptable parts, both process variation and measurement error should be considered in design and manufacturing tolerance allocation.

Measurement tolerance on a finished part affects the specification of product design. From the viewpoint of quality control, it also influences the inspection strategy. Hahn (1982) discussed the problem of a product satisfying a lower specification limit when the data were subject to measurement error. It is assumed that the measurement errors are normally distributed with variance σ_M^2 and that the measurement errors are distributed independently as $N(0,\sigma_M^2)$. Furthermore, only finished components are inspected. No inspection is made for intermediate processes of components fabrication, although the proposed model is expandable to include intermediate inspections. To satisfy the customers and reduce the rejects, the manufacturer needs to take the measurement errors into consideration. Easterling et al. (1991) proposed rational approaches to determining measurement tolerances and controlling the attendant risks. Given a measurement variance (σ_M^2) and process variance (σ_P^2), the total variance can be obtained as follows:

$$\sigma_C^2 = \sigma_P^2 + \sigma_M^2 \tag{10.17}$$

where σ_C is the standard deviation of a component dimension.

Mee (1984) introduced several methods for computing bounds for the proportion of product in conformance with a given specification limit when the observed data were subject to measurement errors. There are two possible ways, which are based on the availability of σ_M^2, to find the measurement bounds. He also addressed the issues for both known σ_M^2 and unknown σ_M^2, where repeated measurements were taken to obtain a sample estimate for the unknown σ_M^2.

10.4.3   Model

Based on the analyses above, a mathematical model for concurrent tolerance optimization is proposed as follows:

$$\text{minimize } G = \text{minimize } \sum_{i=1}^{n} \sum_{j=1}^{m_i} \sum_{k=1}^{ij} X_{ijk} g_{ijkijk}(\delta_{ijk}, C_{pijk}) \qquad (10.18)$$

subject to

$$\sum_{i=1}^{n} \left\{ \left[\sum_{k=1}^{Pim_i} X_{im_ik} \left(\frac{\delta_{im_jk}}{3C_{Pim_ik}} \right)^2 \right] + \sigma_M^2 \right\} \leq \left(\frac{\delta_R}{3C_{PR}} \right)^2$$

$$\left(\sum_{k=1}^{p_{ij}} X_{ijk} \frac{\delta_{ijk}}{3C_{Pijk}} \right)^2 + \left(\sum_{k=1}^{p_{ij}} X_{i(j-1)k} \frac{\delta_{i(j-1)k}}{3C_{Pi(j-1)k}} \right)^2 \leq \sigma_{Z_{ij}}^2$$

$$\sum_{k=1}^{p_{ij}} X_{ijk} = 1$$

$$\delta_{ijk}^l \leq \sigma_{ijk} \leq \delta_{ijk}^u$$

$$\sigma_{ijk}^l \leq \frac{\delta_{ijk}}{3C_{Pijk}} \leq \sigma_{ijk}^u$$

$$C_{Pijk} \geq 0$$

where n = number of components in the assembly
m_i = number of operations (stages) required for producing component i
p_{ij} = number of alternative processes (machines) for the jth operation to produce component i
G = total manufacturing cost of the assembly
g_{ijk} = manufacturing cost of utilizing the kth alternative machine in the jth operation for producing component i
C_{Pijk} = capability of process k in the jth operation for producing component i

C_{PR} = pseudo process capability for the resultant tolerance; it can be calculated from the required acceptance rate (a known parameter) of the resultant tolerance

σ_M = standard deviation of measurement error

$\sigma_{Z_{ij}}$ = standard deviation of stock removal in the jth operation for producing component i

σ^l_{ijk} = lower bound of standard deviation of process k in the operation to produce component i

σ^u_{ijk} = upper bound of standard deviation of process k in the operation to produce component i

δ_R = required resultant tolerance

δ_{ijk} = tolerance of process k in the jth operation for producing component i

δ^l_{ijk} = lower bound of tolerance of process k in the jth operation to produce component i

δ^u_{ijk} = upper bound of tolerance of process k in the jth operation to produce component i

X_{ijk} = if process k is chosen for the jth operation to produce component i; 0, otherwise

A multivariate nonlinear discrete optimization model is proposed for the concurrent tolerance optimization problem. The objective is to minimize the total manufacturing cost by optimally selecting the available processes and their parameters, which include process capability and manufacturing tolerance. Meanwhile, the product assemblability and yield are ensured.

10.5 SOLUTION PROCEDURE

10.5.1 Genetic Algorithms

Genetic algorithms (GAs) are a class of search procedures based on the mechanics of natural genetics and natural selection. GAs are different from traditional search methods encountered in engineering optimization problems in the following ways: GAs work with a coding of the design variables, as opposed to the variables themselves—continuity of parameter space is not a requirement as GAs search from a population of points, not a single point—parallel processing of points reduces the chance of being trapped into a local optimum; GAs use probabilistic transition rules, not deterministic transition rules, which leads to high-quality solutions; and GAs require only the objective function values—minimal requirements broaden GAs' application.

In most GAs, binary-coded chromosomal strings of ones (1's) and zeros (0's) are used to describe the values for each solution. In a multiple-design-variable optimization problem, the individual variable coding is usually concatenated into a complete string. To decode a string, bit strings with specified string length are extracted successively from the concatenated string and the substrings are then decoded and mapped into the value in the corresponding search space.

There are three main GA operators: reproduction, crossover, and mutation. The reproduction operator allows the highly productive chromosomes (strings) to live and produce offspring in the next generation. The crossover operator, used with a specified probability, exchanges genetic information by splitting two chromosomes at a random site and joining the first part of one chromosome with the second part of another chromosome. Mutation introduces occasional changes of a random string position with a specified mutation probability.

The general procedure for genetic algorithms in optimization problems is described as follows:

1. An appropriate chromosome representation should be defined to represent the combinations of design variables that correspond to the fitness or objective function values. To have a normal coding and decoding process, the representation should be a one-to-one mapping.
2. The probabilities of crossover and mutation are specified. The population size and maximum number of generation are selected. An initial population in the genetic system is generated.
3. Evaluate the objective function value or fitness of each solution in the current generation.
4. Apply the three operators (reproduction, crossover, and mutation) to those solutions in the current generation to produce a new population.
5. Repeat steps 3 and 4 until the maximum number of generations is reached.

10.5.2 GAs for Mixed-Discrete Nonlinear Optimization Model

A combinatorial nature exists in the cost–tolerance–process capability optimization model since discrete variables are taken into considerations; the solution space may be much more complicated than that of the nonlinear problem, with solely continuous variables and the existence of many local minima in the solution space. Significant modifications of the traditional genetic algorithms are made to solve mixed-discrete nonlinear optimization problems. These modifications are made to the five components of genetic algorithms: variable representation, genetic parameters, initial population, genetic operators, and evaluation function. In the following sections we discuss the implementation of genetic algorithms as well as modifications made for mixed-discrete nonlinear optimization problems.

Variable Representation. Since various types of variables may be involved in a mixed-discrete optimization problem, some modifications and treatment are necessary in the traditional genetic algorithms. The hybrid string representation is used in this research. The bit-string representation in hybrid GAs uses 0 to 9 for each bit position instead of 0 and 1. This reduces the chromosome length dramatically and simplifies the string processing, which benefits multiple-design-variable problems.

The variable encoding and decoding scheme has been modified to represent both discrete and continuous variables. In genetic algorithms, the chromosome length should be defined first. For each variable, its substring length m is calculated based on its lower and upper bound and the variable type. m is assigned the smallest integer that satisfies the following relationship:

$$10^m \geq \frac{U - L}{R} + 1 \tag{10.19}$$

where U and L are the upper and lower bounds of the search interval and R is the desired resolution for a variable. R is set to unity for discrete variables. It can be seen that the finer the resolution of R, the larger the value of m.

For an integer variable, X_i, the encoded substring is decoded directly and transferred into the search space by using the equation

$$X_i = M(X, U - L) + L \tag{10.20}$$

where X is a decoded base 10 integer, $U - L$ is the search interval, and $M(X, U - L)$ denotes the remainder of the decoded integer divided by the interval of search space. The decoded integer is transferred and mapped into the search space by using equation (10.20). This method occasionally results in an uneven distribution, where some numbers in the design space obtain more representations and others less. One remedy is the *excessive-and-discarded scheme*, which simply discards those numbers within the range of a nonintact representation space.

For binary (zero–one) variables, the encoding and decoding scheme is identical to that for integer variables except that the upper bound is always 1 and the lower bound is always 0. To represent a continuous variable, a numerical operation is applied to decode a substring:

$$X \frac{U - L}{10^m - 1} + L \tag{10.21}$$

where X is a base 10 integer, U and L are the upper and lower bounds of the search interval, and m is the substring length.

Genetic Parameters. The definition of the population pool includes a specification of the total number of strings in the population, the length of each string, and the maximum number of generations. Additional parameters are probabilities of crossover and mutation in the evolution process. Studies conducted by De Jong (1975) and Grefenstette (1986) (see Table 10.1) can be used as guidelines for choosing the GA parameters.

Initial Population. The initial population is randomly generated, with the restriction that no two individuals are allowed to have the same chromosome, in order to get maximal variety. Each chromosome is checked by a validation procedure (see Section 10.5.3) which makes sure that only the feasible solution can be placed in the initial population.

Genetic Operators. The basis of genetic algorithms is their operators. A more comprehensive understanding of the three basic transformations corresponding to reproduction, crossover, and mutation may be obtained by describing a simplistic implementation of each.

Reproduction. There are many ways to achieve effective reproduction (Goldberg, 1989; Freeman et al., 1990). In this research the roulette-wheel selection is employed. The better-fitted chromosomes obtain larger slots on the wheel and have higher probabilities of being selected for producing in the next generation. Another strategy, the *steady-state-without-duplicates scheme,* is also used in this research. This scheme works by replacing only a certain number of chromosomes at a time rather than all individuals in the population. This scheme also ensures that each chromosome is distinctive. In other words, this scheme is processed by copying several better-fitted chromosomes to the next generation and generating the rest from an old population. This prevents the potential source of loss by GA operators when the generational reproduction is used.

Crossover. For a multiple-variable optimization problem, a concatenated string (chromosome) is produced, and with the aid of the string representation scheme, each substring represents a variable. Many different crossover techniques were reported (Goldberg, 1989; Syswerda, 1989). Among those crossover techniques, *one-point crossover* is adapted in this research. This scheme randomly chooses a locus (crossover site), which applies to both chromosomes when performing a crossover operation. This operation results in two

TABLE 10.1. Genetic Algorithm Parameters

	Population Size	Crossover Rate	Mutation Rate
De Jong	50–100	0.6	0.001
Grefenstette	30	0.95	0.01

new chromosomes by cutting two chromosomes at the crossover site and connecting the first part of one chromosome to the second part of another chromosome. This operation produces new values for a certain number of variables. The number of variables whose values change is dependent on the crossover site. In terms of an optimization process, this operator takes a solution point to a new position and forms a new design solution.

Mutation. Mutation is performed on a single gene with a mutation probability when a new chromosome is formed. In hybrid GA, he allele may be one of any integer between zero (0) and nine (9). When the mutation operation is applied to the gene, its allele is changed to a randomly determined number other than itself.

Evaluation Function. The evaluation function needed to obtain a chromosome's fitness is the objective function of the optimization problem. GAs are normally used to solve optimization problems with positive objective function values if roulette-wheel selection is used. In this study the fitness techniques have been altered to solve all-range objective function optimization problems. This is done by obtaining their fitness, with the aid of a normalization process, and computing the probability of each string for reproduction. The normalization process is applied to handle both minimization problems and negative objective function values. The normalization process is indicated by the following relationship:

$$f'(x_i) = f_{max} - f(x_i) \qquad (10.22)$$

where $f(x_i)$ is the objective function value with design vector x_i, f_{max} the maximal objective function value, $f'(x_i)$ the normalized fitness value with design vector x_i, and i the string index. In other words, the fitness is altered by subtracting the fitness from the generational maximum fitness. This normalization process also achieves the windowing effect, which gives the least-fitted member no chance for reproduction.

10.5.3   Treatment of Constraints

Most engineering optimization problems are constrained. For implementing the genetic algorithms, two techniques are often used to handle constraints. One is to restrict the solution space to solutions that conform to the constraints. The other is to allow solutions that violate the constraints at the expense of a suitably defined penalty function. In this study the former is called a variable restriction method and the latter, a penalty function method. In the *variable restriction method,* a new chromosome is generated and the solutions decoded from that chromosome are checked against the constraints. If the constraints are satisfied, the new chromosome is accepted and is placed into the population pool for the next generation; otherwise, the chromosome is discarded and a new chromosome is generated and checked. Therefore, some of the chromosomes generated are unusable and thus a portion of the computational time is not productive.

In the *penalty function method,* all chromosomes are used. The acceptance of a particular chromosome, however, depends on the magnitude of its fitness. Chromosomes that violate constraints are expected to be rejected by its penalized fitness. The larger the violation of the constraints, the higher the probability of the chromosome being rejected. This penalty function approach is likely to lead to a simpler reproduction process. However, caution must be exercised in the selection of the penalty function. A poorly defined penalty function leads to a worse final solution or makes the solution infeasible. Experience and some computational experiments are needed to find an appropriate penalty function for a specific problem. In this study, the variable restriction method is employed for its simplicity.

A validation procedure using the variable restriction method is performed for each chromosome generated before placing it in the solution population. This validation procedure ensures feasible solutions and optimal design.

10.6   EXAMPLE

In this section an example (see Figure 10.3) taken from the literature (Zhang et al., 1992) is provided to illustrate the integrated tolerance optimization model and genetic algorithm solution procedure. The example problem is to

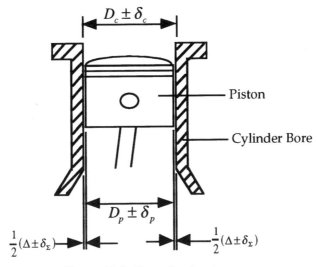

Figure 10.3. Example of a clearance.

determine the optimal design and manufacturing tolerances for an engine assembly fabrication. The assembly is composed of a piston and a cylinder bore. A description of the problem is given in Table 10.2. The design specification of this assembly is that the clearance between the piston and cylinder bore should be kept within 0.0022 ± 0.0008 in. The tolerances under study are the major diameter tolerances for the piston and cylinder bore.

The number of components in the assembly is $n = 2$ and the number of manufacturing stages for these two components are $m_1 = 4$ and $m_2 = 4$, respectively.

Results and Discussions. This example problem was solved by using a genetic algorithm over a number of runs. A tolerance allocation result is shown in Table 10.3. Table 10.4 shows a tolerance allocation result obtained by experience-based calculation.

Table 10.3 shows the result of design tolerance synthesis and manufacturing tolerance allocation for the piston and cylinder bore with the GA optimization procedure. For instance, for piston fabrication, machine 1 is selected at stage 1, with manufacturing tolerance allocated to 0.04 and process capability 1.40, while at stage 4, machine 3 is selected with manufacturing tolerance allocated to 0.00057 and process capability 1.91. The optimal design tolerance on the piston diameter is 0.00057, while the optimal design tolerance on the cylinder bore is 0.00059. The minimum manufacturing cost is $500.75.

Table 10.4 shows the results from experience-based calculation, which normally utilizes the tolerance charting procedure. As indicated before, tolerance charting and calculation was a general way to allocate manufacturing tolerances before optimization methods were introduced. The experience-based calculation method results in a tolerance of 0.00058 on the piston diameter and 0.00061 on the cylinder bore. The total cost obtained by this method is $682.97.

It can be seen from the results that the GA optimization method leads to a significant manufacturing cost reduction. A 26.68% reduction in manufac-

TABLE 10.2. Description of the Example Problem

Piston		Cylinder Bore	
Process Route	Number of Alternative Machines	Process Route	Number of Alternative Machines
Rough turn	2	Drill	2
Finish turn	3	Bore	3
Rough grind	3	Finish bore	3
Finish grind	3	Grind	2

TABLE 10.3. Optimal Tolerance Allocation Result [δ_{ijk} and (C_{pijk})] for the Example Problem

Manuf.	Piston: Alternative Machine			Cylinder bore: Alternative Machine		
Stage	1	2	3	1	2	3
1	0.04 (1.40)	—	N.A.[a]*	0.04 (1.00)	—	N.A.
2	0.01 (1.20)	—	—	0.0147 (1.00)	—	—
3	0.00379 (1.26)	—	—	—	—	0.00293 (1.00)
4	—	—	0.00057 (1.91)	0.00059 (1.70)	—	N.A.
Optimal design tolerance:	0.00057			0.00059		
Cost:	$500.75					

[a] N.A., not applicable.

turing cost can be achieved by employing the optimization method over the manual calculation method. Three optimization runs are performed and the objective function values (i.e., manufacturing costs) are shown in Table 10.5. It can be seen from the computational result that the minimum objective values obtained through the three runs are very close. The genetic algorithms

TABLE 10.4. Experience-Based Tolerance Allocation Result [δ_{ijk} and (C_{pijk})] for the Example Problem

Manuf. Stage	Piston: Selected Machine 1	Cylinder Bore: Selected Machine 1
1	0.02705 (1.00)	0.01890 (1.00)
2	0.00604 (1.00)	0.08228 (1.00)
3	0.00391 (1.00)	0.00297 (1.00)
4	0.00058 (1.00)	0.00061 (1.00)
Design tolance:	0.00058	0.00061
Cost:	$682.97	

TABLE 10.5. Minimum Costs Obtained in Experiments

	Experiment		
	1	2	3
Minimum cost	$500.75	$500.38	$501.53

appear to be robust in solving the nonlinear cost–tolerance–process capability model.

10.7 CONCLUSIONS

An optimization model has been developed and solved for the process selection and tolerance allocation problem. As indicated, a substantial amount of research has been carried out regarding optimal tolerance allocation using cost–tolerance functions. The cost–tolerance–process capability model presented in this study employs nonlinear programming techniques and was solved by GAs. Through this model, not only the optimal tolerance for each component dimension, but also the optimal process capability can be found. Furthermore, the optimal process selection for each manufacturing process can be allocated.

Several conclusions evolve from this study:

- The optimization methodology was able to evaluate alternative processes and to allocate optimal manufacturing tolerances and process capabilities. The objective was to achieve the most economical manufacturing with ensured manufacturability. The optimal manufacturing tolerances and process capabilities could be used as a guideline for selecting machine tool, fixture, tool, and setup methods.
- The advantage of the optimal tolerance allocation method over the manual approach is obvious in cost savings. A significant cost reduction can be achieved if the optimization method is employed.
- The results obtained from the genetic algorithm optimization method are independent of the initial solutions in the population. It provides a robust way to solve tolerance allocation problems.
- The cost–tolerance–process capability model plays a key role in the optimum tolerancing problem. Although it can be obtained from the historical production data or derived from cost prediction for new processes, the cost model may be different for various manufacturing processes. More research work is needed to develop a systematic approach to the cost modeling.

• This study can be used as a quantitative tool in design for manufacturability and concurrent engineering.

REFERENCES

Chase, K. W., Greenwood, W. H., Loosli, B. G., and Hauglund, L. F., 1990, Least Cost Tolerance Allocation for Mechanical Assemblies with Automated Process Selection, *Manuf. Rev.,* Vol. 3, No. 1, March, pp. 49–59.

De Jong, K. A., 1975, An Analysis of the Behavior of a Genetic Adaptive System, Dissertation Abstracts International, University Microfilms, Michigan, Vol. 41, No. 9, p. 3503B.

Dong, Z., and Soom, A., 1990, Automatic Optimal Tolerance Design for Related Dimension Chains, *Manuf. Rev.,* Vol. 3, No. 4, December, pp. 262–271.

Easterling, R. G., Johnson, M. E., Bement, T. R., and Nachtsheim, C. J., 1991, Statistical Tolerancing Based on Consumer's Risk Considerations, *J. Qual. Technol.,* Vol. 23, pp. 1–23.

Freeman, L. M., Krishnakumar, K., Karr, C. L., and Meredith, D. L., 1990, Tuning Fuzzy Logic Controllers Using Genetic Algorithms: Aerospace Applications, *Proceedings of the Aerospace Applications on Artificial Intelligence Conference,* Dayton, OH, October.

Goldberg, D. E., 1989, *Genetic Algorithm in Search, Optimization, and Machine Learning,* Addison-Wesley, Reading, MA.

Grefenstette, J. J., 1986, Optimization of Control Parameters for Genetic Algorithms, *IEEE Trans. Syst. Man Cybernet.,* Vol. 16, No. 1, pp. 122–128.

Hahn, G. J., 1982, Removing Measurement Error in Assessing Conformance to Specifications, *J. Qual. Technol.,* Vol. 14, pp. 117–121.

Harry, M. J., and Stewart, R., 1988, *Six Sigma Mechanical Design Tolerancing,* Schaumburg, IL. Motorola, Inc.

Hauglund, L. F., 1987, Combining Manufacturing Process Selection and Optimization, ADCATS Report 87-1, M.S. thesis, Brigham Young University, Provo, UT, January.

Lee, W., and Woo, T. C., 1990, Tolerances: Their Analysis and Synthesis, *J. Eng. Ind.,* Vol. 112, May, pp. 113–121.

Mee, R. W., 1984, Tolerance Limits and Bounds for Proportions Based on Data Subject to Measurement Error, *J. Qual. Technol.,* Vol. 16, pp. 74–79.

Nielsen, H. S., 1992, Uncertainty and Dimensional Tolerances, *Quality,* May, pp. 25–29.

Ostwald, P. F., and Huang, J., 1977, A Method for Optimal Tolerance Selection, *J. Eng. Ind.,* Vol. 99, August, pp. 558–565.

Parkinson, D. B., 1982, The Application of Reliability Methods to Tolerancing, *J. Mech. Des.,* Vol. 104, pp. 612–618.

Speckhart, F. H., 1972, Calculation of Tolerance Based on a Minimum Cost Approach, *J. Eng. Ind.,* Vol. 94, May, pp. 447–453.

Spotts, M. F., 1973, Allocation of Tolerances to Minimize Cost of Assembly, *J. Eng. Ind.,* Vol. 95, August, pp. 762–764.

Syswerda, G., 1989, Uniform Crossover in Genetic Algorithms, *Proceedings of the Third International Conference on Genetic Algorithms,* San Mateo, CA, J. David Schaffer (ed.), Morgan Kaufmann, San Francisco.

Wang, H. P., and Li, J. K., 1991, *Computer-Aided Process Planning,* Elsevier, New York.

Zhang, C., 1993, Tolerance Analysis and Synthesis for Design and Manufacturing, Ph.D. dissertation, The University of Iowa, Ames, IA, August.

Zhang, C., and Wang, H. P., 1993, Optimum Sequence Selection and Manufacturing Tolerance Allocation, *J. Des. Manuf.,* Vol. 3, pp. 135–146.

Zhang, C., Wang, H. P. and Li, J., 1992, Simultaneous Design and Manufacturing Tolerance Design with Process Selection, *Ann. CIRP,* Vol. 41, No. 1, pp. 569–572.

11

SLICING: PROCEDURE FOR TOLERANCE EVALUATION OF MANUFACTURED PARTS USING CMM MEASUREMENT DATA

YU WANG, SHILENDRA GUPTA, and SRINAVAS RAO

University of Maryland
College Park, Maryland

11.1 INTRODUCTION

Coordinate measuring machines (CMMs) are now being used extensively in industry to take measurements on manufactured parts. Coordinate points on the part surface are often measured for two purposes: (1) to determine if the part meets the designed tolerance specifications and (2) to provide information for improvement in the process control. For either application the CMM data have to be analyzed to obtain a "substitute" geometry model such that the parameters of the substitute geometry can be used to compare with the product specifications and/or to make a change in manufacturing processes.

To obtain the substitute geometric model, the measured coordinate points must be related to the parameters of the geometry model in the coordinate system where the substitute geometry is uniquely defined by the data points. This can be formulated as an optimization problem in which an objective function related to the deviations of the measured points from the model geometry is minimized with respect to the model parameters. The resulting

Advanced Tolerancing Techniques, Edited by Hong-Chao Zhang
ISBN 0-471-14594-7 © 1997 John Wiley & Sons, Inc.

model parameters and pointwise deviations of the data from the model can be used for inspection and/or process control applications.

Among many practical factors, success of the CMM application depends largely on how appropriately the substitute geometric model is chosen and how economically the CMM data are measured and processed. Various standards [e.g., the ANSI Geometric and Tolerancing Standard Y14.5M (ASME, 1982)] specify substitute models for determining form tolerances. Recently, methods based on solid model variations and parameterizations have been proposed for tolerance analysis (Guilford and Turner, 1993; Rivest et al., 1994) and for manufacturing process assessment (Wang et al., 1995a, b). Within the framework of a tolerance evaluation system, it is an important economic factor that the pointwise deviations of the measured points are calculated efficiently. This task can be quite challenging when the manufactured part being evaluated has a complex geometric shape and its geometric features depart from nominal size, location, orientation, and form.

In this chapter we present a general procedure, called *slicing,* for CMM data measurement and processing for tolerance evaluation of manufactured parts. In the proposed procedure, the measured coordinate points are required to be organized such that they lie within a set of planes, usually parallel. When intersecting with the substitute geometric model of the part, these slicing planes yield a set of cross sections. The data-processing procedure makes use of the built-in data construction structure to evaluate deviations of the measurement points from their corresponding cross section. The reliance on planar slices permits an efficient determination of pointwise derivations and therefore efficient facilitation of the CMM applications for automated inspection and manufacturing process improvement.

The primary goal of this chapter is to discuss the basic formulation and to demonstrate the accuracy of the approach proposed. To that end, four examples of cylindricity evaluation are examined with both the slicing method proposed here and the conventional orthogonal distance method, using two different optimization algorithms (i.e., the least squares and minimax algorithms). An industrial problem is used to further illustrate the slicing method, when it is applied in determination of a manufactured part model for an aluminum extrusion of an automotive spaceframe structure (Wang et al., 1995).

11.2   BACKGROUND

The literature involving issues of geometric tolerancing and automated inspection falls into two general categories (Feng, 1991). There are the efforts to formalize tolerance standards and to formulate tolerance representation schemes (ASME, 1982; Requicha, 1983; Voclcker, 1993) and the efforts to develop numerical procedures for measurement data processing and tolerance evaluation (Murthy and Abdin, 1980; Hocken et al., 1993). The ANSI standard Y14.5M (ASME, 1982) specifies requirements for tolerances of geometric form features but does not specify the methods by which these tolerances are to be evaluated using automatic measuring systems (Feng and Hopp, 1991).

Most current CMM systems employ optimization algorithms or other curve- and surface-fitting techniques for tolerance evaluation (Feng and Hopp, 1991; Voelcker, 1993). Least squares fitting is the most common technique (Murthy and Abdin, 1980; Shunmugam, 1987; Elmaraghy et al., 1990). Minimax algorithms, in which the maximum deviation of measured points is minimized, have been used particularly in evaluating form tolerances to guarantee the minimum zone requirement of the standard (e.g., Wang, 1992). Other techniques include Monte Carlo simulation (Murthy and Abdin, 1980), heuristic methods (Shunmugam, 1987; Elmaraghy et al., 1990), and direct search techniques (Traband et al., 1989; Le and Lee, 1991). Issues of statistical representation of tolerances using discrete measurement data are discussed in (Menq et al., 1990; Sweet et al., 1985). It is generally accepted that part tolerances are reasonably evaluated by a sufficient number of data points.

For any of the tolerance evaluation techniques, pointwise deviations of the measured points from the ideal model must be defined and calculated. The current trend is to evaluate the orthogonal distance of a data point perpendicular to the model feature surface, which happens to be the minimum distance in the Euclidean sense. A method based on this distance calculation is referred to as an *orthogonal distance method* [e.g., the orthogonal distance regression method in the statistical analysis (Dowling et al., 1993)]. For two-dimensional geometric features (e.g., lines and circles) and simple three-dimensional features (e.g., planes and cylinders) an orthogonal distance method usually works well, due to the fact that orthogonal distances for these features can be determined analytically (Etesami, 1988; Fent and Hopp, 1991; Sahoo and Menq, 1991).

For a complex three-dimensional part, such as space extrusions and castings used in an automotive body structure (Hulting, 1995; Wang et al., 1995a), it may be computationally expensive to evaluate the orthogonal distance, since a general complex geometry is often defined with higher-order spline surfaces (e.g., NURBS) in a geometric modeling system. The orthogonal distance has to be evaluated numerically, usually in an iterative manner (Sahoo and Menq, 1991). The literature review indicates that computation of pointwise deviations of measurement points, particularly for long and complex-shaped parts, is a problem largely unaddressed by researchers in this area. The work described in this chapter attempts to address this problem and proposes an alternative method for evaluating deviations of measurement points within the context of tolerance analysis of manufactured parts using CMM data.

11.3   ORTHOGONAL DISTANCE DEVIATIONS

To evaluate a geometric feature, measurements on the actual part surface must be compared with an ideal or substitute feature such that all the deviations of the surface from the ideal feature are within the tolerance zone. The ideal feature to be established first from the actual measurements can be expressed as

$$\mathbf{S} = \mathbf{S(u)} \tag{11.1}$$

where $\mathbf{u}$ denotes feature parameters describing the position, orientation, and size of the feature. In the case of a CMM system, a number of discrete points on the surface of the feature are usually measured. The geometric deviation for a sampled point can be expressed in a general form as

$$d_n(\mathbf{u}_i) = |\mathbf{p}_i - \mathbf{r}(\mathbf{u}_i)| \tag{11.2}$$

where $\mathbf{p}_i$ is the measurement coordinate of point i and $\mathbf{r}(\mathbf{u}_i)$ is the corresponding point on the ideal feature $\mathbf{S}$, all in vector forms. The geometric error is usually defined in the Euclidean sense as the shortest distance from the measurement point to the ideal form (Menq et al., 1990), as denoted by $|\cdot|$. This deviation, d_n, is known as the *orthogonal distance deviation*. For simple features such as lines and planes, parameter $\mathbf{u}_i$ may not be explicit, usually yielding an algebraic expression for d_n. For complex geometric forms, however, the value of $\mathbf{u}_i$ may not be directly available, and the orthogonal distance may have to be determined by using an iterative numerical procedure (Sahoo and Menq, 1991; Wang, 1992).

In a tolerance analysis, the form error of the feature can be determined by solving a tolerance zone optimization problem. Existing solution techniques are essentially based on either a least squares formulation or a minimax formulation. The least squares problem can be formulated as

$$\text{LSQ: min} \sum_{i=1}^{m} (d_n)_i^2 \tag{11.3}$$

where m is the total number of data points, while the minimax formulation is defined as

$$\text{minimax: min}(\max_i (d_n)_i) \quad \text{for } i = 1,2,...,m \tag{11.4}$$

Both techniques have been examined extensively in the literature (e.g., El-maraghy et al., 1990; Wang, 1992). Within the same framework of optimization, the procedure proposed here employs a different determination of pointwise deviations of the measurement data points, discussed below.

11.4   SLICING METHOD

It is always desirable to use large sample size of CMM measurements to increase confidence in the accuracy of tolerance evaluation. However, CMM measurement is time consuming, and processing a large amount of data could be computationally expensive. In addition to the method of numerical optimization, efficiency of a CMM application depends largely on (1) the selection of measurement locations and density and (2) the technique for evaluating deviation distance.

Calculation of orthogonal distance may require a surface model described in terms of feature parameters of the manufactured part. For complex-shaped parts, part surfaces are often represented with high-order splines (e.g., NURBS). In such a case the distance calculation may become computationally expensive, as it has to be evaluated by a numerical iterative procedure. A slicing strategy for data sampling and deviation evaluation is discussed here, aimed at reducing the cost associated with CMM data measurement and processing.

11.4.1   CMM Measurement by Slicing

In contrast to the common strategy of uniform sampling, in the proposed procedure measurement locations on a part surface are selected at certain *slicing planes*. During sampling the CMM is commanded to move on each of the planes and a *slice* of the workpiece is taken. The slice consists of the set of CMM points (usually discrete) at the intersection of the slicing plane with the workpiece as illustrated in Figure 11.1. Each set of CMM slice data points is coplanar. Only one slice is shown in the figure.

The slicing strategy is not only intuitive but also cost-effective. In fact, machined parts are often measured in planes in industrial practice of CMM measurement, particularly for a long and slender workpiece. By "structuring" measurement data this way, it is usually easier to program a CMM and faster for the CMM to execute sampling motions.

For convenience the set of slicing planes may be determined to be parallel. However, this is not necessary. In industrial practice, a practitioner often first determines a plane to be perpendicular to the surface being measured by touching the surface at three locations. Then a set of coordinate data points on the part surface are collected within the established plane. This procedure is particularly common for long, slender workpieces such as space extrusions. In general, the slicing planes may not be necessarily parallel, nor are they actually perpendicular to the surface being measured.

Clearly, accuracy of tolerance estimation of a manufactured part depends on sample size and sample location (Dowling et al., 1993). In the slicing strategy, location of a sample point is determined by the location and orientation of the selected slicing plane, in addition to the position of the point on

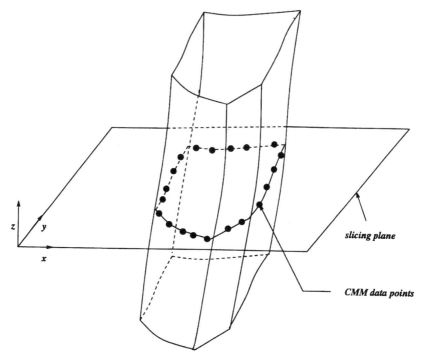

Figure 11.1. CMM measurement by slicing.

the plane. Effect of the orientation of slicing planes on the accuracy of tolerance estimation is examined briefly in Section 11.5 for a cylindricity case.

11.4.2   Planar Pointwise Deviation

Another advantage of the slicing measurement strategy is its potential for reducing computational expenses associated with data processing in tolerance evaluation. Since the speed of a tolerance evaluation algorithm depends on the efficiency of deviation distance evaluation for each data point, one can now make use of the coplanar structure inherent in the sliced measurement data for more efficient distance evaluation.

In a way similar to slicing in CMM measurement, each slicing plane can be intersected with the surfaces of the geometric model (nominal or substitute), yielding a *cross-sectional contour* of the model. This cross-section contour can now be used to approximate the pointwise deviation for each data point measured with the slicing plane. The deviation is defined to be the distance of the data point to the cross-sectional contour measured on the slicing plane. This distance is referred to as *planar distance, d_p*. As shown in Figure 11.2, the planar distance d_p is measured along a line perpendicular to the cross-section contour and passing through sample point **p**, compared

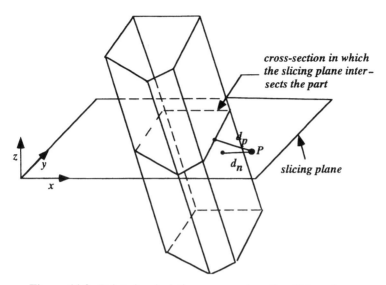

Figure 11.2. Pointwise deviation measured on the slicing plane.

to the orthogonal distance d_n, which is measured along a line perpendicular to the model surface and passing through point **p**.

Instead of using the conventional orthogonal distance, one can use the planar distance to formulate the objective function to be minimized in an optimization algorithm for tolerance evaluation of a manufactured part. Calculation of the planar distance does not necessarily require a surface model as in the case of orthogonal distance. Therefore, use of the alternative distance may make it possible to reduce the computational complexity in deviation calculation. For simple surface features such as planes and cylinders, the resulting computational change is generally minimal, since explicit algebraic expressions are available for both orthogonal and planar distance evaluation. For parts with complex surfaces (e.g., NURBS), however, the advantage of the planar distance may become significant. Particularly, if the part is a long, slender workpiece such as a space-frame extrusion used in aluminum automotive structures (Wang et al., 1995a), use of the slicing technique makes both theoretical and practical sense.

11.4.3 Calculation of the Planar Distance

It is convenient to evaluate the planar distance by establishing a coordinate system in each slicing plane such that a principal axis (e.g., z-axis) is perpendicular to the plane (Figure 11.2). Then the equation of the slicing plane is given simply by

$$z = 0 \qquad (11.5)$$

The equations of each workpiece surface can also be expressed in the same coordinate system. The general implicit equation of the surface is given as

$$f(x,y,z) = 0 \qquad (11.6)$$

Combining these two equations, the cross-sectional contour of the surface on the slicing plane is defined as

$$f(x,y) = 0 \qquad (11.7)$$

For each measured point $\mathbf{p}_i(x_i,y_i)$ that belongs to this slicing plane, its planar distance to the cross-sectional contour is now to be evaluated in the two-dimensional (planar) space. In general, calculating the planar distance is easier than in the three-dimensional space as in the case of the orthogonal distance.

The planar distance may also be approximated by its algebraic counterpart (Sampson, 1982; Sahoo and Menq, 1991). Substituting the coordinates (x_i,y_i) of the data point $\mathbf{p}_i$ into the implicit contour equation (11.7) and then normalizing it with respect to the gradient of the function, an approximation of the planar distance is obtained as

$$d_p = \frac{f(x_i,y_i)}{|\nabla f(x_i,y_i)|} \qquad (11.8)$$

In the case of straight lines, this algebraic distance yields the exact Euclidean distance.

In many engineering applications, contour lines of a surface are often required for other purposes. For example, it is usually necessary to generate parallel contour lines for numerically controlled (NC) machining in computer-aided manufacturing (Satterfield and Rogers, 1985). If accuracy of the contour lines is acceptable, the contour lines may be used directly for planar distance evaluation in the CMM application.

11.5   ACCURACY OF TOLERANCE ZONE EVALUATION

It is obvious that the various approaches to evaluate deviation distances will yield different tolerance evaluation results. For a reliable application of the proposed method, it is imperative to understand the significance of the differences of the results obtained by the slicing method from those obtained by using the orthogonal distance definition. Especially in the case of form tolerance evaluation, the fundamental concern is the reliability of the tolerance zone value obtained.

The slicing approach uses a new distance evaluation method in which the algebraic distances are evaluated. Thus one needs to examine carefully the accuracy of tolerance zones obtained using algebraic distances. By definition, the planar distance d_p of a data point is always greater than or equal to the orthogonal distance d_n of the same data point; that is,

$$d_p \geq d_n \qquad (11.9)$$

Therefore, it is expected that tolerance zones obtained using the planar distance may be greater than those obtained using the traditional orthogonal distance; that is,

$$\text{tolerance zone } (d_p) \geq \text{tolerance zone } (d_n) \qquad (11.10)$$

Consider the trivial case of a workpiece that has no deviations from its ideal model (i.e., a perfect workpiece). If the measurement noise is discarded, the deviation of any measured point from the ideal model would always be equal to zero. Hence the tolerance zones obtained using either of the two methods should yield the same results. Therefore, when an actual part has small deviations from its model, tolerance assessment results made by these two different approaches must not be significantly different.

A general solution explaining the differences in tolerance zones would be useful to predict correctly the influence of the planar distance approximations on tolerance zones. An analytical characterization of the differences is unavailable presently. An empirical analysis of the numerical results of an example is presented below to gain a better understanding of the accuracy of the proposed slicing method.

11.5.1   Tolerance Evaluation of Cylinders

The position and orientation of a cylinder are characterized by the position and orientation of its center axis (as in Menq et al., 1990; Wang, 1992). The axis passes through a point (x_0, y_0, z_0) and has orientation given by the parameters (u, v, w), as seen from the measurement coordinate frame (Figure 11.3). Its radius is denoted by R. The equation of the center axis can be written as

$$\frac{x - x_0}{u} = \frac{y - y_0}{v} = \frac{z - z_0}{w} \qquad (11.11)$$

It is convenient to specify $z_0 = 0$ and $w = 1$.

Measurements on a cylindrical object are taken at different cross sections. While taking measurements, the z-rail of the CMM is held fixed at a particular position and measurements on the surface of the cylinder are taken in the $x–y$ plane around the periphery of the cylinder (Figure 11.3). Then the z-rail is moved to a different height and measurements are taken again in the $x–y$ plane, and so on for different heights. If the cylinder is fixed in a position other than vertical, these measured points would lie around the periphery of

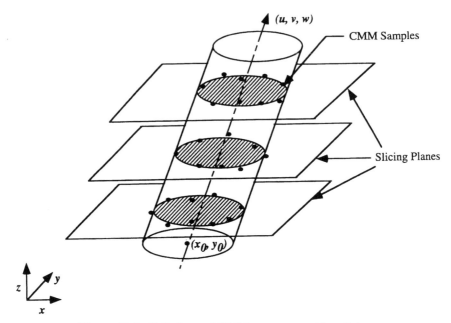

Figure 11.3. Cylinder and CMM measurement by slicing.

a set of ellipses. For the first cylindrical object, four slices of 60 points in total are taken, as shown in Figure 11.4.

In the conventional orthogonal distance evaluation, the deviation distance of a measurement point (x_i, y_i, z_i) to the cylinder is given by (Wang, 1992)

$$d_{ni} = \left\{ \frac{(x_i - x_0 - uz_i)^2 + (y_i - y_0 - vz_i)^2 + [v(x_i - x_0) - u(y_i - y_0)]^2}{1 + u^2 + v^2} \right\}^{1/2}$$
$$- R \tag{11.12}$$

The planar distance would be the distance of the point from the ellipse resulting from the intersection of the slicing plane with the ideal cylinder (Figure 11.5). The equation of the ellipse is given by

$$\frac{(x_i - x_c)^2}{a^2} + \frac{(y_i - y_c)^2}{b^2} = 1 \tag{11.13}$$

where a and b are the semimajor axis and the semiminor axis, respectively, and (x_c, y_c) are the coordinates of the center of the ellipse. As discussed in Section 11.4, the planar distance of the point (x_i, y_i) from this ellipse can be approximated by the algebraic distance

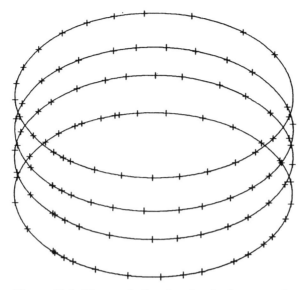

Figure 11.4. Measured slice data for the first example.

$$d_{pi} = \frac{Q(x_i, y_i)}{|\nabla(Q(x_i, y_i))|} \qquad (11.14)$$

where Q is given by

$$Q(x_i, y_i) = \frac{(x_i - x_c)^2}{a^2} + \frac{(y_i - y_c)^2}{b^2} - 1$$

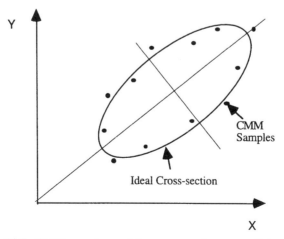

Figure 11.5. CMM samples and the cross section on a slicing plane.

TABLE 11.1. Cylindricity Evaluation Results for the First Example

	LSQ (Orthogonal)	LSQ (Planar)	Minimax (Orthogonal)	Minimax (Planar)
x_0	5.9977E-2	5.9977E-2	6.0866E-2	6.0864E-2
y_0	6.2708E-2	6.2708E-2	6.2802E-2	6.2801E-2
u	−8.0140E-5	−8.0070E-5	1.3094E-4	−1.1280E-4
v	−2.3889E-4	−2.3898E-4	−6.2771E-4	−6.2714E-4
R	9.1160E-1	9.1160E-1	9.1159E-1	9.1159E-1
Zone	2.3460E-3	2.3640E-3	2.1718E-3	2.1718E-3

and x_c, y_c, a, and b can easily be expressed in terms of the five parameters (x_0, y_0, u, v, R) of the cylinder. The magnitude of the gradient is given by

$$|\nabla Q(x_i, y_i)| = 2 \left(\frac{(x_i - x_c)^2}{a^4} + \frac{(y_i - y_c)^2}{b^4} \right)^{1/2}$$

11.5.2 Tolerance Zone Results of Cylinders

The minimization for cylindricity tolerance can be carried out with respect to these parameters. To compare the results of the proposed slicing method and the conventional orthogonal distance method, both least squares and minimax evaluations are performed using the IMSL (IMS, 1989) and CFSQP (Zhou and Tits, 1991) optimization routines, respectively. Four examples of cylindricity evaluation are examined.

The tolerance evaluation results for the first example are shown in Table 11.1, where four slices of 60 points in total were taken in measurement. The very small value of tolerance zone for the example indicates that the cylinder is nearly perfect. Table 11.2 shows results of the second example with a cylinder similar to the one used in the first example.

TABLE 11.2. Cylindricity Evaluation Results for the Second Example

	LSQ (Orthogonal)	LSQ (Planar)	Minimax (Orthogonal)	Minimax (Planar)
x_0	8.1788	8.1912	8.1708	8.1616
y_0	−13.1756	−13.1842	−13.2086	−13.2057
u	−0.0534	−0.0483	−0.0569	−0.0608
v	0.0710	0.0677	0.0586	0.0549
R	0.7043	0.7040	0.7267	0.7266
Zone	0.1361	0.1392	0.1280	0.1309

TABLE 11.3. Cylindricity Evaluation Results for the Third Example

	LSQ (Orthogonal)	LSQ (Planar)	Minimax (Orthogonal)	Minimax (Planar)
x_0	0.5137	0.5826	0.3639	0.4697
y_0	0.6004	0.5998	0.7630	0.7590
u	0.8352	0.6703	0.6868	0.7442
v	−0.2679	−0.1041	0.0063	0.0764
R	2.5469	2.6330	2.4605	2.3589
Zone	2.0338	2.2229	1.7741	1.8466

An example discussed by Shunmugam (1987) and Wang (1992) is analyzed as the third example of cylindricity analysis. The measurement consists of 24 data points measured in three slices. The results of the example are presented in Table 3, and a high value of tolerance zone suggests that the part is not well machined. In fact, this is an artificial example made for a testing purpose only (Shunmugam, 1987).

The last cylindricity example is to illustrate the effect of the orientation of the slicing planes on the accuracy of tolerance estimation. A measured cylinder is inclined from the vertical position by approximately 22°. The results in Table 11.4 show that the planar distance evaluation yields accuracy comparable with that of the orthogonal distance method.

From the results of these examples, one can make the following observations:

1. The tolerance zone obtained by using planar distances is slightly larger than that obtained by using orthogonal distances. This is true for both least squares and minimax evaluations. For the first two examples, the tolerance zone of orthogonal distance is about 98% of that of the planar distance. Even for the third example of largely exaggerated deviations, the tolerance zone of orthogonal distance is 90% of that of the planar

TABLE 11.4. Cylindricity Evaluation Results for the Fourth Example

	LSQ (Orthogonal)	LSQ (Planar)	Minimax (Orthogonal)	Minimax (Planar)
x_0	7.6007	7.5330	7.6305	7.7029
y_0	−3.8954	−3.5367	−3.0347	−3.4985
u	−0.1649	−0.1732	−0.1627	−0.1549
v	0.1052	0.1440	0.1985	0.1513
R	0.6623	0.6594	0.6587	0.6566
Zone	0.1131	0.1147	0.0880	0.0890

TABLE 11.5. Differencesin Tolerance Zones

	LSQ			Minimax		
Example	Orthogonal	Planar	Difference (%)	Orthogonal	Planar	Difference (%)
1	0.02346	0.02364	0.767	0.02172	0.02178	0.276
2	0.1361	0.1392	2.278	0.1280	0.1309	2.265
3	2.0338	2.2239	9.345	1.7741	1.8466	4.080
4	0.1131	0.1147	1.415	0.0880	0.0890	1.136

distance. The differences in tolerance zones are summarized in Table 11.5 for the least squares and minimax optimization solution methods, respectively.

2. The minimax technique yields smaller tolerance zone than the least squares method. This is observed for both the orthogonal and planar distance evaluations. It is consistent with previous findings of Wang (1992) and Shunmugam (1987).

11.6 APPLICATION FOR ALUMINUM SPACE-FRAME EXTRUSIONS

The slicing method is applied to a process for characterizing critical geometric deviations in a space-frame extrusion used in automotive structures. The extrusion part has a rectangular cross section and three bends, marked *left bend, center bend,* and *right bend* in Figure 11.6, produced on a rotary-draw bender. Slices are taken in each of the straight segments, as shown by lines in the figure.

The critical shape parameters of primary interest in process control are the three bend angles and the relative positions of the bends. These parameters are estimated from the measurement data by a process called *manufactured*

FRONT OF EXTRUSION

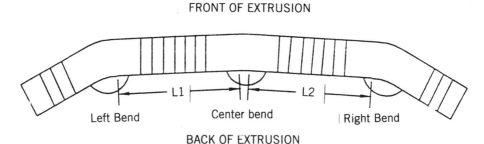

BACK OF EXTRUSION

Figure 11.6. Automotive extrusion part with three bends.

part model analysis. Details of the estimation technique have been described by Wang et al. (1995a).

These critical parameters are estimated with both planar and orthogonal distance evaluations. The results obtained are presented in Table 11.6, where the nominal values of the parameters are the design specifications. An excellent agreement is found between these two estimation methods. As expected, the sum of squared distances (SSD) for the slicing method is larger than that of the orthogonal distance method. The manufactured part model analysis reveals that in the actual part the center bend and the right bend are smaller, and the center distance l_2 is made substantially smaller (by 11.78 mm) than specified.

11.7 CONCLUSIONS

In this chapter we have described a general method for tolerance evaluation of a manufactured part, which uses a slicing procedure for sampling and processing CMM data. In the procedure proposed, the measured coordinate points are required to be organized such that they lie within a set of selected planes. This inherent structure is then utilized for the evaluation of pointwise deviations of the measurement points. The procedure, called *slicing* for obvious reasons, is particularly effective for a long, slender workpiece.

The planar distances can be used to formulate the objective function to be minimized in an optimization algorithm for form tolerance evaluation or other applications. It is illustrated, through several examples, that the accuracy of estimated tolerance is comparable to that of the conventional orthogonal distance method. It is shown that the difference in tolerance zone value using planar and orthogonal distances is in the range 1 to 3% for the realistic cases. Even in the third example, comprised of randomly generated measurement points, the difference is about 10%.

The proposed slicing approach raises an issue related to statistical analysis of data sampling and the quality of tolerance estimation (Dowling et al., 1993). It appears that a slicing sample plan may render the mean function in a statistical model used for data analysis to be independent of the data observed. This possibility would enable us to formulate a fixed-regression model

TABLE 11.6. Critical Parameter Values of the Extrusion

	Left Bend	l_1	Center Bend	l_2	Right Bend	SSD[a]
Nominal	29.0947	376.8173	1.8163	376.8173	29.0947	—
Slicing	29.0974	378.4636	1.2991	365.0333	29.3468	18.3902
Orthogonal	29.0974	378.4675	1.2998	365.0294	29.3463	18.3895

[a]SSD, sum of squared distances.

and use regression theory to address statistical issues. It is worthy of an investigation.

ACKNOWLEDGMENTS

This material is based on work supported by the National Science Foundation and the Alcoa Technical Center.

REFERENCES

ASME, 1982, *Dimensioning and Tolerancing for Engineering Drawings,* ANSI Y14.5M-1982, American Society of Mechanical Engineers, New York.

Dowling, M., et al., 1993, *Statistical Issues in Geometric Tolerance Verification Using Coordinate Measuring Machines,* Technical Report, School of Industrial and Systems Engineering, Georgia Institute of Technology, Atlanta, GA.

Elmaraghy, W. H., Elmaraghy, H. A., and Wu, Z., 1990, Determination of Actual Geometric Deviations Using Coordinate Measuring Machine Data, *Manuf. Rev.,* Vol. 3, No. 1, pp. 32–39.

Etesami, F., 1988, Tolerance Verification Through Manufactured Part Modeling, *J. Manuf. Syst.,* Vol. 7, No. 3, pp. 223–232.

Feng, S. C., and Hopp, T. H., 1991, *A Review of Current Geometric Tolerancing Theories and Inspection Data Analysis Algorithms,* Technical Report NISTIR-4509, National Institute of Standards and Technology, Gaithersburg, MD.

Guilford, J., and Turner, J., 1993, Advanced Analysis and Synthesis for Geometric Tolerances, *Manuf. Rev.,* Vol. 6, No. 4, pp. 305–313.

Hocken, R., et al., 1993, Sampling Issues in Coordinate Metrology, *Manuf. Rev.,* Vol. 6, No. 4, pp. 282–294.

Hulting, F., 1995, An Industry View of Coordinate Measurement Data Analysis, *Stat. Sinica,* Vol. 5, No. 1, pp. 19–31.

IMS, 1989, *IMSL Math/Library User's Manual,* IMS, Houston, TX.

Le, V.-B., and Lee, D. T., 1991, Out-of-roundness problem revisited, *IEEE Trans. Pattern Anal. Mach. Intell.,* Vol. 13, No. 3, pp. 217–223.

Menq, C.-H., Yau, H.-T., and Lai, G.-Y., 1990, Statistical Evaluation of Form Tolerances Using Discrete Measurement Data, in *Advances in Integrated Product Design and Manufacturing,* P. H. Cohen and S. B. Joshi (eds.), American Society of Mechanical Engineers, New York, pp. 135–149.

Murthy, T. S. R., and Abdin, S. Z., 1980, Minimum Zone Evaluation of Surfaces, *Int. Mach. Tool Des. Res.,* Vol. 20, pp. 123–136.

Requicha, A. A. G., 1983, Towards a Theory of Geometric Tolerance, *Int. J. Robot. Res.,* Vol. 2, No. 4, pp. 45–60.

Rivest, L., Fortin, C., and Morel, C., 1994, Tolerancing a Solid Model with Kinematic Formulation, *Comput.-Aid. Des.,* Vol. 26, No. 6, pp. 465–476.

Sahoo, K. C., and Menq, C.-H., 1991, Localization of 3d Objects Using Tactile Sensing and Surface Description, *J. Eng. Ind.,* Vol. 113, pp. 85–92.

Sampson, P. D., 1982, Fitting Conic Sections to Very Scattered Data: An Iterative Refinement of the Bookstein Algorithm, *Comput. Graph. Image Process.*, Vol. 18, pp. 97–108.

Satterfield, S. G., and Rogers, D. F., 1985, A Procedure for Generating Contour Lines from a b-Spline Surface, *IEEE Comput. Graph. Appl.*, Vol. 5, pp. 71–75.

Shunmugam, M. S., 1987, New Approach for Evaluating Form Errors of Engineering Surfaces, *Comput.-Aid. Des.*, Vol. 19, No. 7, pp. 368–374.

Sweet, A. L., Noller, D., and Lee. S.-H., 1985, Statistical Design for the Location of Planes and Circles When Using a Probe, *Precis. Eng.*, Vol. 7, No. 4, pp. 187–194.

Traband, M. T., et al., 1989, Evaluation of Straightness and Flatness Tolerances Using the Minimum Zone, *Manuf. Rev.*, Vol. 2, No. 3, pp. 189–195.

Voelcker, H., 1993, A Current Perspective on Tolerancing and Metrology, *Manuf. Rev.*, Vol. 6, No. 4, pp. 258–268.

Wang, Y., 1992, Minimum Zone Evaluation of Form Tolerances, *Manuf. Rev.*, Vol. 5, No. 3, pp. 213–220.

Wang, Y., et al., 1995a, Manufactured Part Modeling for Characterization of Geometric Errors of Aluminum Automotive Space-Frames, in *Manufacturing Science and Engineering—1995,* American Society of Mechanical Engineers, New York, pp. 1051–1063.

Wang, Y., Hulting, F., and Fussell, P., 1995b, Discovery of Part Shape from Surface Measurement Data, *Proceedings of the NSF Design and Manufacturing System Conference,* LaJolla, CA, Jan. 5–9, 1995, Society of Manufacturing Engineers, New York, pp. 377–378.

Zhou, J. L., and Tits, A. L., 1991, *User's Guide for fsqp Version 2.4, a Fortran Code for Solving Optimization Problems, Possibly Minimax, with General Inequality Constraints and Linear Equality Constraints, Generating Feasible Iterates,* Technical Report TR90-60b, Systems Research Center, University of Maryland, College Park, MD.

12

TOLERANCE ANALYSIS AND SYNTHESIS IN VARIATIONAL DESIGN

Jian (John) Dong*

University of Connecticut
Storrs, Connecticut

Ying Shi

University of Connecticut
Storrs, Connecticut

12.1 INTRODUCTION

Tolerance synthesis and analysis during design stages play a critical role in the entire design and manufacturing processes. Tolerances assigned to assembly components must meet the functional requirements, permit part interchangeability, and led to economical manufacturing and assembly. A tolerance that is too loose may cause product malfunction, and one that is too tight will definitely raise the cost of production (Dong and Hu, 1991). In many cases the tolerances cannot be determined precisely due to the lack of methodology. The current practice regarding tolerance assignment is to use experience-based handbooks, which are error prone. Especially when components form an assembly, the tolerances of each component will accumulate, which may cause a failure in assembly operations (Lee and Yi, 1994).

*Dr. Dong is currently with Boeing North American, Inc., Downey, CA 90242

Advanced Tolerancing Techniques, Edited by Hong-Chao Zhang
ISBN 0-471-14594-7   © 1997 John Wiley & Sons, Inc.

Tolerance specifications can be categorized into *dimensional tolerances* (also called *conventional tolerances* or *local tolerances*) and *geometric tolerances* (perpendicularity, parallelism, etc.). *Tolerance synthesis* (also called *tolerance allocation*) allocates tolerances to individual design constraints based on one or more critical design functions. The design constraints can be all in one part or in different components of an assembly. *Tolerance analysis* verifies that critical design functions meet the design requirements based on given individual design tolerances.

Much research has been done in this area. Bjorke (1989) developed the tolerance chain approach to analyzing the tolerance accumulation in an assembly and assigning dimensional tolerance to minimize the cost of manufacture. Michael and Siddall (1981) proposed a vector-space formulation for tolerance synthesis. Parkinson (1985) developed a mathematical programming formulation with product cost as the objective function and standard deviations of the design variable as decision variables for tolerance synthesis. Lee and Woo (1987) formulated the concept as a combinatorial optimization problem by treating manufacturing cost as an objective function and stacked-up conditions in assembly as the constraints. Dong and Soom (1986) developed a unidirectional tolerance chain analysis system for axisymmetric rotational parts in a two-dimensional CAD system. Chase and Greenwood (Chase et al., 1990; Greenwood and Chase, 1987, 1988) presented a discrete optimization scheme that deals with the combinatorics resulting from alternative manufacturing processes. Zhang and Wang (1993) used sequential quadratic programming (SQP) and a simulated annealing algorithm to deal with the same problem.

Since the objective of tolerance synthesis is to minimize the overall manufacturing cost, identifying the correct relationship between the cost of a specific manufacturing process and the value of a tolerance becomes very important. Some empirical cost–tolerance models have been developed (Spotts, 1973; Dong et al., 1995, 1997; Speckhart, 1994; Wu et al., 1992). However, all these models can be used only for dimensional tolerances.

Constraint-based variational design concepts were first proposed by Hillyard (1978), and extended by Gossart et al. (1988), Lin (1981), and Light and Gossard (1982). In their research a variational geometry (VG) model is generated from a solid boundary model by applying variations to the coordinates of each vertex. The vertex coordinates are treated as a collection of model variables. The vertices determine the faces and edges. Turner (1987, 1990) developed an alternative strategy in which a variational model is generated from a solid boundary model by applying variations to the surface associated with each surface. The coefficients of surface equations are treated as a collection of model variables. Three methods—linear programming, Monte Carlo, and least squares—have been used by Turner (1987, 1990) to perform dimensional tolerance synthesis and analysis.

Most research work in tolerance allocation is limited to dimensional tolerance allocations or one-dimensional tolerancing problems. No suitable method has been developed for optimally allocating both dimensional and geometric tolerances to respect the requirements of the design functions (Roy et al., 1991). Optimal allocation of both dimensional and geometric tolerances is much more difficult than the allocation of dimensional tolerances alone. First, the assumption that all design constraints can be produced independently is no longer valid when geometric constraints are considered. Second, there is no cost–tolerance function available for geometric tolerances. Third, because geometric constraints often represent global relationships between geometric entities, the relationships among dimensional and geometric constraints are often in nonlinear formats, which make it difficult to capture how one constraint variation (tolerance) affects other constraints. But tolerance synthesis without considering geometric tolerances results in allocating larger-than-required tolerances to individual design constraints, which may reduce the cost of manufacturing as well as causing problems in assembly processes and product quality. Tolerance analysis without considering geometric tolerances may result in accepting failed designs (Dong and Shi, 1995, 1997).

To overcome these difficulties, research in using variational geometry concepts as a theoretical basis for optimal tolerance allocation and evaluation was carried out at the Concurrent Design and Manufacturing System (CDMS) Laboratory of the University of Connecticut. In this chapter we present the concepts of tolerance allocation and evaluation in a variational design environment and some research results obtained at the CDMS Lab.

12.2   VARIATIONAL GEOMETRY THEORY FOR TOLERANCE SYNTHESIS AND ANALYSIS

12.2.1   Geometric Model

Under variational geometry (VG) concepts, any geometry is defined by a set of characteristic points. These characteristic points may represent a vertex, the center of a circle, a start point or endpoint of an arc, part of a cubic spine, and so on. All higher-order entities (edges, surfaces, and dimensions) are defined with respect to those characteristic points. As showed in Figure 12.1, a two-dimensional object is decided by six characteristic points. The complete set of characteristic points is described by a geometry vector $\bar{\chi}$ containing their Cartesian coordinates:

$$\bar{\chi} = \{x_1, y_1, z_1, \ldots, x_n, y_n, z_n\}^{\mathrm{T}}$$

$$\bar{\chi} = \{x_1, y_1, z_1, \ldots, x_n, y_n., z_n\}^T$$

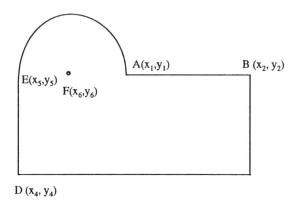

D (x_4, y_4)

Figure 12.1. Characteristic points for a two-dimensional figure.

12.2.2   Types of Constraints

In a VG-based design system, although the geometry is defined by a set of points, it is not directly constructed by explicit coordinates of every point but by a set of geometry constraints: for example, the distance between two points, the angle between two lines, or the parallelism of the two lines. Generally, constraints can be divided into two types. The first is *dimensional constraint* (also known as *explicit constraint*), which constrains, for example, the distance between two points. The second type is *geometric constraint* (or *implicit constraint*), which defines the geometric relationship between geometric entities, such as the parallelism between two lines. For each type of constraint, there is a corresponding nonlinear equation. For example, two points in three dimensional space, $A(x_1, y_1, z_1)$ and $B(x_2, y_2, z_2)$, the equation

$$D = \sqrt{(x_2 - x_1)^2 + (y_2 - y_1)^2 + (z_2 - z_1)^2}$$

constrains the distance between the two points. The dimensional constraints usually include angular dimension, linear dimension, and radial dimension. The equations are listed in Table 12.1.

The geometric constraints usually include coincidence between two points, collinearity, parallelism, and perpendicularity between two lines. Their corresponding equations are listed in Table 12.2.

Since a point in two- or three-dimensional space has two (three) degrees of freedoms, any point in two(three) dimensions needs two (three) equations

TABLE 12.1. Dimensional Constraints

Constraints	Dimension Name	Entities Constrained	Equation
Angular	Angular dimension	p_1p_2, p_3p_4 (Figure 12.2)	$\dfrac{\lvert p_1p_2 \times p_2p_4 \rvert}{p_1p_2 \cdot p_3p_4} - \tan A = 0$ or $\dfrac{(x_2 - x_1)(y_4 - y_3) - (y_2 - y_1)(x_4 - x_3)}{(x_2 - x_1)(x_4 - x_3) + (y_2 - y_1)(y_4 - y_3)} - \tan A = 0$
Linear	Point to point	p_1, p_2 (Figure 12.3)	$(x_2 - x_1)^2 + (y_2 - y_1)^2 - D^2 = 0$
	Point to line	p_1, p_2p_3 (Figure 12.4)	$\bar{u} = \dfrac{x_3 - x_2}{\lvert p_2p_3 \rvert}\,i + \dfrac{y_3 - y_2}{\lvert p_2p_3 \rvert}\,j,\quad \bar{v} = (x_2 - x_1)i + (y_2 - y_1)j$ $\lvert \bar{u} \times \bar{v} \rvert - D = 0$
	Line to line	p_1p_2, p_3p_4 (Figure 12.5)	(1) $\dfrac{y_2 - y_1}{x_2 - x_1} - \dfrac{y_4 - y_3}{x_4 - x_3} = 0$ (2) $(x_3 - x_1)^2 + (y_3 - y_1)^2 - D^2 = 0$
Radical	Horizontal	p_1, p_2 (Figure 12.6)	$x_1 - x_2 - D = 0$
	Vertical	p_1, p_2 (Figure 12.7)	$y_1 - y_2 - D = 0$
	Radical dimension	p_1, $c1$ (Figure 12.8)	$(x - x_1)^2 + (y - y_1)^2 - R^2 = 0$

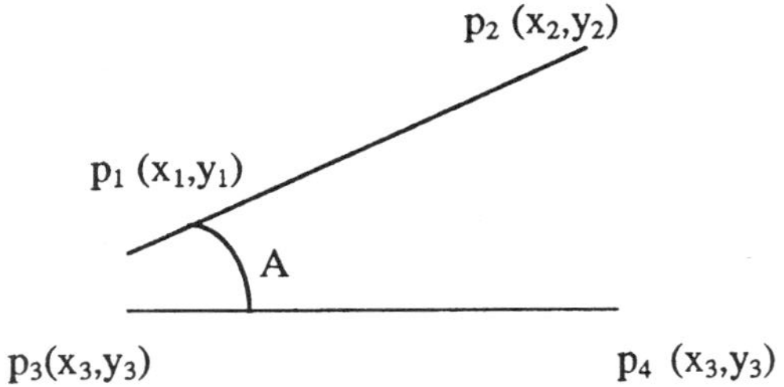

$p_2\,(x_2,y_2)$

A

$p_1\,(x_1,y_1)$

$p_3(x_3,y_3)$

$p_4\,(x_3,y_3)$

Figure 12.2. Angular constraint.

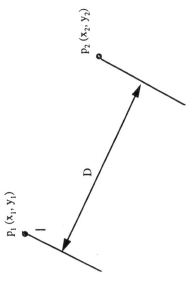

$p_1\,(x_1,y_1)$

$p_2\,(x_2, y_2)$

D

Figure 12.3. Distance from point to point.

305

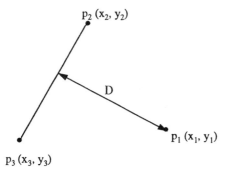

Figure 12.4. Distance between point and line.

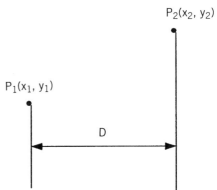

Figure 12.6. Horizontal distance between two points.

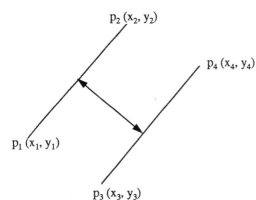

Figure 12.5. Distance from line to line.

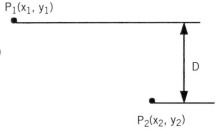

Figure 12.7. Vertical distance between two points.

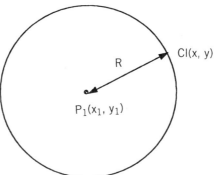

Figure 12.8. Radical constraint.

TABLE 12.2. Geometric Constraints

Constraint	Entities Constrained	Equation
Coincidence	p_1, p_2	$(x_2 - x_1)^2 + (y_2 - y_1)^2 = 0$
Collinearity	p_1p_2, p_3p_4	(1) $\dfrac{y_2 - y_1}{x_2 - x_1} - \dfrac{y_4 - y_3}{x_4 - x_3} = 0$
		(2) $(x_3 - x_1)^2 + (y_3 - y_1)^2 = 0$
Parallelism	p_1p_2, p_3p_4	$p_1p_2 \times p_3p_4 = 0$
		$\dfrac{y_2 - y_1}{x_2 - x_1} - \dfrac{y_4 - y_3}{x_4 - x_3} = 0$
Perpendicularity	p_1p_2, p_3p_4	$p_1p_2 \cdot p_3p_4 = 0$
		$\dfrac{y_2 - y_1}{x_2 - x_1} \cdot \dfrac{y_4 - y_3}{x_4 - x_3} = 1$

to constrain it. A geometry with n characteristic points requires $2n$ ($3n$) constraint equations to constrain itself totally. For simplicity, a two-dimensional rectangle is used as an example to illustrate this concept (Figure 12.9). Constraints for a rectangle are listed as follows:

(1) $x_1 = 0$

(2) $y_1 = 0$

(3) $\dfrac{y_2 - y_1}{x_2 - x_1} - \dfrac{y_4 - y_3}{x_4 - x_3} = 0$

(4) $\dfrac{x_4 - x_1}{y_4 - y_1} - \dfrac{x_3 - x_2}{y_3 - y_2} = 0$

(5) $y_3 - y_2 = 0$

(6) $x_2 - x_1 = 0$
(7) $(x_4 - x_1)^2 + (y_4 - y_1)^2 - a^2 = 0$

(8) $(x_2 - x_1)^2 + (y_2 - y_1)^2 - b^2 = 0$

Because there are four vertices for a rectangle, there must be a total of eight (2×4) constraint equations. The first two constraints define the explicit position of point A. Otherwise, the geometry can translate on this surface. The third and fourth constraints define the parallelism between two sets of lines, $AB\|CD$ and $AD\|BC$. Constraint 5 defines BC as a horizontal line. Constraint 6 defines line AB is a vertical line; otherwise, the rectangle can rotate

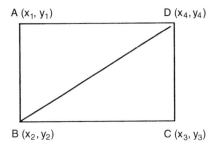

Figure 12.9. Rectangular XY surface.

about point A. Constraints 7 and 8 define the length and the width of a rectangle, $AD = a$, $AB = b$.

A properly dimensional geometry can be defined as one where the position of all graphic entities can be determined from the complete set of general constraints. That is, for any geometry with n characteristic points in two or three dimensions, there are $2n$ (or $3n$) independent constraint equations to describe it. If a geometry is specified by more than $2n$ or $(3n)$ constraints, it is said to be overconstrained. For example, in Figure 12.10, the distance between AB and DC is overconstrained (defined twice by $b + d$ and c, respectively). On the other hand, if the number of the constraints in a geometry is less than $2n$ or $3n$, the geometry is considered to be less constrained.

12.2.3 Constraint Variations (Tolerances) and Sensitivity Analysis

In a variational geometry, tolerances are represented as constraint variations. When a constraint varies, other dimensions may be affected. As showed in

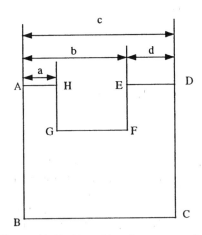

Figure 12.10. Example of overconstraint.

Figure 12.11, if the sixth constraint (length AB) changes to $x_2 - x_1 = \Delta v$ (v is called the variation of a constraint or tolerance), this variation will bring point B to a new position, B'. If BD, the diagonal of this rectangle, is a critical dimension in this mechanical design, the variation of the sixth constraint will definitely affect BD. How and to what degree the variation affects BD is discussed next.

Generally for a three-dimensional geometry, if there are n characteristic points, this geometry requires $3n$ constraint equations:

$$f_1(x_1,y_1,z_1,...,x_n,y_n,z_n) = 0$$

$$f_2(x_1,y_1,z_1,...,x_n,y_n,z_n) = 0 \qquad (12.1)$$

$$\vdots$$

$$f_{3n}(x_1,y_1,z_1,...,x_n,y_n,z_n) = 0$$

Every constraint is an explicit function with respect to the coordinates of the characteristic points, as mentioned before. As a critical dimension, T can be written

$$T = F(f_1,f_2,...,f_{3n}) \qquad (12.2)$$

In equation (12.2), if Taylor's series expression is used, the variation of critical dimension ∇T can be written as

$$\nabla T = \frac{\partial F}{\partial f_1}\nabla f_1 + \frac{\partial F}{\partial f_2}\nabla f_2 + \cdots + \frac{\partial F}{\partial f_{3n}}\nabla f_{3n} + \frac{1}{2}\frac{\partial^2 F}{\partial f_1 \partial f_2}\nabla f_1 \nabla f_2 + \cdots \quad (12.3)$$

If only the first derivative is used, and let $k_1 = \partial F/\partial f_i$, equation (12.3) can be simplified to

$$\nabla T = k_1 \nabla f_1 + k_2 \nabla f_2 + \cdots + k_{3n}\nabla f_{3n} \qquad (12.4)$$

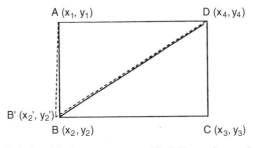

Figure 12.11. Relationship between unspecified dimension and constraints.

This equation defines the relationship between the variation of critical dimension and the variation of constraints. Any small variation in constraint f_i will cause the critical dimension T to vary k_i times of the constraint variation. Along with the $v_l \leq \nabla f_i (i = 1,2,...,3n) \leq v_u$ (v_l and v_u are the lower and upper limits for the variation), Equation (12.4) defines the tolerance design space. $k_i = \partial F / \partial f_i$ represents the sensitivity of critical dimension T to the constraint f_i. To obtain $\partial F / \partial f_i$, the following approach is used. Because f_1 ,f_2,...,f_{3n} are the explicit functions of $x_1,y_1,z_1,...,x_n,y_n,z_n$, the function (12.2) can also be written as

$$T = F(x_1,y_1,z_1,...,x_n,y_n,z_n) \tag{12.5}$$

So the partial derivatives $\partial F / \partial x_i$, $\partial F / \partial y_i$, and $\partial F / \partial z_i$ can be written as

$$\frac{\partial F}{\partial x_1} = \frac{\partial F}{\partial f_1}\frac{\partial f_1}{\partial x_1} + \frac{\partial F}{\partial f_2}\frac{\partial f_2}{\partial x_1} + \cdots + \frac{\partial F}{\partial f_{3n}}\frac{\partial f_{3n}}{\partial x_1}$$

$$\frac{\partial F}{\partial y_1} = \frac{\partial F}{\partial f_1}\frac{\partial f_1}{\partial y_1} + \frac{\partial F}{\partial f_2}\frac{\partial f_2}{\partial y_1} + \cdots + \frac{\partial F}{\partial f_{3n}}\frac{\partial f_{3n}}{\partial y_1}$$

$$\frac{\partial F}{\partial z_1} = \frac{\partial F}{\partial f_1}\frac{\partial f_1}{\partial z_1} + \frac{\partial F}{\partial f_2}\frac{\partial f_2}{\partial z_1} + \cdots + \frac{\partial F}{\partial f_{3n}}\frac{\partial f_{3n}}{\partial z_1} \tag{12.6}$$

$$\vdots \qquad \vdots \qquad \vdots \qquad \ddots \qquad \vdots$$

$$\frac{\partial F}{\partial z_n} = \frac{\partial F}{\partial f_1}\frac{\partial f_1}{\partial z_n} + \frac{\partial F}{\partial f_2}\frac{\partial f_2}{\partial z_n} + \cdots + \frac{\partial F}{\partial f_{3n}}\frac{\partial f_{3n}}{\partial z_n}$$

Written in matrix format, we have

$$
\begin{bmatrix}
\dfrac{\partial F}{\partial x_1} \\
\dfrac{\partial F}{\partial y_1} \\
\dfrac{\partial F}{\partial z_i} \\
\vdots \\
\dfrac{\partial F}{\partial z_n}
\end{bmatrix}
=
\begin{bmatrix}
\dfrac{\partial f_1}{\partial x_1} & \dfrac{\partial f_2}{\partial x_1} & \dfrac{\partial f_3}{\partial x_1} & \cdots & \dfrac{\partial f_{3n}}{\partial x_1} \\
\dfrac{\partial f_1}{\partial y_1} & \dfrac{\partial f_2}{\partial y_1} & \dfrac{\partial f_3}{\partial y_1} & \cdots & \dfrac{\partial f_{3n}}{\partial y_1} \\
\dfrac{\partial f_1}{\partial z_1} & \dfrac{\partial f_2}{\partial z_1} & \dfrac{\partial f_3}{\partial z_1} & \cdots & \dfrac{\partial f_{3n}}{\partial z_1} \\
\vdots & \vdots & \vdots & \ddots & \vdots \\
\dfrac{\partial f_1}{\partial z_n} & \dfrac{\partial f_2}{\partial z_n} & \dfrac{\partial f_3}{\partial z_n} & \cdots & \dfrac{\partial f_{3n}}{\partial z_n}
\end{bmatrix}
\begin{bmatrix}
\dfrac{\partial F}{\partial f_1} \\
\dfrac{\partial F}{\partial f_2} \\
\dfrac{\partial F}{\partial f_3} \\
\vdots \\
\dfrac{\partial F}{\partial f_{3n}}
\end{bmatrix}
\tag{12.7}
$$

where the $3n \times 3n$ matrix is a Jacobian matrix, because f_i is an explicit constraint equation with respect to x_i, y_i, z_i. Each element $\partial f_i / \partial x_i$, $\partial f_i / \partial y_i$, $\partial f_i / \partial z_i$ of this matrix is known. The inverse Jacobian matrix is

$$
\begin{bmatrix} \dfrac{\partial F}{\partial f_1} \\[2ex] \dfrac{\partial F}{\partial f_2} \\[2ex] \dfrac{\partial F}{\partial f_3} \\[2ex] \vdots \\[2ex] \dfrac{\partial F}{\partial f_{3n}} \end{bmatrix}
=
\begin{bmatrix}
\dfrac{\partial f_1}{\partial x_1} & \dfrac{\partial f_2}{\partial x_1} & \dfrac{\partial f_3}{\partial x_1} & \cdots & \dfrac{\partial f_{3n}}{\partial x_1} \\[2ex]
\dfrac{\partial f_1}{\partial y_1} & \dfrac{\partial f_2}{\partial y_1} & \dfrac{\partial f_3}{\partial y_1} & \cdots & \dfrac{\partial f_{3n}}{\partial y_1} \\[2ex]
\dfrac{\partial f_1}{\partial z_1} & \dfrac{\partial f_2}{\partial z_1} & \dfrac{\partial f_3}{\partial z_1} & \cdots & \dfrac{\partial f_{3n}}{\partial z_1} \\[2ex]
\vdots & \vdots & \vdots & \ddots & \vdots \\[2ex]
\dfrac{\partial f_1}{\partial z_n} & \dfrac{\partial f_2}{\partial z_n} & \dfrac{\partial f_3}{\partial z_n} & \cdots & \dfrac{\partial f_{3n}}{\partial z_n}
\end{bmatrix}^{-1}
\begin{bmatrix} \dfrac{\partial F}{\partial x_1} \\[2ex] \dfrac{\partial F}{\partial y_1} \\[2ex] \dfrac{\partial F}{\partial z_1} \\[2ex] \vdots \\[2ex] \dfrac{\partial F}{\partial z_n} \end{bmatrix}
\tag{12.8}
$$

From equation (12.8), the sensitivities of each constraint to the variation of critical dimension T can be calculated.

12.3 VARIATIONAL GEOMETRY-BASED TOLERANCE ANALYSIS

12.3.1 Conventional Tolerance Analysis Versus VG-Based Tolerance Analysis

As mentioned before, tolerance analysis is to verify whether the variation of a critical dimension is small enough so that no quality problem will occur later (i.e., given the tolerances for all relevant design constraints, the variation of the critical dimension should be guaranteed in the allowable range). The two most commonly used tolerance analysis methods are the worst-case and root sum of squares (statistical method) (Fortini, 1967). The worst-case method makes no assumptions of how the relevant design constraints vary within the tolerance zone, and the statistical method assumes that any of the relevant design constraint tolerances has a statistical distribution with mean value μ and standard deviation σ. If we consider only one critical dimension, tolerance analysis can be represented as (in the worst case)

$$
\begin{aligned}
|\eta|_{vg} &= |k_1|v_1 + |k_2|v_2 + \cdots + |k_n|v_n \\
&= (|DT_1|v_{DT1} + |DT_2|v_{DT2} + \cdots + |DT_s|v_{DTs}) + (|GT_1|v_{GT1} \\
&\quad + |GT_2\{v_{GT2} + \cdots + |GT_r|v_{GTr}) \\
&\geq (|DT_1|v_{DT1} + |DT_2|v_{DT2} + \cdots + |DT_s|v_{DTs}) = [|DT|][v_{DT}] \\
&= |\eta|_{conv}
\end{aligned}
\tag{12.9}
$$

$$
s + r = 3n
$$

where s = number of dimensional constraints

r = number of geometric constraints

$DT_i(i = 1,...,s)$ = sensitivity of dimensional constraint i to the critical dimension

GT_j $(j = 1,...,r)$ = sensitivity of geometric constraint j to the critical dimension

$|\eta|_{vg}$ = variation of the critical dimension resulted from VG-based tolerance analysis

$|\eta|_{conv}$ = variation of critical dimension resulted from conventional tolerance analysis

n = number of the total characteristic points

The statistical method can be represented as follows: The mean value of the critical dimension variation

$$
\begin{aligned}
E[\eta] &= \mu_{vg} \\
&= E[k_1 v_1 + k_2 v_2 + \cdots + k_n v_n] \\
&= k_1 \mu_1 + k_2 \mu_2 + \cdots + k_n \mu_n \\
&= (DT_1 \mu_{DT1} + DT_2 \mu_{DT2} + \cdots + DT_s \mu_{DTS}) \\
&\quad + (GT_1 \mu_{GT1} + GT_2 \mu_{GT2} + \cdots + GT_r \mu_{GTr}) \\
&= \mu_{DT} + \mu_{GT} \\
&= \mu_{conv} + \mu_{GT} \qquad (12.10)
\end{aligned}
$$

and the standard deviation of the critical dimension

$$
\begin{aligned}
\sigma_{vg} &= \sqrt{k_1^2 \sigma_1^2 + k_2^2 \sigma_2^2 + \cdots + k_n^2 \sigma_n^2} \\
&= \sqrt{(DT_1^2 \sigma_1^2 + \cdots + DT_s^2 \sigma_s^2) + (GT_1^2 \sigma_1^2 + \cdots + GT_r^2 \sigma_r^2)} \quad (12.11) \\
&\geq \sqrt{DT_1^2 \sigma_1^2 + \cdots + DT_s^2 \sigma_s^2} = \sigma_{conv}
\end{aligned}
$$

With either the statistical or worst-case approach, conventional tolerance analysis will give more optimistic results which may result in accepting a failed design (Dong and Shi, 1994, 1995).

12.3.2 Example

Figure 12.12 is a simple assembly used for tolerance analysis, which includes two parts, part 1 and part 2. If the assembly dimension T is critical to this assembly, and allowable variation for this critical dimension is $\pm \eta/2$, then after assigning each constraint (including both dimensional and geometric tolerance) a tolerance, and by using the VG-based tolerance analysis method, we can find the actual variation zone of this critical dimension. Comparing the actual variation zone with allowable variation $\pm \eta/2$, we can see whether this tolerance allocation is suitable. This process is usually called *tolerance analysis* or *tolerance evaluation.*

With the data in Table 12.3 we use both the statistical and worst-case methods in (12.9)–(12.11) to see whether the required variation, $\pm \eta/2$, of the critical dimension is satisfied. In this example, the required variation of the critical dimension $\pm \eta/2$ is ± 0.02. The result of this evaluation is listed in Table 12.4.

From Table 12.4 we can see that the tolerance allocation (Table 12.3) satisfies the design requirement by using the statistical method $[0.0114 - (-0.0114) = 0.0228 < 0.04 = 0.02 - (-0.02)]$. However, the tolerance allocation does not satisfy the design requirement $[0.0319 - (-0.0319) = 0.0638 > 0.04]$ by using the worst-case method. When the conventional method is used (i.e., the geometrical variations are not considered), the tolerance allocation is as listed in Table 12.5.

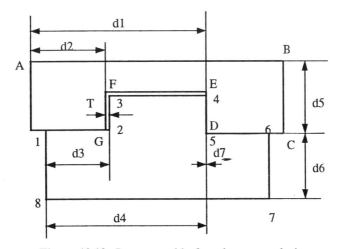

Figure 12.12. Parts assembly for tolerance analysis.

TABLE 12.3. Assigned Tolerances and Their Contributions to the Critical Dimension[a]

	d1	d2	d3	d4	d7	Pa11	Pa17	Pa21	Pa4	Pa6
Nomi.	1.8	1.0	1.0	1.75	0.0	0.0	0.0	0.0	0.0	0.0
Up. Tol	+0.004	+0.004	+0.004	+0.004	+0.004	+0.002	+0.002	+0.002	+0.002	+0.002
Lo. Tol	−0.004	−0.004	−0.004	−0.004	−0.004	−0.002	−0.002	−0.002	−0.002	−0.002
Sensi.	1.0	−1.0	1.0	−1.0	1.0	1.227	0.844	0.847	0.286	0.286
Contri. (%)	19.9	19.9	19.9	19.9	4.97	7.49	3.54	3.54	0.41	0.41

[a]Nomi., nominal dimension; Up. Tol, upper tolerance assigned to a dimension; Lo. Tol, lower tolerance assigned to a dimension; Sensi., sensitivity of a constraint to this critical dimension; Contri., contribution (%) of a constraint's variation to the critical dimension's variation; Pa, parallelism between two lines, which is as follows (see Figure 12.12): Pa11, parallelism between A–H and 4–5; Pa17, parallelism between A–H and 1–8; Pa21, parallelism between 1–8 and 4–5; Pa4, parallelism between A–H and F–G; Pa6, parallelism between A–H and E–D.

TABLE 12.4. Evaluation Results (with VG-Based Approach) for Critical Dimension T^a

Nomi	Up. Tol	Low. Tol	Stat. Tol	Worst Tol
0.05	+0.02	−0.02	±0.0114	±0.0319

[a]Nomi, nominal dimension of the critical dimension; Up. Tol, upper tolerance required for this critical dimension; Low. Tol, lower tolerance required to the critical dimension; Stat. Tol, using the statistical method to calculate the actual tolerance zone for this critical dimension, the statistic tolerance value in Table 12.4 is $6\sigma_{vg}$; Worst Tol, using the worst-case method to calculate the actual tolerance zone for this critical dimension.

From Table 12.6, the tolerance allocation using conventional methods satisfies the design requirements [0.0089 − (−0.0089) = 0.0178 < 0.04 and 0.02 − (0.02) = 0.04 ≤ 0.04]. It is clear to see that because the geometrical tolerances are not taken into consideration, the critical tolerance zone calculated becomes tighter and the acceptable rate is higher than that when using the VG-based methods.

12.4   VARIATIONAL GEOMETRY-BASED OPTIMAL DIMENSIONAL TOLERANCE ALLOCATIONS

12.4.1   Mathematical Models for Optimal Tolerance Allocations

As mentioned before, with one or more given critical dimensions in a mechanical assembly, the tolerance allocation is to distribute tolerances among relevant design constraints to achieve the lowest overall manufacturing cost. The following is an optimization model for tolerance allocations, in which multimanufacturing processes are assumed to be available to produce each constrain, and process selection is therefore involved in the model (Dong and Shi, 1994, 1995).

TABLE 12.5. Dimensional Tolerances Assigned and Their Contributions to Critical Dimension^a

	d1	d2	d3	d4	d7
Nomi.	1.8	1.0	1.0	1.75	0.0
Up. Tol	+0.004	+0.004	+0.004	+0.004	+0.004
Low. Tol	−0.004	−0.004	−0.004	−0.004	−0.004
Sensi.	1.0	−1.0	1.0	−1.0	1.0
Contri. (%)	23.53	23.53	23.53	23.53	5.88

[a]See the footnote to Table 12.3 for explanations of the terms.

TABLE 12.6. Evaluation Results (with Conventional Methods)

Nomi.	Up. Tol	Low. Tol	Stat. tol	Worst Tol
0.05	+0.02	−0.02	±0.0089	±0.02

[a]See the footnote to Table 12.4 for explanations of the terms.

$$\text{minimize } C_q(v) = \text{minimize } \sum_{i=1}^{n} \sum_{j=1}^{pi} \delta_{ij} s_i c_{ij}(v_{ij}) \tag{12.12}$$

subject to (tolerance design space):

Worst case:

$$\sum_{i=1}^{n} [\delta_{ij} k_{ij} c_{ij}(v_{ij})] \leq |\eta|_q \tag{12.13}$$

or

statistical method:

$$\sum_{i=1}^{n} [\delta_{ij} k_{ij}^2 \sigma_{ij}^2] \leq \sigma_q^2 \tag{12.14}$$

$$\mu_q = 0 \qquad |\eta|_q = 6\sigma_q$$

$$v_{u_i} \geq v_{ij} \geq v_{l_i} \qquad (i = 1,...,n, \quad j = 1,...,pi)$$

where $|\eta|_q$ = functional requirement to the critical dimension

$C_q(v)$ = overall manufacturing cost to achieve the allowable variation $|\eta|_q$ of the critical dimension

$c_{ij}(v_{ij})$ = cost for manufacturing constraint i with process j

$v_{ij}(i = 1,...,n)$ = tolerance (variation) for design constraint i with process j

σ_{ij} = deviation for the variation of constraint i with process j

σ_q = allowable deviation for the qth critical dimension (here we assume that $|\eta|_q = 6\sigma_q$)

$k_{ij}(i = 1,...,n, j = 1,...,p) =$ sensitivity of constraint i to critical dimension q with process j

$\delta_{ij}(i = 1,...,n, j = 1,...,p) =$ 1, if process j is chosen to produce constraint i; 0, otherwise

$n =$ total number of relevant design constraints to the qth critical dimension

$pi =$ total number of available processes for producing constraint i

$q =$ number to represent the qth critical dimension; if there are m critical design functions, $q = 1,...,m$

$v_{ui} =$ upper tolerance limit for constraint i

$v_{li} =$ lower tolerance limit for constraint i

For simplicity the worst-case approach is used and only one critical design function and one manufacturing process per design constraint are considered. Model (12.12) can therefore be written as

$$\text{minimize } C(v) = \text{minimize } [s_1 c_1(v_1) + s_2 c_2(v_2) + \cdots + s_n c_{3n}(v_{3n})]$$

$$= \text{minimize } \sum_{i=1}^{s} [s_1 C_{DT11}(v_{DT11}) + \cdots + s_s C_{DTs}]$$

$$+ \sum_{i=1}^{r} [s_{1i} C_{GT1}(v_{GT1}) + \cdots + s_r C_{GTr}]$$

$$= \text{minimize } [C_{DT}(v_{DT}) + C_{GT}(v_{GT})] \tag{12.15}$$

subject to the worst case:

$$|\eta|_{vg} \geq |k_1|v_1 + |k_2|v_2 + \cdots + |k_{3n}|v_{3n}$$

$$= (|DT_1|v_{DT1} + |DT_2|v_{DT2} + \cdots + |DT_s|v_{DTs})$$

$$+ (|GT_1|v_{GT1} + |GT_2|v_{GT2} + \cdots + |GT_r|v_{GTr})$$

$$= [|DT|][v_{DT}] + [|GT|][v_{GT}] \tag{12.16}$$

$$v_{li} \leq v_i \leq v_{ui} \qquad i = 1,2,...,3n \tag{12.17}$$

where $C_{DT}(v_{DT})$ is the overall manufacturing cost for obtaining all dimensional tolerances (DT) and $C_{GT}(v_{GT})$ the overall manufacturing cost for obtaining all geometric tolerances (GT).

In conventional tolerance synthesis, geometric constraints are not considered [$C_{DT}(v_{DT})$ = 0 and [|GT|][v_{GT}] = 0], and |DT| is assumed to be 1 for all dimensional constraints (Chase et al., 1990; Zhang and Wang, 1993). Conventional tolerance synthesis is a very special case in the VG-based tolerance synthesis, which rarely happens in practical applications.

12.4.2 VG-Based Manufacturing Cost–Tolerance Models for Optimal Dimensional Tolerance Allocation

The relationship between dimensional tolerance and manufacturing cost has been studied by many researchers. Some experiments have been done to obtain the empirical cost–tolerance data for typical production processes, including turning on lathe, face milling and drilling, hole machining, and so on (Wu et al., 1992). A lot of research has gone into formulating empirical cost–tolerance data into nonlinear cost–tolerance formulas. Formulas proposed include an exponential formula (Speckhart, 1994), a reciprocal squared formula (Wu, Elmaraghy, 1988), a reciprocal powers formula, and a powers and exponential hybrid formula (Dond and Zoom, 1986). Dong, Hu and Xue (1991) tested various types of mathematics formulas and built five new formulas for cost–tolerance relation with better fittings:

- Combined reciprocal powers and exponential function (combined RP-E):

$$c(v) = a_0 + a_1 v^{-a_2} + a_3 e^{-a_4 v} \tag{12.18}$$

- Combined linear and exponential function (combined L-E):

$$c(v) = a_0 + a_1 v + a_2 e^{-a_3 v} \tag{12.19}$$

- Cubic polynomial (cubic-p):

$$c(v) = a_0 + a_1 v + a_2 v^2 + a_3 v^3 \tag{12.20}$$

- Fourth-order polynomial (4th-p):

$$c(v) = a_0 + a_1 v + a_2 v^2 + a_3 v^3 + a_4 v^4 \tag{12.21}$$

• Fifth-order polynomial (5th-p):

$$c(v) = a_0 + a_1v + a_2v^2 + a_3v^3 + a_4v^4 + a_5v^5 \qquad (12.22)$$

Because of the smallest fitting errors, the fifth-order polynomial formula is selected as the cost–tolerance function $C_i(v)(i = 1,2,...,I)$ in our research, where I represents the total number of relevant dimensional constraints.

12.4.3 Example

Figure 12.13 is a simplified intersection of the slideway assembly used for tolerance analysis. The slideway was designed by the Precision Manufacturing Center at the University of Connecticut, and has two subassemblies, a slider subassembly and a saddle subassembly. To simplify this problem, we treat each subassembly as a single mechanical part.

To allow lubricant oil flow, very thin clearances x_1 and x_2 are required when these two subassemblies are put together. Clearances x_1 and x_2 are very critical to the product, as they directly determine the stability and assemblability of the product. The nominal dimension of the rectangular slot in the slider subassembly is 1.000 in., and the nominal dimension of the slab in the saddle subassembly is 0.999 in. So the total nominal clearance between these two subassemblies is just 0.001 in. From the point of view of the product's functionality, a clearance between 0.0005 and 0.0015 in. will satisfy the function's requirement very well; that is, this critical dimension is required to be 0.001 ± 0.0005 in. Although clearances x_1 and x_2 are very important to the

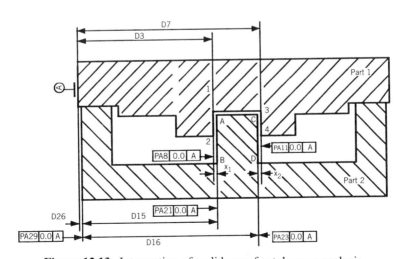

Figure 12.13. Intersection of a slideway for tolerance analysis.

functionality of the slideway, they cannot be controlled directly. They are affected by many other dimensions, such as the dimension of the slot, the dimension of the slab, the parallelism of the slot and slab, and so on. In this example, x_1 and x_2 are chosen as the critical dimensions. Using the VG-based algorithm, we can identify the relevant constraints and the relationships between variations of the critical dimensions and those of other relevant constraints (Hurt, 1994, SDRC, 1995). From the tolerance analysis results listed in Table 12.7, it is very clear to see that six constraints influence critical dimensions x_1 and x_2: d3, d7, d15, d16, and d26 are dimensional constraints and Pa8, Pa11, Pa21, and Pa29 are geometric constraints, as illustrated in Figure 12.13. Their relationships are represented in equations (12.23) and (12.24). The clearance between the slot and the slab is

$$|\text{distance between } 1–2 \text{ and } 3–4| - |\text{distance between } A–B \text{ and } C–D|$$

$$= |\text{distance between } 1–2 \text{ and } A–B| + |\text{distance between } 3–4 \text{ and } C–D|$$

$$= x_1 + x_2$$

The relevant constraints that will affect critical dimensions x_1 and x_2 were listed in Table 12.7. If only the worst case is considered, from equation (12.16) we have

$$|v_1| + |v_2| + |v_3| + |0.9140\, v_4| + |0.5v_5| + |0.253v_6|$$

$$+ |v_1'| + |v_2'| + |1.24520v_3'| + |0.5v_4'| \leq 0.0005 \quad (12.23)$$

where v_1, v_2, v_3, v_1', and v_2' stand for the dimensional variations of d3, d15, d26, d16, and d7, and v_4, v_5, v_6, v_3', v_4' stand for the geometric variations Pa21, Pa8, Pa29, Pa11, and Pa23.

Because cost–tolerance functions are currently available only for dimensional constraints, tolerance for geometric constraints such as parallelism and perpendicularity are not considered in this model (this issue will be studied

TABLE 12.7. Results for x_1 and x_2

Relevant Constraints for x_1			Relevant Constraints for x_2		
Name	Nominal	Sensitivity	Name	Nominal	Sensitivity
d3	5.41	−1.0	d16	6.39	1.0
d15	5.39	1.0	d7	6.41	−1.0
d26	0.015	−1.0	d26	0.015	1.0
Pa21	0.0	0.9140	Pa11	0.0	1.2452
Pa8	0.0	0.5	Pa23	0.0	0.5
Pa29	0.0	0.253	Pa29	0.0	0.253

further and the results will be published elsewhere, and all these tolerances are therefore set to zero). The functional equation (12.23) becomes

$$|v_1| + |v_2| + |v_1'| + |v_2'| \leq 0.0005 \tag{12.24}$$

The complete optimization model for critical dimension x_1 is

$$\text{minimize } C(v) = \text{minimize } [S_1 C_1(v_1) + S_2 C_2(v_2)$$

$$+ S_1' C_1'(v_1') + S_2' C_2'(v_2')] \tag{12.25}$$

$$|v_1| + |v_2| + |v_1'| + |v_2'| \leq 0.0005 \tag{12.26}$$

and

$$0.00005 \leq v_i \leq 0.001 \qquad i = 1,2,3 \tag{12.27}$$

In this model the cost–tolerance functions for milling, surface grinding, and lapping are as follows:

$$c(v) = 11.08 + 8.51 \times 10^3 v - 1.65 \times 10^5 v_2 + 1.22 \times 10^6 v^3$$

$$- 4.03 \times 10^6 v^4 + 5.03 \times 10^6 v^5 \tag{12.28}$$

$$c(v) = 98.86 - 3.678 \times 10^3 v + 1.568 \times 10^5 v^2 - 3.536 \times 10^6 v^3$$

$$+ 3.919 \times 10^7 v^4 - 1..647 \times 10^8 v^5 \tag{12.29}$$

$$c(v) = 104.4 - 2.34 \times 10^3 v - 5.05 \times 10^4 v^2 + 5.85 \times 10^6 v^3$$

$$- 7.06 \times 10^7 v^4 - 1.17 \times 10^9 v^5 \tag{12.30}$$

Here 0.00005 and 0.01 in. refer to the upper and lower manufacturing capabilities; and $S_1 = 4.25$ in^2, $S_2 = 8.64$ in^2, $S_1' = 4.25$ in^2, and $S_2' = 8.64$ in^2.

The optimization problem is solved with IMSL subroutine ICONF. Given the iteration criterion $\epsilon < 1.0E-6$, after hundreds of iterations the optimal tolerances values for d3, d15, d7, and d16 are obtained. The manufacturing processes selected and the optimal tolerances for these dimensions are listed in Table 12.8.

TABLE 12.8. Manufacturing Processes and Optimal Tolerance Allocations

	Optimal Tolerance for x_1		Optimal Tolerance for x_2	
Dimension	d3	d15	d7	d16
Nominal	5.41	5.39	6.41	6.39
Optimum tolerance	0.00005	0.0002	0.00005	0.0002
Process[a]	F and L	F and SG	F and L	F and SG

[a]F, face milling; SG, surface grinding; L, lapping.

With tolerance–cost functions, the lowest manufacturing cost can be calculated. For critical dimension x_1, with the data given we can get the total cost:

$$C(v) = 188.45$$

For critical dimension x_2, the total cost will be

$$C(v) = 188.45$$

The relative cost to achieve 0.001 ± 0.0005 in. as the critical clearance between the slab and the slot is therefore

$$C(v_i) = \Sigma[S_i C_i(v_i) + S_i' C_i'(v_i')] \qquad i = 1,2$$
$$= 188.45 + 188.45 = 376.90 \tag{12.31}$$

Without using the VG-based manufacturing tolerance–cost model, the engineers used experience-based handbooks and chose the manufacturing processes and assigned tolerances as listed in Table 12.9. With the manufacturing process and tolerances, we can get the total cost $C(v_i) = 434.24$.

The manufacturing cost difference for the two sets of tolerances is

$$\nabla C(v) = C(v) - C(v)_{VG} = 434.24 - 376.90 = 57.34$$

TABLE 12.9. Selected Manufacturing Process and Tolerance Allocations (Experience-Based Approach)

	Optimal Tolerance for x_1		Optimal Tolerance for x_2	
Dimension	d3	d15	d7	d16
Nominal	541	5.39	6.41	6.39
Tolerance	0.0001	0.0001	0.0001	0.0001
Process	F and L	F and L	F and L	F and L

The saving is

$$\frac{\nabla C(v)}{C(v)} = \frac{57.34}{434.24} = 0.132 = 13.2\% \text{ per slideway}$$

Considering the large quantities involved in manufacturing, the savings can be significant.

12.5 CONCLUSIONS AND FUTURE RESEARCH

The concepts of tolerance analysis and synthesis in a variational design environment have been presented. The newly developed algorithms have been applied successfully to tolerance evaluation and optimal allocation of tolerances for the components selected. The optimal tolerance allocations achieved not only meet the requirements of minimizing manufacturing cost but also meet the functionality requirements for both components and the assembly system. This optimal tolerance allocation algorithm has great potential to be developed as a computer tool for concurrent design and manufacturing.

Only the local dimensional tolerances were considered in the tolerance synthesis approach presented. Global geometric tolerances such as variations in parallelism and perpendicularity are not taken into account. These global tolerances are usually affected not only by the manufacturing processes, but also by the way a machining process is set up and the fixtures used. Further research is under way to study the issue.

REFERENCES

Bjorke, O., 1989, *Computer-Aided Tolerancing,* Tapir, Trondheim, Norway.

Chase, K. W., Greenwood, W. H., Loosli, B. G., and Hauglund, L. F., 1990, Least Cost Tolerance Allocation for Mechanical Assemblies with Automated Process Selection, *Manuf. Rev.,* Vol. 3, No. 1, March, pp. 211–218.

Dong, Z., and W. Hu, 1991, Optimal Process Sequence Identification and Optimal Process Tolerance Assignment in Computer Aided Process Planning, *Comput. Ind.,* Vol. 17, pp. 19–32.

Dong, J. and Shi, Y., 1997, Tolerance Sensitivity Analysis in a Variational Design Environment, International Journal of Vehicle Design, Vol. 18, No. 5, pp. 474–486.

Dong, J., and Shi, Y., 1995, Variational Geometry Theory for Tolerance Sensitivity Analysis, presented at the 1995 ASME Winter Congress and Exposition, Geometry and Tolerance Conference, 95-WA/DE-7, San Francisco, November 12–17.

Dong, Z., and Soom, A., 1986, Automatic Tolerance Analysis from a CAD Database, *Proceedings of the* 86-DET-36, *Design Engineering Conference,* Columbus, Ohio, Oct. 5–8, 1986, American Society of Mechanical Engineers, New York.

Dong, Z., W. Hu, and Xue, D., 1994, New Production Cost-Tolerance Models for Tolerance Synthesis, *J. Eng. Ind.,* Vol. 116, pp. 199–206.

Fortini, E. T., 1967, *Dimensioning for Interchangeable Manufacturing,* Industrial Press, New York.

Gossard, D. C., Zuffante, R. P., and Sakurai, H., 1988, Representing Dimensions, Tolerances, and Features in MCAE Systems, *IEEE Comput. Graph. Appl.,* March, pp..

Greenwood, W. H. and Chase, K. W., 1987, A New Tolerance Analysis Method for Designers and Manufacturers, *J. Eng. Ind.,* Vol. 109, May, pp. 112–116.

Greenwood, W. H., and Chase, K. W., 1988, Worst Case Tolerance Analysis with Nonlinear Problems, *J. Eng. Ind.,* Vol. 101, August, pp. 232–235.

Hillyard, R. C., 1978, Dimensions and Tolerances in Shape Design, Ph.D. dissertation, University of Cambridge, Cambridge, England.

Hillyard, R. C., and Braid, I. C., 1978, Analysis of Dimensions and Tolerances in Computer-Aided Mechanical Design, *Comput.-Aid. Des.,* Vol. 10, No. 3, May, pp. 161–166.

Hurt, J., 1994, *The Gaussian Assumption in Ideas for Tolerance Analysis,* Technical Report, Structural Dynamics Research Corporation,

Lee, S., and Yi, C., 1994, Tolerance Analysis for Assembly Planning, 0-8186-6510-6/ 94, *Proceedings of the 4th International Conference on Computer Integrated Manufacturing and Automation Technology,*

Lee, W., and Woo, T. C., 1987, *Optimum Tolerance Selection of Discrete Tolerances,* Technical Report 87-34, University of Michigan, Ann Arbor, MI.

Light, R., and Gossard, D., 1982, Modification of Geometric Models Through Variational Geometry, *Comput.-Aid. Des.,* Vol. 14, No. 4, July, pp. 209–214.

Lin, Y. C., 1981, Variational Geometry in Computer Aided Design, M.S. thesis, MIT, Cambridge, MA.

Michael, W., and Siddall, J. N., 1981, The Optimization Problem with Optimal Tolerance Assignment and Full Acceptance, *J. Mech. Des.,* Vol. 103, October, pp. 842–848.

Parkinson, D. B., 1985, Assessment and Optimization of Dimensional Tolerances, *Comput.-Aid. Des.,* Vol. 17, No. 4, May, pp. 191–199.

Roy, U., Liu, C. R., and Woo, T. C., 1991, Review of Dimensioning and Tolerancing: Representation and Processing, *Comput.-Aid. Des.,* Vol. 23, No. 7, pp. 466–483.

SDRC, 1995, *Tolerance Analysis Manual,* Structural Dynamics Research Corporation,

Speckhart, F. H., 1994, Calculation of Tolerance Based on a Minimum Cost Approach, *J. Eng. Ind.,* Vol. 94, No. 2, pp. 447–453.

Spotts, M. F., 1973, Allocation of Tolerances to Minimize Cost of Assembly, *J. Eng. Ind.,* August, pp. 762–764.

Turner, J. U., 1987, Tolerances in Computer Aided Geometric Design, Ph.D. thesis, Rensselaer Polytechnic University, Troy, NY.

Turner, J. U., 1990, Exploiting Solid Models for Tolerance Computations, in *Geometric Modeling for Product Engineering,* M. Wozony, J. Turner, and K. Preiss (eds.), North-Holland, New York.

Wu, Z., Elmaraghy, W. H., and Elmaraghy, H. A., 1992, Evaluation of Cost-Tolerance Algorithm for Design Tolerance and Synthesis, *Manuf. Rev.,* Vol. 1, No. 3, June, pp. 168–179.

Zhang, C., and Wang, H.-P., 1993, The Discrete Tolerance Optimization Problem, *Manuf. Rev.,* Vol. 6, No. 1, March, pp. 60–70.

III
APPLICATIONS AND INDUSTRIAL PRACTICES

13

DIMENSIONING AND TOLERANCING FOR FUNCTION

Roland D. Weill

Technion—Israel Institute of Technology
Haifa, Israel

13.1   INTRODUCTION

When conceiving a new product, a designer has to take into account both the functional requirements for its utilization and the requirements for its manufacture, assembly, and inspection. In other words, the definition of the product geometry, material choice, rough blank choice, and many other decisions have to satisfy simultaneously the functional characteristics of the final product and the manufacturing feasibility, including economic considerations. This means that in most practical cases, a compromise must be found between different, sometimes conflicting, requirements.

In most practical cases, however, functionality requirements are dominant and have to be observed as a priority, although in some special cases, such as mass production, product design is influenced directly by production conditions. In general, the design of the product (i.e., its shape, weight, size, material, dimensional accuracy, etc.) determines the production and inspection facilities used for the product. The general technical documentation helps in finding suitable production conditions according to the product design (Halevi and Weill, 1995).

Basically, products are described in great detail by technical drawings, which serve as the source of information for their definition. An important

Advanced Tolerancing Techniques, Edited by Hong-Chao Zhang
ISBN 0-471-14594-7   © 1997 John Wiley & Sons, Inc.

part of the description in the drawing concerns the dimensions and tolerances of the components, which define a part completely and without ambiguity. By dimensioning and tolerancing a part precisely on a drawing, its manufacturing processing method (sequence of operations, tooling and machines required, inspection instruments, etc.) is basically defined. If dimensioning and tolerancing are not precise enough, it will be difficult to define a suitable process plan and problems will appear during fabrication. A definition of too small tolerances, for example, will be difficult to satisfy in production, or will at least increase the cost of fabrication, without necessarily contributing to better functionality. Therefore, the importance of dimensioning and tolerancing cannot be overemphasized and is examined in detail in this chapter.

Dimensioning and tolerancing by a designer should satisfy the following basic conditions (Ropion, 1968):

1. Define without ambiguity a product suitable for an application.
2. Confer the widest manufacturing tolerances possible that enable functionality and interchangeability.
3. Allow manufacturing personnel the freedom to choose the most suitable and economical production facilities.

The adoption of the preceding principles is called *dimensioning and tolerancing for function* (DTF) or *functional dimensioning*. It represents the basic approach to design for manufacturing and has been largely accepted in actual manufacturing. Farmer calls its *tolerance analysis* and has given a precise definition of its meaning (Farmer and Gladman, 1986), with a view toward computerized tolerance analysis:

1. Describe and quantify the functional requirements of the design.
2. Identify datums, features, and geometric relations between the features that will influence functional and assembly requirements.
3. Develop functional equations with the functional requirements as dependent variables and the sizes and tolerances of the dimensions affecting the functional requirements as independent variables.
4. Determine an economic solution to these equations.

At this stage, the importance of respecting as a priority the functionality of a product should be emphasized strongly. In some cases a designer could be inclined to define dimensions easy to control by inspection or more suitable for certain types of tooling. However, this approach does not consider the fact that fabrication and inspection means are different from enterprise to enterprise and that satisfaction of the same requirements in design can be obtained in many different ways. Against the argument that some dimensions might be difficult to inspect, it can be remarked that functional dimensions are generally part of a mechanism and are in direct relation with a counterpart in the mechanism. Therefore, the functional gage to check a dimension can be derived from the design of the conjugate part and its feasibility is warranted. In this case the inspection of the part is "the image of the function" (Ropion, 1968).

Therefore, it is reasonable to conclude that the functional approach, the DTF approach, is optimal from a design point of view and satisfies most of the conditions required during production and inspection. This approach can be applied in a very general way to all types of functions (e.g., mechanical strength, power transmission, guidance in translation and rotation, sealing, etc.). A special case is represented by assemblability, which is analyzed in more detail later.

Before reviewing applications of the DTF principle, it will be useful to give precise definitions of functional and nonfunctional dimensions, well described in the Australian standard (Farmer, 1993) as follows:

> A distinction is made between functional and non-functional dimensions. The term functional dimension is described in AS1100 101-1992 as a dimension which affects directly the functioning of the product.

> Dimensions shall be selected and arranged to suit the function and mating relationship of a part and shall not be subject to more than one interpretation. Such dimensions are termed functional dimensions.

> The drawing should define a part without specifying construction and inspection methods. Thus, only the diameter of a hole is given without indicating whether it is drilled, reamed, punched, or made by any other operation.

It is appropriate to add that international standards (from ISO, the International Organization for Standardization) or national standards (from ANSI, the American National Institute of Standards) are essentially based on considerations of functionality, particularly on considerations of assemblability. It is therefore strongly recommended that the existing standards be used in mechanical design to guarantee respect for the principle of DTF.

13.2 DTF CONCEPT AND ITS LIMITATIONS

13.2.1 Simple Examples

To give an idea of the significance of the DTF concept, some simple examples will be reviewed. The distinction between functional and nonfunctional dimensions is well illustrated by the example given in the Australian standard (Farmer, 1993), shown in Figure 13.1. It represents a sliding panel guide assembly, where the dimensioning of the external cylindrical sections of the axle can be defined by functional or nonfunctional dimensions, as shown in Figure 13.2. For example, dimension A can be defined by contact of the shoulder with the guide frame so that the mating cylinder of the axle will

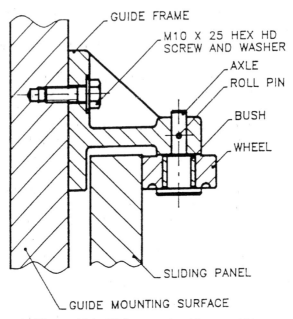

Figure 13.1. Sliding panel guide assembly.

protrude from the hole in the frame. Therefore, dimension *A* has to be just longer than the length of the mating hole in the frame. In the same fashion, dimension *B* is defined to guarantee free rotation of the wheel on the axle, and dimension *C* is defined as a width sufficient to withstand end loading. If nonfunctional dimensions, as shown in Figure 13.2, had been chosen (dimensions *D, E,* and *F*), they would not directly fulfill the functions prescribed. A transfer of tolerances would be necessary between the functional and nonfunctional dimensions, which are the dimensions executed in manufacturing. As a consequence, the tolerances of dimensions, *D, E,* and *F* would have to be restricted to satisfy the given function. Obviously, such transfer is not desirable for economical reasons and should be avoided.

This simple example can be used to define some general, very important rules concerning DTF, such as:

1. The DTF concept implies that the datums for manufacturing and inspection should be identical with the datums in design.

2. The functional dimensions are defined by the end surfaces that are in contact with other surfaces (i.e., the joints, which can be fixed as in assemblies, or mobile as in kinematic chains). Other dimensions in the composing links are nonfunctional and should not be considered in defining the dimensional chains that satisfy the functionality.

3. The dimensional chains defining a functional characteristic should be minimal, so as not to generate stackup of tolerances.

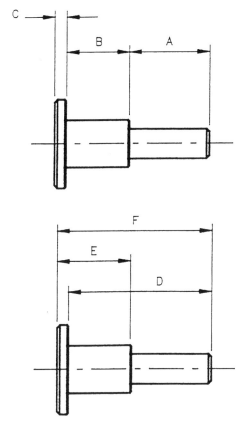

Figure 13.2. Functional and non-functional length dimensions of axle.

The assembly shown in Figure 13.3 (Ropion, 1968) can serve as an illustration of these basic principles. The functional quantity j, which is the gap between the end surfaces, could be defined by the dimensions of the composing parts, as follows:

$$j = -C_1 - B_1 - B_2 - A_4 + A_6 - A_1 - A_2 \qquad (13.1)$$

with the consequence that the high number of dimensions involved will produce a stackup of tolerances and difficulties to respect the imposed tolerance for j (0.3 mm). On the contrary, by limiting the number of dimensions to the really functional ones (i.e., the dimensions between the junction surfaces of the components), the dimensional chain becomes much simpler and the worst-case tolerance will be defined as follows (Figure 13.4):

$$\text{Tol } j = \text{Tol } C_1 + \text{Tol } B_5 + \text{Tol } A_7 = 0.3 \text{ mm} \qquad (13.2)$$

By taking reasonable tolerances in manufacturing for the dimensions C_1, B_5,

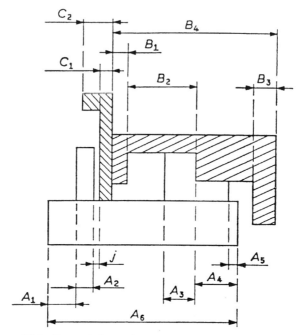

Figure 13.3. General dimensioning and tolerancing of clearance j.

and A_7, suitably distributed according to the capabilities of production means, one can reach an economic optimum. In this special case, the tolerances for C_1 and A_7 can be taken equal to 0.06 mm, obtained by grinding, because a good surface finish is required for these surfaces. A large tolerance is left for B_5 (0.18 mm), which can be obtained by a less precise process.

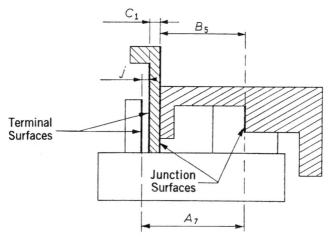

Figure 13.4. Definition of functional dimensions.

The basic principles just defined for linear chains can easily be generalized for inclined surfaces, such as the sliding guide surface of part A in Figure 13.5. The functional position of this surface should ensure correct motion of plunger B, so that the dogs H and H_1 are displaced alternatively. It means that the functional surface between A and B should be positioned in relation to the contact surfaces F and F_1 of the support M. By defining the position of this surface by dimensions $a1$, 2, and angle α as shown in Figure 13.6b, a stackup of tolerances is produced. On the contrary, if the functional dimensions, as defined in Figure 13.6a are angle α and the distance 1, a gain in accuracy is obtained. In this case also, the functional dimension 1 is related to a joint surface (i.e., intersection of the dihedral formed by surfaces F and F_1). Practically, the physical intersection, which is not a precise reference, has not really to be considered, as shown in Figure 13.7a and b, where the setting in the production stage and the checking in the inspection stage are related to two functional surfaces F and F_1 by using a precise jig having the shape of part A. As shown, the dimensions used in manufacturing and in inspection are the angle α and the distance 1. A simple plug gage will check the position of the gliding surface in relation to the datum surfaces. The principle of the identity of the datums in design, production, and inspection is well respected in this example.

Of course, the definition of functional dimensions in the simple examples given before is not very difficult. For industrial practical cases, the choice of relevant functional dimensions is not as easy in the general case. For example, in the mechanism represented in Figure 13.8, the lever P has an inclined end surface F_1 which should be at a distance equal to 2.25 $\pm$ 0.25 mm from the conjugate surface F_2 when both levers are in a blocked position on anvils B_1 and B_2. The functional surfaces here are obviously the bore A_1 and the contact surface on anvil B_1. A possible functional dimensioning is given in Figure 13.9, consisting in a defined orientation of surface F_1 and a given distance F from a point L as represented. The datums of the inspection device (Figure

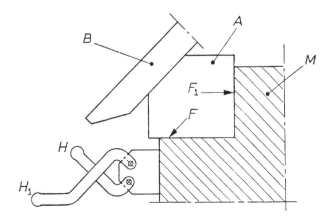

Figure 13.5. Case of inclined functional surface.

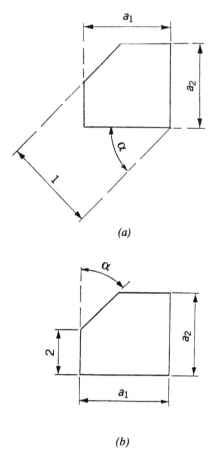

Figure 13.6. Dimensioning by (*a*) distance 1 and angle α and (*b*), distances *a*1 and 2 and angle α.

13.10) and of the manufacturing workholder should, of course, be identical to the functional datums defined in Figure 13.9.

13.2.2   Two Special Applications of the DTF Principle

The principle of DTF is used implicitly in the concept of the MMC condition. The MMC principle considers the case of mating parts and allows enlargement of a geometrical tolerance when the feature considered is not at the maximum material size (ISO Standard 2692, 1988), therefore providing a reduction in the manufacturing cost. In addition, application of the MMC principle can be checked by simple gages, which is also an economic and technical gain. Respecting the functions of mating completely, the MMC principle possesses the advantages just mentioned and can be applied successfully to complex mating parts that are dimensioned by *composite tolerancing*.

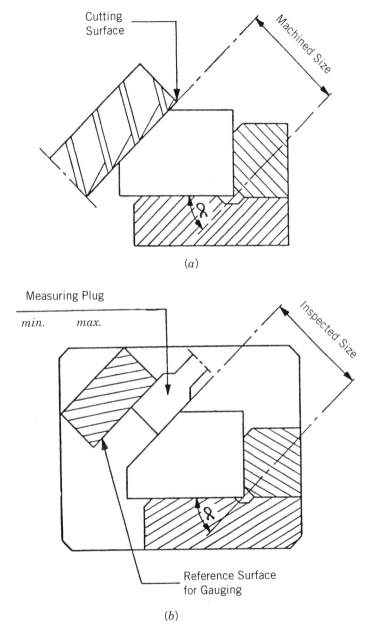

Figure 13.7. Functional dimensioning in (*a*) manufacturing and (*b*) inspection.

A good example of composite tolerancing is given in the Standard ISO 5459 (1981) on datum definitions for geometrical tolerances. As shown in Figure 13.11, the tolerancing uses the group of four holes, called *C*, as a datum for the group of three holes while defining their position tolerances

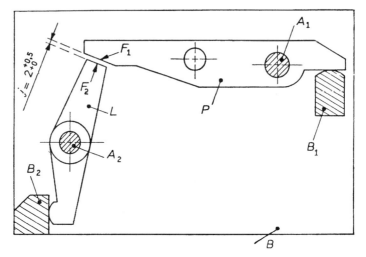

Figure 13.8. Locking lever.

with the MMC condition, also applied for the four holes of pattern C. In addition, the feature positioning of the three holes is defined by a smaller tolerance, also using the MMC condition. As a consequence, the positions of the equally spaced three holes benefit from two bonuses: one is the result of the feature-related tolerance of 0.05 mm and the other is the result of the pattern-related tolerance to the group C, equal to 0.15 mm.

A problem similar to pattern tolerancing has been raised by the tolerancing of profiles, for example, in the case of a cam controlling the motion of a follower to given tolerances, or the tolerancing of the shape of a cover plate that must fit in a recessed conjugate part. The function and assembly of such components are dependent on correct geometric tolerancing of the functional requirements and on knowledge about errors in production processes, fixturing, and positioning, as well as in inspection by fixed gages or by coordinate measuring machines (CMMs). The inspection of positioning in composite

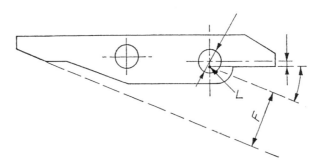

Figure 13.9. Functional dimensioning of the locking lever.

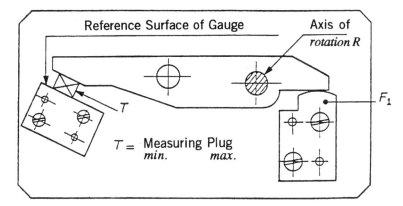

Figure 13.10. Fixture for functional inspection of a locking lever.

tolerancing can be carried out by a functional gage which is a simulation of the mating part, taking into account the bonuses from position tolerances. Special programs have also been developed to check hole or pin patterns to ensure proper mating (Lehtihet and Gunasena, 1991).

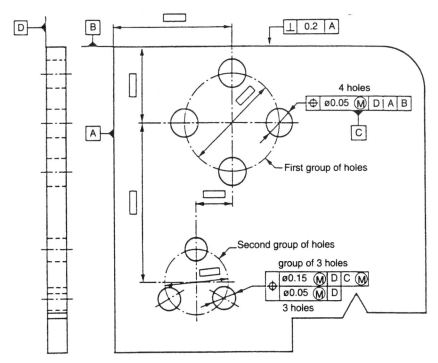

Figure 13.11. Composite tolerancing of a pattern of holes.

13.2.3 Limitations of DTF

Limitations Due to Production Conditions. Although as discussed before, the DTF principle is of very general application, there are a number of circumstances where its application is difficult, or not desirable, primarily for reasons of productivity. In principle, functional dimensioning and tolerancing, as specified by design requirements, should be transferred directly to manufacturing dimensioning and tolerancing. By so doing the dimensions are manufactured as direct dimensions and their tolerances can reach the highest values, which is, of course, the aim of economical production.

However, for different reasons, due mainly to constraints in the available equipment or to difficulties in resetting of parts and generating losses of time, direct transfer is not advisable and, as a consequence, a reduction in tolerances must be accepted. If the imposed tolerances are not too narrow, this procedure is acceptable and the functional dimensions can be obtained, but at the expense of tightening the manufacturing tolerances and, eventually, rejecting parts that are functionally correct.

The problem just mentioned can easily be illustrated by the simple example shown in Figure 13.12, representing the machining of a slot. The functional dimensions are a and b, connecting faces A and B to face C. However, for reasons of commodity of fabrication, the production dimensions chosen are b and c, connecting faces A and B, easy to obtain by a milling operation. The dimension C will be after transfer of tolerances;

$$c = 30.15 \pm 0.05 \text{ mm}$$

which means a reduction from an initial tolerancing of 0.2 mm to 0.1 mm. Inspection during production will check the size of b with a snap gage and the size of c with a plug gage. The parts passing this inspection will be functionally correct. However, some parts, rejected in production, are nevertheless functionally correct. For example, a part having the dimensions $b =$

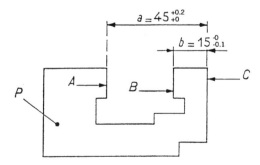

Figure 13.12. Milling of a slot.

14.9 mm and $c = 30.3$ mm will be rejected, although it is functionally correct, the dimension a being $a = 45.2$ mm.

To compensate for the loss of parts due to a wrong sorting, one could perform an additional check based on functional dimensions, but there will be an additional cost for production. Another way to avoid losses of good parts consists in applying statistical criteria to dimensional inspection. At the expense of a small risk of accepting bad parts, statistical inspection will substantially enlarge part tolerances and therefore reduce production costs (Halevi and Weill, 1995).

The problem of transfer of tolerances just described happens very frequently in production. Many routines have been developed to find an optimum definition of production dimensions and tolerances guaranteeing functional dimensions (Fainguelernt et al., 1986), at the same time taking in account limitations in manufacturing facilities and other manufacturing constraints, such as chip thickness, cutting forces, and thermal deformations. Efficient strategies in tool and part setting have also been developed to ensure optimal working conditions (Mathieu and Weill, 1991).

When a transfer of tolerances is not possible, mainly because the tolerances have to be restricted to very small values that are difficult or impossible to obtain in production, manufacturing facilities generally use one of two alternative approaches:

1. In the first approach, the components of a unit are not manufactured independently with restricted tolerances, difficult to observe, and then assembled. On the contrary, subsets of the complete unit are prepared in advance and the functional dimensions are executed respecting the required accuracy on the subset. This approach, called *unit assembly approach,* is illustrated in Figure 13.13 for the assembly of parts A and B. Because the tolerances for dimensions a and b would be difficult to reach to satisfy the functional tolerance of dimension f, which is very small, a subassembly of A and B is constructed and the functional dimension f is executed directly with the required accuracy. This approach is also called *partial interchangeability.*

2. In the second approach, the components of a dimensional chain are produced with a given scatter that allows sorting of suitable parts with dimensions mating the conjugate part. A variant of this approach consists in assembling the components in the final assembly by a suitable setting mechanism. Of course, this approach, called *selective assembly procedure,* implies that the population of parts contains a sufficient number of parts of the size required.

Other Limitations. Application of the DTF principle can also be limited by factors of a more general nature. In particular, the actual trend in industrial design of referring more and more to factors such as maintainability, relia-

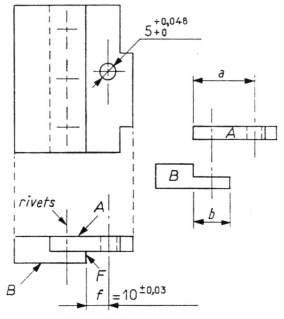

Figure 13.13. Unit assembly.

bility, product life cycle, servicing, environmental conditions, marketability, ergonomics, and so on, influences the design decisions and, consequently, DTF very deeply. Although indirect in nature, the impact of these factors is very crucial in design for function.

Additionally, a particularly important role in designing a product is related to the cost of manufacturing conditions. Unfortunately, information sources on costing in production are difficult to obtain and are seldom very reliable. The problem is complicated by the reluctance of companies to provide costing information and by the fact that costing is related to general company policies, which vary widely. Obviously, manufacturing cost has no absolute value, and therefore it has been proposed that relative values be used to compare the merits of different processes before making a decision (Chisholm, 1973).

13.3 APPLICATION OF THE DTF PRINCIPLE TO A COMPLEX MECHANICAL ASSEMBLY

To show how the DTF principle can be applied to a real mechanical assembly, where a number of functions have to be respected simultaneously, it is interesting to refer to the analysis performed by Farmer (Farmer and Gladman, 1986; Farmer et al., 1991). Taking as an example a gear pump assembly

(Figure 13.14), Farmer defines all the functions to be fulfilled to establish correct operating conditions (e.g., pressure, flow rate, etc.) for the pump.

To this end, two conditions have been stipulated with respect to the functional requirements:

1. The gear peripheries must not rub on the pump body.
2. The maximum clearance between gear peripheries and pump body recesses must not exceed 0.4 mm.

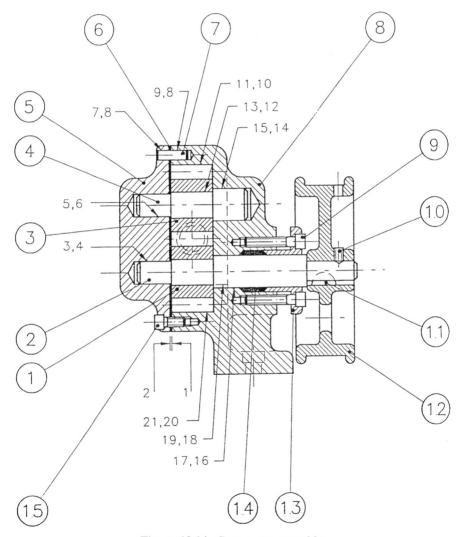

Figure 13.14. Gear pump assembly.

These conditions suggest designing the driving gear and the driven gear accordingly. Considering only driven gear (3), the design has to allow rotation on shaft (4), which is press fitted into pump body (8). The two functional conditions are related to the following dimensions:

1. The diameters of the pump body recess and the gear periphery (features 11 and 10).
2. The diameters of the pump body bore and the shaft (15, 14).
3. The diameters of the gear bore and the shaft (13, 12).

In addition, the fit conditions between the shaft and the hole for the gear (13, 12) and the press fit of the shaft into the pump body will also influence the functional conditions. Applying the envelope principle (Halevi and Weill, 1995), the sizes of the gear hole, the shaft, and the pump body hole are well defined for function and assembly without requiring other geometrical tolerances: roundness, for example.

Considering the rubbing and clearance conditions between gear and body, it is necessary to define the diameters of the pump body recess D_1 and the gear periphery D_2. The maximum clearance C_{max} occurs when both diameters are at their least material conditions and is equal to

$$C_{max} = (D_1 + d_1) - (D_2 - d_2) \qquad (13.3)$$

with d_1 and d_2 being the tolerances of D_1 and D_2. The minimum clearance C_{min} occurs when the pump body recess and gear periphery diameters are at their maximum material conditions and is equal to

$$C_{min} = D_1 - D_2 \qquad (13.4)$$

The envelope principle is also applied in this case to the diameters D_1 and D_2 to ensure proper assembly, regardless of geometric form errors. However, additional errors of axis offsets can also influence the clearance conditions. This can be the case of the fit (13, 12) between gear hole diameter and shaft diameter when they are at their least material conditions. For the interference fit (15, 14), it is assumed that no offset will result. For the internal and external diameters of the gear, for the diameters of the shaft and the diameters of the pump body, an error of concentricity can result. Finally, all these errors are combining to form a relation between the concentricities, the diameters, and their tolerances that defines the condition of nonrubbing (Farmer and Gladman, 1986). A suitable choice of sizes, feature tolerances, and position tolerances, taken primarily from existing standards, will satisfy the functional requirements. Fortunately, in most cases, the manufacturing capabilities, assembly constraints, and inspection techniques will also be met easily because the industrial standards are based on relevant industrial experience.

The gear pump example proposes another good analysis of tolerancing in the case of the assembly of the pump cover (5) on the pump body (8) (Figure 13.14). Here the flat surface (2) (Figure 13.15) and the two location holes (7) are the datum features to position the cover on the pump body. The two locating holes are drilled and reamed and the flat surface is milled and ground. Afterward, these features are datums for manufacturing the two bearing holes 3 and 5 (Figure 13.15). The inspection gage, represented in Figure 13.16, is designed to check the position of the two bearing holes, applying the principles of composite tolerancing and maximum material condition. Therefore, the identity of datums for function, manufacturing, and inspection is well respected. If, for practical reasons, this identity cannot be respected, a transfer of tolerances will result with negative consequences, as explained before.

It is important in this case to note that other conditions also have to be respected to guarantee proper assembly and functioning. For example, the concentricity of diameter 4 (see fit 3,4) in relation with diameter 16 (see fit 17,16) for the driving shaft and of diameter 6 (5,6) in relation with diameter 14 (15,14) of the driven shaft (Figure 13.14) plays an important role in assembly and function. Other constraints concern the flatness of surface 2 and the roundness and straightness of the shafts and the mating holes, all of which influence proper functioning of the shafts in the pump body.

Analysis of the gear pump assembly has emphasized some basic approaches that can be regarded as universal. As shown, the functional requirements can be satisfied in many cases by using existing standard procedures, such as composite tolerancing. In principle, it is possible to develop an algorithm to satisfy functional requirements, but in many cases the necessary data are lacking. Therefore, designers have to rely on their personal experience and on empirical data for process capabilities and relative production costs, preferred dimensions from standards, and so on. Some information concerning relations between tolerance grades and process capabilities, as

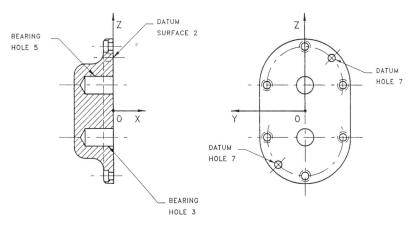

Figure 13.15. Gear pump cover.

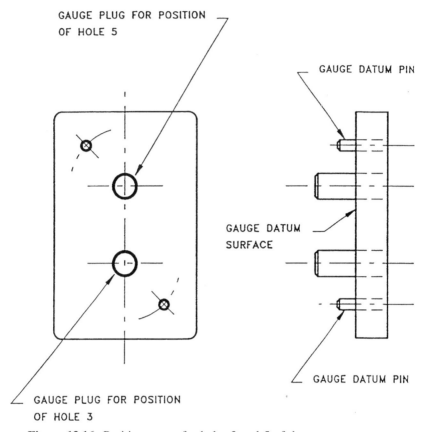

GAUGE PLUG FOR POSITION
OF HOLE 5

GAUGE DATUM PIN

GAUGE DATUM
SURFACE

GAUGE DATUM PIN

GAUGE PLUG FOR POSITION
OF HOLE 3

Figure 13.16. Position gauge for holes 3 and 5 of the gear pump cover.

well as relations between geometric tolerances and machine capabilities or relations between surface finish and functional requirements such as wear, corrosion, fatigue, noise, load-carrying capacity, and sealing, can be found in the literature. But few complete and reliable references exist.

It is obvious that application of the DTF principle requires that the designer be well qualified in many fields, such as standards, process capabilities, and inspection techniques. It is therefore doubtful if the automation of design introduced by the use of computers will be able to compete with human design capabilities. This aspect of dimensioning and tolerancing is examined next in more detail.

13.4   COMPUTER TECHNIQUES IN DIMENSIONING AND TOLERANCING FOR FUNCTION

13.4.1   General Problem of Integration of the Computer in DTF

Problems raised by computer introduction in design activities such as dimensioning and tolerancing have been examined by various authors. A good analysis has been given by Farmer (Farmer, 1989), who concludes that despite the development of very fine computer-aided design (CAD) systems in recent years, there is still a wide lack of capabilities in these systems concerning dimensioning and tolerancing. In general, the conceptual aspects of design are difficult, if not impossible, to carry out by computer. For example, the search for functional dimensions and their definition, which is fundamental for tolerance analysis, seems at present to be out of reach of the capabilities of a computer in a CAD system.

For other computer utilization, however, in particular for applications of a more systematic nature, the repetitive capacities of a computer can be of great help. The storage of information used by designers, such as preferred sizes, preferred limits and fits, standard parts such as screws and threads, process capabilities concerning accuracy and surface finish, operations costs, and tolerance recommendations, can be very useful. By using databases containing this information, a comparative evaluation of different alternatives can be carried out easily. Although absolute costs are difficult to define, it is possible to evaluate changes in designs on the basis of relative costs. By retrieving this information interactively, designers can use computers to speed up their work and to obtain reliable information.

In tolerancing, it is well known that typical simple dimensional chains occur frequently and their tolerance analysis can be integrated by computer. When similar situations arise, it is convenient to be able to retrieve earlier solutions.

Modules for the transfer of design dimensions and tolerances to production dimensions and tolerances have already been developed for linear chains of dimensions (Fainguelernt et al., 1986). The extension of tolerance analysis to three-dimensional chains is less advanced but is under investigation. Computer-aided process planning (CAPP) programs have also been developed in recent years and can be used to evaluate the feasibility of proposed process plans or to suggest modifications in design. Up to now, however, the available programs have not been advanced enough to solve process plan problems for real industrial parts.

A domain in which use of the computer is now entranched is the domain of coordinate measuring machines (CMMs). Compared with conventional inspection, where auxiliary reference datum surfaces had to be used, and therefore problems of stackup of tolerances have occurred, CMM machines can use directly functional datums as defined in the design. However, the results given by CMM machines raise basic problems of accuracy because of inconsistencies due to error definitions which are not completely resolved (Sona and Farmer, 1988).

It can be concluded that introduction of the computer in product design and manufacture will not fundamentally change the conditions required to achieve correct observation of functional requirements. It can, however, be a powerful help for designers in being intelligent and creative in their work. The best approach to developing computer-aided programs for design will certainly consist in employing the consistent system of dimensioning and tolerancing developed over the years by competent experts and embodied generally in industrial standards. A current effort toward this end has been undertaken in development of the TTRS (Technologically and Topologically Related Surfaces) system, which is intended to define functional dimensions and tolerances (Charles et al., 1989).

13.4.2 The TTRS Approach

The purpose of the TTRS approach is to develop a computer-aided functional tolerancing module applied to a given function, such as assembly, sealing, circular guidance, linear guidance, or power transmission. To illustrate this approach, the function of assembly has been chosen and applied to the example shown in Figure 13.17. The casing (1) and cover (2) have to be assembled and secured by the screws (3).

The meaning of TTRS can be explained by the example in Figure 13.18, representing a set of two parallel planes, toleranced by two tolerance zones of width e_1 and e_2. The minimal and maximal distances of two real planes will be D_{min} and $D_{max,}$ which will define maximal and minimal clearances in the assembly with another set of planes.

In general, a TTRS is defined by a set of surfaces situated on the same part having given tolerances of position. The TTRS module carries out the following tasks:

1. It identifies functional surfaces on the part of a mechanism (contact surfaces) (e.g., S21 and S22 in Figure 13.17).

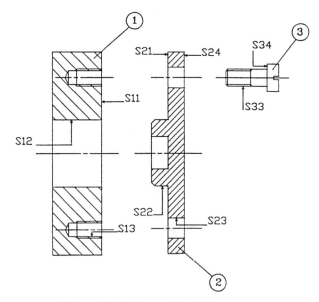

Figure 13.17. Assembly of three parts.

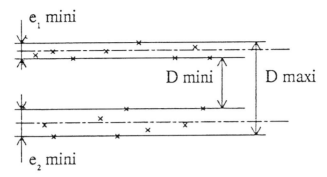

Figure 13.18. TTRS consisting of two parallel planes.

2. It identifies functional surface associations (e.g., S11S12 on part 1), (i.e., a TTRS, now called TTRS 1).
3. It proposes the functional relevant parameters (dimensional and positional) for each TTRS according to the parameters listed in the table shown in Figure 13.19.
4. It assists in defining tolerances for the parameters according to the manufacturing capabilities and inspection techniques.

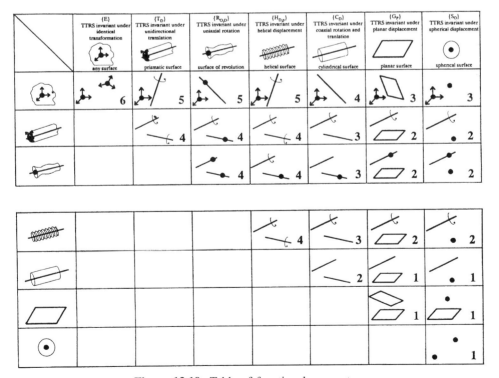

Figure 13.19. Table of functional parameters.

5. It verifies the feasibility of assembly as a function of the tolerances chosen (minimum clearance).

6. It computes the maximum clearance to assess the quality of the assembly.

Of course, the choices are made according to industrial standardization and the capabilities of CAD systems (e.g., CSG representations). The system also has to be compatible with manufacturing and inspection principles.

Tasks 1 and 2 are performed automatically by a computer program using the kinematic link diagram and the liaison graph of the mechanism, which gives the contact surfaces and related surfaces (TTRS). Figure 13.20 shows such a graph with the contact surfaces and their association as TTRS surfaces for the parts of Figure 13.17 (SATT = TTRS). The graph represents the three components (1, 2, and 3) of Figure 13.17, connected by three independent assembly loops between the parts (Figure 13.20b) (e.g., between 1 and 2 by TTRS1 and TTRS2). It also identifies related surfaces (TTRS) such as S11 and S12, now called TTRS1. The other TTRS can be associated, and finally,

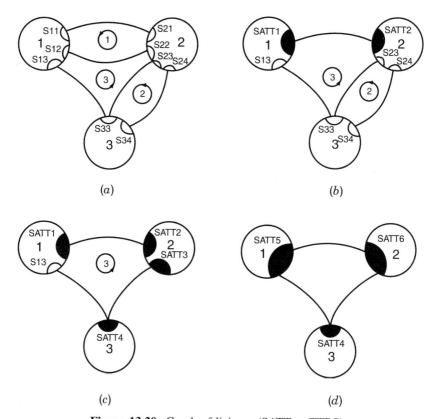

Figure 13.20. Graph of liaisons (SATT = TTRS).

Figure 13.20*d* represents the functional liaison of TTRS 6 (TTRS2 and TTRS3) with TTRS 5 and TTRS4, as well as the liaison between TTRS5 and TTRS4 [i.e., the total liaison $(1 + 2 + 3)$].

For task 3 it is necessary to refer to the table of parameters (Figure 13.19), which shows 28 classes of couples of surfaces, based on seven elementary types of mechanical surfaces which can be reproduced by translation or rotation (planes, cylinders, etc.), as required in assembly. The table also gives the number of functional parameters defining the relative position of the surfaces; for example, the association of a plane and a cylinder is defined by one parameter only, their relative inclination or perpendicularity.

For tasks 4, 5, and 6 the system proposes tolerancing modules consisting of cost evaluations, manufacturing capabilities, and inspection procedures. Soft gages, taking in account virtual sizes for MMC and LMC conditions, are implemented in the computer to verify the feasibility of assemblies (minimum clearance) or their quality (maximum clearance). They simulate the solid functional gages used in conventional inspection techniques.

13.4.3 Tolerancing of the Assembly of Figure 13.17

To illustrate the application of the TTRS principles, we now analyze the assembly of the cover (2) with the casing (1) secured by the bolts (3). There are three TTRS elements to analyze for part (2): TTRS2, TTRS3, and TTRS 6 (Figure 13.20). TTRS2, consisting of the association of a plane and a cylinder, is defined by one parameter, in this case by the perpendicularity tolerance of 0.1 for a diameter of 39.85 ± 0.05 at MMC. Therefore, the soft gage should have a diameter of 40.00 for checking the perpendicularity. The same analysis is pertinent for the tolerancing of TTRS3.

For the tolerancing of TTRS6 (union of TTRS2 and TTRS3), the table of parameters in Figure 13.19 indicates for the case of two surfaces of revolution (considering at first only one hole of the pattern of four holes, shown in Figure 13.21) that four parameters are necessary. Translated into tolerance standard language, the parameters are shown in Figure 13.21 as tolerances of perpendicularity to *A* of diameter 0,1 and of perpendicularity to *B* of diameter 0.1, as well as a position tolerance of 0.3 of the hole relative to cylinder *C,* and a distance of 10 ± 0.2 between planes *A* and *B*. Since the axes and planes of TTRS2 and TTRS3 are parallel, only two positional parameters are used here: the distance between the axes of the cylinders and the distance between the planes. When the pattern of the four holes is considered, it is tolerated in relation with cylinder *C*. Therefore, all the features are completely functionally toleranced. The nonfunctional surfaces (e.g., the external surfaces of the part) are not considered because they do not play any role in the function of assembly of parts. The tolerancing of part (2) is carried out in the same way, already having parts (3) fixed to it for simplicity.

Finally, assembly of the four parts will be verified by functional gages, taking in account the virtual conditions due to use of the MMC convention. Instead of physical gages, soft gages at an MMC condition corresponding to

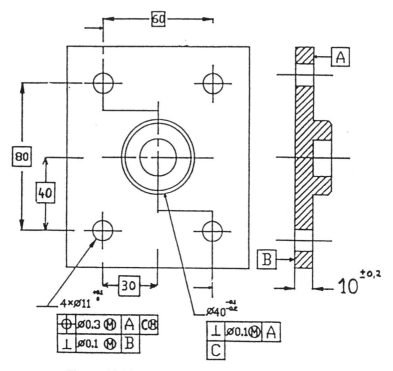

Figure 13.21. Geometrical tolerancing of part (1).

TTRS5 and TTRS6 can be used. As shown in Figure 13.22, the result of inspection of part (2) on a CMM machine and its fitting in the soft gages is represented. The methods used to check the fitting of the parts in the soft gage are well known and can be used to determine minimal and maximal clearances in the assembly (Bourdet and Clement, 1988).

The TTRS approach has been applied to most of the basic functions of mechanical systems (power transmission, linear guidance, etc.). This approach is described in detail in a handbook for the dimensioning of three-dimensional mechanical systems (Clement et al., 1994).

13.5   CONCLUSIONS

The concept of dimensioning and tolerancing for function (DTF) has been discussed and illustrated by examples, starting with simple cases and ending with practical applications. The limitations of DTF caused by manufacturing and inspection constraints have also been analyzed and illustrated by a complex mechanical assembly. Attempts to introduce the DTF concept in computerized design systems have been reviewed and critically assessed.

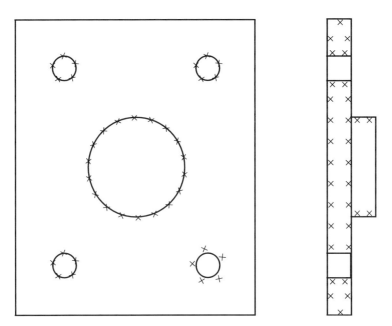

Figure 13.22. Checking of manufactured parts by soft gages.

To use the DTF concept efficiently, a designer must possess extended knowledge in several fields, such as standards, process capabilities and costs, and inspection techniques. The designer should have comprehensive training in these fields and be constantly striving to enrich his or her technical expertise. A good way to educate a proficient designer is to oblige the designer to spend considerable time in production.

To improve designers' qualifications, it is logical to make use of the powerful capabilities of modern computer systems. However, it is doubtful if computers can replace human designers in all functions, in particular for creative activities. This does not mean that computers cannot be very helpful in repetitive tasks, such as storing relevant technical data and retrieving them when necessary. The development of technological databases is therefore a fundamental need in advanced computerized design systems.

It is clear that efficient application of the DTF concept will require designers who have broad expertise in all aspects of mechanical systems design. Every practical design problem is, finally, a specific case that needs a special, relevant approach. Obviously, at this stage, design is more an art than a science.

REFERENCES

Bourdet, P., and Clement, A., 1988, A Study of Optimal Criteria Identification Based on the Small Displacement Screw Model, *Ann. CIRP,* Vol. 37, No. 1, pp. 503–506.

Charles, B., Clement, A., et al., 1989, Toward a Computer Aided Functional Tolerancing Model, *Proceedings of the First CIRP Seminar on Computer Aided Tolerancing,* Jerusalem, Israel, December 11–12.

Chisholm, A. W. J., 1973, Design for Economic Manufacture, *Ann. CIRP,* Vol. 22, No. 2, pp. 243–247.

Clement, A., Riviere, A., and Tennenbaum, M., 1994, *Cotation tridimensionnelle des systèmes mécaniques, théorie et pratique,* Pyc Edition, Paris.

Fainguelernt, D., Weill, R., and Bourdet, P., 1986, Computer Aided Tolerancing and Dimensioning in Process Planning, *Ann. CIRP,* Vol. 35, No. 1, pp. 381–386.

Farmer, L. E., 1993, *Dimensioning and Tolerancing to AS 1100.101–1992 and AS1100.201–1992,* Standards Australia, Canberra, Australia.

Farmer, L. E., 1989, Tolerancing for Function in a CAD/CAM Environment, *Proceedings of the First CIRP Seminar on Computer Aided Tolerancing,* Jerusalem, Israel, December 11–12.

Farmer, L. E., and Gladman, C. A., 1986, Tolerance Technology: Computer Based Analysis, *Ann. CIRP,* Vol. 36, No. 1, pp. 7–10.

Farmer, L. E., Gladman, C. A., and Edensor, K. H., 1991, The Scope of Tolerancing Problems in Engineering, CIRP International Working Seminar on Computer Aided Tolerancing, Penn State University, University Park, PA, May 16–17.

Halevi, G., and Weill, R. D., 1995, *Principles of Process Planning: A Logical Approach,* Chapman & Hall, London.

Lehtihet, E. A., and Gunasena, N. U., 1991, On the Composite Position Tolerances for Patterns of Holes, *Ann. CIRP,* Vol. 40, No. 1, pp. 495–498.

Mathieu, L., and Weill, R., 1991, A Model for Machine Tool Setting as a Function of Positioning Errors, CIRP International Working Seminar on Computer Aided Tolerancing, Penn State University, University Park, PA, May 16–17.

Ropion, R., 1968, *La Cotation fonctionnelle des dessins techniques,* Dunod, Paris.

Sona, C. M., and Farmer, L. E., 1988, The Inspection of Holes with Co-ordinate Measuring Machines, *Proceedings of the 4th International Conference on Manufacturing Engineering,* Brisbane, Australia.

14

WORKPIECE ACCURACY AND TOLERANCE ANALYSIS IN MODULAR FIXTURES

A. Y. C. NEE AND A. SENTHIL KUMAR

National University of Singapore
Republic of Singapore

14.1 INTRODUCTION

Fixtures are tooling devices used to locate, support, and hold workpieces during machining, assembly, inspection, and other manufacturing processes. The lead time required of a product is determined not only by the time required to design and fabricate the part but also by the time required to design and build the fixtures. For products made in small batch sizes, the time spent on designing and building fixtures becomes a fairly large proportion of the total production time, and it is therefore essential to try to shorten this necessary but "unproductive" time as much as possible.

Fixtures used to be very dedicated devices (i.e., specifically tailored for making only a single part according to its specifications and requirements). If the part is no longer in production, investment in the fixture is written off completely since it is not possible to use the same fixture for other parts. However, many companies are reluctant to throw away old fixtures and end up shelving them in large quantities, which adds to the cost of storage and inventory.

With the development of new manufacturing techniques in the last three to four decades and the diminishing batch size of products, the use of dedicated fixtures for workholding is becoming counterproductive (Hoffman,

Advanced Tolerancing Techniques, Edited by Hong-Chao Zhang
ISBN 0-471-14594-7   © 1997 John Wiley & Sons, Inc.

1987). The idea that fixtures should be flexible was conceived by John Warton in the early 1940s. He introduced the concept of modular fixtures, and the first modular system was introduced at the Bristol Aero Company. The concept of modular fixtures is similar to that of children's Lego sets, where a finite number of components, through suitable combinations, are able to produce almost an infinite number of configurations to suit workpieces of various shapes and dimensions.

Manufacturers find modular fixtures particularly useful for prototyping purposes, where parts have not been finalized and are still going through design refining and iteration processes as well as for one-off parts and parts of small batch sizes. Many commercial modular fixturing systems have been developed. The basic principles of operation are quite similar, although they can be broadly classified into the T-slot and the dowel-pin systems, or a combination of both. Precision elements for locating, supporting, and clamping the workpiece can be built on top of base plates of various shapes and sizes. As the elements are assembled together, sometimes reaching rather unwieldy shapes and heights in the final configuration, it has been a major concern to designers and users whether such composites are still within the tolerance specifications originally intended for the individual elements and how part accuracy would be affected.

In this chapter we address the specific problem outlined above. For a particular set of modular fixturing elements where the individual dimensional and form tolerances are given, a computer program has been developed to analyze the final configuration of a fixture in terms of the cumulative tolerances of the elements and the effect on workpiece accuracy. The program is able to recommend customized elements should a combination of standard elements exceed the tolerance specified on a workpiece. The present approach is, however, limited to static analysis based on the specifications of fixture elements and the workpiece. The approach also excludes any inaccuracy that may be caused by movements of the machine tool slides and spindles. Workpiece distortion and deflection during machining has been analyzed by Nee et al. (1995) and there is an ongoing project to address on-line monitoring of clamping and supporting forces to minimize workpiece deformation.

14.2 RELATED WORK IN MACHINING ACCURACY AND FIXTURE DESIGN

Although there has been a very large amount of work in tolerance analysis and synthesis, geometric dimensioning and tolerancing (GD&T), association of tolerance information with computer-aided design (CAD) models, and so on, there is relatively little reported work on fixture tolerancing and analysis and the effect on workpiece accuracy. The earliest works on the effect of fixture design in a fixture–workpiece system were reported by Shawki and Abdel-Aal (1965,1966a,b) and Shawki (1967). To achieve maximum workpiece dimensional accuracy, the rigidity of the fixture was concluded to be a prime factor. They also analyzed the rigidity of locating elements and clamps both experimentally and theoretically. In his concluding paper (Shawki, 1967), a comprehensive list of fixture design guidelines was provided to enable better fixture design so that workpiece accuracy can be improved. This series of four papers represented some of the earliest and most useful work in the analysis of fixture–workpiece interaction.

In their survey on fixture-design automation, Trappey and Liu (1990), covered a number of major automated and semiautomated fixture design systems up to the late 1980s. They reported a lack of treatment of dimensioning and tolerancing information to guide the preference of operation sequences, which were largely feature based.

Boerma (1990) examined the possible causes of errors that may occur during a manufacturing process. Since a workpiece is mounted on a fixture, which in turn is fastened to a pallet and possibly a rotary table and then to the machine table, there could be considerable error buildup as far as workpiece positioning is concerned. This is compounded further by the fact that the tool may have spindle error, clamping error, tool-positioning error, and so on, as shown in Figure 14.1. Errors can be additive or cancel each other, depending on the interaction, and hence it will be extremely difficult to arrive at a complete and accurate solution of how these errors will ultimately affect workpiece accuracy.

Cogun (1992) analyzed the effect of the sequence of applying clamping forces on workpiece accuracy both theoretically and experimentally. He concluded that a rational selection of clamping sequence is important in controlling workpiece displacement but that it is difficult to propose a particular

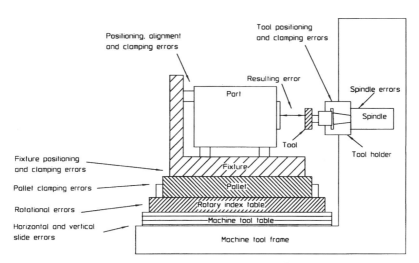

Figure 14.1. Possible geometrical errors between the components of a machining center. (From Boerma, 1990.)

preferred clamping sequence such that all types of deviations can be optimized simultaneously.

Bai and Rong (1993) reported the use of a datum-machining surface relationship graph (DMG) to perform a tolerance chain analysis of dimension variations. They consider fixturing errors to comprise two categories: first, deterministic errors, which are caused primarily by the locating errors of the fixture, such as the position error of fixture locating components, datum variations of the workpiece, and tool-fixture alignment errors; and second, random errors, caused by clamping deformation, cutting force, and thermal deformation during the machining process. Their solution, however, arises primarily from using the approach of tolerance chains to determine dimensional and form tolerances.

King and de Sam Lazaro (1994) discussed the fact that tolerance considerations have generally been handled inadequately in fixture design, due to the unavailability of such information from CAD models. In their approach, tolerance information has been incorporated in the design algorithm. This is achieved by identifying the datums from which all dimensions are toleranced. A tolerance attribute table (tolerance spreadsheet) consisting of both surface and tolerance information is used to describe tolerance relationships of workpiece features.

Halevi and Weill (1995) pointed out that when a part is positioned in a fixture, its real positions and orientation in space are difficult to determine due to errors in the locators and form errors in the surface of the part. They also remarked that those errors are most important, as they are accumulated since the design stage. Due to errors in the locators, only a theoretical datum is assumed; the actual reference datum has been modified. Their observation is also the objective of the present approach in assessing the "worst" situation by summing up tolerances in locators as well as form tolerances of surfaces where the locators are in contact. If the worst situation already exceeds the tolerance specification of a workpiece, not to mention the other random errors mentioned by Boerma (1990), it will be quite unlikely that the workpiece can be machined within tolerance requirements.

Rong et al. (1995) presented the effects of location error on machining accuracy. A locating reference surface model is developed to determine the deviation of locating surface from an arbitrary point on the machining surface. This is used to estimate geometric inaccuracies such as parallelism, perpendicularity, and angularity errors produced in a machining process due to inaccurate locator positions. Their method, did not, however, consider the effect of tolerance buildup on the location accuracy.

14.3   TWO-DIMENSIONAL TOLERANCE ANALYSIS OF GEOMETRIC TOLERANCE

14.3.1   Tolerance Buildup

In modular fixture design, a number of fixturing components can be stacked one over another to form the necessary support and clamping towers, and hence the analysis of tolerance buildup is important to maintain part accuracy within specified values. A simple modular fixture based on component stackup is shown in Figure 14.2.

Tolerance buildup can be calculated using the method of extremes (MOE) (Bennett and Gupta, 1970). It is a technique where the maximum and minimum tolerances are considered to determine the effect of tolerance buildup. This buildup relates the resultant variation of modular fixturing elements to the workpiece tolerances. The MOE ensures that if all the component variables are within their allowable tolerance ranges, the assemblies composing of such components will have their resultant tolerance within the functional requirement, assuming that the components are rigid and there is no appreciable elastic or plastic deformation due to clamping and machining forces.

Tolerance buildup in a modular fixture design can be calculated by considering the tolerance on the part and the tolerance difference between any two of the available locating chains when they are at their extremes (maximum and minimum tolerance). The two locating chains chosen for the analysis are termed a *loop*. A datum locating chain is chosen initially from several available locating chains. It is chosen in such a way that the number of elements in that particular chain is minimal. Next, the buildup is allocated between the chains under consideration. Once the allocations are made, the next chain is analyzed with the datum chain. The tolerance allocation is again made for this loop. When all the chains are exhausted, a cross-check is made between the chains, apart from the datum chain, to ensure that the component is within the limits. If excess tolerance (i.e., tolerance in excess to that specified) is present during rechecking, the upper tolerance limit in that loop is always corrected.

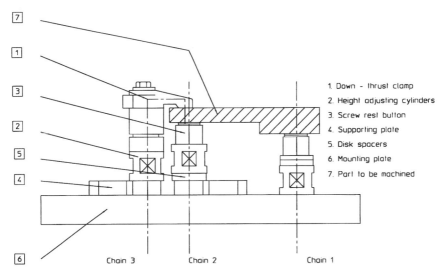

Figure 14.2. Simple modular fixture.

The tolerance buildup in Figure 14.2 can be calculated as follows. As the tolerance effect of the clamping elements is less significant than the effect of the locating elements, only the latter is considered, as shown in Figure 14.3. If the profile of the part is irregular, the part is divided into several primitive parts. The divided part can also be considered as a modular element, but the tolerance on this fixed elementary part (FEP) will be considered for calculating the tolerance buildup. The process of dividing the part for analysis is explained in detail in the appendix.

The profile of the part shown in Figure 14.3 is not regular, so it has been divided into two elementary parts, each having a tolerance of ± 0.01 mm. A profile is said to be *irregular* if different locating pins are not in the same plane. Using the MOE, the following expression is obtained:

$$\text{Chain } 2_{\max} = E_{\max} + F_{\max} + G_{\max} + H_{\max}$$

$$= 50.01$$

$$\text{Chain } 1_{\max} = A_{\min} + B_{\min} + C_{\min} + D_{\min} + \text{FEP}_{\min}$$

$$= 49.95$$

The distance between chains 1 and 2 is given as 50 mm. Hence the angle of inclination of the plane joining the locating elements is

$$\phi_1 = 0.068754°$$

(The angle above may increase if more modular components are stacked.) From Figure 14.4, the maximum inclination of the part, neglecting the FEP (between chains 1 and 2) is to be within the tolerance limit of

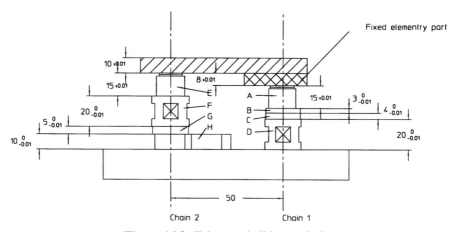

Figure 14.3. Tolerance buildup analysis.

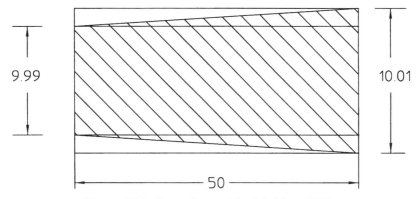

Figure 14.4. Part to be machined (without FEP).

$$\phi_2 = 0.022918°$$

For the part to be within the limits, the angle of inclination ϕ_1 should be less than or equal to ϕ_2. In this case the excess tolerance buildup (i.e., the tolerance above the workpiece tolerance) due to the modular components on the part is found to be 0.04. This is obtained by considering the maximum and minimum values of the chains with the part tolerance. In this case it is obtained by taking the chain 2 maximum, chain 1 minimum, and part tolerance. This relationship can be represented by an equation:

$$\text{excess tolerance} = \text{chain 2 maximum} - \text{chain 1 minimum} -$$

$$\text{part tolerance maximum}$$

$$+ \text{part tolerance minimum}$$

$$= 50.01 - 49.95 - 0.01 - 0.01$$

$$= 0.04$$

This holds for both unilateral and bilateral tolerances. Suitable algorithms are developed to obtain the excess tolerance automatically considering the other possible combination (i.e., chain 1_{max} and chain 2_{min}). The maximum excess tolerance obtained is then allocated as explained in subsequent sections of this chapter.

14.3.2 Tolerance Allocation

Once the tolerance buildup is calculated, the subsequent step is to allocate the excess tolerance so that the tolerance of the part after machining can be held within the tolerance zone. The first stage is to reorganize the modular fixture in such a way that the tolerance buildup is minimized. The reorganization is done by replacing say, two 20-mm-height adjusting cylinders by a 40-mm-height adjusting cylinder, or vice versa. If reorganization of the modular element is not possible with the available elements, the locating elements are repositioned such that the final configuration satisfies the tolerance requirements. The general algorithm for reorganizing the modular elements follows.

```
Identify_contact_points
    Check_link_continuity
for i = 1 to m
        Identify_linking elements (chain(i))
        Identify_end_element
End for
        if (end_element.EQ.fixing_element)
        ignore_chain
/*The effect of tolerance buildup on the locating
```

elements is more important than on fixing elements*/

```
        else
            continue
        endif
        Check_tolerance_buildup
    Choose_chain
        Check_for_same_group_elements
        if (elements.EQ.same)
            replace_with_new_elements
            check_tolerance_buildup
        endif
        if (new_ang.GT.old_ang)
            perfer_old_setup
        else
            new_setup
        endif
```

The algorithm can be illustrated with a simple example. Consider a slot to be machined according to the dimensions shown in Figure 14.5. The standard modular fixture components available in the database are shown in Table 14.1 (only relevant modular fixture components are shown).

In the example above, the algorithm checks initially for the number of contact points on the part. The number of contact points in this case is 3, as shown in Figure 14.6. Next, the continuity of the link is checked. If the continuity does not exist, an error message will be displayed. This is to ensure that all the components are linked together. The subsequent step is to check the clamping chain, of which only one is shown in Figure 14.6. Locating

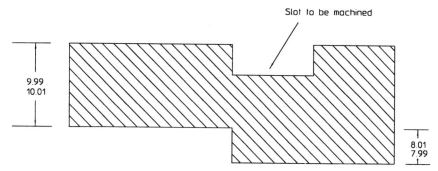

Figure 14.5. Part to be machined (with FEP).

elements in chain 1 will be checked. If suitable elements are available in the database, the algorithm attempts to replace the same group of elements with a single piece and computes the new tolerance buildup. If the new tolerance buildup is more than the preceding one, the algorithm reverts to the original setup. The procedure is repeated for the other chains.

In this way, the tolerance buildup at the extremes can be considerably reduced by exploiting the components available in the database. In the example shown, chain 1 is taken for analysis. Using the algorithm it was found that no reorganization of the supporting elements is possible with the available components in the database. Chain 2 is considered next. The total height of 50 mm can be replaced by two components to the same height using a screw rest button and a height-adjusting cylinder of 10 mm and 40 mm, respectively. Here chains 1 and 2 are considered as a loop. Chain 3 is ignored, as its purpose is not to locate but to clamp the part from moving while machining. The effect of tolerance buildup due to this replacement is again calculated for this loop, using the MOE. Referring to Figure 14.7, we have

TABLE 14.1. Components Available to the Database

Type	Dimensions and Tolerance Available	
Screw rest button	$10^{+0.01}_{-0.01}$	$15^{+0.01}_{-0.01}$
	$25^{+0.01}_{-0.01}$	
Down thrust clamp	Minimum possible	
	height is 30 mm	
Height-adjusting cylinder	$20^{+0}_{-0.01}$	$40^{+0}_{-0.01}$
	$80^{+0}_{-0.01}$	
Supporting plate	$10^{+0}_{-0.01}$	
Disk spacers	$3^{+0}_{-0.01}$	$4^{+0}_{-0.01}$
	$5^{+0.01}_{-0.01}$	

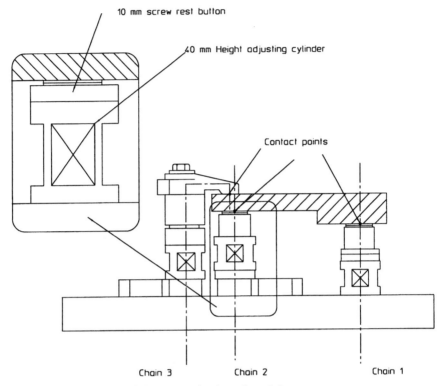

Figure 14.6. Reorganization of modular components.

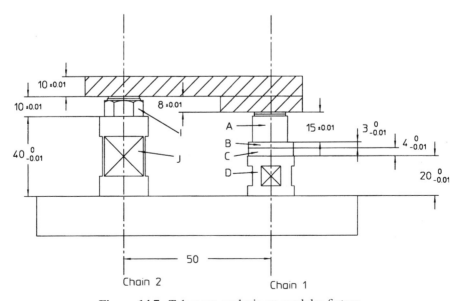

Figure 14.7. Tolerance analysis on modular fixture.

$$\text{Chain } 2_{max} = I_{max} + J_{max}$$

$$= 50.01$$

$$\text{Chain } 1_{min} = A_{min} + B_{min} + C_{min} \, D_{min} + \text{FEP}_{min}$$

$$= 49.95$$

The distance between chains 1 and 2 is given as 50 mm. The angle of inclination of the part due to the difference in height between chains 1 and 2 is
found to be 0.068754°. Similarly, the other possibilities can be checked:

$$\text{Chain } 2_{min} = I_{min} + J_{min}$$

$$= 49.93$$

$$\text{Chain } 1_{max} = A_{max} + B_{max} + C_{max} + D_{max} + \text{FEP}_{max}$$

$$= 50.02$$

The angle of inclination of the part due to the difference in heights between
chains 1 and 2 is found to be 0.04583°. The maximum angle of inclination
is considered in this case. Further replacement of elements is not possible
with the components available in the database.

Even though the angle of inclination is equal to the previous arrangement,
it is preferable to replace E, F, G, and H by I and J as the number of
components is reduced. This is because the liability of tolerance getting accumulated due to four components is more than for two components. For
higher precision, the excess tolerance buildup has to be suitable allocated to
the modular fixture elements. Using the MOE, the excess tolerance can be
distributed suitably on the available modular fixture elements so that the tolerance of the part is within the limits. The technique adopted can be explained
using the following algorithm.

```
Calculate_excess tolerance
Choose_chain
   if (lower_tolerance.NE.0) then
      new_tolerance_1 = excess_tolerance + lower_tolerance
      if (new_tolerance_1.LT.0) then
         new_tolerance = new_tolerance_1
         calculate_new_excess_tolerance
      else
         new_tolerance = 0
         excess_tolerance = new_tolerance_1
         continue_allocation
   else
         continue_allocation
```

```
      endif
   end
     if (upper_tolerance.NE.0) then
        new_tolerance_1 = excess_tolerance - upper_tolerance
        if (new_tolerance_1.GT.0) then
           new_tolerance = new_tolerance_1
           calculate_new_excess_tolerance
        else
           continue_allocation
      else
        continue_allocation
      endif
   end
     if (excess_tolerance.NE.0) then
        choose_another_chain
        continue_allocation
     endif
     check_for_tolerance-buildup
     calculate_excess_tolerance
     if (excess_tolerance . LE . 0) then
        stop
     else
        repeat_procedure
     endif
   end
```

Referring to Figure 14.7, the excess tolerance on the modular component over the part is found to be 0.04. This excess tolerance is obtained by considering the chain 2 maximum, chain 1 minimum, and part tolerances. The excess tolerance obtained is allocated to the chain that has minimum height within the loop considered. It is done such that the lower tolerance limit of the first component in that chain is given the entire tolerance. If that particular limit is zero, the next component's lower tolerance limit will be adjusted until all the excess tolerance is distributed. Once again the excess tolerance is calculated and the method above is repeated until the excess tolerance is less than or equal to zero. If all the lower tolerance limits in a particular chain are zero, the other chain in the loop is considered. In such cases the upper tolerance limit is taken for analysis and the foregoing procedure is repeated.

The excess tolerance obtained is distributed, and the new tolerance limits obtained by this method are shown in Figure 14.8. Table 14.2 shows the distribution of excess tolerance and the excess tolerance for each stage. The flowchart shown in Figure 14.9 depicts the procedure followed in tolerance allocation. The CAD model and the modular element database are given as an input to the system. The tolerance information associated with the geom-

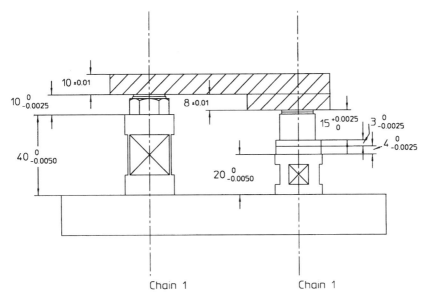

Figure 14.8. Distributed new tolerance limits.

etry and the feature are stored in the CAD model as attributes. They can be extracted from the CAD model for calculating the tolerance buildup. C routines have been developed to reorganize the modular fixture and compute tolerance buildup automatically.

TABLE 14.2. Tolerance Distribution on Modular Elements

| Stages | Modular Component | | | | | | Excess Tolerance |
	A	B	C	D	I	J	
1	$15^{+0}_{-0.01}$	$3^{+0}_{-0.01}$	$4^{+0}_{-0.01}$	$20^{+0}_{-0.01}$	$10^{+0.01}_{-0.01}$	$40^{+0}_{-0.01}$	0.04
2	$15^{+0.01}_{-0.}$	$3^{+0}_{-0.01}$	$4^{+0}_{-0.01}$	$20^{+0}_{-0.01}$	$10^{+0.01}_{-0.01}$	$40^{+0}_{-0.01}$	0.03
3	$15^{+0.01}_{-0.}$	$3^{+0}_{-0.01}$	$4^{+0}_{-0.01}$	$20^{+0}_{-0.01}$	$10^{+0}_{-0.01}$	$40^{+0}_{-0.01}$	0.02
4	$15^{+0.01}_{-0.}$	$3^{+0}_{-0.005}$	$4^{+0}_{-0.005}$	$20^{+0}_{-0.005}$	$10^{+0}_{-0.01}$	$40^{+0}_{-0.01}$	0.005
5	$15^{+0.01}_{-0}$	$3^{+0}_{-0.0025}$	$4^{+0}_{-0.0025}$	$20^{+0}_{-0.005}$	$10^{+0}_{-0.01}$	$40^{+0}_{-0.01}$	0.0
Cross checking:							
6	$15^{+0.01}_{-0}$	$3^{+0}_{-0.0025}$	$4^{+0}_{-0.0025}$	$20^{+0}_{-0.005}$	$10^{+0}_{-0.01}$	$40^{+0}_{-0.01}$	0.02
7	$15^{+0.005}_{-0}$	$3^{+0}_{-0.0025}$	$4^{+0}_{-0.0025}$	$20^{+0}_{-0.005}$	$10^{+0}_{-0.01}$	$40^{+0}_{-0.01}$	0.015
8	$15^{+0.005}_{-0}$	$3^{+0}_{-0.0025}$	$4^{+0}_{-0.0025}$	$20^{+0}_{-0.005}$	$10^{+0}_{-0.005}$	$40^{+0}_{-0.005}$	0.005
9	$15^{+0.0025}_{-0}$	$3^{+0}_{-0.0025}$	$4^{+0}_{-0.0025}$	$20^{+0}_{-0.005}$	$10^{+0}_{-0.005}$	$40^{+0}_{-0.005}$	0.0025
10	$15^{+0.0025}_{-0}$	$3^{+0}_{-0.0025}$	$4^{+0}_{-0.0025}$	$20^{+0}_{-0.0025}$	$10^{+0}_{-0.0025}$	$40^{+0}_{-0.005}$	0.0
Cross checking:							
11	$15^{+0.0025}_{-0}$	$3^{+0}_{-0.0025}$	$4^{+0}_{-0.0025}$	$20^{+0}_{-0.0025}$	$10^{+0}_{-0.0025}$	$40^{+0}_{-0.005}$	0.0

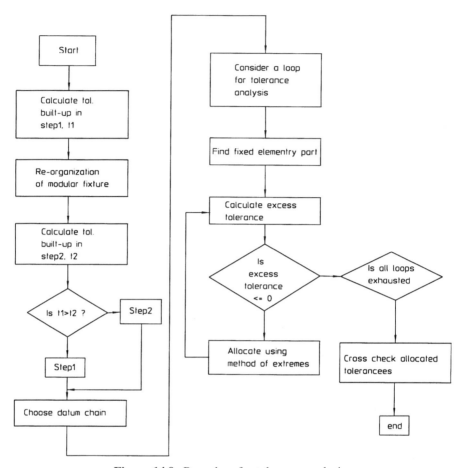

Figure 14.9. Procedure for tolerance analysis.

14.4   THREE-DIMENSIONAL TOLERANCE ANALYSIS OF FORM AND POSITIONAL TOLERANCE

In this section both form and positional tolerances due to the stackup of modular fixturing elements are analyzed. Supporting elements affect the form tolerance of a feature while the locating elements affect its positional tolerance. Hence a careful study has to be made to identify the errors accumulated due to the stackup of modular fixturing elements.

14.4.1   Reduction of Tolerance Buildup by Recombination or Replacement of Fixture Elements

The first step in identifying the tolerance error is to determine the tolerance specification of each feature. The tolerance accumulated due to the modular element stackup is determined using the method of extremes (Bennett and Gupta, 1970). An attempt is made to reduce the tolerance buildup by replacing two or more elements with a single element. An algorithm for recombining the modular elements is shown below.

Recombination Algorithm

```
Identify the contact points{
   Check the link continuity
   Identify the elements (for each link)
}
Choose a chain {
   Check for same group of elements
   Check if combination of elements possible
   If possible {
      Replace
      Compute tolerance buildup
      If computed angle > old angle {
         Prefer old setup
      }
      else {
         New setup
      }
   }
}
```

If the tolerance buildup is still not within the limits, the excess tolerance obtained will be distributed on a custom-made fixturing element. This is to ensure that the workpiece is within the specified tolerance limits. An algorithm for analyzing form and positional toleranced features is shown below.

Tolerance Analysis Algorithm.

Form Tolerance
```
Identify normal vector on the reference feature
Identify the form feature tolerance zone {
   Determine the normal vector of the form feature
}
Identify the tolerance zone due to elements {
   Determine geometric tolerance zone
   Determine the form tolerance zone
   Combined effect of the tolerance zone on the workpiece
}
Transfer tolerance from a reference plane
Compute the excess tolerance
```

Remove the excess tolerance {
  Recombination
  Tolerance distribution
  Custom-made part {
    Tolerance determination
  }
}

Positional Tolerance
  Identify normal vector on the reference feature
  Identify the form feature tolerance zone {
    Determine the normal vector of the form feature
  }
  Identify the tolerance zone due to elements {
    Identify angle formed by the positioning elements
  }
  Transfer tolerance from a reference plane
  Remove excess tolerance {
    Recombination
    Custom-made fixturing element
  }

The algorithms above are explained with respect to the example shown in Figure 14.10. Considered first is a hole to be machined according to the dimensions shown in Figure 14.11. The standard molecular elements available in the database and their tolerance values are shown in Table 14.1. The first step is to identify the form and positional tolerances on the workpiece and the normal vector on the reference surface. The normal vector on the reference surface in this case is along the z-axis, which is also the approach direction of the feature. Next the form feature tolerance zone is identified, which is 0.015 mm.

The supporting points on the workpiece can be joined to form a plane. The geometric and form tolerance of each element is determined, and the combined effect will form a total tolerance zone containing a worst plane (a plane having the maximum permissible errors). In the present approach an exhaustive search is made to determine the worst inclined plane. This worst plane, which is also the maximum inclined plane, can be defined by a point and a normal vector. The angle formed between the normal vector of the modular element plane and the reference plane is determined and in the present case is found to be

$$\theta_w = 0.02868°$$

The maximum inclination of the feature is shown in Figure 14.12 and the maximum angle of inclination is found to be

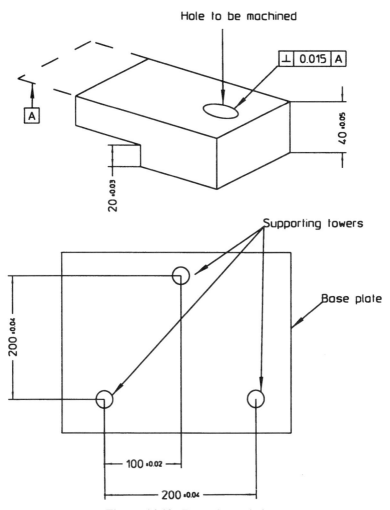

Figure 14.10. Example workpiece.

$$\theta_f = 0.03968°$$

The inclination of the workpiece is greater than the inclination of the feature, which may result in workpiece inaccuracy during machining. This can be avoided if the angle of inclination of the workpiece is less than the permissible angle of inclination of the feature (θ_f).

The next step is to bring the angle of inclination of the workpiece to within the feature inclination by replacing two or more modular elements with a single element. The heights of the new supporting towers are shown in Figure 14.13. The angle of inclination of the workpiece is now found to be

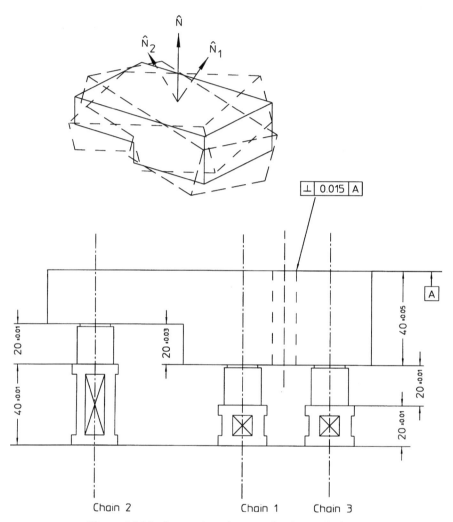

Figure 14.11. Supporting elements for the workpiece.

$$\theta_w = 0.0334°$$

which is again greater than the feature inclination. Hence the excess tolerance present must be distributed to a custom-made fixturing element. The excess tolerance is computed using the difference in angle, any two supporting points, and the z-vector. The general equation for finding the angle between the vectors is

$$az = |a \cdot |z| \cos \theta$$

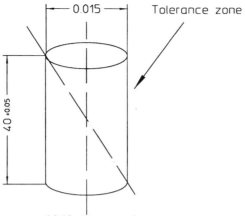

14.12. Feature tolerance zone.

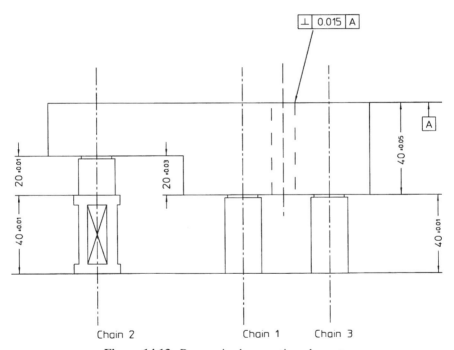

Figure 14.13. Reorganized supporting elements.

In this equation, the values of θ, the z-vector, and any two supporting points are known and hence the equation of the plane can be constructed. Using the x,y coordinates of the third point, the z-value, which governs the height of the supporting tower, can be found. In the present case, it is found to be 0.014 mm, and hence the excess tolerance is

$$\text{excess tolerance} = \text{cumulative tolerance on the supporting tower} -$$
$$\text{computed tolerance}$$
$$= 0.02 - 0.014$$
$$= 0.007$$

This can be distributed to the modular element towers such that a custom-made element can be designed to replace the tower. The tolerance is distributed in Figure 14.14.

14.4.2   Example

A simple workpiece together with its fixture is shown in Figure 14.15. A hole is to be drilled using a three-axis vertical machining center. The first step in analyzing the tolerance is to identify the tolerance on the feature. In the present case, both the form and positional tolerances are specified on the feature.

Form Tolerance Analysis. The maximum possible angle formed by the axis of the feature within the given tolerance zone without error is found to be

$$\theta_f = \tan^{-1} \frac{0.01}{9.99}$$
$$= 0.05735°$$

The angle between the normal vector on the plane formed by the three supporting pins and the reference vector (z-axis) is found to be

$$\theta_w = 0.06316°$$

In the present situation $\theta_f < \theta_w$, which implies that the possibility of an error occurring on the feature after machining is high. This can be avoided only if $\theta_f > \theta_w$. This can be achieved by:

- Recombining two or more modular elements to one, so that the tolerance accumulation can be reduced
- Replacing the modular elements with a custom-made part

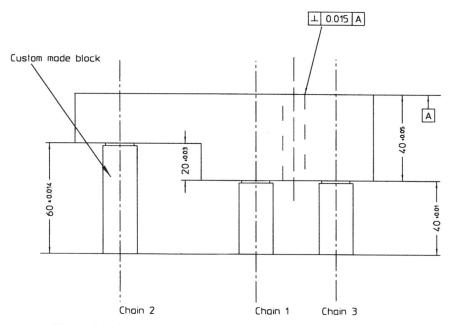

Figure 14.14. Distributed new tolerance on the custom-made block.

The excess tolerance present is computed by considering the normal vector, difference in angle ($\theta_w - \theta_f$), and locations of any two supporting points. In the present case it is found to be 0.005°. This can be removed by:

- Machining the base plate of the modular element (flatness should be respecified to 0.015°)
- Making a custom-made part to replace a supporting tower (height of 20 ± 0.015 mm)

Positional Tolerance Analysis. The positional tolerance is largely affected by the locating elements on the secondary and tertiary locating planes. The positional tolerance zone of the feature to be machined is given to be 0.028 mm, which implies that any displacement of the workpiece should not exceed 0.028 mm. This is computed using the tolerance zones available on the locating elements; however, it should also be noted that any change in the tolerance of any one point will affect the rest of the points. This may cause the workpiece to rotate and hence affect the position of the feature. In the present case it is fund to be 0.0352 mm, which can be reduced by:

- Changing the tolerance zone on the locating element
- Designing a custom-made part

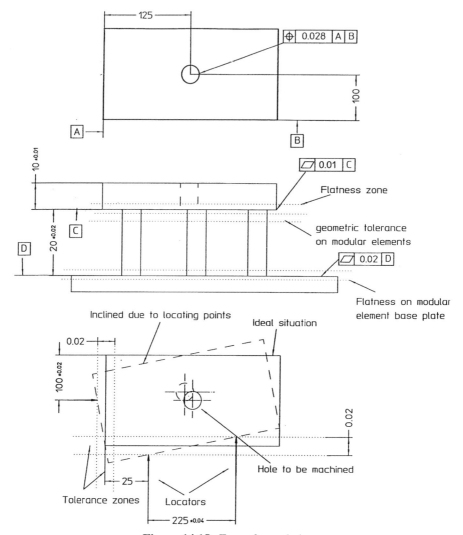

Figure 14.15. Example workpiece.

The excess tolerance is computed to be 0.0072 mm and can be removed from any of the locating elements. The final configuration of the fixture is shown in Figure 14.16.

14.5   CONCLUSIONS

Tolerance analysis of modular fixtures ensures a fixture designer that the machining errors of the part features due to tolerance buildup of the fixturing

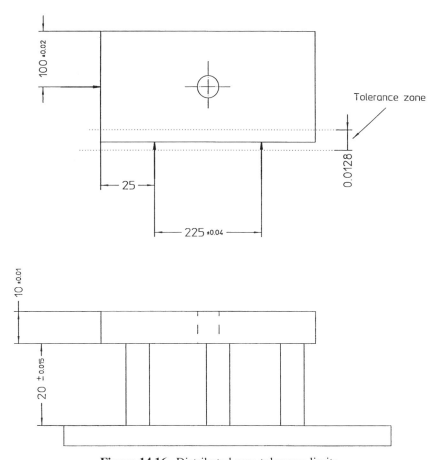

Figure 14.16. Distributed new tolerance limits.

elements are reduced, at least at the fixture synthesis stage. In this chapter, computational geometry-based techniques to solve the tolerance accumulation problem of modular fixtures is studied. Both the two-dimensional (planar) and three-dimensional (GD&T) approaches to solve the tolerance accumulation problem are analyzed by considering the dimensional, form, and positional tolerances on a workpiece. It is also important to note that the form tolerance is generally affected by the supporting points and the approach direction of a feature, while the positional tolerance of a feature is affected by the locating points. The work described considers flatness, perpendicularity, and positional tolerance of the workpiece. However, the concept developed could be further extended to solve the tolerance accumulation problem associated with cylindricity, angularity, circular runout, total runout, concentricity, and so on.

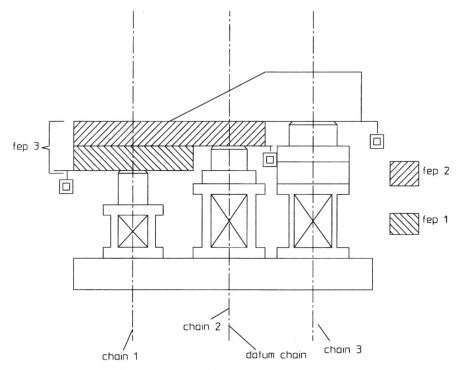

Figure 14.17. Dividing fixed elementary part.

REFERENCES

Bai, Y., and Rong, Y., 1993, Machining Accuracy Analysis for Computer-Aided Fixture Designs, *Manufacturing, Science and Engineering,* PED-Vol. 64, American Society of Mechanical Engineers, New York, pp. 507–512.

Bennett, G., and Gupta, L. C., 1970, Least-Cost Tolerance, Parts I and II, *Int. J. Prod. Res.,* Vol. 8, No. 1, pp. 65–74, and Vol. 8, No. 2, pp. 169–181.

Boerma, J. R., 1990, The Design of Fixtures for Prismatic Parts, Ph.D. thesis, Laboratory of Production Engineering, University of Twente, Enschede, The Netherlands.

Cogun, C., 1992, The Importance of the Application Sequence of Clamping Forces on Workpiece Accuracy, *J. Eng. Ind.,* Vol. 114, pp. 539–543.

Halevi, G., and Weill, R. D., 1995, *Principles of Process Planning: A Logical Approach,* Chapman & Hall, London.

Hoffman, E. G., 1987, *Modular Fixturing,* Manufacturing Technology Press, Lake Geneva, Wis.

King, D. A., and de Sam Lazaro, A., 1994, Process and Tolerance Considerations in the Automated Design of Fixtures, *Trans. ASME,* Vol. 116, pp. 480–486.

Nee, A. Y. C., Whybrew, K., and Senthil Kumar, A., 1995, *Advanced Fixture Design for FMS,* Springer-Verlag, New York.

Rong, Y., Li, W., and Bai, Y., 1995, Locating Error Analysis for Computer-Aided Fixture Design and Verification, *Proceedings of the ASME Conference on Computers in Engineering,* Boston, September 17–21, pp. 825–832.

Shawki, G. S. A., 1967, Rigidity Considerations in Fixture Design: Contact Rigidity for Eccentric Clamping, *Int. J. Mach. Tool Des. Res.,* Vol. 7, pp. 195–209.

Shawki, G. S. A., and Abdel-Aal, M. M., 1965, Effect of Fixture Rigidity and Wear on Dimensional Accuracy, *Int. J. Mach. Tool Des. Res.,* Vol. 5, pp. 183–202.

Shawki, G. S. A., and Abdel-Aal, M. M., 1966a, Rigidity Considerations in Fixture Design: Contact Rigidity at Locating Elements, *Int. J. Mach. Tool Des. Res.,* Vol. 6, pp. 31–43.

Shawki, G. S. A., and Abdel-Aal, M. M., 1966b, Rigidity Considerations in Fixture Design: Rigidity of Clamping Elements, *Int. J. Mach. Tool Des. Res.,* Vol. 6, pp. 207–223.

Trappey, J. C., and Liu, C. R., 1990, A Literature Survey of Fixture-Design Automation, *Int. J. Adv. Manuf. Technol.,* Vol. 5, pp. 240–255.

APPENDIX

A part is said to be irregular if different locating pins are not in the same plane. This does not include V-blocks and locators used to locate the sides of a part. The locating pins that are in contact with the bottom surface of the part are considered here. The general rule in dividing a part into fixed elementary parts (FEPs) is: If the part is irregular, draw a horizontal plane from the surface of the locating pins under consideration such that the part between the two planes form the fixed elementary part. Care should be taken since the FEP changes with the loop under consideration.

Consider Figure 14.17; here chain 2 is taken as the datum chain. First consider the loop containing chains 1 and 2. Draw a horizontal plane for each chain (i.e., plane A and plane B). The distance between them form the FEP (fep 1) for analysis. Next, consider chains 2 and 3 to obtain the FEP (fep 2). While checking, planes C and A are considered and the distance between A and C forms the FEP (fep 3). In this way, the FEPs are found and they are considered as modular components. However, their tolerance specifications are not altered while allocating the excess tolerance.

15

TOLERANCE AND ACCURACY ANALYSIS IN COMPUTER-AIDED FIXTURE DESIGN

YIMING (KEVIN) RONG

Southern Illinois University at Carbondale
Carbondale, Illinois

15.1 INTRODUCTION

Manufacturing accuracy depends on the relative positions of the machining tool and the workpiece (Rong et al., 1988). Fixtures are used to locate and hold a workpiece in the proper position during machining processes. Development of computer-aided fixture design (CAFD) systems is becoming increasingly important within flexible manufacturing systems (FMS) and computer-integrated manufacturing systems (CIMS) (Thompson and Gandi, 1986). Basically, two major approaches exist in CAFD. The first is rule-based (or knowledge-based) automated fixture design, where geometric reasoning, kinematics analysis, or screw theory may be applied (Pham and de Sam Lazaro, 1990; Trappey and Liu, 1990a; Chou et al., 1989). The second is group technology (GT)-based search and retrieval of existing fixture designs (Grippo et al., 1987; Rong and Zhu, 1992). The former is ideal for total automation but can usually only be applied to simple workpiece geometry because of the difficulties in geometric modeling and rule extraction. The latter is practical for industrial applications due to the use of existing knowledge in fixture designs.

Advanced Tolerancing Techniques, Edited by Hong-Chao Zhang
ISBN 0-471-14594-7   © 1997 John Wiley & Sons, Inc.

Once a fixture is designed by CAFD, its performance needs to be evaluated. Fixture design performance may include the locating accuracy needed to ensure the tolerance requirements of a product design, clamping and machining stability, fixturing stiffness to resist fixture component deformations, and an interference-free tool path (Menassa and DeVries, 1991). In previous CAFD research, possible interference between the cutting tool and fixture components was checked visually (Barry, 1982), force equilibrium of machining and clamping was verified (Trappey and Liu, 1989), clamping stability was evaluated automatically (Rong et al., 1994a), locating rigidity was considered and supporting position was optimized (Menassa and DeVries, 1990), and fixture component deformation was studied (Zhu et al., 1993). Actually, locating accuracy is the most important performance measure because the major purpose of CAFD is to provide a fixture design that can ensure the machining quality in manufacturing processes. Unfortunately, omitting dimensioning and tolerancing (D&T) analysis is very common in CAFD research (Trappey and Liu, 1990b). Very few papers take D&T information into account when flexible fixturing issues are discussed, including a simple case study of tolerance buildup in modular fixture design (Senthil kumar and Nee, 1990) and monitoring and diagnosis of fixturing failure detection in autobody assembly (Ceglarek and Shi, 1994).

Tolerance analysis has been an important problem in mechanical design, process planning, assembly, and fixture design. The computer-aided tolerancing has become one of the key issues in concurrent engineering (CE) and CIMS (Roy et al., 1991; Zhang and Huq, 1992). As computer numerical control (CNC) machine tools and machining centers are widely utilized in industry, fixture design functions have been changed to a simple structure, high accuracy, and single setup for multiple operations (Rong et al., 1994b). Therefore, accuracy analysis becomes more important, and the method used should be adaptable to these changes. Multioperation under a single setup becomes popular in modern manufacturing, where fixture design plays important roles in realization of machining process design and NC programming. To verify fixturing accuracy, machining errors need to be analyzed.

15.2 MACHINING ACCURACY ANALYSIS

A lot of research has been carried out to implement computer-aided tolerancing in manufacturing systems. Most current tolerance-related research is concerned with the operational and assembly tolerance chain analysis in computer-aided process planning (CAPP) and assembly, including tolerance modeling and analysis for satisfying clearance conditions of mating parts (Lee and Woo, 1990); tolerance allocation for mechanical assembly with automated process selection (Greenwood and Chase, 1987; Chase et al., 1990); studies on optimal tolerance assignment problems in CAPP (Dong and Soom, 1990; Manivanna et al., 1989); and operational tolerance analysis of rotational workpiece with setup effects, where the dependence of dimension variations in different operations was first considered (Zhang et al., 1991). Boerma and Kals studied fixture-setup planning from the view of ensuring tolerances of the workpiece (1988). In their work, different kinds of tolerances are converted into nondimensional values representing the maximum possible rotational error. These nondimensional values, called *tolerance factors,* can be compared. The process of fixture planning is to determine the locating (as well as clamping) positions and directions, which is determined primarily by and based on tolerance factors to ensure fixturing accuracy.

Analysis and synthesis of design and operational dimensions and tolerances are two aspects of computer-aided tolerancing. Tolerance charting is a tool representing relationships among dimensions and tolerances in different operations and setups. In previous studies a trace method was developed for tolerance charting with consideration of form and position tolerance effects (He and Lin, 1992; He and Gibson, 1992); a graph representation and linear programming method was applied to the tolerance charting problem for allocating tolerances among individual machining cuts (Mittal et al., 1991); a tree approach with a linear programming model was proposed for tolerance assignment in the tolerance charting (Ji, 1993a,b); a routed tree representation and relationship matrix method was developed for tolerance chart balancing (Ngoi, 1992, 1993); a dimensional tree method was utilized to determine operational dimensions and tolerances (Li and Zhang, 1989); the machining process effect was considered in stackup analysis of tolerances (tool wear and setup selection may affect machining errors) (Mei and Zhang, 1992); manufacturing cost effects were considered in tolerance synthesis (Dong and Soom, 1990; Dong and Hu, 1991), especially in optimum selection of discrete tolerances (Lee and Woo, 1989; Zhang and Wang, 1993); and by using descriptive rules and reasoning algorithms, expert systems were developed for tolerance assignment and verification in CAPP (Abdou and Cheng, 1993; Panchal et al., 1992; Janakiram et al., 1989). Most of this research studied linear dimension problems, assumed independent relationships among dimension variations, and did not consider setup and fixture design effects in the tolerance chain analyses, especially for nonrotational parts machined with machining centers.

15.2.1 Machining Error Analysis

Machining error is an opposite measure of machining accuracy. Dimensional tolerance is an allowable variation range of a dimension. A product designer provides tolerances to limit the range of machining errors. The deviation of the actual dimension from the theoretical dimension is the machining error. In a machining process, the errors include cutter-fixture relative alignment errors, tool wear, motion errors of a machine table, force and thermal effects, and vibration. These undesirable operating conditions are inevitable. Therefore, the dimensions generated cannot be exactly equal to the theoretically desired dimensions. Figure 15.1 shows some general descriptions of machining errors, which are the differences between the actual dimension and the theoretical dimension, where (a) is the linear dimensional error, ΔX, and (b), (c), and (d) are angular dimensional errors, $\Delta \alpha$ [referred to as orientation errors (Foster, 1982)], including parallel, perpendicular, and angular errors.

A fixture is applied to locate the workpiece relative to the cutter, therefore to ensure product quality because the dimensional accuracy depends primarily on the relative position of the workpiece and cutter in the machining process. When the locating datum in fixture design differs from the measuring datum in product design, an operational tolerance chain is formed and needs to be analyzed to estimate machining errors. In this chapter, dependent relationships of operational dimensions are analyzed. It is assumed that the workpiece size is not too large so that the tool wear within one setup is constant, although more than one cut may be involved, and the motion error of the machine table is random and uncontrollable without a specially designed compensator. Therefore, variations in workpiece dimensions (linear and angular) are not considered as functions of time.

Machining Errors of Linear Dimensions. Machining errors can be divided into two components (Rong et al., 1988), deterministic and random components:

$$\Delta X = \Delta X_d + \Delta X_r \tag{15.1}$$

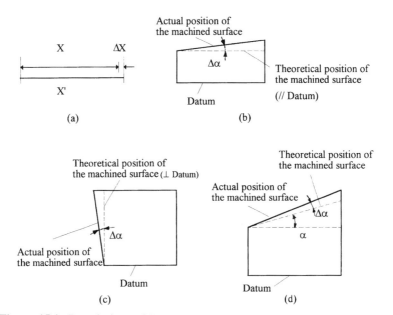

Figure 15.1. Descriptions of linear and angular dimension machining errors.

where ΔX is the error of dimension X; ΔX_d and ΔX_r are the deterministic and random components of ΔX, respectively, which could be positive or negative corresponding to a figure larger or smaller than the nominal dimension X.

The deterministic and random components of machining errors can be recognized based on analysis of machining processes. For example, in the study of variations of different dimensions generated in a machining center with a single setup and using the same cutting tool, the following discussion is true. Generally speaking, the deterministic machining errors are determined primarily by the locating errors of the fixture, including the position error of fixture-locating components and datum variations of the workpiece, errors caused by tool-fixture alignment errors and tool wear, and other deterministic errors in the machining process; that is,

$$\Delta X_d = \Delta X_1 + \Delta X_t + \Delta X_0 \qquad (15.2)$$

where ΔX_1 is an error component due to the fixture locating errors, ΔX_t an error component due to tool alignment errors and tool wear, and ΔX_0 other deterministic errors. The random errors are determined primarily by the clamping deformation, cutting force and thermal deformation during the machining process, and other random components of machining errors.

Each of the error components listed above can be estimated either by theoretical calculations or from empirical data. However, in a machining process, not every dimension can be generated directly. Some dimensions are formed as resultant dimensions by several relevant dimensions (Li and Zhang, 1989). The errors of relevant dimensions may affect the resultant dimension. The deterministic machining errors may be added or canceled with different fixturing methods. The widely used method to stack errors is the dimension and tolerance (D&T) chain theory which assumes an independence of the variations of the relevant dimensions (Bjorke, 1989). The variations in the relevant dimensions are summed up as the variation of the resultant dimension. This theory cannot be used directly to analyze the chain of dimensions which are machined in CNC machine tools or machining centers because the dimensions are dependent (Zhang et al., 1991). In this chapter we discuss a more general D&T chain issue that is suitable not only for turning but also for milling and other machining operations.

Figure 15.2 shows a machining case performed on a vertical CNC machine tool, where all machining and locating surfaces are indexed by numbers. If the subscripts are defined to present the dimension between the two surfaces, dimensions X_{3-5}, X_{3-7} and X_{9-8}, X_{9-6} are obtained in the first setup. Dimension X_{5-1} is obtained in the second setup. Dimensions X_{5-7}, X_{3-1} are ensured indirectly by dimensions X_{3-5}, X_{3-7} and X_{3-5}, X_{5-1}, and dimension X_{6-8} is ensured indirectly by dimensions X_{9-8}, X_{9-6}.

Because dimensions X_{3-5}, X_{5-1} are machined in different setups with different cutting tools, they are independent of each other because variations of these dimensions are dominated by the relative distance between the locating

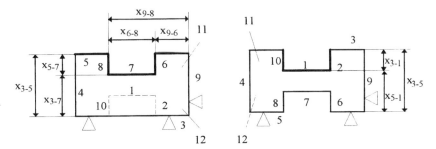

Figure 15.2. Workpiece machined in a vertical machining center.

datum and cutting tool as well as other random factors. The possible variation of resultant dimension X_{3-1} can be calculated by the D&T chain method, where the worst cases could be considered in error estimation, that is,

$$\Delta X_{3-1} = \Delta X_{3-5} + \Delta X_{5-1} = (\Delta X_{3-5d} + \Delta X_{3-5r}) + (\Delta X_{5-1d} + \Delta X_{5-1r}) \quad (15.3)$$

where ΔX_{3-1} is a possible variation of dimension X_{3-1}.

Dimensions X_{3-5}, X_{3-7} are machined in the same setup with the same cutting tool, so their variations are dependent. The two machining surfaces 5 and 7 have the same normal directions. The tool error (alignment error and tool wear) and locating errors are in the same direction, with identical values for both X_{3-5} and X_{3-7} (ΔX_{t2} and ΔX_{12} in Figure 3). When calculating the possible variations of X_{5-7}, ΔX_{3-5d} and ΔX_{3-7d} will cancel each other (including tool alignment and wear errors and fixturing locating errors). Therefore, the variation of the resultant dimension X_{5-7} becomes

$$\Delta X_{5-7} = \Delta X_{3-5r} + \Delta X_{3-7r} \quad (15.4)$$

In this case the variation of the resultant dimension is only the summation of the random components of the relevant dimensions.

However, if dimensions X_{3-5} and X_{3-7} are machined in the same setup with different cutting tools, the variation of the resultant dimension X_{5-7} becomes

$$\Delta X_{5-7} = (\Delta X_{3-5t} + \Delta X_{3-5r}) + (\Delta X_{3-7t} + \Delta X_{3-7r}) \quad (15.5)$$

Only the locating errors may be canceled, while the tool alignment error and tool wear effect may be different and cannot cancel each other.

In the case of the resultant dimension X_{6-8}, which is a result from dimensions X_{9-8} and X_{9-6}, the two machining surfaces 8 and 6 are obtained from the same setup and with the same tool and have opposite normal directions. The tool errors of X_{9-8} and X_{9-6} have the same value but opposite direction (ΔX_{t1} in Figure 3) (i.e., $\Delta X_{9-8t} = -\Delta X_{9-6t}$). Therefore, the variation of resultant X_{6-8} should be calculated as

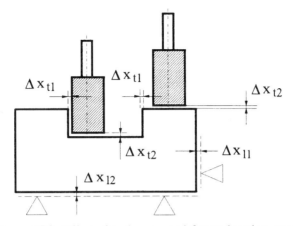

Figure 15.3. Effect of tool errors and fixture locating errors.

$$\Delta X_{6-8} = 2\Delta X_t + \Delta X_{9-8r} + \Delta X_{9-6r} \qquad (15.6)$$

where $\Delta X_t = \Delta X_{9-8t} = -\Delta X_{9-6t}$.

From the discussion above it can be seen that different setup planning may result in different combinations of the relevant dimensions in forming the resultant dimension, where the calculation of the variation of the resultant dimension should be different. The same conclusions can be drawn from the angular dimension analysis.

Machining Errors of Angular Dimensions. When a part design contains an accuracy requirement of orientations (parallelism, perpendicularity, or angularity), the machining errors can be analyzed and calculated in terms of an angular dimension chain. If more than one operation is concerned, the variation of the resultant angular dimension can also be discussed by examining the dependencies between the relevant angular dimensions. Because the tool alignment error and tool wear may affect the parallelism, perpendicularity, or angularity differently with the effects on linear dimensions, the analysis discussed above should be modified according to the analysis of angular dimension chains.

Similar to the linear dimension errors, the angular dimension errors can also be divided into deterministic ($\Delta\alpha_d$) and random ($\Delta\alpha_r$) components:

$$\Delta\alpha = \Delta\alpha_d + \Delta\alpha_r \qquad (15.7)$$

Because it is assumed that the workpiece size is not very large and the tool wear effect is not significant in one workpiece machining, the tool alignment error and tool wear will not affect the parallelism, perpendicularity, or angularity errors. Therefore, the deterministic machining errors are composed primarily of fixture locating errors ($\Delta\alpha_1$) and other deterministic errors ($\Delta\alpha_0$):

$$\Delta\alpha_d = \Delta\alpha_1 + \Delta\alpha_0 \tag{15.8}$$

To illustrate the dependent relationships between angular dimensions, the part in Figure 15.2 is taken as an example, where the parallelisms between surfaces 1 and 3, 5 and 7, and the perpendicularity between surfaces 2 and 3 are considered. Because, surfaces 1 and 3 are processed in different setups (surface 3 is assumed as a premachined surface), the parallelism between these two surfaces is ensured through surface 5. Therefore, the parallelism error between surfaces 1 and 3 can be estimated as

$$\Delta\alpha_{1-3} = \Delta\alpha_{3-5} + \Delta\alpha_{5-1} = (\Delta\alpha_{3-5d} + \Delta\alpha_{3-5r}) + (\Delta\alpha_{5-1d} + \Delta\alpha_{5-1r}) \tag{15.9}$$

By using the same principle, the perpendicularity error between surfaces 2 and 3 can be expressed as

$$\Delta\alpha_{3-2} = (\Delta\alpha_{3-5d} + \Delta\alpha_{3-5r}) + (\Delta\alpha_{5-2d} + \Delta\alpha_{5-2r}) \tag{15.10}$$

Surfaces 5 and 7 are obtained in the same setup. Their common datum is surface 3. Therefore, the parallelism error can be calculated as

$$\Delta\alpha_{5-7} = \Delta\alpha_{3-5} + \Delta\alpha_{3-7} = \Delta\alpha_{3-5r} + \Delta\alpha_{3-7r} \tag{15.11}$$

where the deterministic errors, $\Delta\alpha_{3-5d}$ and $\Delta\alpha_{3-7d}$, cancel each other.

Because the tool alignment error and tool wear do not contribute to the angular errors, equation (15.11) is valid for the machining error estimation under one setup, whether or not the normal directions of the surfaces are the same, and whether or not the surfaces are machined with the same or different cutting tools.

Summary of Machining Error Analysis. Based on the analysis above, the following five models can be summarized and defined.

Dimensional Variation Relationship Model 1. The variation of dimensions between a locating datum and machining surfaces contains deterministic and random components. If a dimension of the machining surface can be generated and measured directly from the locating surface, its variation can be calculated using equation (15.1) for linear dimensions or (15.7) for angular dimensions.

Dimensional Variation Relationship Model 2. If a resultant dimension is generated indirectly by two relevant dimensions that are machined in the same setup with the same tool, and the two surfaces associated with the resultant dimension have the same normal direction, the variation of the resultant dimension is only the summation of the random components of the relevant dimensions and can be calculated by equation (15.4) for linear dimensions or (15.11) for angular dimensions.

Dimensional Variation Relationship Model 3. If a resultant dimension is determined indirectly from two relevant dimensions that are obtained in the same setup with the same tool and the two surfaces forming the resultant dimension have opposite normal directions, the variation of the resultant dimension will be the summation of the deterministic and random components of the relevant dimensions, where their deterministic components caused by the tool alignment error and tool wear are the same and added up and the deterministic components resulting from the fixture locating error are canceled. The variation in the resultant dimension can be calculated by equation (15.6) for linear dimensions. In this case (15.11) is also valid for angular dimensions because the tool alignment error and tool wear do not affect the angular errors.

Dimensional Variation Relationship Model 4. If a resultant dimension is formed indirectly by two relevant dimensions that are obtained in a same setup with different cutting tools, the variation of the resultant dimension will be the summation of the deterministic and random components of the relevant dimensions, where their deterministic components caused by tool alignment error and tool wear are added up and the deterministic components caused by the fixture locating error are canceled. The variation of the resultant dimension can be calculated by equation (15.5) for linear dimensions. In this case, (15.11) is also valid for angular dimensions.

Dimensional Variation Relationship Model 5. When a resultant dimension is generated by two relevant dimensions that are obtained in different setups with different tools, the machining errors involved in obtaining the relevant dimensions are independent. The variation of the resultant dimension will be the summation of the deterministic and random components of the relevant dimensions, and can be calculated by equation (15.3) for linear dimensions, or equation (15.9) [equation (15.10)] for angular dimensions.

The variation of other dimensions can be considered as a combination of these five basic models. For example, the variation between surfaces 1 and 7 in Figure 15.2 is a combination of model 1 between surfaces 1 and 5 and model 2 between surfaces 5 and 7, discussed further below.

In fixture design, selecting different locating datums will cause different machining errors. To achieve high machining accuracy, relationship models 1 and 2 should be chosen with priorities. In the case of relationship models 3 and 4, reducing the effects of tool alignments and tool wear is more important than the position error of the fixture locating elements. In the analysis, form and location tolerances, as well as their effects on the dimensional errors, are not included, which may lead to a more complicated discussion.

15.2.2   Datum–Machining Surface Relationship Graph

Since the dimensional variation relationship models have been developed, how to determine the relationship models of the dimensions and tolerance (D&T) chain automatically is crucial in the analysis and verification of the locating accuracy in CAFD. A traditional tolerance chart can only be used to represent the linear dimensional relationship in one direction; it is not valid for angular dimension relationships. A datum-machining surface relationship graph (DMG) is developed to solve this problem.

DMG can be defined as a set of relationship graphs $G = \{G_i\}$, $i = 1, 2, ..., M$, where M is the number of setups. G_i represents the relationship between datum and machining surfaces in setup i:

$$G_i = \{D_{ij}, M_{ik}, A_{ijk}, T_{ijk}\} \qquad j = 1, 2, ..., N, \quad k = 1, 2, ..., N, \quad j \neq k \quad (15.12)$$

where N is the surface number of the workpiece; D_{ij} the set of nodes representing the locating surface j in setup i; M_{ik} the set of nodes representing the machining surface j in setup i; A_{ijk} the connections between nodes D_{ij} and M_{ik} or between M_{ij} and M_{ik}, representing manufacturing relationships; and T_{ijk} the set of attributes of the connection A_{ijk}. It is defined as follows: If $T_{ijk} = 0$, there is a linear dimension and parallelism relationship determined between the surfaces j and k in setup i (i.e., $D_{ij} \parallel M_{ik}$ or $M_{ij} \parallel M_{ik}$), which are with same normal directions; if $T_{ijk} = 180$, there is a linear dimension and parallelism relationship determined between surfaces j and k in setup i, but with opposite normal directions; if $T_{ijk} = 90$, there is a perpendicularity relationship determined between surfaces j and k in setup i (i.e., $D_{ij} \perp M_{ik}$ or $M_{ij} \perp M_{ik}$) and not a linear dimension relationship determined in this setup; in general, $T_{ijk} = \alpha$ represents an angular relationship of surfaces j and k; and if $T_{ijk} = 1$, there is only a linear dimension relationship determined between surfaces j and k in setup i.

Figure 15.4 shows an example of the DMG of a machining process for the workpiece illustrated in Figure 15.2. When the DMG is constructed, each geometric surface, such as plane, internal or external surface of a cylinder, and so on, is first assigned by an index number, as shown as in Figure 15.2. The datum surfaces and machining surfaces in each setup are represented by the nodes in DMG according to the process planning and fixture design results. These nodes are connected by lines with attributes to represent different relationships among these surfaces.

In Figure 15.4 two setups are required to process the workpiece, where the 3–2–1 locating method is applied. In the first setup, surfaces 3, 9, and 11 are the locating (or datum) surfaces, and surfaces 5, 6, 7, and 8 are the machining surfaces. Similarly, in the second setup, surfaces 5, 9, and 12 are the locating surfaces, and surfaces 1, 2, and 10 are the machining surfaces. These surfaces are represented by the nodes in the figure. The connecting lines between a locating surface and a machining surface within one setup represent the model 1 relationships with attributes that specify the dimension-determining rela-

surface number	First setup		Second setup	
	locating surfaces	machining surfaces	locating surfaces	machining surfaces

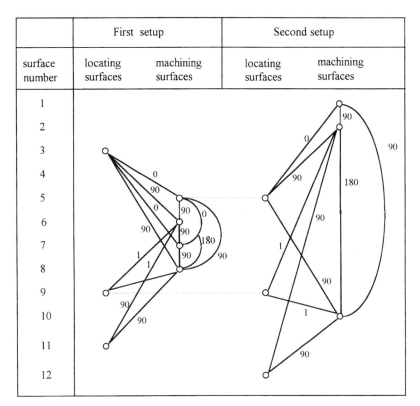

Figure 15.4. Datum–machining surface relationship graph for the example workpiece.

tionships. For example, in the second setup, the relationship between surfaces 5 and 1 is a model 1 with dimension-determining relationship because surface 1 is generated directly when surface 5 is the locating datum, where equation (15.1) should be used to estimate the machining errors of the dimension between surfaces 5 and 1. If the attribute value is 90, there is no linear dimension relationship between the datum and machining surfaces (e.g., the relationship between surfaces 5 and 2).

When the connecting line is between two machining surfaces, it represents a model 2 relationship (with an attribute of 0) or a model 3 relationship (with an attribute of 180), where variation in the resultant dimension is affected by relevant dimensions and the corresponding formulas should be applied. Examples of the relationships in Figure 15.10 include the relationship between surfaces 2 and 10 in the second setup. Nodes with the same surface numbers are considered as one node, connected by a dashed line between two setups (e.g., surfaces 5 and 9).

Once a verification requirement is specified for two surfaces of a workpiece, a search for the shortest path with a specific attribute between the surfaces is conducted from the last setup in DMG (from right to left). The shortest path is defined by a minimum number of nodes used to connect the two surfaces, which reflects the final dependent relationship of the surfaces (dimensions) during machining processes. The manufacturing relationship is determined and corresponding formulas are used to estimate the machining errors. For example, to analyze variation in the linear dimensions between surfaces 1 and 7 in Figure 15.10, the shortest path with attributes 1 or 0 is found as

$$\text{surface 1–surface 5–surface 7}$$

where surface 5 is the locating datum for machining surface 1 in the second setup, while surface 3 was the locating datum for machining surfaces 5 and 7 in the first setup. Therefore, the variation of the dimension between surfaces 1 and 7 becomes

$$\Delta X_{1-7} = \Delta X_{1-5} + \Delta X_{5-7} \tag{15.13}$$

where ΔX_{1-5} obeys the model 1 relationship and should be calculated using equation (15.1); ΔX_{5-7} obeys the model 2 relationship and should be calculated by using (15.4). Finally, the variation of the dimension between surfaces 1 and 7 is calculated by

$$\Delta X_{1-7} = \Delta X_{1-5} + (\Delta X_{3-5r} + \Delta X_{3-7r}) \tag{15.14}$$

The discussion above describes linear dimensions. Actually, angular dimensions can be analyzed in the same way. In the DMG representation of dimension-dependency relationships, datum surfaces are separated from machining surfaces, which is more convenient for relationship tracing. It should be noted that the DMG method can be applied not only in machining error analysis with NC machine tools or machining centers where multi-operations under a single setup are concerned, but also in problems with traditional machine tools where the DMG becomes wide and shallow (i.e., more setups in DMG and fewer machining surfaces in one setup).

15.2.3 Relationship Search: Matrix Approach

DMG is a graph representation of datum and machining surface relationships. To achieve an automated relationship search and automated calculation of variations between surfaces, a computer representation of DMG and a relationship search algorithm need to be developed. In this chapter a matrix reasoning approach is introduced for this purpose.

Basic-Relationship Matrix. Every setup J corresponds to two matrices, $A^{IJ} = \{a_{ij}^{IJ}\}$ ($I = 1,2$, $J = 1,2,...,M$; M is the number of setups). A^{IJ}, presenting the relationships between machining surfaces in setup J, is constructed based on the following rule:

$$a_{ij}^{IJ} = \begin{cases} 0 & \text{if machining surfaces } i \text{ and } j \text{ are parallel and have the} \\ & \quad \text{same normal direction} \\ \alpha & \text{if machining surfaces } i \text{ and } j \text{ have an angular relationship} \\ -1 & \text{otherwise} \end{cases}$$

where i and j are surface indexes $(i,j = 1,2,...,N;$ N is the number of surfaces).

A^{2J}, presenting the relationships between datum and machining surfaces in setup J, is constructed based on the following rules:

$$a_{ij}^{2J} = \begin{cases} 0 & \text{if datum } i \text{ and machining surface } j \text{ are parallel and have} \\ & \quad \text{the same normal direction} \\ \alpha & \text{if there is an angular relationship between datum } i \\ & \quad \text{and machining surface } j \\ 1 & \text{if there is a linear dimension relationship between datum } i \\ & \quad \text{and machining surface } j \\ -1 & \text{otherwise} \end{cases}$$

In matrix A^{2J}, $a_{ij}^{2J} = 1$ or 0 indicates a linear dimension relationship and $a_{ij}^{2J} = \alpha$ shows an angular dimension relationship between the datum and machining surfaces. When the angle α is 0 or 180°, the parallel relationship of surfaces i and j is presented, which have the same or opposite normal directions, and when α is 90°, the perpendicular relationship is presented. Therefore, orientation machining errors (parallelism, perpendicularity, and angularity) can be evaluated. A^{IJ} is defined as basic-relationship matrix presenting the DMG. It shows basic relationships between datum and machining surfaces. Matrices A^{11}, A^{21}, A^{12}, and A^{22} are the basic-relationship matrices of a machining process for the workpiece illustrated in Figures 15.2 and 15.4. These matrices are shown in Figure 15.5. They can be transferred to specific matrices according to the type of machining errors (linear or angular dimension errors) that need to be estimated.

Linear-Dimension Relationship Matrix. Basic-relationship matrices include information on both linear and angular dimension relationships. When linear dimension errors need to be estimated, a set of linear dimension-relationship matrices B^{IJ} should first be developed from their corresponding basic matrices A^{IJ}. The component, b_{iij}^{IJ}, in B^{IJ} can be generated from a_{ij}^{IJ} in A^{IJ} based on the following rules:

$$\text{if } a_{ij}^{IJ} = 0 \text{ or } 1, \qquad b_{ij}^{IJ} = 1$$

$$\text{if } a_{ij}^{IJ} = 180, \qquad b_{ij}^{IJ} = -1$$

$$\text{otherwise,} \qquad b_{ij}^{IJ} = 0$$

When $b_{ij}^{IJ} = 1$ or -1, there is a linear dimension relationship between

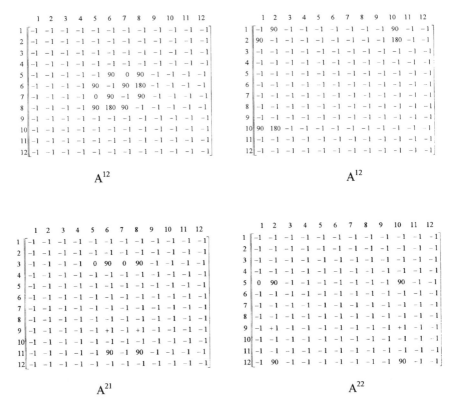

Figure 15.5. Basic relationship matrices.

surfaces i and j with the same or opposite normal directions. The angular relationships, including parallelism and perpendicularity, are represented by c_{ij}^{IJ} in matrix C^{IJ}, which is presented in the next section. The meaning of the matrix B^{IJ} is similar to A^{IJ}, where B^{IJ} presents the linear dimension relationship between machining surfaces in setup J and corresponds to models 2 and 3; B^{2J} presents the linear dimension relationship between datum and machining surfaces in setup J and corresponds to model 1. When two surfaces are machined based on the same datum but in different setups, the search process of the dimensional variation relationships needs a third matrix, $B^{3J} = (B^{2J})^{\mathrm{T}}$ (where T denotes the transposed matrix), which can be seen in case 3 of the example shown below. Therefore, three sets of matrices are formed and expressed as B_{IJ}, $I = 1,2,3$; $J = 1,...,M$.

Once B^{IJ} is constructed, a linear dimension chain relationship between specified surfaces (say, m and n) can be determined by checking the corresponding nonzero element (b_{mn} or b_{nm}) in the relationship-search matrix B based on a search strategy. If the two surfaces form a datum-machining surface relationship or are finally processed in the same setup, the nonzero element can be found in a single matrix, B_{IJ}, representing the relationships in the setup (see case 1 in the example). When the relationship spread into different setups, it can be found as a nonzero element in a matrix obtained through matrix multiplication (see case 2 in the example). The relationship search starts from right to left until a nonzero element is found in a related location of the matrix. In the search process, a shortest path is to be identified in DMG because only the relationship associated with the last operations represents a final relationship between the surfaces specified.

A search algorithm has been designed to identify a dimensional relationship between two specified surfaces. The input of the algorithm includes a set of matrices representing linear dimensional relationships for a workpiece in all setups (i.e., B^{IJ}) and the surface indexes of a specified linear dimension (m and n). The procedure for the relationship search is as follows.

1. Nonzero elements b_{im}^{2J} and b_{jn}^{2J} (where i and j are surface indexes) are sought in matrix $B^{2J}(J = M, M - 1,...,1)$ to determine the last setups in which surfaces m and n are machined. It can be assumed in a general sense that the setups are H and G, and that $G \geq H$.

2. If one surface is machined in setup G and the other is the locating datum in the same setup ($i = n$ or $j = m$, i.e., b_{mn}^{2G} or $b_{nm}^{2G} \neq 0$ in B^{2G}), a direct relationship of datum and machining surfaces can be identified, with the relationship represented by model 1. The search is finished.

3. If surfaces m and n are machined in the same setup ($H = G$), the locating datum should be the same ($i = j$) where dimensional relationship model 2 or 3 may be applied. In the corresponding matrix B^{2G}, it can be found that b_{mn}^{1G} or $b_{nm}^{1G} \neq 0$.

4. If surfaces m and n are machined in different setups, b_{mn}^{IR} or b_{nm}^{2G} is examined in a newly formed matrix $B = B^{IR} \times B^{2G}$, $I = 1,2,3$; $R = G - 1,...,1$, until a nonzero value is found, which is in the form $b_{mp}^{IR} \times b_{pm}^{2G} \neq 0$ (or $b_{np}^{IR} \times b_{pm}^{2G} \neq 0$), where p represents a surface that is machined in setup R and used as a datum in setup G. Once such a nonzero element is found and surface p is not machined in any setup between R and G (i.e., $b_{ep}^{2L} = 0$, $R < L < G$; $e = 1,2,...,N$), the dimension relationship can be determined with different relationship models, which can be seen in the examples that follow. In this case two related setups are involved.

5. If a dimension is generated within three related setups, a nonzero element b_{mn} or b_{nm} can be determined from one of the following matrices:

$$B = B^{IQ1} \times B^{2Q2} \times B^{2G} \quad \text{or} \quad B = B^{3Q2} \times B^{IQ1} \times B^{2G}$$

where $I = 1,2,3$; $Q_1, Q_2 = G - 1, G - 2,...,1$; and $Q_1 < Q_2$. With the rules described in step 4, the dimensional relationship of surfaces m and n can be identified.

6. When four related setups are necessary in the dimension chain, the nonzero element b_{mn} or b_{nm} can be determined from one of the matrices:

$$B = B^{IQ1} \times B^{2Q2} \times B^{2Q3} \times B^{2G}$$

or

$$B = B^{3Q2} \times B^{IQ1} \times B^{2Q3} \times B^{2G}$$

or

$$B = B^{3Q3} \times B^{IQ1} \times B^{2Q2} \times B^{2G}$$

or

$$B = B^{3Q3} \times B^{3Q2} \times B^{IQ1} \times B^{2G}$$

where $I = 1,2,3$; $Q_1,Q_2,Q_3 = G - 1, G - 2,....,1$; and $Q_1 < Q_2 < Q_3$.

7. When there are even more relevant dimensions, a general search step needs to be developed. If k relevant dimensions are involved, the nonzero element b_{mn} or b_{nm} can be found in the matrix:

$$B = B^{IQ1} \times B^{2Q2} \times B^{2Q3} \times \cdots \times B^{2Qk-1} \times B^{2G}$$
$$\vdots$$
$$B = \prod_{u=1}^{U} B^{3Qiu} \times B^{IQ1} \times \prod_{v=1}^{V} B^{2Qjv} \times B^{2G}$$
$$i_u > i_{u+1}, j_v < j_{v+1},$$
$$U + V = k - 2,$$
$$i_u \neq j_v$$
$$\vdots$$
$$B = B^{3Qk-1} \times B^{3Qk-3} \times \cdots \times B^{3Q2} \times B^{IQ1} \times B^{2G}$$

where $I = 1,2,3$; $Q_1,Q_2,...,Q_{k-1} = G - 1, G - 2,....,1$; and $Q_1 < Q_2 < Q_3 < \cdots < Q_{k-1}$.

As long as such a relationship exists, a nonzero element b_{mn} or b_{nm} can be found by following the foregoing procedure. Therefore, the dimensional relationship of the surfaces can be identified. It should be mentioned that although many setups could be designed for part production, the number of relevant dimensions in most dimension-chain analysis problems for machining operations is less than four, especially for operations with CNC machines and machining centers. The relationship search procedure can be much simplified.

For example, the matrices A^{11}, A^{21}, A^{12}, and A^{22} are the basic-relationship matrices for the machining case in Figure 15.2, and B^{11}, B^{21}, B^{12}, and B^{22} are the linear dimension relationship matrices generated from matrices A^{11}, A^{21}, A^{12}, and A^{22}. The elements of the B matrices are shown in Figure 15.6.

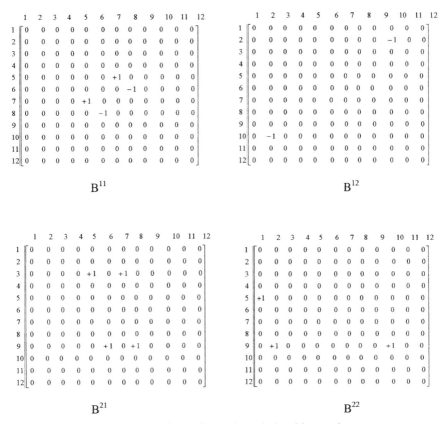

Figure 15.6. Linear dimension relationship matrices.

Case 1. If the variation in dimensions between surfaces 3 and 5 needs to be estimated, the first step is to examine their relationship. According to the relationship search algorithm, the search procedure is to find a nonzero element from the B matrices which refers to surfaces 3 and 5, in the sequence B^{IJ} ($I = 1,2; J = M,...,1$) and manipulations of B matrices, that is,

$$B^{22} - B^{21}$$

In the relationship search matrix B^{21}, it is found that $b_{35} = 1$. This result shows that the dimension between surfaces 3 and 5 is ensured directly in setup 1 because the subscript $J = 1$ in the relationship search matrix. The relationship between surfaces 3 and 5 is represented by model 1 since the subscript $I = 2$, which indicates the relationship between datum and machining surface. (When $I = 1$, model 2 should be applied if $b_{mn} = 1$, or model 3 applied if $b_{mn} = -1$.) Hence equation (15.1) is used to calculate the dimension variation between surfaces 3 and 5.

Case 2. When the variation of dimension between surfaces 1 and 7 is estimated, the search procedure is

$$B^{22}\text{--}B^{11} \times B^{22}$$

In matrix $B^{12} \times B^{22}$ we find that $b_{71} = b_{75}^{11} \times b_{51}^{22} = 1$. That means that the dimension between surfaces 1 and 7 is ensured indirectly by dimensions X_{7-5} and X_{7-1}. The relationship between surfaces 7 and 5 is presented by model 2 because $b_{75}^{11} = 1$, and the machining error relationship between surfaces 5 and 1 is specified by model 1 since $b_{51}^{22} = 1$ appears in matrix B^{22}. The machining error between surfaces 1 and 7 is a summation of the errors of the foregoing two dimensions and becomes

$$\Delta X_{1-7} = \Delta X_{7-5} + \Delta X_{5-1} \tag{15.15}$$

Case 3. When the variation of a linear dimension between surfaces 2 and 8 is to be estimated, the search procedure is

$$B^{22}\text{--}B^{11} \times B^{22}\text{--}B^{21} \times B^{22}\text{--}B^{31} \times B^{22}$$

In matrix $B^{31} \times B^{22}$ it is found that $b_{28} = b_{89}^{31} \times b_{92}^{22} = 1$. Therefore, the linear dimension between surfaces 2 and 8 is ensured indirectly by dimensions X_{9-8} and X_{9-2}. The dimension relationships between surfaces 9 and 8 and 9 and 2 are presented by model 1 because it is found that $b_{89}^{31} = 1$ and $b_{92}^{22} = 1$. The machining error between surfaces 2 and 8 is a summation of the errors of the two relevant dimensions:

$$\Delta X_{2-8} = \Delta X_{9-8} + \Delta X_{9-2} \tag{15.16}$$

These examples illustrate that the search for linear dimension relationships is implemented by manipulating the B matrices, which can be done automatically through programming. A similar search procedure is valid for angular dimensions.

Angular-Dimension Relationship Matrix. When the stackup of angular dimension error (parallelism, perpendicularity, and angularity) is estimated, the family of angular-dimension relationship matrices C^{IJ} can be generated from their corresponding matrices A^{IJ}. The matrix element c_{ij}^{IJ} is determined based on the following rules:

$$\text{If } a_{ij}^{IJ} = \begin{cases} \pm 1 & c_{ij}^{IJ} = 0 \\ \text{otherwise} & c_{ij}^{IJ} = 1 \end{cases}$$

C^{1J} presents the angular-dimension relationship between machining surfaces in setup J, and C^{2J} presents the angular-dimension relationship between datum and machining surface in setup J. Similarly, a third matrix is defined as $C^{3J} = (C^{2J})^T$, for convenience in the relationship search.

The case of stacking-up angular dimensional errors is more complicated than the case of linear dimensional errors since the angular dimensional errors are concerned with variations in angular dimensions. Before we can stack up the angular dimensional errors, we need to verify if it is meaningful to stack up these angular dimensional errors. Actually, we can only stack up the angular dimensional errors when their measuring planes are the same or parallel to each other. Figure 15.7 shows an example in which surfaces 1 and 3 are machined in terms of a datum, surface 2. Although the perpendicularity between surfaces 1 and 2 and surfaces 3 and 2 might be generated in the machining operation, the relationship between surfaces 1 and 3 (parallelism or perpendicularity) may be considered only when the normal directions of their measuring planes are the same. The measuring plane of two surfaces (planes or cylinders) can be defined by a normal vector, $\mathbf{d}_{ij}$, which is determined through a cross-product of feature vectors of these two surfaces (normal vector for planes and vector in axis direction for cylinders); that is,

$$\mathbf{d}_{ij} = \mathbf{n}_i \times \mathbf{n}_j \tag{15.17}$$

In the case of Figure 15.7, when the relationship between surfaces 1 and 3 is examined, the cross-product of measuring surface vectors should be considered as

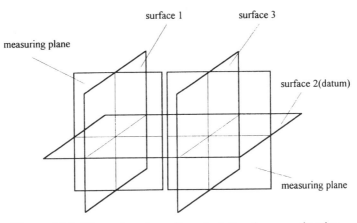

Figure 15.7. Measuring planes for orientational error estimation.

$$\mathbf{d} = \mathbf{d}_{12} \times \mathbf{d}_{23} \qquad (15.18)$$

Only when the product of two normal vectors of the measuring planes is zero are the two measuring planes parallel so that the angular dimensional errors can be stacked up. Therefore, this test becomes a criterion of evaluating relationships of geometric dimensions. The search strategy for the angular dimension relationship between surfaces m and n is the same as the strategy for the linear dimensional relationship shown earlier.

The example shown in Figure 15.2 can also be used for the angular dimension-relationship analysis. Angular relationship matrices C^{11}, C^{21}, C^{12}, and C^{22} are generated from the basic relationship matrices A^{11}, A^{21}, A^{12}, and A^{22}. The C matrices are shown in Figure 15.8. To estimate the variation of angular dimension between surfaces 2 and 3, their relationship is first searched. The search procedure is to find a nonzero element representing the relationship of the two surfaces from the following matrices:

$$C^{22}\text{--}C^{11} \times C^{22}\text{--}C^{21} \times C^{22}$$

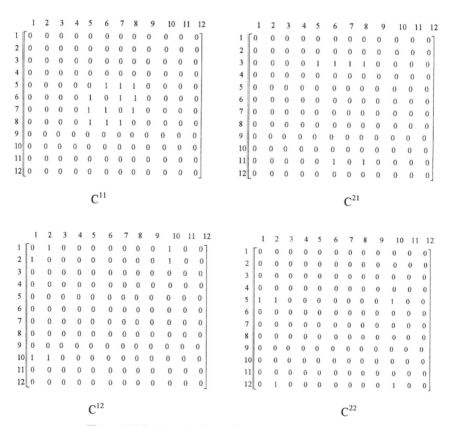

Figure 15.8. Angular dimension relationship matrices.

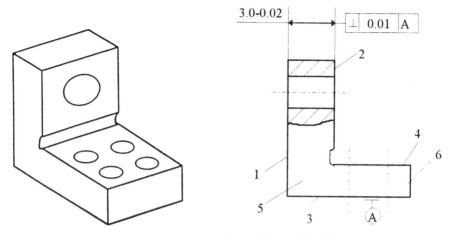

Figure 15.9. Workpiece to be machined.

In matrix $C^{21} \times C^{22}$, $c_{32} = c_{35}^{21} \times c_{52}^{22} = 1$. Verification of the error stackup condition shows that $\mathbf{d}_{3,5} = 0$, so that $\mathbf{d}_{3,5} \times \mathbf{d}_{5,2} = 0$. Therefore, the angular dimension between surfaces 2 and 3 is ensured indirectly by angular dimensions between surfaces 2 and 5 and surfaces 5 and 3. The error relationship between surfaces 2 and 3 is the summation of angular dimension errors between surfaces 2 and 5 and surfaces 5 and 3, where the relationship model 1 is applied twice.

19.2.4 Tolerance Analysis in Setup Planning

Figure 15.9 shows a workpiece to be machined on a vertical machining center, where all the major machining surfaces are indexed by numbers and two accuracy requirements are to be ensured with tolerance specifications and design datum. Fixturing plans are made with a CAPP system, which requires three setups, as shown in Figure 15.10. The fixturing plans are shown in Table 15.1. A common datum (surface 1) is employed in setups 2 and 3 to ensure

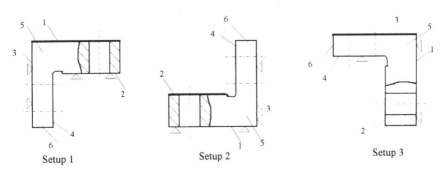

Figure 15.10. First fixturing plan for the workpiece.

TABLE 15.1. Machining and Locating Surfaces in Each Setup with Different Fixturing Plans

	First Fixturing Plan			Second Fixturing Plan		
	Setup 1	Setup 2	Setup 3	Setup 1	Setup 2	Setup 3
Machining surface	1	2	3	3	1	2
Primary locating surface	2	1	1	4	3	3
Secondary locating surface	3	3	6	1	6	1
Tertiary locating surface	6	6	4	6	2	6

linear dimension tolerance between surfaces 1 and 2, and angular dimension (perpendicularity) accuracy between surfaces 2 and 3. After machining error analysis, the estimated variations in the linear and angular dimensions are listed in Table 15.2. If the design requirement regarding perpendicularity cannot be satisfied where machining errors in two setups are added up, the fixturing plan needs to be modified (i.e., different locating datum may be selected). Figure 15.11 shows a second fixture plan where surface 3 is a locating datum in both setups 2 and 3 where surfaces 1 and 2 are machined (see Table 15.1). The results of the machining error analysis is also listed in Table 15.2. Comparison of columns 2 and 3 in Table 15.2 shows the improvement in ensuring the perpendicularity requirement (fewer error terms are involved).

15.3   LOCATING ERROR ANALYSIS

Fixture design activities include three steps: setup planning, fixture planning, and fixture configuration design (Rong and Bai, 1995). The objective of the setup planning is to determine the number of setups needed, orientation of the workpiece in each setup, and the machining surfaces in each setup. The setup planning could be a subset of process planning. Fixture planning is to determine the locating, supporting, and clamping points on workpiece sur-

TABLE 15.2. Comparison of Machining Errors in Different Fixturing Plans

	First Fixturing Plan	Second Fixturing Plan
Variation of linear dimension between surfaces 1 and 2	$\Delta X_{2-1} = \Delta X_{2-1d} \pm \Delta X_{2-1r}$	$\Delta X_{2-1} = \Delta X_{2-1d} \pm \Delta X_{2-1r}$
Variation of angular dimension between surfaces 2 and 3.	$\Delta \alpha_{3-2} = \Delta \alpha_{2-1d} \pm \Delta \alpha_{2-1r}$ $\pm \Delta \alpha_{3-1d} \pm \Delta \alpha_{3-1r}$	$\Delta \alpha_{3-2} = \Delta \alpha_{3-2d} \pm \Delta \alpha_{3-2r}$

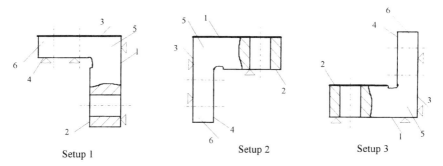

Setup 1   Setup 2   Setup 3

Figure 15.11. Second fixturing plan for the workpiece.

faces. The task of fixture configuration design is to select fixture elements and place them into a final configuration to locate and clamp the workpiece. Locator position errors contribute to locating accuracy, yielding geometric tolerances in product designs. Analysis of locator position variation effects provides information for the selection of locating surface and points and for fixture design verifications.

Fixture design involves locating a workpiece in the proper position and maintaining the position during one or more manufacturing operations. In view of the kinematics, fixture design is to restrict the six degrees of freedom (DOF) of a free-body (workpiece) motion. To ensure manufacturing accuracy, fixture design is to select a proper locating surface (datum) as well as distributions of locating points. For a complete locating design, six locators (or equivalent) are required to constrain all the DOFs of a workpiece. The 3–2–1 locating method, shown in Figure 15.12, is the method used most frequently for machining prismatic workpieces (Hoffman, 1991). Locators 1, 2, and 3

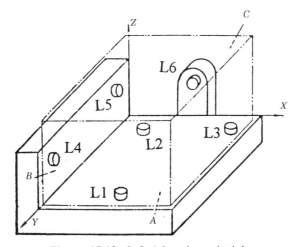

Figure 15.12. 3–2–1 locating principle.

restrict possible linear motion in the Z-direction and angular motions about the X and Y axes. Locators 4 and 5 restrict linear motion in the X-direction and rotation about the Z-axis. Finally, locator 6 constrains motion in the Y-direction. Although there may be other locating methods, they are variations of the 3–2–1 locating principle, where six locating points are necessary and distributed in different ways. Figure 15.13 shows examples of the variations. The function of locators in the fixture is to provide workpieces with right location, including position and orientation, so that the relative relationship between cutting paths (or positions of cutting tools) and the workpiece can be ensured. When the positions of the six locators are inaccurate, this relationship may be changed. Therefore, manufacturing errors may be generated because of the locating errors.

According to the error source, the manufacturing process error, Δ_p, related to the locating datum can be expressed as

$$\Delta_p = \Delta_m + \Delta_l \tag{15.19}$$

where Δ_m and Δ_l are the machining error and locating error, respectively. These two terms can be divided into a summation of deterministic and random components and can be expressed further according to the error sources as

$$\Delta_m = \Delta_{md} + \Delta_{mr} = (\Delta_{tw} + \Delta_{mod}) + \Delta_{mr} \tag{15.20}$$
$$\Delta_l = \Delta_{ld} + \Delta_{lr} = (\Delta_{wv} + \Delta_{pv}) + \Delta_{lr}$$

where Δ_{md} and Δ_{ld} = deterministic components of Δ_m and Δ_l

Δ_{mr} and Δ_{lr} = random components of Δ_m and Δ_l

Δ_{tw} = tool wear and alignment error

Δ_{mod} = other deterministic machining errors

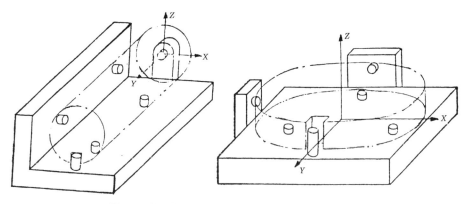

Figure 15.13. Variations of the 3–2–1 principle.

Δ_{wv} = workpiece datum error caused by workpiece size variations

Δ_{pv} = locating-point position variation

Machining errors may arise from tool alignment errors and tool wear, cutting force and thermal deformation during the machining process, and the machine table's motion errors. As indicated earlier, processing errors were analyzed by considering the dependency relationships among operational dimensions. There are three types of relationships between machined features or between machined features and the locating datum: errors between machining–machining (M-M) surfaces in one setup, between datum–machining (D-M) surfaces in one setup, and between M-M or D-M surfaces in multiple setups. When only locating error effects are considered, the error in the first relationship is zero. The complex relationships in a multiple setup can be decomposed into errors generated in each individual setup by using DMG analysis. The effects on manufacturing accuracy of imperfect locator position (Δ_{pv}) and workpiece size variation (Δ_{wv}) associated with different locating methods are analyzed to provide information for CAFD and fixture design verification.

15.3.1 Effect of Locator Position Error

When the locator positions are inaccurate, the position and orientation of the workpiece may change, affecting manufacturing process accuracy. Manufacturing process errors may be generated due to this change. Because the position change of even a single locator has a complex effect on workpiece position and orientation, a variation in the workpiece position or orientation due to locating error is first analyzed compared with the ideal situation, which implies a change in the locating datum. The position change of any relevant point on the machining surface of the workpiece is then determined. The manufacturing errors caused by the locating errors can thus be estimated. The estimation results can be used to verify and improve the fixture design.

Locating Reference Surface Modeling. To formulate a change in workpiece position and orientation due to inaccurate locator positions, three locating reference planes are established, according to the actual potions of the locators. Based on the 3–2–1 locating principle, three orthogonal planes are established as the locating reference surfaces. Change in locator position results in change of the plane equations. A general plane equation can be expressed as

$$Ax + By + Cz + D = 0 \tag{15.21}$$

where A, B, and C are direction cosines of the normal vector of the plane, D

is a constant, and x, y, and z are the coordinates of an arbitrary point on the plane.

The three locating reference planes are expressed as:

$$\text{primary plane:} \quad A_1x + B_1y + C_1z + D_1 = 0$$

$$\text{secondary plane:} \quad A_2x + B_2y + C_2z + D_2 = 0 \quad (15.22)$$

$$\text{tertiary plane:} \quad A_3x + B_3y + C_3z + D_3 = 0$$

When the coordinates of six locating points are known, the reference planes can be determined by calculating the direction cosines (i.e., A_i, B_i, and C_i, $i = 1,2,3$) and D_i values. The primary locating plane is defined by locators L1, L2, and L3 (see Figure 15.12) with their coordinates (x_1,y_1,z_1), (x_2,y_2,z_2), and (x_3,y_3,z_3). Therefore, its plane equation is

$$\begin{vmatrix} x - x_1 & y - y_1 & z - z_1 \\ x_2 - x_1 & y_2 - y_1 & z_2 - z_1 \\ x_3 - x_1 & y_3 - y_1 & z_3 - z_1 \end{vmatrix} = 0 \quad (15.23)$$

The three locators cannot be placed along a straight line; otherwise, it is true that

$$\frac{x_3 - x_1}{x_2 - x_1} = \frac{y_3 - y_1}{y_2 - y_1} = \frac{z_3 - z_1}{z_2 - z_1} \quad (15.24)$$

which leads to nonunique solutions of equation (15.23). Thus A_1, B_1, C_1, and D_1 can be obtained as

$$A_1 = \begin{vmatrix} y_1 - y_2 & z_2 - z_3 \\ y_2 - y_3 & z_1 - z_2 \end{vmatrix} \quad B_1 = -\begin{vmatrix} x_1 - x_2 & z_2 - z_3 \\ x_2 - x_3 & z_1 - z_2 \end{vmatrix}$$

$$\quad (15.25)$$

$$C_1 = \begin{vmatrix} x_1 - x_2 & y_2 - y_3 \\ x_2 - x_3 & y_1 - y_2 \end{vmatrix} \quad D_1 = -x_1A_1 - y_1B_1 - z_1C_1$$

The secondary locating plane is defined by locators L4 and L5 with their coordinates (x_4,y_4,z_4), and (x_5,y_5,z_5) and perpendicular to the first plane. The plane equation is

$$\begin{vmatrix} x - x_4 & y - y_4 & z - z_4 \\ x_4 - x_5 & y_4 - y_5 & z_4 - z_5 \\ A_1 & B_1 & C_1 \end{vmatrix} = 0 \quad (15.26)$$

The two locators cannot be placed along a vertical line relative to the first plane; otherwise, it becomes true that

$$\frac{x_4 - x_5}{A_1} = \frac{y_4 - y_5}{B_1} = \frac{z_4 - z_5}{C_1} \tag{15.27}$$

which leads to nonunique solutions of equation (15.26). Therefore, A_2, B_2, C_2, and D_2 can be calculated as

$$A_2 = \begin{vmatrix} y_4 - y_5 & z_4 - z_5 \\ C_1 & B_1 \end{vmatrix} \qquad B_2 = -\begin{vmatrix} x_4 - x_5 & z_4 - z_5 \\ C_1 & A_1 \end{vmatrix}$$

$$C_2 = \begin{vmatrix} x_4 - x_5 & y_4 - y_5 \\ B_1 & A_1 \end{vmatrix} \qquad D_2 = -x_4 A_2 - y_4 B_2 - z_4 C_2 \tag{15.28}$$

The tertiary locating plane equation can be determined by locator L6 with coordinates (x_6, y_6, z_6) and keeps the perpendicular relationships with the first and second locating planes; that is,

$$\begin{vmatrix} x - x_6 & y - y_6 & z - z_6 \\ A_1 & B_1 & C_1 \\ A_2 & B_2 & C_2 \end{vmatrix} = 0 \tag{15.29}$$

Similarly, A_3, B_3, C_3, and D_3 become

$$A_3 = \begin{vmatrix} B_1 & B_2 \\ C_1 & C_2 \end{vmatrix} \qquad B_3 = -\begin{vmatrix} A_1 & A_2 \\ C_1 & C_2 \end{vmatrix}$$

$$C_3 = \begin{vmatrix} A_1 & A_2 \\ B_1 & B_2 \end{vmatrix} \qquad D_3 = -x_6 A_3 - y_6 B_3 - z_6 C_3 \tag{15.30}$$

The position and orientation of the workpiece are determined by the three reference planes, while the planes are determined by the positions of the six locators. When the positions of the locators deviate from the nominal positions, these reference planes need to be reexamined.

It should be mentioned that even if the locators L1, L2, and L3 are not in the same plane, the primary plane can still be determined according to coordinates (x_1, y_1, z_1), (x_2, y_2, z_2), and (x_3, y_3, z_3). A similar situation applies to locators L4 and L5. Details on how to derive equations to establish the locating reference planes can be found in the literature (Li, 1995). If other locating methods are utilized in a fixture design, three corresponding locating reference planes can still be established according to the locating information. For example, if two pins (one round and one diamond-shaped) are used with locating holes to restrict three DOFs beyond the three restricted by the locators under the bottom surface of the workpiece, the locating reference planes are chosen as the three orthogonal planes shown in Figure 15.14, where the primary locating plane is determined by the positions of the three locators, the secondary locating plane is constructed to contain the center axes of the two

(a)

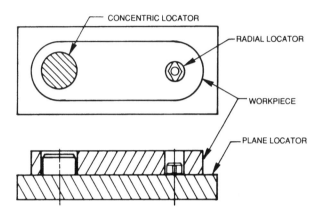

(b)

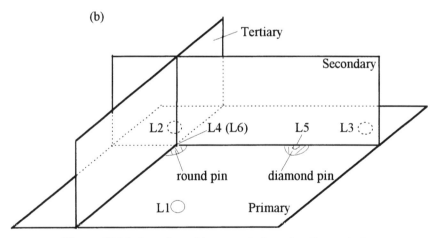

Figure 15.14. Locating with a round pin and a diamond pin.

locating pins and perpendicular to the primary plane, and the tertiary locating plane is the plane perpendicular to the first two and contains the center axis of the round pin. It is obvious that the spatial positions of the reference planes are determined by the positions of locators. Therefore, the plane equations can be built up in the same way as shown above, where $x_4 = x_6$, $y_4 = y_6$, and $z_4 = z_6$.

Position Variation of a Point on the Workpiece Surface. The variation of one locator's position may not result simply in a movement of the workpiece in the direction the locator is applied to restrict. A slip motion in another direction may exist due to the effects of other locators. This variation may cause a complex change in the workpiece orientation. Therefore, the deviation of a point on the workpiece surface needs to be defined and identified in terms of the locating reference planes.

An arbitrary point (say, M with coordinates x, y, and z) on a machining surface of the workpiece can be defined by the distances of the point to the three locating planes. These distances are constant in the product design. When the distances are defined by a, b, and c, respectively, the following equations are deduced:

$$A_1 x + B_1 y + C_1 z + D_1 = a\sqrt{A_1^2 + B_1^2 + C_1^2}$$
$$A_2 x + B_2 y + C_2 z + D_2 = b\sqrt{A_2^2 + B_2^2 + C_2^2} \qquad (15.31)$$
$$A_3 x + B_3 y + C_3 z + D_3 = c\sqrt{A_3^2 + B_3^2 + C_3^2}$$

where the signs of a, b, and c should be the same as that of the left-hand terms in the foregoing equations. In matrix form it can be expressed as

$$\begin{bmatrix} A_1 & B_1 & C_1 \\ A_2 & B_2 & C_2 \\ A_3 & B_3 & C_3 \end{bmatrix} \begin{bmatrix} x \\ y \\ z \end{bmatrix} = \begin{bmatrix} a\sqrt{A_1^2 + B_1^2 + C_1^2} - D_1 \\ b\sqrt{A_2^2 + B_2^2 + C_2^2} - D_2 \\ c\sqrt{A_3^2 + B_3^2 + C_3^2} - D_3 \end{bmatrix} \qquad (15.32)$$

Thus the coordinates of M can be resolved as follows:

$$\begin{bmatrix} x \\ y \\ z \end{bmatrix} = \begin{bmatrix} A_1 & B_1 & C_1 \\ A_2 & B_2 & C_2 \\ A_3 & B_3 & C_3 \end{bmatrix}^{-1} \begin{bmatrix} a\sqrt{A_1^2 + B_1^2 + C_1^2} - D_1 \\ b\sqrt{A_2^2 + B_2^2 + C_2^2} - D_2 \\ c\sqrt{A_3^2 + B_3^2 + C_3^2} - D_3 \end{bmatrix} \qquad (15.33)$$

If the inverse matrix exists in equation (15.33), the following determinant cannot be zero:

$$\det = \begin{vmatrix} A_1 & B_1 & C_1 \\ A_2 & B_2 & C_2 \\ A_3 & B_3 & C_3 \end{vmatrix} \neq 0 \qquad (15.34)$$

By considering the relationships in equation (15.30), the value of (15.34) becomes

$$\det = (B_1 C_2 - B_2 C_1)^2 + (A_1 C_2 - A_2 C_1)^2 + (A_1 B_2 - A_2 B_1)^2 \qquad (15.35)$$

Because the primary locating surface is not parallel to the secondary locating surface, the three terms in equation (15.35) cannot be zero at the same time. Therefore, equation (15.33) can be used to determine the coordinates of a point on the workpiece surface in terms of locating reference planes, while the planes are described by the locators' coordinates. When a variation in the locators' positions is identified, the locating reference planes are reevaluated. The coordinates of relevant points on the workpiece surface are determined. Finally, these coordinates are compared with those before the variation and geometric error generated in the machining process can be recognized, which resulted from the variation of locators' positions.

Locator Position Error Effects on Geometric Accuracy. Manufacturing errors in the workpiece geometry include dimensional errors, form errors, orientation errors, position errors, and surface roughness. The shape errors and surface quality are usually dependent on the tool geometry, the motion of the machine table, and/or the interaction process of the tool and workpiece (e.g., tool wear and vibration effects). The locating errors may affect the dimensional errors, orientation errors such as parallelism, perpendicularity, and angularity, and the true position errors. For example, Figure 15.15 shows a simple case in which the parallelism error of surface P1 and perpendicularity error of surface P2 are to be examined relevant to the bottom plane, which is assumed to be the primary locating plane. Once the coordinates of vertices on the machining surfaces are calculated in terms of three locating references in perfect condition planes (i.e., *XOY, YOZ,* and *XOZ* planes) and with locator deviations, the orientation errors due to the locating inaccuracy can be estimated as follows:

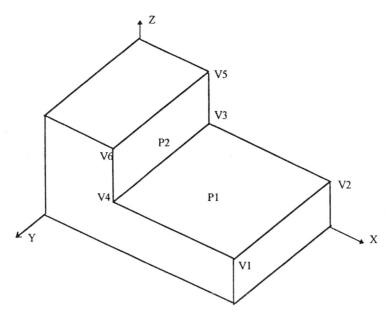

Figure 15.15. Locating error effect on workpiece geometric accuracy.

$$E_{\parallel} = \max[|(z_i - z'_i) - (z_j - z'_j)|, \, i, j = 1,2,3,4; \, i \neq j]$$

$$E_{\perp} = \max[|(x_i - x'_i) - (x_j - x'_j)|, \, i, j = 3,4,5,6; \, i \neq j]$$

(15.36)

where $E_{\parallel}$ and $E_{\perp}$ denote the parallelism error and perpendicularity error, respectively; x_i, y_i, z_i and x'_i, y'_i, z'_i are the vertex coordinates of machining surfaces without and with a position deviation of the locators, respectively.

Equation (15.36) can only be used to estimate the orientation errors caused by inaccurate locator positions in a specific condition. To evaluate the locating error effect, the following procedure can be applied:

1. Establish locating reference planes based on the designed locator positions.
2. Estimate the locating reference planes with locator position deviations.
3. Determine the geometric error items to be evaluated, as well as relevant machining surfaces.
4. Determine the coordinates of the vertices of the machining surfaces (or endpoints of the feature axis) before and after the locator deviations, based on the fact that their distances to the reference planes are constant.
5. Estimate the geometric errors according to differences in the position variations of these points.

When a coordinate system is selected properly, this procedure can be much simplified. Figure 15.16 shows a block diagram of the procedure for locating error effect analysis. To illustrate this procedure, a virtual part, shown in Figure 15.17, is taken as an example. Two setups are required to machine this part. The locating and machining surfaces in the two setups are shown in Table 15.3.

In setup 1, the locating error effect on the parallelism errors of surfaces P3 and P12 and the perpendicularity errors of surface P4 are considered relevant to the locating surface P5. First, a single deviation of each individual locator is assumed to result in different effect sensitivities to the parallelism and perpendicularity. The results are listed in Table 15.4, where the locator deviations are assumed to be 0.02 in. (in locating directions) and the geometric errors are calculated by applying Eq. (15.36) (also in inches). The orientation errors listed in the table are relative to the locating plane (i.e., P5). To evaluate the parallelism between surface P3 and P12, an error-stackup method can be applied. It can be seen that the inaccurate locators positions have different effects on parallelism and perpendicularity errors. The scale-level difference of the parallelism errors of P3 and P12 has resulted from the different areas of the planes. The same test is repeated with different locator coordinates to examine the effects of the locator layout. Table 15.5 shows four cases where the locators' coordinates are given in the first column. Finally, when a uniform tolerance is given to all the locators, the maximum

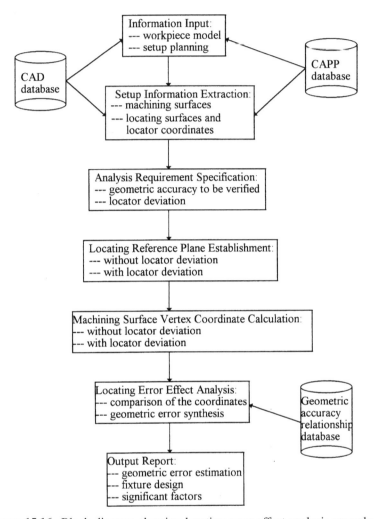

Figure 15.16. Block diagram showing locating error effect analysis procedure.

effect on parallelism and perpendicularity is estimated and has been shown in Table 15.6.

In setup 2, the locating error effect on the perpendicularity errors of surface P7, angularity errors of surface P10, and parallelism errors of hole H1 are considered relevant to the locating surface P3. Table 15.7 lists locator coordinates and geometric error values under single deviations of each locator.

It should be noted that the results shown in the tables come from a specific workpiece and specific setup planning. To evaluate the locating error effects quantitatively, the effective areas of machining surfaces should be taken into account. The example above shows the method and functions of a locating

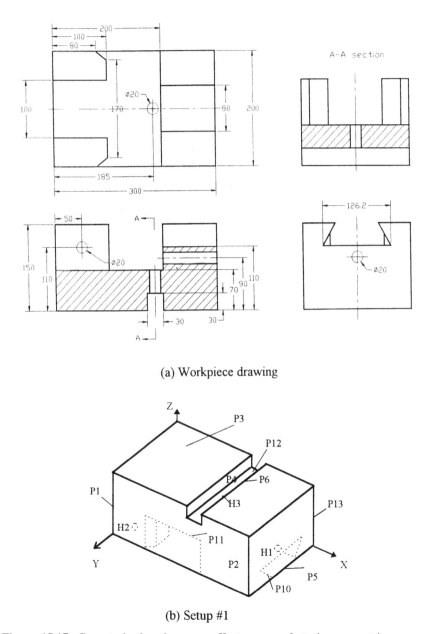

(a) Workpiece drawing

(b) Setup #1

Figure 15.17. Case study: locating error effect on manufacturing geometric accuracy.

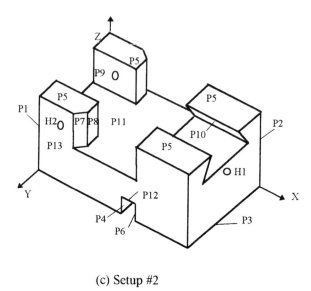

(c) Setup #2

Figure 15.17. (*Continued*)

error effect analysis system. Further studies are undertaken by applying the system to study general conclusions of the locating error effects. This method can be applied to more complex tolerance problems with multiple setups where the resultant geometric errors are decomposed into those produced from individual setups. The information from locating error effect analysis is valuable and important for locating datum selection and verification in computer-aided process planning and fixture designs.

15.3.2 Locating Method Effects

Although there are a variety of fixture design configurations, the locating methods are quite limited. Workpiece surfaces that can be used as locating datums include planes, holes, and external profile surfaces (Rong et al., 1993). When different locating methods are applied, different combinations of these surfaces are utilized (Pollack, 1988). For example, in the 3–2–1 locating

TABLE 15.3. Setup Planning for the Sample Part

	Setup 1	Setup 2
Primary locating surface	P5	P3
Secondary locating surface	P13	P2
Tertiary locating surface	P1	P1
Machining surfaces considered	P3, P4, P12	P7, P10, H1

TABLE 15.4. Geometric Errors Due to Individual Deviations of Locator Positions

Locator Coordinates	Single Deviation	$E_\parallel - P3$	$E_\perp - P4$	$E_\parallel - P12$
L1: 66, 66, 0	$\Delta z = 0.02$	0.0767	0.0018	0.0606
L2: 66, 134, 0	$\Delta z = 0.02$	0.0767	0.0018	0.0606
L3: 234, 100, 0	$\Delta z = 0.02$	0.0357	0.0036	0.0036
L4: 40, 0, 75	$\Delta y = 0.02$	0	0.0182	0
L5: 260, 0, 75	$\Delta y = 0.02$	0	0.0182	0
L6: 0, 100, 50	$\Delta x = 0.02$	0	0	0

TABLE 15.5. Comparison of Geometric Errors Under Different Locator Layouts

Locator Coordinates	Single Deviation	$E_\parallel - P3$	$E_\perp - P4$	$E_\parallel - P12$
Case 1				
L1: 66, 90, 0	$\Delta z = 0.02$	0.2179	0.0018	0.2018
L2: 66, 110, 0	$\Delta z = 0.02$	0.2179	0.0018	0.2018
L3: 234, 100, 0	$\Delta z = 0.02$	0.0357	0.0036	0.0036
L4: 40, 0, 75	$\Delta y = 0.02$	0	0.0182	0
L5: 260, 0, 75	$\Delta y = 0.02$	0	0.0182	0
L6: 0, 100, 50	$\Delta x = 0.02$	0	0	0
Case 2				
L1: 100, 66, 0	$\Delta z = 0.02$	0.0888	0.0030	0.0618
L2: 100, 134, 0	$\Delta z = 0.02$	0.0888	0.0030	0.0618
L3: 200, 100, 0	$\Delta z = 0.02$	0.0600	0.0060	0.0060
L4: 40, 0, 75	$\Delta y = 0.02$	0	0.0182	0
L5: 260, 0, 75	$\Delta y = 0.02$	0	0.0182	0
L6: 0, 100, 50	$\Delta x = 0.02$	0	0	0
Case 3				
L1: 150, 66, 0	$\Delta z = 0.02$	0.0588	0	0.0588
L2: 100, 134, 0	$\Delta z = 0.02$	0.0894	0.0060	0.0354
L3: 200, 134, 0	$\Delta z = 0.02$	0.0894	0.0060	0.0354
L4: 40, 0, 75	$\Delta y = 0.02$	0	0.0182	0
L5: 260, 0, 75	$\Delta y = 0.02$	0	0.0182	0
L6: 0, 100, 50	$\Delta x = 0.02$	0	0	0
Case 4				
L1: 66, 66, 0	$\Delta z = 0.02$	0.0767	0.0430	0.0606
L2: 66, 134, 0	$\Delta z = 0.02$	0.0767	0.0430	0.0606
L3: 234, 100, 0	$\Delta z = 0.02$	0.0357	0.0036	0.0036
L4: 100, 0, 30	$\Delta y = 0.02$	0	0.0400	0
L5: 200, 0, 100	$\Delta y = 0.02$	0	0.0400	0
L6: 0, 100, 50	$\Delta x = 0.02$	0	0	0

TABLE 15.6. Geometric Errors Under a Given Uniform Locating Tolerance

Locator Coordinates	Uniform Tolerance	$E_\parallel - P3$	$E_\perp - P4$	$E_\parallel - P12$
L1: 66, 66, 0	$\Delta x = \pm 0.02$	0.1533	0.0435	0.1212
L2: 66, 134, 0	$\Delta y = \pm 0.02$			
L3: 234, 100, 0	$\Delta z = \pm 0.02$			
L4: 40, 0, 75				
L5: 260, 0, 75				
L6: 0, 100, 50				

method, three plane surfaces (or equivalent) are used as the locating datum, and one plane and two holes may be used in another locating method. Due to the variation in workpiece size or clearance between the workpiece and fixture locating components, the position of the locating datum may vary in a certain range, which will contribute to the machining errors. The variation in the locating datum position is defined as a datum-position error, which is a part of the fixturing locating error (ΔX_1 or $\Delta \alpha_1$), defined earlier.

Locating Method Effect on Linear Dimensions. If planes are used as the locating datum, the effect of datum-position error requires a simple expression and the variation of the locating datum positions is constant once the fixture is constructed. The variation of workpiece geometry does not affect the datum position in this case. However, when a hole is used as a locating datum, the variation of locating datum position is determined by the difference between the maximum hole dimension and the minimum pin dimension, as shown in Figure 15.18. The ideal locating datum is the geometric center of hole. The maximum variation of the datum is

$$\Delta_1 = T_D + T_d + \Delta_{\min} + \Delta_{\text{pin}} \tag{15.37}$$

where T_D is the hole tolerance of the workpiece, T_d the pin tolerance of the fixture locating component, $\Delta_{\min}$ the minimum clearance between the hole and pin, and Δ_{pin} the position error of the locating pin, which could be eliminated through calibration and adjustment during the operation.

TABLE 15.7. Geometric Errors Due to Individual Deviations of Locator Positions

Locator Coordinates	Single Deviation	$E_\perp - P7$	$E_\angle - P10$	$E_\parallel - H1$
L1: 66, 66, 0	$\Delta z = 0.02$	0.0217	0.0166	0.0060
L2: 66, 134, 0	$\Delta z = 0.02$	0.0160	0.0166	0.0060
L3: 234, 100, 0	$\Delta z = 0.02$	0.0057	0.0060	0.0119
L4: 40, 0, 75	$\Delta y = 0.02$	0.0023	0.0079	0
L5: 260, 0, 75	$\Delta y = 0.02$	0.0023	0.0079	0
L6: 0, 100, 50	$\Delta x = 0.02$	0	0	0

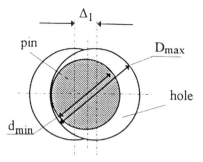

Figure 15.18. Datum-position error in pinhole.

If the workpiece is placed in a vertical orientation, because the gravity always pushes the workpiece contacting the locating pin in one direction, the variation of the locating datum becomes

$$\Delta_1 = \frac{T_{D6} + T_d}{2} \qquad (15.38)$$

A V-block is another typical locating component, where the geometric center of the cylindrical surface of the workpiece is desired to be the locating datum. Figure 15.19 shows a sketch of the locating datum position with V-block locating. The maximum variation of the datum can be calculated by

$$\Delta_1 = \frac{T_d}{2 \sin(\alpha/2)} \qquad (15.39)$$

where T_d is the tolerance of the workpiece cylinder and α is the V-block angle (Figure 15.18).

In the estimation of fixturing related machining errors, the locating method effects should be included. Figure 15.20 shows an example of machining error estimation, where one plane and two holes are used as the locating datum in

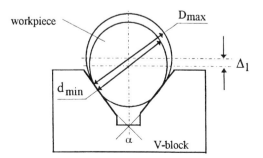

Figure 15.19. Datum position in V block.

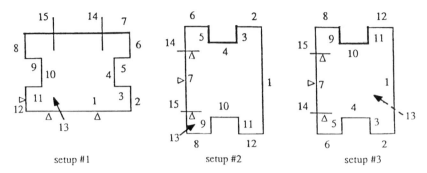

Figure 15.20. Example of locating workpieces with pinholes and a plane.

setups 2 and 3. The locating datum (plane and holes) is generated in setup 1. It should be noted that one hole is located by a round pin (fitted with hole 14 in setup 1 and hole 15 in setup 2) and the other is located by a diamond pin.

If the machining error of the dimension between surfaces 4 and 10 is to be estimated, the dimension X_{4-10} is determined by relevant dimensions X_{14-4} and X_{15-10}, while the dimension X_{14-15} is generated in setup 1. Therefore, the variation of the dimension between surfaces 10 and 4 becomes

$$\Delta X_{4-10} = \Delta X_{15-10} \pm \Delta X_{15-14} \pm \Delta X_{14-4} \qquad (15.40)$$

where ΔX_{15-10} and ΔX_{14-4} are the model 1 relationships and including the datum-position errors; ΔX_{15-14} is the model 2 relationship and is calculated by

$$\Delta X_{15-14} = \Delta X_{12-15r} \pm \Delta X_{12-14r} \qquad (15.41)$$

where surface 12 is the locating datum for holes 15 and 14 machined in setup 1. In the calculation of ΔX_{15-10} and ΔX_{14-4}, the locating error should be evaluated by using equation (15.38) because locating holes and pins are used.

Locating Method Effect on Angular Dimensions. When different locating methods are utilized, the variation in locating datum position will also affect the parallelism, perpendicularity, or angularity accuracy. In the application of 3–2–1 locating method, where three planes are used as locating surfaces, the angular errors caused by locating datum position errors is influenced by the variation in locator heights. Figure 15.21 shows an angular-dimension variation in the secondary locating plane, where two locators are used to restrict the rotational degree of freedom. The angular locating error can be estimated by

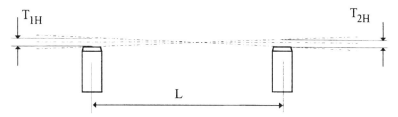

Figure 15.21. Locating error with plane locating method.

$$\Delta_1' = 2 \tan^{-1} \frac{T_{1H} + T_{2H}}{2L} \tag{15.42}$$

where T_{1H} and T_{2H} are the height tolerances of the first and second locating pins, respectively, and L is the distance between the two locating pins.

Figure 15.22 shows the locating errors with another common locating method, where one plane and two holes are used as locating surfaces (a diamond pin is used with the right hole). Due to the clearances between the locating pins and the holes (one of the pins is diamond), the angular locating error caused by the variation in locating datum position can be calculated as

$$\Delta_1' = 2 \tan^{-1} \frac{T_{1D} + T_{1d} + \Delta_{1min} + T_{2D} + T_{2d} + \Delta_{2min}}{2L} \tag{15.43}$$

where T_{1D} = diameter tolerance of the workpiece hole fitted with the round locating pin

T_{1d} = diameter tolerance of the round pin

$\Delta_1{}^{min}$ = minimum clearance between the round pin and the hole

T_{2D} = diameter tolerance of workpiece hole fitted with the diamond locating pin

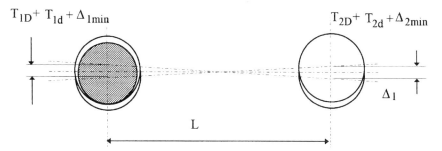

Figure 15.22. Locating errors with two-hole and one-plane locating method.

T_{2d} = diameter tolerance of the diamond pin

Δ_2^{min} = minimum clearance between the diamond pin and the hole

L = distance between the two holes

Figure 15.23 shows an example of using V-blocks as the primary locating components. Variations in workpiece diameters will affect the accuracy of the angular dimension. The deviation of the workpiece central axis from the desired locating axis can be calculated as

$$\Delta_1' = 2 \tan^{-1} \frac{1}{4L} \left[\frac{T_{2d}}{\sin(\alpha_2/2)} + \frac{T_{1d}}{\sin(\alpha_1/2)} \right] \qquad (15.44)$$

where T_{1d} and T_{2d} are the tolerances of the workpiece diameters and α_1 and α_2 are the V-block angles.

Equations (15.42)–(15.44) can be derived from the geometric relationships shown in Figures 15.20 to 22, where a maximum variation of the error angle is taken into account. In the example of Figure 15.20, equation (15.43) should be applied in estimating the fixture locating errors if the parallelism, perpendicularity, or angularity must be ensured.

15.4 LOCATING ACCURACY VERIFICATION OF CAFD

Once a fixture planning is conducted through a CAFD system, the locating accuracy needs to be verified to ensure machining quality. Machining errors are analyzed and decomposed into individual setups according to process planning information and DMG analysis. Dimension variations caused by locating datum variation are estimated based on fixture configuration design

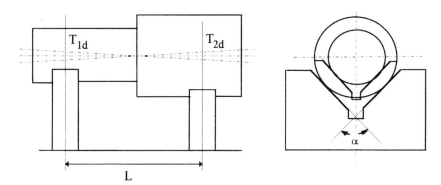

Figure 15.23. Locating errors with two-V-block locating method.

and workpiece geometry, where a fixture component database is used to assess the information regarding locator position deviations. Machining process errors resulting from vibration, thermal deformation, tool wear, and other process factors need to be estimated based on a machining process model and machining process database. The procedure of implementing the locating accuracy verification is shown in Figure 15.24. The input information includes workpiece geometry, geometric features and tolerances from the part design; locating method and datum from process planning and fixture design, and the fixturing tolerances and machining process error estimations from a manufacturing planning database. The system is applied to calculate the relationships between machining surfaces and locating surfaces where the machining error of a specified dimension is decomposed into machining errors in each individual setup, to calculate fixturing errors through a tolerance chain analysis, and to synthesize the possible maximum machining errors. The output of the system is an estimation of maximum machining errors and their most significant components, which is verified with the part design requirement.

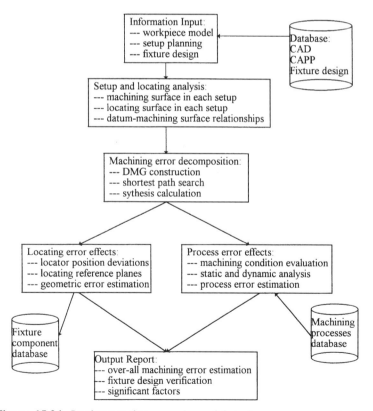

Figure 15.24. Implementation procedure of fixturing accuracy verification.

15.5 SUMMARY

A machining accuracy analysis is presented for fixture design verifications, where the dependency of resultant dimension variations on the variations of relevant dimensions are studied, including linear and angular dimensions. Five basic dimension relationship models of locating datum and machining surfaces are given to estimate the machining error under different setup conditions. The locating method effects on fixturing accuracy analysis are discussed. A datum–machining surface relationship graph (DMG) has been developed to construct a tolerance chain analysis. A matrix representation and reasoning algorithm is developed for automatic search and evaluation of the dimension relationships of the datum and machining surfaces. Although this method is general and may be applied to other tolerance chain analyses, general form tolerance (as well as their effects on dimensional variations) is not included in this chapter.

An analysis of locating error effects on manufacturing accuracy is also presented in this chapter. Inaccurate locator positions may cause a change in the workpiece location and orientation. Therefore, after machining operations, workpiece geometric errors may be generated, such as parallelism, perpendicularity, and angularity errors. Based on the locator positions, locating reference planes are established to determine workpiece location and orientation. The coordinates of vertices of machining surfaces (or endpoints of a feature axis) are calculated relevant to the locating reference planes. When locator positions deviate from their accurate positions, the locating reference planes will change as well as the vertices of machining surface. According to this change, geometric errors due to the inaccurate locator positions can be estimated. A case study example illustrates the procedure of the analysis. Once the analysis system is integrated with the tolerance analysis system developed before, more complex multiple-setup problems can be analyzed. The information from the analysis is useful and important in computer-aided process planning and computer-aided fixture design and verification.

ACKNOWLEDGMENTS

This research is partially funded by the National Science Foundation. Encouragement and support from Ingersoll Milling Machine Company and Bluco Cooperation are appreciated by the author. Thanks are extended to Yong Bai and Weizhen Li, who were involved in the work presented in this chapter.

REFERENCES

Abdou, G., and Cheng, R., 1993, TVCAPP: Tolerance Verification in Computer-Aided Process Planning, *Int. J. Prod. Res.,* Vol. 31, No. 2, pp. 393–411.

Barry, D. C., 1982, Application of CAD/CAM to Fixture Design, *Proceedings of the First Biennial International Machine Tool Technology Conference,* Chicago, September, pp. 43–66.

Bjorke, O., 1989, *Computer-Aided Tolerancing,* 2nd ed., ASME Press, New York.

Boerma, J. R., and Kals, H. J. J., 1988, FIXES: A System for Automatic Selection of Setup and Design of Fixtures, *Ann. CIRP,* Vol. 37, No. 1, pp. 443–446.

Ceglarek, D., and Shi, J., 1994, Fixture Failure Diagnosis for the Autobody Using Pattern Recognition, *Proceedings of the ASME Winter Annual Meeting,* Chicago, November 6–11, pp. 263–275.

Chase, K. W., Greenwood, W. H., Loosli, B. G., and Hauglund, L. F., 1990, Least Cost Tolerance Allocation for Mechanical Assemblies with Automated Process Selection, *Manuf. Rev.,* Vol. 3, No. 1, pp. 49–59.

Chou, Y. C., Chandru, V., and Barash, M. M., 1989, A Mathematical Approach to Automatic Configuration of Machining Fixtures: Analysis and Synthesis, *J. Eng. Ind.,* Vol. 111, pp. 299–306.

Dong, Z., and Hu, W., 1991, Optimal Process Sequence Identification and Optimal Process Tolerance Assignment in Computer-Aided Process Planning, *Comput. Ind.,* Vol. 17, pp. 19–32.

Dong, Z., and Soom, A., 1990, Automatic Optimal Tolerance Design for Related Dimension Chains, *Manuf. Rev.,* Vol. 3, No. 4, pp. 262–271.

Foster, L. W., 1982, *Modern Geometric Dimensioning and Tolerancing with Workbook Section,* 2nd ed., National Tooling and Machining Association, Baltimore, MD.

Greenwood, W. H., and Chase, K. W., 1987, A New Tolerance Analysis Method for Designers and Manufacturers, *J. Eng. Ind.,* Vol. 109, pp. 112–116.

Grippo, P. M., Gandi, M. V., and Thompson, B. S., 1987, The Computer-Aided Design of Modular Fixturing Systems, *Int. J. Adv. Manuf. Technol.,* Vol. 2, No. 2, pp. 75–88.

He, J. R., and Gibson, P. R., 1992, Computer-Aided Geometrical Dimensioning and Tolerancing for Process-Oriented Planning and Quality Control, *Int. J. Adv. Manuf. Technol.,* Vol. 7, pp. 11–20.

He, J. R., and Lin, G. C. I., 1992, Computerized Trace Method for Establishing Equations for Dimensions and Tolerances in Design and Manufacture, *Int. J. Adv. Manuf. Technol.,* Vol. 7, pp. 210–217.

Hoffman, E. G., 1991, *Jig and Fixture Design,* 3rd ed., Delmar, Albany, NY.

Janakiram, D., Prasad, L. V., and Rao, U. R. K., 1989, Tolerancing of Parts Using an Expert System, *Int. J. Adv. Manuf. Technol.,* Vol. 4, pp. 157–167.

Ji, P., 1993a, A Linear Programming Model for Tolerance Assignment in a Tolerance Chart, *Int. J. Prod. Res.,* Vol. 31, No. 3, pp. 739–751.

Ji, P., 1993b, A Tree Approach for Tolerance Charting, *Int. J. Prod. Res.,* Vol. 31, No. 5, pp. 1023–1033.

Lee, W. J., and Woo, T. C., 1989, Optimum Selection of Discrete Tolerances, *J. Mech. Transm. Autom. Des.,* Vol. 111, pp. 243–251.

Lee, W. J., and Woo, T. C., 1990, Tolerances: Their Analysis and Synthesis, *J. Eng. Ind.,* Vol. 112, pp. 113–121.

Li, J. K., and Zhang, C., 1989, Operational Dimensions and Tolerances Calculation in CAPP Systems for Precision Manufacturing, *Ann. CIRP,* Vol. 38, pp. 403–406.

Li, W., 1995, Locating Error Analysis for Computer-Aided Fixture Design and Verification, M.S. thesis, Southern Illinois University, Carbondale, IL, June.

Manivanna, S., Lehtihet, A., and Egbelu, P. J., 1989, A Knowledge Based System for the Specification of Manufacturing Tolerances, *J. Manuf. Sys.,* Vol. 8, No. 2, pp. 153–160.

Mei, J., and Zhang, H. C., 1992, Tolerance Analysis for Automated Setup Selection in CAPP, *Proceedings of the ASME Winter Annual Meeting,* Anaheim, CA, November 8–13, PED-Vol. 59, pp. 211–220.

Menassa, R. J., and DeVries, W. R., 1990, A Design Synthesis and Optimization for Fixtures with Compliant Element, *Proceedings of the ASME Winter Annual Meeting,* PED-Vol. 47, pp. 203–218.

Menassa, R. J., and DeVries, W. R., 1991, Optimization Methods Applied to Selecting Support Positions in Fixture Design, *J. Eng. Ind.,* Vol. 113, pp. 412–418.

Mittal, R. O., Irani, S. A., and Lehtihet, E. A., 1991, Tolerance Control in Machining of Discrete Components, *J. Manuf. Sys.,* Vol. 9, No. 3, pp. 233–246.

Ngoi, B. K. A., 1992, Applying Linear Programming to Tolerance Chart Balancing, *Int. J. Adv. Manuf. Technol.,* Vol. 7, pp. 187–192.

Ngoi, B. K. A., 1993, A Complete Tolerance Charting System, *Int. J. Prod. Res.,* Vol. 31, No. 2, pp. 453–469.

Panchal, K., Raman, S., and Pulat, P. S., 1992, Computer-Aided Tolerance Assessment Procedure (CATAP) for Design Dimensioning, *Int. J. Prod. Res.,* Vol. 30, No. 3, pp. 599–610.

Pham, D. T., and De Sam Lazaro, A. 1990, Autofix: An expert CAD System for Jig and Fixtures, *J. Mach. Tools Manuf.,* Vol. 30, No. 3, pp. 403–411.

Pollack, H. W., 1988, *Tool Design,* 2nd ed., Prentice Hall, Upper Saddle River, NJ.

Rong, Y., and Bai, Y., 1995, Automated Generation of Modular Fixture Configuration Design, *Proceedings of the ASME Design Automation Conference,* Boston, September 17–21, pp. 681–688.

Rong, Y., and Zhu, Y., 1992, Application of Group Technology in Computer-Aided Fixture Design, *Int. J. Syst. Autom. Res. Appl.,* Vol. 2. No. 4, pp. 395–405.

Rong, Y., Ni, J., and Wu, S., 1988, An Improved Model Structure for Forecasting Compensatory Control of Machine Tool Errors, *Proceedings of the ASME Winter Annual Meeting,* Chicago, November 27–December 2, PED Vol. 33, pp. 175–181.

Rong, Y., Zhu, J., and Li, S., 1993, Fixturing Feature Analysis for Computer-Aided Fixture Design, *Proceedings of the ASME Winter Annual Meeting,* New Orleans, LA, November 28–December 3, PED-Vol. 64, pp. 267–271.

Rong, Y., Chu, T., and Wu, S., 1994a, Automated Verification of Clamping Stability in Computer-Aided Fixture Design, *Proceedings of the ASME Conference on Computers in Engineering,* Minneapolis, MN, September 11–14, pp. 421–426.

Rong, Y., Li, S., and Bai, Y., 1994b, Development of Flexible Fixturing Technique in Manufacturing Industry, *Proceedings of the 5th International Symposium on Robotics and Manufacturing,* Maui, HI, August 15–18, pp. 661–666.

Roy, U., Liu, C. R., and Woo, T. C., 1991, Review of Dimensioning and Tolerancing: Representation and Processing, *Comput.-Aid. Des.,* Vol. 23, No. 7, pp. 466–483.

Senthil kumar, A., and Nee, A. Y. C., 1990, Tolerance Analysis in Modular Fixture Design, *J. Inst. Eng.,* Singapore, Vol. 30, No. 1, pp. 40–50.

Thompson, B. S., and Gandi, M. V., 1986, Commentary on Flexible Fixturing, *Appl. Mech. Rev.,* Vol. 39, No. 9, pp. 1365–1369.

Trappey, J. C. A., and Liu, C. R., 1989, An Automated Workholding Verification System, *Proceedings of the 4th International Conference on Manufacturing Science and Technology of the Future,* Stockholm, June, pp. 23–34.

Trappey, J. C. A., and Liu, C. R., 1990a, Automatic Generation of Configuration for Fixturing an Arbitrary Workpiece Using Projective Spatial Occupancy Enumeration Approach, *Proceedings of the ASME Winter Annual Meeting,* Dallas, November 25–30, PED-Vol. 47, pp. 191–202.

Trappey, J. C. A., and Liu, C. R., 1990b, A Literature Survey of Fixture-Design Automation, *Int. J. Adv. Manuf. Technol.,* Vol. 5, No. 3, pp. 240–255.

Zhang, H. C., and Huq, M. E., 1992, Tolerancing Techniques: The State-of-the-Art, *Int. J. Prod. Res.,* Vol. 30, No. 9, pp. 2111–2135.

Zhang, C., and Wang, H. P., 1993, The Discrete Tolerance Optimization Problem, *Manuf. Rev.,* Vol. 6, No. 1, pp. 60–71.

Zhang, H. C., Mei, J., and Dudek, R. A., 1991, Operational Dimensioning and Tolerancing in CAPP, *Ann. CIRP,* Vol. 40, No. 1, pp. 419–422.

Zhu, Y., Zhang, S., and Rong, Y., 1993, Experimental Study on Fixturing Stiffness of T-Slot Based Modular Fixtures, *Proceedings of NAMRC XXI,* Stillwater, OK, May, pp. 231–235.

16

TOLERANCE ANALYSIS FOR SETUP PLANNING IN CAPP

SAMUEL H. HUANG

EDS/Unigraphics
Cypress, California

HONG-CHAO ZHANG

Texas Tech University
Lubbock, Texas

16.1 INTRODUCTION

Tolerance analysis for process planning plays a crucial role in precision manufacturing. It ensures that resultant part dimensions and tolerances do not violate design requirements. However, most existing computer-aided process planning (CAPP) systems lack the ability of automated tolerance analysis and rely on manually calculated tolerance insertions (Wade, 1990). In recent years many researchers have realized this problem and are interested in integrating tolerance analysis into CAPP systems (Ngoi and Fang, 1994; Tang et al., 1994; Abdou and Cheng, 1993; Ji, 1993a,b; Ngoi and Ong, 1993a,b; Ngoi, 1992; Dong and Hu, 1991; Dong and Soom, 1990; Mittal et al., 1990; Bjorke, 1989; Irani et al., 1989; Li and Zhang, 1989; Manivannan et al., 1989; Weill, 1988; Xiaoqing and Davies, 1988; Ahluwalia and Karolin, 1986; Fainguelernt and Weill, 1986; Farmer and Gladman, 1986).

The foundation for the research noted above is tolerance synthesis and tolerance chart analysis. A tolerance chart analysis (Wade, 1967, 1983; Johnson, 1963; Eary and Johnson, 1962; Gadzala, 1959) shows how individual machining cuts combine to produce each blueprint dimension. It yields a set of linear algebraic expressions showing the relationship between each desired

Advanced Tolerancing Techniques, Edited by Hong-Chao Zhang
ISBN 0-471-14594-7   © 1997 John Wiley & Sons, Inc.

blueprint dimension and the individual cuts that contribute to it. Since the dimension produced by a specific cut is a stochastic variable dependent on factors such as machine tool capability, raw material, cutting tool, and so on, production is then faced with the task of distributing tolerances among individual cuts so that when these cuts combine, blueprint tolerance specifications are not exceeded. With the help of a tolerance chart, a process planner will be able to determine mean values and tolerances of the operational dimensions in a systematic way.

However, a tolerance chart can be built only after all the initial engineering decisions have been made concerning the process plan: the selection of setups, datums, operation sequences, and so on (Wade, 1983). When these decisions are not made properly, a tolerance chart analysis may require the use of machine tools with higher accuracy than necessary. As a result, the manufacturing cost will be higher than necessary, which is undesirable.

Tolerance chart analysis can only reactively check existing tolerance stackup problems rather than proactively minimizing tolerance stackup by selecting an appropriate setup plan. This problem is overlooked in current CAPP research primary because setup planning is one of the least studied areas in automated process planning (Sakurai, 1991). Recently, researchers are beginning to consider the issue of automated setup planning and fixture design (King and de Sam Lazaro, 1994; Chen, 1993; Lee and Cutkosky, 1991; Sakurai, 1991; Menassa and DeVries, 1990; Chu et al., 1989; Lee and Haynes, 1987). The research, however, has been focused on geometric analysis, kinematic analysis, force analysis, and deformation analysis. Tolerance analysis is the least explored issue in setup planning (Huang and Gu, 1994; Boerma and Kals, 1988).

Setup planning and fixture design are two closely related tasks in process planning. To set up a workpiece is to locate the workpiece in a desired position on the machine table. A fixture is then used to provide a clamping mechanism to maintain the workpiece in position and to resist the effects of gravity and/or operational forces (Karash, 1962). The purpose of setup and fixturing is to ensure the stability and, more important, the precision of machining processes. Therefore, an important guideline for setup planning and fixture design is design tolerance requirements.

Tolerance analysis for setup planning and fixture design is a relatively unexplored research issue. Although some rules of thumb for setup planning based on tolerance analysis have been proposed (Huang and Gu, 1994; Mei and Zhang, 1992; Boerma and Kals, 1988), no systematic approach is available in the literature. To automate (computerize) the task of setup planning and fixture design, one needs to understand the problem at hand fully and then solve it using an algorithmic approach. Although setup planning and fixture design are two closely related tasks in CAPP, in this chapter we focus on setup planning. It is assumed that conventional fixtures (e.g., three-jaw chuck for rotational parts and vise for prismatic parts) are used in the machining process.

16.2   SETUP PLANNING

The purpose of setup is to locate and fix a part in a definite manner on a machine tool so that machining can take place. Usually, a part needs to be manufactured in more than one setup. Therefore, the task of setup planning includes (1) grouping features into setups, (2) selecting setup datums for the setups, and (3) sequencing the setups.

Boerma and Kals (1988) reported on a system called FIXES which can perform the task of setup selection. Two main objectives of the setup selection procedure were:

1. To reduce the number of critical tolerances in the geometrical relations between features belonging to different setups.
2. To keep the number of setups as small as possible

Huang and Gu (1994) also provided two rules for setup selection:

1. A setup should be selected such that maximum number of features can be machined synchronously.
2. A setup should be selected such that minimum locating errors can be achieved.

The researchers noted above provided some guidelines for setup planning based on tolerance analysis. However, no systematic approach for setup planning is available in the literature. There is a need to develop a systematic approach for setup planning based on tolerance analysis.

In numerically controlled (NC) machining, three setup methods are used to obtain a dimension between two surface features within a part. These methods are machining the two features in the same setup, using one feature as the locating datum and machine the other, and using an intermediate locating datum to machine the two features in different setups, denoted as setup methods I, II, and III, respectively.

When locating and clamping (setting up) a workpiece on a fixture, a setup error occurs. A *setup error* includes workpiece-locating error, workpiece-clamping distortion, geometrical and dimensional inaccuracy of the fixture, and so on. After a workpiece has been located and clamped, the setup error remains a constant until the workpiece is removed from the fixture. In addition to setup errors, there exist machine motion errors. A *machine motion error* is the deviation of a machine movement from its ideal position due to the inaccuracy of the machine tool, thermal deformation of the machine tool, and so on (Huang and Zhang, 1996).

When using setup method I, the setup error is not included in the dimension obtained. The geometric relationship of the features machined in the same setup depends primarily on the geometry built into the machine tool. The dimensional relationship, such as the distance between two parallel surfaces, is determined mainly by the accuracy of the machine control unit (MCU), which is a built-in capability of the machine tool. Unlike manual machining, no tolerance stackup problem is encountered for dimensions obtained using setup method I. Dimensions obtained using setup method I consist of the least manufacturing errors. Whenever possible, this setup method should be used to facilitate tolerance control.

When setup method II is used, the setup error is included in the dimension obtained. To control the tolerance of the dimension, the accuracy of setting up the part must be considered, and it becomes a major part of the tolerance stack. Dimensions obtained using setup method II are usually less accurate than those obtained using setup method I. However, it is regarded as a good setup method when two features cannot be machined in the same setup.

Setup method III is the least desired setup method. A dimension (tolerance) chain is formed for every dimension obtained using setup method III. As a result, the tolerances will stack up. When the required tolerance of a dimension is tight, setup method III should be avoided. If this setup method is used to obtain a dimension with tight tolerance, although tolerance chart analysis can be used to calculate the operational dimensions and tolerances, the undesired stackup of tolerances might make its manufacture impossible even with the best machine tool.

The discussion above shows that tolerance control can be achieved proactively via setup planning. Features with tight tolerance relationship should be arranged into the same setup whenever possible (setup method I). When two features with a tight tolerance relationship cannot be machined in the same setup, they should be mutually datumed (setup method II). Only when the tolerance relationship between two features is not tight can an intermediate datum be used (setup method III). Refer to Huang and Zhang (1996) for more details.

16.3 GRAPH-BASED APPROACH TO SETUP PLANNING

Setup planning is a complex problem. Consider the simple example part shown in Figure 16.1. Features A, B, C, D, and E are to be machined. Usually, the part is machined in two setups, with features A and B in one setup and features C, D, and E in the other setup. Either setup can be machined first. When machining features C, D, and E, either feature A or feature B can be used as a setup datum. Similarly, any one of features C, D, and E can be used as a setup datum when machining features A and B. Therefore, there are $2\binom{2}{1}\binom{3}{1} = 12$ alternative setup plans (Table 16.1). Given the design tolerance requirements, some of the setup plans are appropriate (i.e., they will reduce the problem of tolerance stackup) and one of them should be selected to ensure the resultant part dimensions. The purpose of setup planning is to select a setup plan that will facilitate tolerance control.

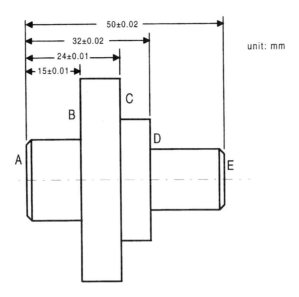

Machined features: A, B, C, D, and E

Possible setup plans: $2\binom{2}{1}\binom{3}{1} - 12$

Figure 16.1. Complex problem of setup planning.

16.3.1 Graph Representation of the Setup Planning Problem

Graph theory (Tulasiraman and Swamy, 1993; Foulds, 1992) is a powerful mathematical tool to the solution of a great number of complex problems. Using the language of graphs one can represent the structure of the setup planning problem in a simple manner. In this section we show how to represent the setup planning problem using the language of graph.

Setup Formation. The first step in setup planning is to group part features into setups: namely, setup formation. Tool approach direction is an important factor in determining setups. The tool approach direction of a feature is an unobstructed path that a tool can take to access the feature (Chang, 1990). Features with the same tool approach direction can be grouped into one setup. A feature may have more than one tool approach direction and hence can be grouped into different setups. The feature might have tolerance relationships with another feature that has only one tool approach direction. If these two features share a common tool approach direction, they should be grouped into the same setup (setup method I) to eliminate setup error. This is the purpose of setup formation.

The problem of setup formation can be represented using the language of graphs as follows. Part features (we are talking about surface features through-

TABLE 16.1. Setup Methods for the Part Shown in Figure 16.1

Setup Plan	Method
1	Machine features C, D, and E using feature A as a setup datum and then machine features A and B using feature C as a setup datum
2	Machine features A and B using feature C as a setup datum and then machine features C, D, and E using feature A as a setup datum
3	Machine features C, D, and E using feature A as a setup datum and then machine features A and B using feature D as a setup datum
4	Machine features A and B using feature D as a setup datum and then machine features C, D, and E using feature A as a setup datum
5	Machine features C, D, and E using feature A as a setup datum and then machine features A and B using feature E as a setup datum
6	Machine features A and B using feature E as a setup datum and then machine features C, D, and E using feature A as a setup datum
7	Machine features C, D, and E using feature B as a setup datum and then machine features A and B using feature C as a setup datum
8	Machine features A and B using feature C as a setup datum and then machine features C, D, and E using feature B as a setup datum
9	Machine features C, D, and E using feature B as a setup datum and then machine features A and B using feature D as a setup datum
10	Machine features A and B using feature D as a setup datum and then machine features C, D, and E using feature B as a setup datum
11	Machine features C, D, and E using feature B as a setup datum and then machine features A and B using feature E as a setup datum
12	Machine features A and B using feature E as a setup datum and then machine features C, D, and E using feature B as a setup datum

out this chapter) and their relationships (in terms of tool approach direction) can be represented as a graph $G_0 = (V,E)$, called a *feature relationship graph*. The features are represented as the vertices $V = \{v_1, v_2, ..., v_n\}$, where n is the number of features. Their relationships (in terms of tool approach direction) are represented as the edges $E = \{e_1, e_1, ..., v_m\}$, where m is the number of edges. If two features v_i and v_j $(i \neq j)$ can be machined within a setup (i.e., they have the same tool approach direction), then $(v_i, v_j) \in E$; if $i = j$, then $(v_i, v_j) \in E$; otherwise, $(v_i, v_j) \notin E$. Let $M_0 = [m_{ij}]$ be G_0's adjacency matrix. Let $\mathbf{r}_i$ and $\mathbf{c}_i$ $(i = 1, 2, ..., n)$ be M_0's row vector and column vector, respectively. Since G_0 is an undirected graph, M_0 is symmetric. Hence, $\mathbf{r}_i = \mathbf{c}_i^T$ $(i = 1, 2, ..., n)$.

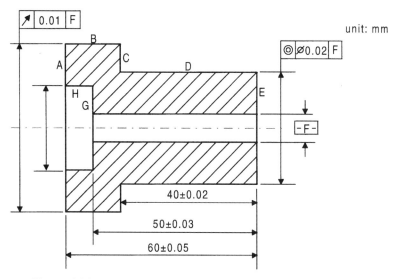

Figure 16.2. Example part (only relevant dimensions are shown).

An example is used for the purpose of illustration. Figure 16.2 shows the example part. The part consists of eight features, denoted as A, B, C, D, E, F, G, and H. Feature B has a circular runout tolerance (0.01) when rotated about the datum axis (feature F). Features D and F have a concentricity tolerance (0.02). The size between features A and E, G and E, and C and E are denoted as d_{AE}, d_{GE}, and d_{CE}, respectively. The tolerances for dimensions d_{AE}, d_{GE}, and d_{CE} are ±0.05, ±0.03, and ±0.02, respectively. This part is to be machined using an NC lathe. The fixture to be used is a three-jaw chuck.

The feature relationship graph G_0 for the example part is shown in Figure 16.3. M_0, the adjacency matrix of G_0, is shown as follows:

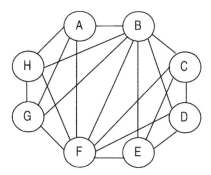

Figure 16.3. Feature relationship graph G_0 of the example part (self-loops are omitted).

$$
\begin{array}{c}
\begin{array}{cccccccc}
 & A & B & C & D & E & F & G & H
\end{array} \\
\begin{array}{c}
A \\ B \\ C \\ D \\ E \\ F \\ G \\ H
\end{array}
\left[
\begin{array}{cccccccc}
1 & 1 & 0 & 0 & 0 & 1 & 1 & 1 \\
1 & 1 & 1 & 1 & 1 & 1 & 1 & 1 \\
0 & 1 & 1 & 1 & 1 & 1 & 0 & 0 \\
0 & 1 & 1 & 1 & 1 & 1 & 0 & 0 \\
0 & 1 & 1 & 1 & 1 & 1 & 0 & 0 \\
1 & 1 & 1 & 1 & 1 & 1 & 1 & 1 \\
1 & 1 & 0 & 0 & 0 & 1 & 1 & 1 \\
1 & 1 & 0 & 0 & 0 & 1 & 1 & 1
\end{array}
\right]
\end{array}
$$

The feature relationship graph consists of a number of complete subgraphs. Each complete subgraph represents a setup. The vertices within a complete subgraph represent the features to be machined in the setup. The purpose of setup formation is to find these subgraph $G_1' = (V_1', E_1')$, $G_2' = (V_2', E_2')$,...,$G_l' = (V_l', E_l')$, such that $V_i' \cap V_j' = \emptyset$, for $i = 1,2,...,l$, $j = 1,2,...,l$, $i \neq j$ and $V_1' \cup V_2' \cup \cdots \cup V_l' = V$.

Examining the matrix M_0, one finds that it consists of three distinct row (column) vectors ([1 1 0 0 0 1 1 1], [0 1 1 1 1 1 0 0], and [1 1 1 1 1 1 1 1]). This means that the features can be categorized into three sets. The physical meaning of the sets is explained as follows. For rotational parts, there are two tool approach directions, the left and the right. A feature can thus be machined (1) only from the left, (2) only from the right, or (3) from either the left or the right. Therefore, the features in a rotational part can be categorized into three sets defined as follows:

$$
v_i \in \begin{cases}
\mathbf{S_L}, & \text{if feature } v_i \text{ can be machined only from the left} \\
\mathbf{S_R}, & \text{if feature } v_i \text{ can be machined only from the right} \\
\mathbf{S_B} & \text{if feature } v_i \text{ can be machined from either the left or the right}
\end{cases}
$$

$$(16.1)$$

in which $i = 1,2,...,n$. In this example one can see that $\mathbf{S_L} = \{A,G,H\}$, $\mathbf{S_R} = \{C,D,E\}$, and $\mathbf{S_B} = \{B,F\}$. Rearranging the matrix M_0 according to the sets, one has a new matrix M_1:

$$
\begin{array}{c}
\begin{array}{cccccccc}
 & A & G & H & B & F & C & D & E
\end{array} \\
\begin{array}{c}
A \\ G \\ H \\ B \\ F \\ C \\ D \\ E
\end{array}
\left[
\begin{array}{cccccccc}
1 & 1 & 1 & 1 & 1 & 0 & 0 & 0 \\
1 & 1 & 1 & 1 & 1 & 0 & 0 & 0 \\
1 & 1 & 1 & 1 & 1 & 0 & 0 & 0 \\
1 & 1 & 1 & 1 & 1 & 1 & 1 & 1 \\
1 & 1 & 1 & 1 & 1 & 1 & 1 & 1 \\
0 & 0 & 0 & 1 & 1 & 1 & 1 & 1 \\
0 & 0 & 0 & 1 & 1 & 1 & 1 & 1 \\
0 & 0 & 0 & 1 & 1 & 1 & 1 & 1
\end{array}
\right]
\end{array}
$$

From the matrix M_1 one can see that the feature relationship graph G_0 consists of at least two complete subgraphs. Since it is desirable to reduce the number of setups in machining (Chang, 1990), the graph G_0 should be partitioned into exactly two complete graphs, denoted $G_1' = (V_1', E_1')$ and $G_2' = (V_2', E_2')$, respectively. It is easy to see the vertices in the set $\mathbf{S}_L$ belong to a complete graph, while the vertices in the set $\mathbf{S}_R$ belong to the other one. Without loss of generality, let $\mathbf{S}_L \subseteq V_1'$ and $\mathbf{S}_R \subseteq V_2'$. Since a vertex in the set $\mathbf{S}_B$, denoted v^*, is adjacent to all the other vertices of the graph G_0, G_1' will still be a complete graph if v^* is assigned to V_1'. Similarly, G_2' will still be a complete graph if v^* is assigned to V_2'. Therefore, the problem of setup formation is transformed to assign vertices in the set $\mathbf{S}_B$ to either V_1' or V_2'. The requirement for such an assignment is that the resultant complete subgraphs (setups) should be optimal in terms of tolerance control.

To find the optimal complete subgraphs, a tolerance graph is constructed based on the feature relationship graph and the design tolerance information. The procedure for constructing the tolerance graph is as follows. First, vertices in the feature relationship graph G_0 are partitioned into three sets as defined in (1). The edge connections of G_0 are then removed, and the design tolerance information is added as new, weighted edges, with the size of the tolerance zone being the weights. The feature relationship graph G_0 evolves to a *tolerance graph* G_1. The tolerance graph G_1 for the example part is shown in Figure 16.4.

The assignment of vertices in the set $\mathbf{S}_B$ to V_1' or V_2' is based on the tolerance graph. The primary concern is that features with tight tolerance relationships should be machined in the same setup (corresponding vertices belong to the same subgraph). The result of this assignment is a *setup graph* G_2. The setup graph G_2 consists of two subgraphs (not complete subgraphs because the setup graph is different from the feature relationship graph) rep-

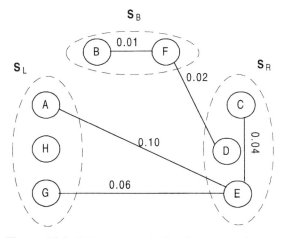

Figure 16.4. Tolerance graph G_1 of the example part.

resenting the two setups, denoted $G_L = (V_L, E_L)$ and $G_R = (V_R, E_R)$, respectively.

In this example (Figure 16.4) one can see that features B and F need to be assigned to either $\mathbf{S_L}$ or $\mathbf{S_R}$. Since feature F has a tolerance relationship with feature D (the size of the tolerance zone is 0.02) and D $\in \mathbf{S_R}$, feature F should be assigned to $\mathbf{S_R}$. After feature F has been assigned to $\mathbf{S_R}$, feature B should also be assigned to $\mathbf{S_R}$ since it has a tolerance relationship with feature F (the size of the tolerance zone is 0.01). As a result, the features within the example part are grouped into two sets, $\mathbf{S_L} = \{A, G, H\}$ and $\mathbf{S_R} = \{B, C, D, E, F\}$. Let $V_L = \mathbf{S_L}$ and $V_R = \mathbf{S_R}$, the two subgraphs G_L and G_R are obtained. The setup graph G_2 for the example part is shown in Figure 16.5.

Datum Selection and Setup Sequencing. After setups have been formed, setup datums need to be selected and the sequence of the setups needs to be decided. Dimensions obtained using setup method I are determined primarily by the built-in machine/process capabilities and have nothing to do with the selection of setup datums and setup sequence. Therefore, their tolerance information is not necessary for datum selection and setup sequencing and can be screened out. Thus the concern at this stage is to apply setup method II to obtain those dimensions that cannot be obtained using setup method I. The first step is proper datum selection to ensure that the dimension with the tightest tolerance is obtained using setup method II. The second step is to sequence the setups so that setup method II may be applied to obtain as many dimensions with specified tolerances as possible.

To fix a rotational part on a machine tool, one (and only one) vertical feature and one (and only one) cylindrical feature are needed. Therefore, vertical features and cylindrical features need to be distinguished. The vertices in the setup graph are partitioned into four subgraphs denoted as $G'_{LV} = (V'_{LV}, E'_{LV})$ (V'_{LV} consists of vertices representing vertical features that need to be machined from the left), $G'_{LC} = (V'_{LC}, E'_{LC})$ (V'_{LC} consists of vertices representing cylindrical features that need to be machined from the left), $G'_{RV} = (V'_{RV}, E'_{RV})$ (V'_{RV} consists of vertices representing vertical features that need to be machined from the right), and $G'_{RC} = (V'_{RC}, E'_{RC})$ (V'_{RC} consists of vertices representing cylindrical features that need to be machined from the right), respectively. As a result, the setup graph G_2 evolves to a *datum graph G_3*. The datum graph G_3 for the example part is shown in Figure 16.6.

Datum selection and setup sequencing are both based on the datum graph. The purpose of datum selection is to find one and only one vertex from each of V'_{LV}, V'_{LC}, V'_{RV}, and V'_{RC}. The vertices found are to be used as the setup datums. The purpose of setup sequencing is to sequence G_L and G_R so that the sequence of the corresponding setups can be determined. Again, the decisions made should facilitate tolerance control.

In this example (Figure 16.6) one can see that the tolerance between features A and E (0.10) and the tolerance between features G and E (0.06) are to be assured. The tighter tolerance is the one between features G and E. Therefore, features G and E are to be used as setup datums. Since the degree

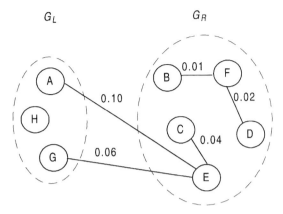

Figure 16.5. Setup graph G_2 of the example part.

of E [$d(E) = 2$] is greater than the degree of G [$d(G) = 1$], feature E needs to be machined in the first setup. In this way, features A and G are both machined using feature E as a setup datum in the finishing process (setup method II is applied). Since there is no tolerance requirement among the cylindrical features, the selection of cylindrical setup datums does not influence the tolerance control. In this case, features H and D are selected as the cylindrical setup datums for ease of fixturing.

16.3.2   Setup Planning Algorithm for Rotational Parts

Section 16.3.1 provided a graph representation of the setup planning problem. It showed that a graph-based approach can be used to advantage in generating

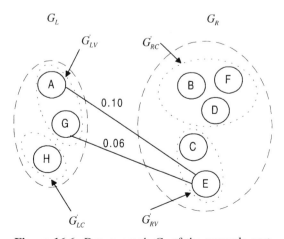

Figure 16.6. Datum graph G_3 of the example part.

appropriate setup plans. In this section a setup planning algorithm for rotational parts is developed. The following assumptions are made:

1. The part is to be machined on an NC lathe.
2. The part is to be set up on the lathe using a three-jaw chuck.
3. Secondary features such as keyways, chamfers, grooves, and threads are not considered.
4. All part features are to be machined.
5. Different types of tolerances are of equal importance.
6. The tool approach direction of each feature is given.
7. Each setup requires one (and only one) vertical feature and one (and only one) cylindrical feature as the setup datums.

The task of setup planning is divided into three steps: (1) setup formation, (2) datum selection, and (3) setup sequencing. The procedure for the graph-based approach to setup planning is shown in Figure 16.7. The input to the algorithm for setup formation is the tolerance graph G_1, whose vertices are categorized into three sets denoted $\mathbf{S}_L$, $\mathbf{S}_R$, and $\mathbf{S}_B$. Let $e_l = (v_i, v_j)$ denote the weight of the edge incident on vertices v_i and v_j. Let r be the number of edges in G_1 that incident on at least one vertex in the set $\mathbf{S}_B$. The algorithm is given as follows:

If $\mathbf{S}_B = \emptyset$, then exit
Else
    Find $e_l = (v_p, v_q) = \min [e_i]$, in which $i = 1, 2, ..., r$, $v_p \in \mathbf{S}_B$, $v_q \notin \mathbf{S}_B$,
    If $v_q \in \mathbf{S}_L$, then assign v_p to $\mathbf{S}_L$, otherwise, assign v_p to $\mathbf{S}_R$.
    Let $\mathbf{S}_B = \mathbf{S}_B - \{v_p\}$
    Repeat

The output of the algorithm is a setup graph G_2. The setup graph G_2 is then transformed into a datum graph G_3 (see Section 16.3.1 for the procedure), which consists of four subgraphs, $G'_{LV} = (V'_{LV}, E'_{LV})$, $G'_{LC} = (V'_{LC}, E'_{LC})$, $G'_{RV} = (V'_{RV}, E'_{RV})$, and $G'_{RC} = (V'_{RC}, E'_{RC})$. The datum graph G_3 is the input for the algorithms for datum selection and setup sequencing.

Let t be the number of edges in $G_3 = (V, E)$. Let D denote the set of features selected to be the setup datums. Initially, $D = \emptyset$. The algorithm for datum selection is given as follows:

If the number of elements in D equals 4, then exit
Else
    Find $e_l = (v_p, v_q) = \min [e_i]$, in which $i = 1, 2, ..., t$.
    Let $D = D \cup \{v_p\}$ if no element in D and v_p belongs to the same subgraph

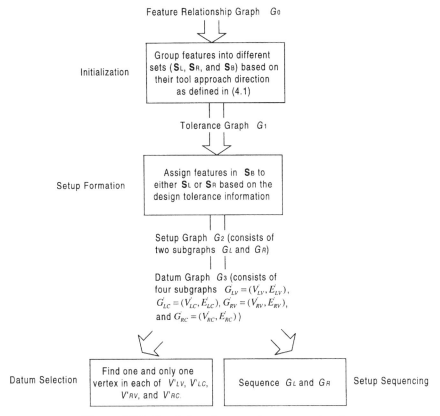

Figure 16.7. Graph-based approach to setup planning.

Let $D = D \cup \{v_q\}$ if no element in D and v_q belongs to the same subgraph

Let $E = E - \{e_m = (v_a, v_b)\}$, in which v_a and v_p belong to the same subgraph while v_b and v_q belong to the same subgraph

Repeat

Let v_i denote the element in D, in which $i = 1, 2, 3, 4$. Let $d(v_j)$ denote the degree of vertex v_j in the datum graph G_3. The algorithm for setup sequencing is given as follows:

Find $d(v_p) = \max[d(v_i)]$, in which $i = 1, 2, 3, 4$

If $v_p \in V_L$, then machine the features in V_L in the first setup; otherwise, machine the features in V_R in the first setup

The detailed setup planning algorithm for rotational parts is shown in Appendix A.

16.3.3 Setup Planning Algorithm for Prismatic Parts

Setting Up a Prismatic Part. A prismatic part can be viewed as a rigid body that has six degrees of freedom in the Cartesian space: three translations along the X, Y, and Z axes denoted as $\overset{\leftrightarrow}{X}$, $\overset{\leftrightarrow}{Y}$, and $\overset{\leftrightarrow}{Z}$, respectively, and three rotations around the X, Y, and Z axes denoted as α, β, and γ, respectively (Figure 16.8). When these degrees of freedom of a workpiece are not all restricted, the workpiece cannot have a definite position in space. In a machining operation, the workpiece must have a definite position on the machine table. Therefore, the degrees of freedom of the workpiece must be properly restricted.

Suppose that the prismatic workpiece shown in Figure 16.9 is to be machined. If points 1, 2, and 3 (not in a straight line) are selected to support the workpiece, then $\overset{\leftrightarrow}{Z}$, α, and β are restricted. Points 1, 2, and 3 establish the primary datum of the workpiece. If points 4 and 5 are then selected to contact the side surface of the workpiece, two more degrees of freedom $\overset{\leftrightarrow}{X}$, and γ are restricted. Points 4 and 5 establish the secondary datum of the workpiece. In addition, if point 6 is selected to contact the rear surface of the workpiece, $\overset{\leftrightarrow}{Y}$ is restricted. Point 6 establishes the tertiary datum of the workpiece. Using these six points, all six degrees of freedom of the workpiece are restricted (i.e., the position of the workpiece is completely defined). It is clear that the six degrees of freedom of a workpiece can be restricted by six properly selected points. This is known as the *3–2–1 principle* or the *six-point locating principle* (Wang and Li, 1991).

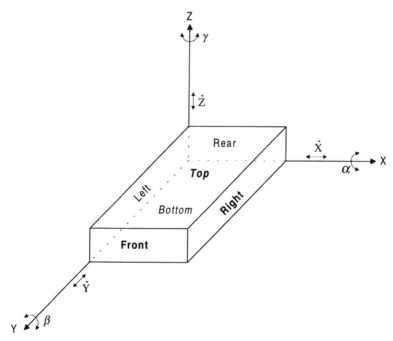

Figure 16.8. Six degrees of freedom of a prismatic part.

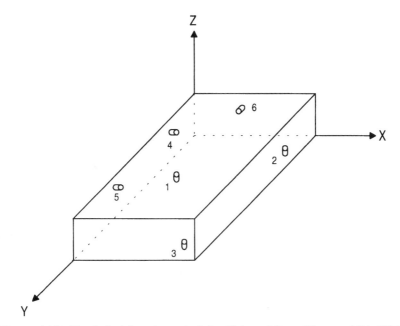

Figure 16.9. The 3–2–1 locating principle. (Adapted from Wang and Li, 1991.)

To set up a part means to locate the part in a definite position and then fix the part at the position located. For prismatic parts, a vise is generally used to provide the required clamping mechanism so that a workpiece can be fixed at the position located. First, the workpiece is placed against the bottom of the vise, which establishes the primary datum. Then the workpiece is pushed toward the stationary jaw of the vise, which establishes the secondary datum. A stop is then used to establish the tertiary datum (Figure 16.10). This procedure locates the workpiece in a definite position. The workpiece can then be fixed by clamping using the moving jaw of the vise. It is then obvious that only the top and front of the workpiece can be accessed by a cutting tool. Referring to Figure 16.8, one can see that in general the top of a prismatic part can be machined together with one and only one of the side features—the front, rear, left, and the right of the part, but not the bottom of the part. Similar relations can be deduced for the bottom, the front, the rear, the left, and the right of a prismatic part. Basically, a feature can only be machined with one other feature that has an orthogonal tool approach direction, but never with a feature that has an opposing tool approach direction.

Illustrative Example and the Algorithm. Setup planning for prismatic parts is similar to setup planning for rotational parts. The only difference is that more setups are involved and the problem is more complex. Before presenting the setup planning algorithm for prismatic parts, an example is used to illustrate the principle of graph-based setup planning for prismatic parts. Consider

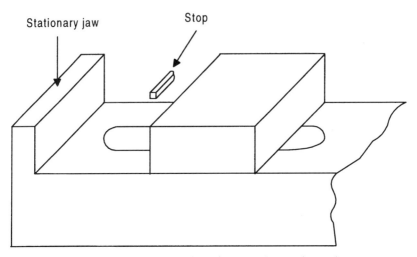

Figure 16.10. Setting up a prismatic part using a vise and a stop.

the prismatic part shown in Figure 16.11. Feature C is a through hole that can be machined from either the top or the bottom. In other words, it has two tool approach directions, the top and the bottom. Features B, H, I, J, and K can be machined from the top. Features D, L, M, N, O, and P can be machined from the bottom. Features A, E, F, and G can be machined from the front, the rear, the left, and the right, respectively. Let S_T, S_B, S_L, S_R, S_F, and S_E denote the set of features that can be machined from the top, the bottom, the left, the right, the front, and the rear of the part. In this example one can see that

$$S_T = \{B,C,H,I,J,K\}$$

$$S_B = \{C,D,L,M,N,O,P\}$$

$$S_L = \{F\}$$

$$S_R = \{G\}$$

$$S_F = \{A\}$$

$$S_E = \{E\}$$

Since feature C presents in both S_T and S_B, it needs to be assigned to either S_T or S_B. The assignment is based on the tolerance graph of the part (Figure 16.12). Since feature C has a concentricity tolerance (0.05) with feature D and feature D can be machined only from the bottom (D $\in$ S_B), feature C is hence assigned to S_B.

Since the dimensions involving features that can be machined from the same tool approach direction are assured automatically by setup method I,

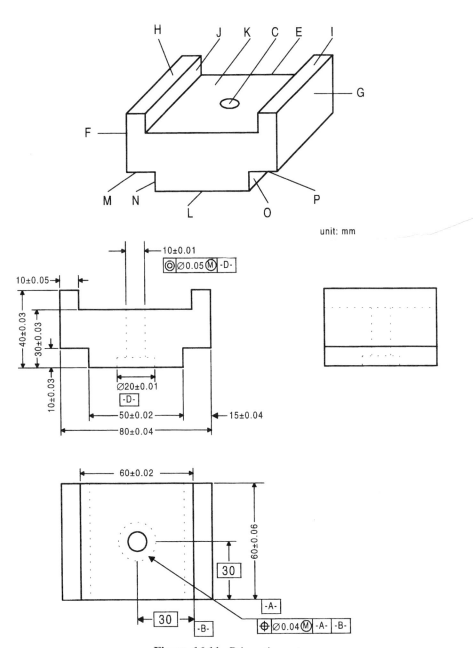

Figure 16.11. Prismatic part.

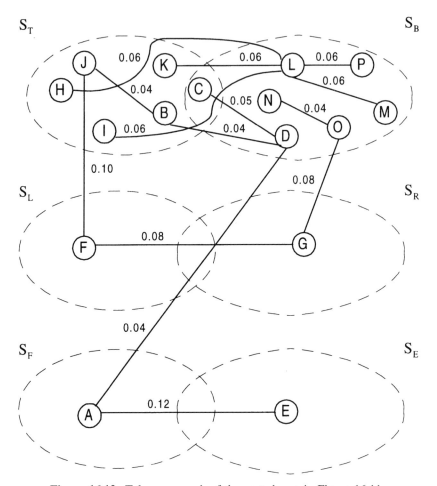

Figure 16.12. Tolerance graph of the part shown in Figure 16.11.

the tolerance requirements for these dimensions do not influence datum selection or setup sequencing and can be screened out. Let $e_l = (v_p, v_q)$ be an
edge of the tolerance graph. The edge e_l is screened out (permanently removed) if one of the following is true:

1. $v_p \in \mathbf{S}_T$ and $v_q \in \mathbf{S}_T$.
2. $v_p \in \mathbf{S}_B$ and $v_q \in \mathbf{S}_B$.
3. $v_p \in \mathbf{S}_L$ and $v_q \in \mathbf{S}_L$.
4. $v_p \in \mathbf{S}_R$ and $v_q \in \mathbf{S}_R$.
5. $v_p \in \mathbf{S}_F$ and $v_q \in \mathbf{S}_F$.
6. $v_p \in \mathbf{S}_E$ and $v_q \in \mathbf{S}_E$.

As has been shown before, features within S_T can be machined together with features within one and only one of S_L, S_R, S_F, and S_E but cannot be machined together with features within S_B. Let e_l represent the tolerance requirement for a dimension involving a feature in S_T and the other feature in S_B. It is then clear that e_l does not influence the formation of setups because features within S_T cannot be machined together with features within S_B. However, e_l does influence the selection of datums. Therefore, e_l can be removed temporarily for the purpose of setup formation but needs to be added back when performing the datum selection. Similar conclusions can be drawn with regard to S_B, S_L, S_R, S_F, and S_E. The edge $e_l = (v_p, v_q)$ is removed temporarily if one of the following is true:

1. $v_p \in S_T$ and $v_q \in S_B$.
2. $v_p \in S_B$ and $v_q \in S_T$.
3. $v_p \in S_L$ and $v_q \in S_R$.
4. $v_p \in S_R$ and $v_q \in S_L$.
5. $v_p \in S_F$ and $v_q \in S_E$.
6. $v_p \in S_E$ and $v_q \in S_F$.

As a result of the operations above, the tolerance graph evolves to a setup graph as shown in Figure 16.13, where the edges represented by dashed lines are edges removed temporarily. From the figure one can see that the tightest tolerance (among those tolerances that are removed neither permanently nor temporarily is the one for the dimension between features A and D (0.04). Therefore, feature A is assigned to S_B. The edge incident on A and D is then removed permanently. In the meantime, the edge incident on O and G is removed temporarily because feature G can no longer be assigned to S_B (S_B already consists features that can be machined from the top and from the front). Similarly, feature F is assigned to S_T, and the edge incident on F and J is removed permanently. Although there is no tolerance relation between features G and E, features G and E are grouped together to reduce the number of setups.

After the setups are formed, edges that were removed temporarily are added back. The setup graph evolves to a datum graph as shown in Figure 16.14. From the figure one can see that the part can be machined in three setups, with features B, F, H, I, J, and K in one setup denoted S_T, features A, C, D, L, M, N, O, and P in another setup denoted S_B, and features G and E in the other setup denoted S_R. Let Δ_T, Δ_B, and Δ_R be the maximum vertex degree of vertices in sets S_T, S_B, and S_R, respectively. In this example, $\Delta_T = 1$, $\Delta_B = 3$, and $\Delta_R = 2$. Since $\Delta_B > \Delta_R > \Delta_T$, the setup sequence can be decided as follows: S_B is the first setup, S_R is the second setup, and S_T is the last setup.

When machining features A, C, D, L, M, N, O, and P (features within S_B), it is clear that the top should be used as the primary setup datum. Since

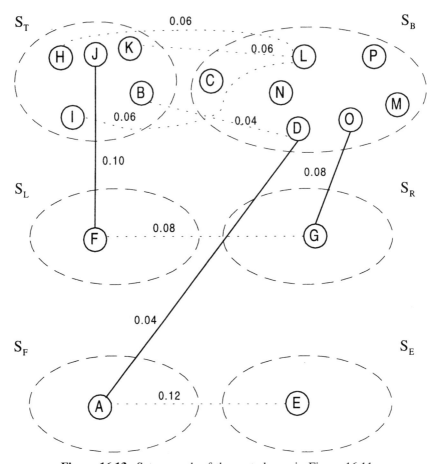

Figure 16.13. Setup graph of the part shown in Figure 16.11.

feature A, whose tool approach direction is the front, is also machined in this setup, it is clear that the rear should be used as the tertiary setup datum. Now consider the secondary setup datum; either the right or the left should be selected. Since feature G, whose tool approach direction is the right, has a tolerance relationship with feature O, the right is selected as the secondary setup datum. Similarly, when machining features G and E, the left is used as the primary setup datum, the bottom is used as the secondary setup datum, and the front is used as the tertiary setup datum. When machining features B, F, H, I, J, and K, the bottom is used as the primary setup datum, the front is used as the secondary setup datum, and the right is used as the tertiary setup datum.

A detailed setup planning algorithm for prismatic parts is given in Appendix B. The following assumptions are made:

1. The part is to be machined on an NC machining center.

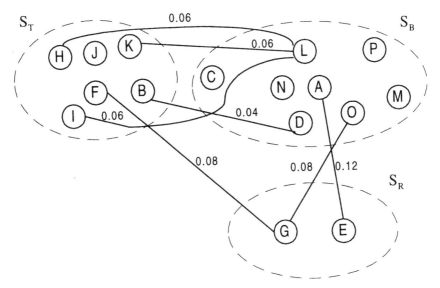

Figure 16.14. Datum graph of the part shown in Figure 16.11.

2. The part is to be set up on the machining center using a vise.
3. All the features of the part are to be machined.
4. Different types of tolerances are of equal importance.
5. The tool approach direction of each feature is given.

16.3.4 Setup Planner: Setup Planning Program

The setup planning algorithms were implemented using C/C++ under the Microsoft Windows environment. The setup planning program is named "Setup Planner." It can generate setup plans for both rotational and prismatic parts based on tolerance analysis.

To run all the functions of the setup planning program successfully, two files, NOTEPAD.EXE and BCWW.DLL, must be available. NOTEPAD.EXE is a text editor supplied with Microsoft Windows. BCWW.DLL is supplied with Borland C++ for Windows (a product of Borland International, Inc.) and should be present in the directory C:\WINDOWS\SYSTEM. Furthermore, a commercial drawing software named VISIO is used to draw and show part designs. VISIO is a product of Shapeware Corporation. It should be installed in the directory C:\VISIOLTE.

To obtain a setup plan for a part, just read the part data file and then click on "Setup Plan" at the menu bar. To read a part data file, click on "File" at the menu bar and then click on "Read" at the pull-down menu. The setup plan generated is stored in a temporary file and shown via NOTEPAD. The user can then save the file using a desired file name. The part drawing can be shown by click on "View" at the pull-down menu of "File" as long as the drawing is saved using a file format supported by VISIO. A part data file can be edited by a click on "Edit" at the pull-down menu of "File." The file formats for rotational parts and prismatic parts are different. The file format for rotational parts is as follows:

SH_PART_R	(header)
n	(number of feature)
$E_1 E_2 \ldots\ldots E_n$	(feature type of each feature)
$A_1 A_2 \ldots\ldots A_n$	(tool approach direction of each feature)
$S_1 S_2 \ldots\ldots S_n$	(existence of each feature on the stock)
$t_{11} t_{12} \ldots\ldots t_{1n}$	(tolerance matrix)
$t_{21} T_{22} \ldots\ldots t_{2n}$	
$\vdots \quad \vdots \quad \text{::::::} \quad \vdots$	
$t_{n1} t_{n2} \ldots\ldots t_{nn}$	

The file format for prismatic parts is as follows:

SH_PART_P	(header)
n	(number of feature)
$A_1 A_2 \ldots\ldots A_n$	(tool approach direction of each feature)
$t_{11} t_{12} \ldots\ldots t_{1n}$	(tolerance matrix)
$t_{21} t_{22} \ldots\ldots t_{2n}$	
$\vdots \quad \vdots \quad \text{::::::} \quad \vdots$	
$t_{n1} t_{n2} \ldots\ldots t_{nn}$	

Definitions of E_i, A_i, S_i, and t_{ij} (in which $i = 1,2,\ldots,n$ and $j = 1,2,\ldots,n$) for rotational parts are given in Appendix A, while definitions of A_i and t_{ij} (in which $i = 1,2,\ldots,n$ and $j = 1,2,\ldots,n$) for prismatic parts are given in Appendix B. The setup planning program was used to generate the setup plans for the example parts shown in Figures 16.2 and 16.11. The results, including the user interface, part drawing, and input and output files, are shown in Figures 16.15 and 16.16, respectively.

16.4 CONCLUSION

Tolerance control is absolutely crucial to achieving production of high-precision parts at low cost. It is desirable to apply tolerance analysis at the early stage of process planning since no great time or dollar loss will occur if a process, setup, or locating datum change is required. Tolerance chart analysis has been the major tool used for tolerance analysis in process plan-

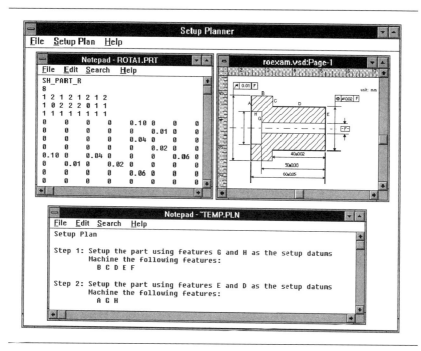

Figure 16.15. Part drawing, data file, and setup plan for the example part shown in Figure 16.2.

ning. However, a tolerance chart can be built only after a setup plan of a part has been generated. Tolerance chart analysis can only passively calculate the operational dimensions and tolerances for an existing process plan. Setup planning, however, can proactively select an appropriate plan so that the influence of tolerance stackup is kept to a minimum.

The major contribution of this research is the development of a graph-based approach to solving the setup planning problem in CAPP. It has been recognized that current approaches to CAPP are knowledge-based in nature and thus hinder the progress of CAPP development. The graph-based approach is a mathematical approach and is an important step toward building a scientifically rigorous base for CAPP research.

Although significant progress has been made, the research is far from complete. Work needs to be done on both algorithm development and system implementation. Although the algorithm for rotational parts is quite mature, the following should be addressed to develop a superior setup planning algorithm for rotational parts:

1. *Fixtures for Rotational Parts.* Although the three-jaw chuck is the most popular fixture for rotational parts, some rotational parts, such as shafts,

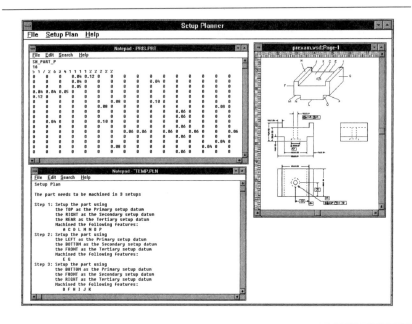

Figure 16.16. Part drawing, data file, and setup plan for the example part shown in Figure 16.11.

are usually machined using center support. Other fixtures for rotational parts, such as the V-block, also need to be considered.

2. *Number of Setups.* A rotational part usually needs to be machined in more than two setups when both roughing processes and finishing processes are involved. Although the setup datums and operation sequence will not be affected, it is desirable to have an algorithm capable of generating the exact number of setups.

3. *Optimization.* The developed setup planning algorithm is heuristic in nature. Although it can generate good setup plans, the optimality of the generated setup plans cannot be guaranteed. The setup planning problem can be formulated as an optimization problem and a superior algorithm can be developed to guarantee the optimality of the solution.

The algorithm for prismatic parts is relatively primitive. Although a vise was a very popular fixture for prismatic parts, modular fixtures are frequently used in today's metal-cutting processes. When considering modular fixtures, the setup planning algorithm will be much more complex than the present one. The issue of optimization will also complicate the problem. Extensive research is required before a mature setup planning algorithm for prismatic parts can be developed.

REFERENCES

Abdou, G., and Cheng, R., 1993, TVCAPP: Tolerance Verification in Computer-Aided Process Planning, *Int. J. Prod. Res.,* Vol. 31, No. 2, pp. 393–411.

Ahluwalia, R. S., and Karolin, A. V., 1986, CATC: A Computer Aided Tolerance Control System, *J. Manuf. Syst.,* Vol. 3, pp. 153–160.

Bjorke, O., 1989, *Computer-Aided Tolerancing,* 2nd ed., ASME Press, New York.

Boerma, J. R., and Kals, H. J. J., 1988, FIXES: A System for Automatic Selection of Set-Ups and Design of Fixtures, *Ann. CIRP,* Vol. 37, No. 1, pp. 443–446.

Chang, T.-C., 1990, *Expert Process Planning for Manufacturing,* Addison-Wesley, Reading, MA.

Chen, C. L. P., 1993, Setup Generation and Feature Sequencing Using Unsupervised Learning Algorithm, *Proceedings of the 1993 NSF Design and Manufacturing Systems Conference,* Society of Manufacturing Engineers, Dearborn, MI, Vol. 1, pp. 981–986.

Chou, Y. C., Chandru, V., and Barash, M. M., 1989, A Mathematical Approach to Automatic Configuration of Machining Fixtures: Analysis and Synthesis, *J. Eng. Ind,* Vol. 111, pp. 299–306.

Dong, Z., and Hu, W., 1991, Optimal Process Sequence Identification and Optimal Process Tolerance Assignment in Computer-Aided Process Planning, *Comput. Ind.,* Vol. 17, pp. 19–32.

Dong, Z., and Soom, A., 1990, Automatic Optimal Tolerance Design for Related Dimension Chains, *Manuf. Rev.,* Vol. 3, No. 4, pp. 262–271.

Eary, D. F., and Johnson, J. E., 1962, *Process Engineering for Manufacturing,* Prentice Hall, Upper Saddle River, NJ.

Fainguelernt, D., and Weill, R., 1986, Computer-Aided Tolerancing and Dimensioning in Process Planing, *Ann. CIRP,* Vol. 35, No. 1, pp. 381–386.

Farmer, L. E., and Gladman, C. A., 1986, Tolerance Technology: Computer-Based Analysis, *Ann. CIRP,* Vol. 35, No. 1, pp. 7–10.

Foulds, L. R., 1992, *Graph Theory Applications,* Springer-Verlag, New York.

Gadzala, J. L., 1959, *Dimensional Control in Precision Manufacturing,* McGraw-Hill, New York.

Huang, X., and Gu, P., 1994, Tolerance Analysis in Setup and Fixture Planning for Precision Machining, *Proceedings of the 4th International Conference on Computer Integrated Manufacturing and Automation Technology,* Rensselaer Polytechnic Institute, Troy, NY, October 10–12, pp. 298–305.

Huang, S. H., and Zhang, H.-C., 1996, Use of Tolerance Charts for NC Machining, *J. Eng. Des. Autom.,* Vol. 2, No. 1, pp. 91–104.

Irani, S. A., Mittal, R. O., and Lehtihet, E. A., 1989, Tolerance Chart Optimization, *Int. J. Prod. Res.,* Vol. 27, No. 9, pp. 1531–1552.

Lee, J. D., and Haynes, L. S., 1987, Finite-Element Analysis of Flexible Fixturing System, *J. Eng. Ind.,* Vol. 109, pp. 134–139.

Lee, S. H., and Cutkosky, M. R., 1991, Fixture Planning with Friction, *J. Eng. Ind.,* Vol. 113, pp. 320–327.

Li, J. K., and Zhang, C., 1989, Operational Dimensions and Tolerances Calculation in CAPP Systems for Precision Manufacturing, *Ann. CIRP,* Vol. 38, No. 1, pp. 403–406.

Ji, P., 1993a, A Linear Programming Model for Tolerance Assignment in a Tolerance Chart, *Int. J. Prod. Res.,* Vol. 31, No. 3, pp. 739–751.

Ji, P., 1993b, A Tree Approach for Tolerance Charting, *Int. J. Prod. Res.,* Vol. 31, No. 5, pp. 1023–1033.

Johnson, A. M., 1963, Tolerance Charts, in *Manufacturing Planning and Estimating Handbook,* McGraw-Hill, New York, pp. 1–14.

Karash, J. I., 1962, Principles of Locating and Positioning, in *Handbook of Fixture Design,* McGraw-Hill, New York, pp. 2-1 to 2-22.

King, D. A., and de Sam Lazaro, A., 1994, Process and Tolerance Considerations in the Automated Design of Fixtures, *J. Eng. Ind.,* Vol. 116, pp. 480–486.

Manivannan, S., Lehtihet, A., and Egbelu, P. J., 1989, A Knowledge-Based System for the Specification of Manufacturing Tolerances, *J. Manuf. Syst.,* Vol. 8, No. 2, pp. 153–160.

Mei, J., and Zhang, H.-C., 1992, Tolerance Analysis for Automated Setup Selection in CAPP, *Concurrent Engineering,* PED-Vol. 59, American Society of Mechanical Engineers, New York, pp. 211–220.

Menassa, R. J., and DeVries, W. R., 1990, A Design Synthesis and Optimization Method for Fixtures with Compliant Elements, *Proceedings of the ASME Symposium on Advances in Integrated Product Design and Manufacturing,* pp. 203–218.

Mittal, R. O., Irani, S. A., and Lehtihet, E. A., 1990, Tolerance Control in the Machining of Discrete Components, *J. Manuf. Syst.,* Vol. 9, No. 3, pp. 233–246.

Ngoi, B. K. A., 1992, Applying Linear Programming to Tolerance Chart Balancing, *Int. J. Adv. Manuf. Technol.,* Vol. 7, No. 2, pp. 187–192.

Ngoi, B. K. A., and Fang, S. L., 1994, Computer-Aided Tolerance Charting, *Int. J. Prod. Res.,* Vol. 32, No. 8, pp. 1939–1954.

Ngoi, B. K. A., and Ong, C. T., 1993a, A Complete Tolerance Charting System, *Int. J. Prod. Res.,* Vol. 31, No. 2, pp. 453–469.

Ngoi, B. K. A., and Ong, C. T., 1993b, Process Sequence Determination for Tolerance Charting, *Int. J. Prod. Res.,* Vol. 31, No. 10, pp. 2387–2401.

Sakurai, H., 1991, Automatic Setup Planning and Fixture Design for Machining, *J. Manuf. Syst.,* Vol. 11, No. 1, pp. 30–37.

Tang, G. R., Kung, R., and Chen, J. Y., 1994, Optimal Allocation of Process Tolerance and Stock Removals, *Int. J. Prod. Res.,* Vol. 32, No. 1, pp. 23–35.

Tulasiraman, K., and Swamy, M. N. S., 1993, *Graphs: Theory and Algorithms,* Wiley, New York.

Wade, O. R., 1967, *Tolerance Control in Design and Manufacturing,* Industrial Press, New York.

Wade, O. R., 1983, Tolerance Control, in *Tool and Manufacturing Engineers Handbook,* Vol. 1, *Machining,* Society of Manufacturing Engineers, Dearborn, MI, pp. 2-1 to 2-60.

Wade, O. R., 1990, The Role of Tolerance Stackup Control in Manual or CAPP Systems, *Proceedings of the Pacific Conference on Manufacturing,* Sydney and Melbourne, Australia, December 17–21, 1990, pp. 1252–1261.

Weill, R., 1988, Integrating Dimensioning and Tolerancing in Computer-Aided Process Planning, *Robot. comput.-Integrat. Manuf.,* Vol. 4, No. 1–2, pp. 41–48.

Xiaoqing, T., and Davies, B. J., 1988, Computer Aided Dimensional Planning, *Int. J. Prod. Res.,* Vol. 26, No. 2, pp. 283–297.

APPENDIX A: DETAILED SETUP PLANNING ALGORITHM FOR ROTATIONAL PARTS

The following notations are adopted:

n number of features within the part

A_i tool approach direction defined as follows

$$A_i = \begin{cases} 1, & \text{if feature } i \text{ can be machined only from the left} \\ 2, & \text{if feature } i \text{ can be machined only from the right} \\ 0, & \text{otherwise} \end{cases}$$

in which $i = 1,2,...,n$

E_i feature type defined as follows

$$E_i = \begin{cases} 1, & \text{if feature } i \text{ is a vertical feature} \\ 2, & \text{if feature } i \text{ is a cylindrical feature} \\ 0, & \text{otherwise} \end{cases}$$

in which $i = 1,2,...,n$

S_i stock shape defined as follows

$$S_i = \begin{cases} 1, & \text{if feature } i \text{ exists on the stock} \\ 0, & \text{otherwise} \end{cases}$$

in which $i = 1,2,...,n$

$T = [t_{ij}]$ tolerance matrix defined as follows

$$t_{ij} = \begin{cases} \delta_{ij}, & \text{if features } i \text{ and } j \text{ have tolerance} \\ & \text{requirement and} \\ & \text{the size of the tolerance zone is } \delta_{ij} \\ 0, & \text{otherwise} \end{cases}$$

in which $i = 1,2,...,n$; $j = 1,2,...,n$

The algorithm is as follows:

Step 1. Setup Formation
   find $t_{pq} = \min[t_{ij}]$, in which $i = 1,2,...n$; $j = 1,2,...n$; $A_i = 0$; $A_j \neq 0$; $t_{ij} \neq 0$
   if no such t_{pq} can be found then
      for $i = 1$ to n do
         if $A_i = 0$, then
            let $l = \sum_{j=1}^{n} x_j$, in which $x_j = \begin{cases} 1, \text{ if } A_j = 1 \\ 0, \text{ otherwise} \end{cases}$
            let $r = \sum_{j=1}^{n} x_j$, in which $x_j = \begin{cases} 1, \text{ if } A_j = 2 \\ 0, \text{ otherwise} \end{cases}$
            if $l < r$, then
               let $A_i = 1$
            else
               let $A_i = 2$
      go to step 2
   else
      let $A_p = A_q$
      let $t_{pq} = 0$; $t_{qp} = 0$
      repeat step 1
Step 2. Datum Selection
   Step 2.0
      let $t_{ij} = 0$ if $A_i = A_j$, in which $i = 1,2,...,n$; $j = 1,2,...,n$
      let $d_{ij} = 0$, in which $i = 1,2$; $j = 1,2$
      let $T' = [t'_{ij}] = T$
   Step 2.1
      find $t'_{pq} = \min[t'_{ij}]$, in which $i = 1,2,...n$; $j = 1,2,...n$; $t'_{ij} \neq 0$
      if no such t'_{pq} can be found, then
         for $i = 1$ to 2 do
      for $j = 1$ to 2 do
         if $d_{ij} = 0$ then

find feature w such that $A_w = i$ and $E_w = j$

let $d_{ij} = 1$; $D_{ij} = w$

go to step 3

else

let $t'_{pq} = 0$; $t'_{qp} = 0$

if $E_p \neq 0$ and $d_{A_p E_p} = 0$, then

let $d_{A_p E_p} = 1$

$D_{A_p E_p} = p$

if $E_q \neq 0$ and $d_{A_q E_q} = 0$, then

let $d_{A_q E_q} = 1$

$D_{A_q E_q} = q$

if $\displaystyle\sum_{i=1}^{2} \sum_{j=1}^{2} d_{ij} = 4$, then

go to step 3

else

go to step 2.1

Step 3. Setup Sequencing

for $i = 1$ to 2 do

for $j = 1$ to 2 do

let $R_{ij} = \displaystyle\sum_{k=1}^{n} x_k$, in which $x_k = \begin{cases} 1, & \text{if } t_{D_{ijk}} \neq 0 \\ 0, & \text{otherwise} \end{cases}$

find $R_{pq} = \max [r_{ij}]$, in which $i = 1,2; j = 1,2$

let $Z = \begin{cases} 1, & \text{if } p = 2 \\ 2, & \text{otherwise} \end{cases}$

Step 4. Generate Setup Plan Based on Stock Shape

let $c = 0$

if $S_{D_{Z1}} \neq 1$, then

find feature f so that $A_f = Z$, $E_f = 1$, and $S_f = 1$

if such a feature cannot be found, then

let $c = 1$

else

let $D_{Z1} = f$

if $S_{D_{Z2}} \neq 1$, then

find feature f so that $A_f = Z$, $E_f = 2$, and $S_f = 1$

if such a feature cannot be found, then

let $c = 1$

else

if $c \neq 1$, then

let $D_{Z2} = f$

Let $S_L = \emptyset$; $S_R = \emptyset$

Let $S_L = S_L \cup \{i\}$ if $A_i = 1$, in which $i = 1,2,...,n$

Let $S_R = S_R \cup \{i\}$ if $A_i = 2$, in which $i = 1,2,...,n$

if $c = 1$, then

  machine features D_{Z1} and D_{Z2}

if $Z = 1$, then

    features within S_R are to be machined in the first setup using features D_{11} and D_{12} as the setup datums; features within S_L are to be machined in the second setup using features D_{21} and D_{22} as the setup datums

else

    features within S_L are to be machined in the first setup using features D_{21} and D_{22} as the setup datums; features within S_R are to be machined in the second setup using features D_{11} and D_{12} as the setup datums

APPENDIX B: DETAILED SETUP PLANNING ALGORITHM FOR PRISMATIC PARTS

The following notations are adopted:

n number of features within the part

A_i tool approach direction defined as follows

$$A_i = \begin{cases} 1, \text{ if feature } i \text{ can be machined from the top} \\ 2, \text{ if feature } i \text{ can be machined from the bottom} \\ 3, \text{ if feature } i \text{ can be machined from the left} \\ 4, \text{ if feature } i \text{ can be machined from the right} \\ 5, \text{ if feature } i \text{ can be machined from the front} \\ 6, \text{ if feature } i \text{ can be machined from the rear} \\ 7, \text{ if feature } i \text{ can be machined from both the top and the bottom} \\ 8, \text{ if feature } i \text{ can be machined from both the left and the right} \\ 9, \text{ if feature } i \text{ can be machined from both the front and the rear} \end{cases}$$

in which $i = 1,2,..., n$

$T = [t_{ij}]$ tolerance matrix defined as follows

$$t_{ij} = \begin{cases} \delta_{ij}, \text{ if features } i \text{ and } j \text{ have tolerance requirement and the} \\ \quad\quad \text{size of the tolerance zone is } \delta_{ij} \\ 0, \quad \text{otherwise} \end{cases}$$

in which $i = 1,2,...,n; j = 1,2,...,n$

P_i Primary setup datum for setup i

D_i Secondary setup datum for setup i

E_i Tertiary setup primary datum for setup i

The algorithm is as follows:

Step 1. Setup Formation
Step 1.0
  let $S_i = 0$, in which $i = 1,2,...6$
  let $c = 0$
Step 1.1
  for $\tau = 7$ to 9 do
    Step 1.1.1
      find $t_{pq} = \min[t_{ij}]$,
        in which $i = 1,2,...n$; $j = 1,2,...n$;
          $A_i = \tau$, $A_j = 2 \times (\tau - 6) - 1$ or $2 \times (\tau - 6)$; $t_{ij} \neq 0$
      if no such t_{pq} can be found, then
        for $i = 1$ to n do
          if $A_i = \tau$, then

$$\text{let } l = \sum_{j=1}^{n} x_j, \text{ in which } x_j = \begin{cases} 1, \text{ if } A_j = 2 \times (\tau - 6) - 1 \\ 0, \text{ otherwise} \end{cases}$$

$$\text{let } r = \sum_{j=1}^{n} x_j, \text{ in which } x_j = \begin{cases} 1, \text{ if } A_j = 2 \times (\tau - 6) \\ 0, \text{ otherwise} \end{cases}$$

            if $l < r$, then
              let $A_i = 2 \times (\tau - 6) - 1$
            else
              let $A_i = 2 \times (\tau - 6)$
      else
        let $A_p = A_q$
        let $t_{pq} = 0$; $t_{qp} = 0$
        repeat step 1.1.1
Step 1.2
  let $t_{ij} = 0$ if $A_i = A_j$, in which $i = 1,2,...,n$; $j = 1,2,...,n$
  let $T' = [t'_{ij}] = T$
  for $\tau = 1$ to 3 do
    if $A_i = 2 \times \tau$ and $A_j = 2 \times \tau - 1$, then
      let $t_{ij} = 0$; $t_{ji} = 0$
        in which $i = 1,2,...,n$; $j = 1,2,...,n$
  let $Z_i = 0$, in which $i = 1,2,3,4,5,6$
Step 1.3
  find $t_{pq} = \min[t_{ij}]$, in which $i = 1,2,...n$; $j = 1,2,...n$

if no such t_{pq} can be found, then
    let $\gamma = -1$
    for $i = 1$ to 4 do
       let $\gamma = \gamma \times (-1)$
       if $Z_i = 0$, then
          let $Z_i = 1$; $c = c + 1$; $S_c = i$
          if (i mod 2) $= 0$, then
             let $e = 1$
          else
             let $e = 2$
          if $\gamma = 1$ then
             let $\alpha = i + e$; $\beta = 6$
          else
             let $\alpha = 6$; $\beta = i + e$
          for $j = \alpha$ to β do
             let $Z_j = 1$

$$\text{let } l = \sum_{k=1}^{n} x_k, \text{ in which } x_k = \begin{cases} 1, & \text{if } A_k = i \\ 0, & \text{otherwise} \end{cases}$$

$$\text{let } r = \sum_{k=1}^{n} x_k, \text{ in which } x_k = \begin{cases} 1, & \text{if } A_j = j \\ 0, & \text{otherwise} \end{cases}$$

        if $l < r$, then
           let $S_c = j \times 10 + i$
           for $u = 1$ to n do
             let $A_u = j$ if $A_u = i$
       else
           let $S_c = i \times 10 + j$
           for $u = 1$ to n do
             let $A_u = i$ if $A_u = j$
  for $i = 5$ to 6 do
    if $Z_i = 0$, then
       let $Z_i = 1$; $c = c + 1$; $S_c = i$
else
  let $Z_{A_p} = 1$; $Z_{A_q} = 1$
  let $c = c + 1$
  for $i = 1$ to n do
    if $A_i = A_p$ or $A_i = A_q$, then
      let $t_{ij} = 0$; $t_{ji} = 0$; in which $j = 1,2,...n$

$$\text{let } l = \sum_{j=1}^{n} x_j, \text{ in which } x_j = \begin{cases} 1, \text{ if } A_j = A_p \\ 0, \text{ otherwise} \end{cases}$$

$$\text{let } r = \sum_{j=1}^{n} x_j, \text{ in which } x_j = \begin{cases} 1, \text{ if } A_j = A_q \\ 0, \text{ otherwise} \end{cases}$$

if $l < r$, then

    let $S_c = A_q \times 20 + A_p$

    for $k = 1$ to n do

        let $A_k = A_q$ if $A_k = A_p$

else

    let $S_c = A_p \times 10 + A_q$

    for $k = 1$ to n do

        let $A_k = A_p$ if $A_k = A_q$

repeat step 1.3

Step 2. Datum Selection

    for $i = 1$ to c

        if $S_i < 10$, then

            if $(S_i \bmod 2) = 0$, then

                let $P_i = S_i - 1$

            else

                let $P_i = S_i + 1$

            let $D_i = ((S_i + 1) \bmod 6) + 1$; $E_i = ((S_i + 3) \bmod 6) + 1$;

        else

            let $Y_j = 0$, in which j = 1, 2, 3, 4, 5, 6

            let $a = \left\lceil \dfrac{S_i}{10} \right\rceil$; $b = S_i \bmod 10$

            if $(a \bmod 2) = 0$, then

                let $P_i = a - 1$;

            else

                let $P_i = a + 1$

            if $(b \bmod 2) = 0$, then

                let $E_i = b - 1$

            else

                let $E_i = b + 1$

            let $Y_a = 1$; $Y_{Pi} = 1$; $Y_b = 1$; $Y_{Ei} = 1$

            find $t'_{pq} = \min[t'_{rs}]$, in which $r = 1,2,...n$; $s = 1,2,...n$; $A_r = a$; $A_s \neq P_i$; $A_s \neq E_i$;

            if no such t_{pq} can be found, then

for $j = 1$ to 6 do
if $Y_j = 0$
let $D_i = j$
else
let $D_i = A_q$

Step 3. Setup Sequencing

let $t'_{ij} = 0$ if $A_i = A_j$, in which $i = 1,2,...,n$; $j = 1,2,...,n$
let $G_i = -1$, in which $i = 1,2,...c$
for $i = 1$ to c do
  if $S_i < 10$, then
    let $a = S_i$
  else

$$\text{let } a = \left[\frac{S_i}{10} \right]$$

  for $j = 1$ to n do
    if $A_j = a$, then
      let $b = 0$
      for $k = 1$ to n do
        if $t'_{jk} \neq 0$
          let $b = b + 1$
      if $b > G_i$, then
        let $G_i = b$

Sort G_i in descending order gives the sequence of the setups.

17

CATCH: *COMPUTER-AIDED TOLERANCE CHARTING*

G. A. BRITTON

Nanyang Technological University
Singapore

K. WHYBREW

University of Canterbury
Christchurch, New Zealand

17.1   INTRODUCTION

Manual tolerance charting has been discussed in Chapters 2 and 3. Manual charting is time consuming and tedious; therefore, recent research has focused on automating dimensioning and tolerancing during process planning. Virtually all of the computer-based methods are variations of manual tolerance charting methods. Some researchers have focused on tolerance optimization, which is discussed later in the chapter. In this introduction we discuss various techniques used to represent dimension chains and to calculate tolerance stackups; for a recent review of tolerancing techniques, refer to Zhang and Huq (1992).

Bjorke's (1978) text on tolerancing is often cited. His work has limited application to tolerance charting because it focuses on tolerancing of assemblies during design. However, his tolerance chain theory can and has been used to calculate tolerance stackups in computer-aided tolerance charting. His theory is based on scalar dimension chains.

In 1982, Sack published a paper describing a stand-alone, interactive, computer-aided, generative, process-planning system. The program was writ-

Advanced Tolerancing Techniques, Edited by Hong-Chao Zhang
ISBN 0-471-14594-7   © 1997 John Wiley & Sons, Inc.

ten specifically for cylindrical parts. Process decision rules and machining data are entered into the program manually. These and a model of the part, defined interactively, are used to generate a tolerance chart (internal to the program) automatically using a variation of Wade's technique. The program has an optimizer to maximize operation tolerances. A process plan is generated as output and this can be modified interactively.

Ahluwalia and Karolin (1984) describe a stand-alone, interactive, tolerance charting program which they call CATC. The user inputs a drawing of the finished workpiece and an outline of the raw material using a graphical interface, together with the manufacturing sequence, and design and raw material dimensions. Stock removal allowances and minimum operation tolerances are retrieved from a database. The authors do not specify the technique they use to identify tolerance chains and to perform the tolerance stackup calculations.

Fainguelernt et al. (1986) present a completely different approach to computer-aided tolerancing; it is not based on manual tolerance charting. They have developed a fundamental theory that differentiates between positioning and machining tolerances and takes into account tool wear and repositioning of the tool during manufacture. (Note that in manual charting these tolerances and deviations are incorporated together as the process tolerance.) The user inputs the finished and initial workpiece dimensions. The program uses a matrix method for calculating the resultant tolerances based on minimum positioning and machining tolerances. These tolerances are then relaxed using a simple optimizing procedure. Finally, the working dimensions are calculated. The program allows the user to input half-limit dimensions to control the stock allowance for first cuts. Their program does not check tolerance stackups against stock removal allowances on succeeding cuts.

Tang and Davies (1988) describe an interactive, tolerance charting program that uses a matrix tree technique to identify the tolerance chains. The dimensions are input in matrix form. Matrix transformations are used to construct basic scalar trees. A scalar tree consists of several dimension chains connected together and may have branches. Each dimension in the tree is a scalar dimension. The basic trees are combined and manipulated to produce valid trees. The entire procedure is extremely complex. The program includes an optimization algorithm to relax tolerances.

Irani et al. (1989) describe a graph-theoretic technique for tolerance charting. The authors note that "the graph representation of tolerance chains is suitable for representation within and extraction from a CAD database" (pp. 1550–1551). Each face is identified by a number. Each machining operation is represented by a directed arc identified by a datum face and a cut face. Directed trees are formed to represent the operations contributing to the tolerance stackups for design specification dimensions and stock allowances. A directed tree consists of several dimension chains connected together and may have branches. Each dimension in the tree is a vector dimension. The trees are cyclic (more than one operation can input to a datum face) because each surface is labeled only once.

A rooted tree graph-theoretic technique has been published by the authors (Whybrew et al., 1990). Our technique differs in one important respect from that of Irani et al. (1989). We differentiate and label each cut on a workpiece face and treat them as separate surfaces. The surfaces resulting from each cut are uniquely labeled. This enables us to represent the entire manufacturing sequence as a single rooted tree, with no cycling. That is, one and only one operation can input into each datum face.

Our labeling technique is the only means for generating a rooted tree (i.e., every cut surface must be uniquely labeled). Our first labeling technique used a two-digit code to label surfaces: a letter and a number (e.g., A2). This is the most efficient code possible. It can represent up to 25 design specification dimensions using the 26 letters of the alphabet and up to nine cuts on a surface. Furthermore, it clearly distinguishes the surface identifier, represented by a letter, and the number of cuts on the surface, represented by a number. Our current code is a four-digit number. This change was necessary because practical charts often have more than 25 design specification dimensions. It is possible to meet this demand with a three-digit code, but we chose a four-digit code (Figure 17.5) because it is easier to read and interpret, which is an essential consideration for interactive tolerance charting. Tolerance stackups and working dimensions can be calculated easily by using different rules to traverse the rooted tree. The tree is a directed tree.

In a later paper (Britton et al., 1996) we describe an industrial implementation of our rooted tree technique. We have named our software CATCH, which is an acronym of computer-aided tolerance charting and is intended to be evocative of catching tolerance stackups before parts are actually manufactured. We argue that the rooted tree graph-theoretic technique is efficient and practical because it is fully compatible with commercial computer-aided design (CAD) systems. Our program, CATCH, is fully interactive, and currently it interfaces directly with EDS's Unigraphic UGII software (a proprietary CAD system).

Another graph-theoretic technique is described by Ping (1993). Like us, the author differentiates surfaces during the manufacturing sequence, but he creates three rooted trees: a scalar design specification dimension tree, a scalar design specification dimension and stock removal tree, and a directed working dimension tree.

Li and Zhang (1989) use a scalar dimension tree technique for calculating dimensions and tolerances. Two types of trees are required: a design specification dimension tree and a stock allowance tree. The authors also describe an algorithm that checks the design specification dimension tree to ensure that all the design specification dimensions have been entered and that there are no redundant dimensions.

He and Gibson (1992) describe a nonlinear optimization technique to minimize the cost of scrap. The manufacturing sequence is input into the optimizing software as a set of constraint equations to control how the tolerance stackups are performed. They use a matrix method to order the raw input sequence for the constraint equations. The software simultaneously calculates the assigned tolerances for each operation, the tolerance stackups, and the working dimensions.

Recently, another approach to tolerance charting was suggested by Tang et al. (1993). The authors propose a list technique. They argue that commercial software uses linked lists and therefore that a list technique is both practical and efficient. Their argument is true with respect to commercial database software. It is not true for CAD software, which is based on tree logic.

It can be seen that there are different computer-aided techniques available for tolerance charting. This raises a crucial issue as to how computer-aided charting should be performed. We propose the following.

1. A graphical (traditional) tolerance chart should be used to represent a manufacturing sequence. The graphical chart is an extremely useful and efficient tool for this purpose and can represent all practical manufacturing processes.

2. Computer-aided tolerancing should be interactive. A major failing of most tolerance optimization programs is that the manufacturing sequence must be planned prior to running the programs. Interactive tolerance charting is a process planning tool, not simply a process checking tool. That is, it is capable of generating alternative manufacturing sequences as well as checking the processes for each sequence. Furthermore, if a direct interface can be provided with existing CAD software, process planners will not have to redraw workpieces or reenter design specification dimensions. In short, interactive charting is much more flexible and efficient than noninteractive charting. This capability is becoming essential in today's concurrent engineering environment.

3. The rooted tree graphic-theoretic technique should be used. This technique has several advantages. First, it is fully compatible with existing commercial CAD software. Second, it is an extremely efficient technique for representing tolerance stacks and performing tolerance stackup calculations. Third, a graphical illustration of the tree clearly highlights critical datum surfaces and is an invaluable aid for planning and modifying a manufacturing sequence.

4. Balance dimensions are not required for computer-aided tolerance charting. Programs using balance dimensions are inefficient, as additional computations are performed unnecessarily. There are also circumstances where the calculations using balance dimensions will yield incorrect results.

5. The most efficient sequence for performing tolerance checking is as follows:

   a. Calculate resultant tolerances and compare with design specifications. The upper and lower resultant dimension limits are compared with the design target limits. The limits must be compared individually because the resultant dimensions may be offset. Design specifications are checked first because the process plan must be changed if they cannot be met.

b. Calculate stock removal tolerance stackups and compare with allowances; adjust allowances for special machining conditions such as grinding. There is no point in calculating working dimensions if allowances have to be changed. Therefore, these are checked next. If allowances cannot be achieved, the process plan must be changed.

c. Calculate working dimensions.

Our software, CATCH, meets these requirements. It and the rooted tree logic are explained in this chapter. CATCH is a practical, computer-aided tolerance charting system developed by the authors in collaboration with Sundstrand Pacific Aerospace Pte. Ltd. in Singapore. Sundstrand manufactures high-precision components for the aerospace industry and has been using a manual tolerance charting technique for many years. Sundstrand has actively supported our work because there are significant productivity gains to be realized through computer-aided tolerancing.

CATCH was derived from early research work at the University of Canterbury by Whybrew, Britton, and Sermsuti-Anuwat (Whybrew et al., 1990; Sermsuti-Anuwat et al., 1991, 1995; Sermsuti-Anuwat, 1992) and prototype interactive tolerance charting software developed at Nanyang Technological University under the supervision of Britton (Kong, 1992; Teng, 1993; Koh, 1993). CATCH incorporates these prototype developments and additional improvements in the tree traversing algorithms.

A tolerance chart generated by CATCH is shown in Figures 17.1 and 17.2. The chart has been edited and subdivided for clarity of exposition. A practical working chart would, in fact, be printed on an A1-size sheet of paper. This chart will be used as an example throughout this chapter. The reader should note that Sundstrand uses Imperial units (inches) for charting.

17.2   CATCH LOGIC

CATCH is based on the tolerance chart layout used by Sundstrand Pacific Aerospace Pte. Ltd. (Singapore), which is shown in Figure 17.3. A two-dimensional profile of the finished workpiece is drawn at the top of the chart. Face lines are drawn down from selected points that represent surfaces, centerlines, or arc centers. Design specification dimensions and the different types of operations are represented by symbols connecting two face lines; see Figure 17.4 for the symbol codes. Design specification dimensions are entered at the bottom of the chart. The nominal distance and bilateral tolerance are recorded on the chart. Resultant dimensions are also entered at the bottom of the chart.

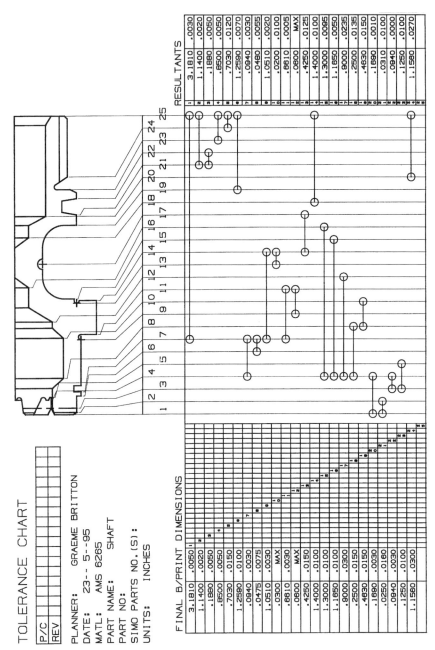

Figure 17.1. Design specification dimensions for example tolerance chart.

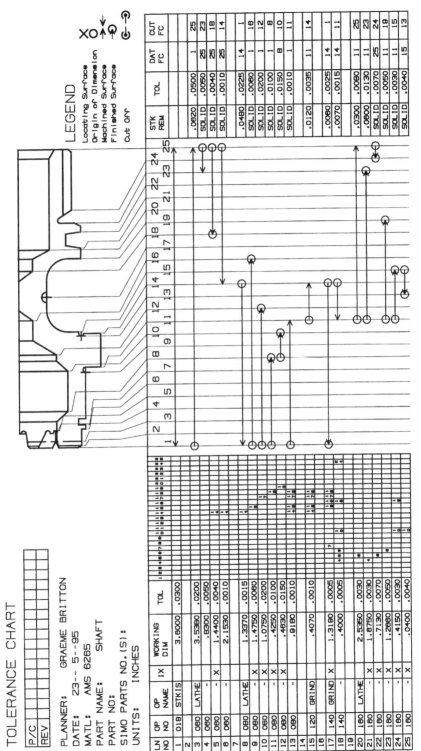

Figure 17.2. Manufacturing sequence for example tolerance chart.

467

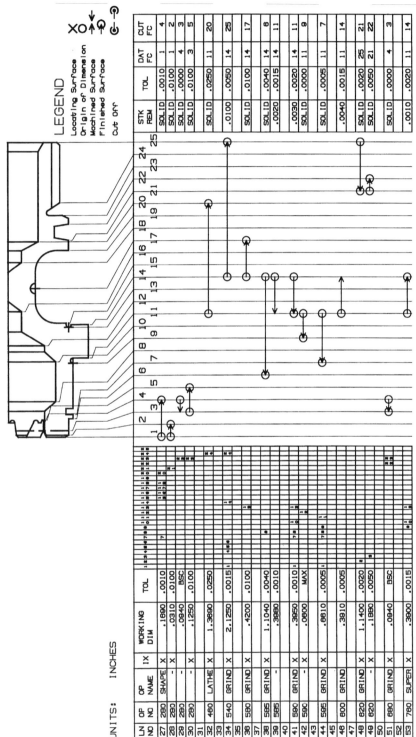

Figure 17.2. (*Continued*)

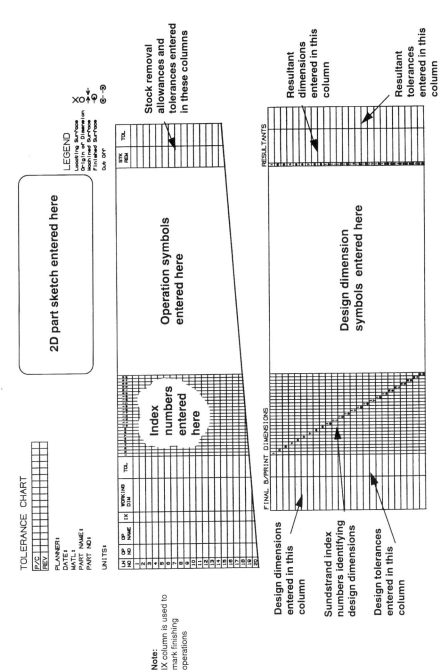

Figure 17.3. Sundstrand tolerance chart layout.

469

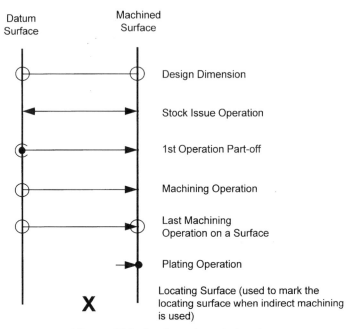

Figure 17.4. Sundstrand symbol codes.

Machining operations are entered in the middle of the chart. Each machining operation is represented by an arrow. The datum surface of the operation is the tail of the arrow and is indicated by a small circle. The cut surface is the head of the arrow. The direction of the arrow is from the datum to the cut surface. Finished (last cut) surfaces are indicated by a small circle over the arrow heads.

Columns are drawn on the chart for recording the following manufacturing information:

- Operation number
- Operation name
- IX: last cut marker (this indicates the last operation to machine a surface)
- Working dimension (nominal dimension only)
- Operation tolerance
- Stock removal allowance
- Stock removal tolerance stackup

In addition to the above, the chart in Figure 17.3 has index columns. Each design specification dimension is identified by an index number. If a machining operation contributes to the tolerance stackup for a given design specification dimension, the index number of the dimension is written in the appropriate column for the operation. This helps the process planner identify important datums and speeds up the calculation of tolerance stackups for manual charting. These index numbers were first used by the Sundstrand Company and we therefore refer to them as Sundstrand index numbers.

A four-digit number is used to label cut surfaces; the convention is defined in Figure 17.5. A finished surface and the prior, preliminary cut surfaces for the finished surface can be considered as a set and given a set identifier. We

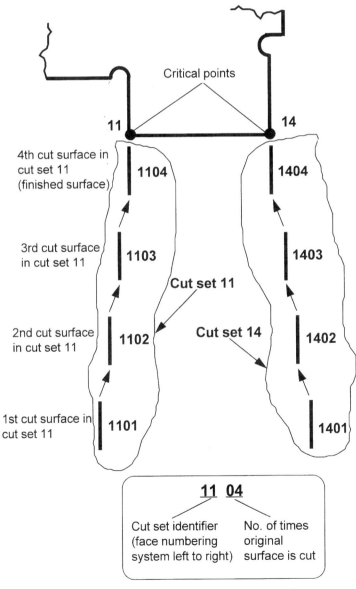

Figure 17.5. Rooted tree representation convention.

refer to these sets as *cut sets*. The finished (last) surfaces of the cut sets are numbered from left to right, corresponding to the order of their coordinate positions from a preset datum, which is the left-hand end of the part. The sets are identified by these two-digit numbers, which are the first two digits of the four-digit labels (e.g., 01 in label 0102, where 01 indicates that the surface is the left-hand end of the part).

The surfaces within each cut set are identified by a two-digit number representing the number of times that the original (starting) surface has been modified by the manufacturing sequence. The first surface in a cut set is labeled 01, the next is labeled 02, and so on. The two digits are appended to the cut-set identification number to produce a unique four-digit number for each cut surface (e.g., 0102). The datum surface of the first operation in the manufacturing sequence is the tree root and is uniquely identified by setting the last two digits to 00 (e.g., 0100 in Figure 17.6). The last (highest numbered) surface in a cut set is a finished surface.

Figure 17.6 is a rooted tree graph for the tolerance chart shown in Figures 17.1 and 17.2. This graph is read from left to right. Operations are placed in sequence order from top to bottom at branch nodes. This assists interpretation of the graph. Each operation can be numbered according to its position in the manufacturing sequence (line number on the tolerance chart). These numbers are shown in smaller typeface for easy cross-reference to the tolerance chart.

There is a considerable amount of information contained in the rooted tree graph.

- The entire graph is a succinct summary of the operation sequence.
- Important datum surfaces can be recognized easily: They are datums for two or more operations (e.g., surface 1402). All datum surfaces are members of datum sets. The members of each datum set are datum surfaces from the same cut set; that is, datum sets are a special subclass of cut sets. These datum sets are extremely important for process planning and are the basis of the automated process planning system developed by Sermsuti-Anuwat (Sermsuti-Anuwat, 1992; Sermsuti-Anuwat et al., 1995).
- The cut surface numbering system clearly shows how many operations have been performed in each cut set.

All of the information above is essential for process planning. It is easily read from the graph because of our labeling technique. A basic requirement of our technique is that the tree must be rooted. All operations must start either from the first datum surface in a manufacturing sequence or from a surface resulting from a machining operation. Any gap in the tree indicates an attempt to locate on a nonexistent surface. This rooted tree constraint is used to check operations for their feasibility when they are put into the system. An operation that violates the constraint is "bad" process planning and is rejected.

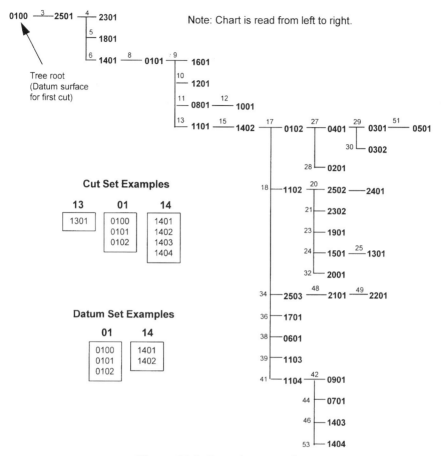

Figure 17.6. Rooted tree graph.

In the CATCH software, string variables and pointers are used to represent the rooted tree. Each operation has a string variable associated with it. This variable (the *reversed path*) is a serial list of the four-digit numbers of the surfaces contained in the path from the tree root to the surface resulting from that operation. For example, consider cut surface 0901 in Figure 17.6. Its reversed path is 0100–2501–1401–0101–1101–1402–1104–0901. The reversed paths are used to identify branch nodes in the tree. Table 17.1 lists the reversed paths for all the operations in the example.

Pointers are used to traverse the tree rapidly. The pointer convention is shown in Figure 17.7. Upward and downward pointers are used to link the surfaces in each cut set. Downward pointers can be used effectively because the sequence of operations in a cut set is unique in both directions. These pointers are used to guide the calculation of working dimensions. Upward pointers are used to link operations back to previous operations. Downward pointers have not been used because the tree may have branches; the reversed

TABLE 17.1. Reversed Paths for Tolerance Chart Example

Operation Line Number	Reversed Path
3	0100–2501
4	0100–2501–2301
5	0100–2501–1801
6	0100–2501–1401
8	0100–2501–1401–0101
9	0100–2501–1401–0101–1601
10	0100–2501–1401–0101–1201
11	0100–2501–1401–0101–0801
12	0100–2501–1401–0101–0801–1001
13	0100–2501–1401–0101–1101
15	0100–2501–1401–0101–1101–1402
17	0100–2501–1401–0101–1101–1402–0102
18	0100–2501–1401–0101–1101–1402–1102
20	0100–2501–1401–0101–1101–1402–1102–2502
21	0100–2501–1401–0101–1101–1402–1102–2302
23	0100–2501–1401–0101–1101–1402–1102–1901
24	0100–2501–1401–0101–1101–1402–1102–1501
25	0100–2501–1401–0101–1101–1402–1102–1501–1301
27	0100–2501–1401–0101–1101–1402–0102–0401
28	0100–2501–1401–0101–1101–1402–0102–0201
29	0100–2501–1401–0101–1101–1402–0102–0401–0301
30	0100–2501–1401–0101–1101–1402–0102–0401–0302
32	0100–2501–1401–0101–1101–1402–1102–2001
34	0100–2501–1401–0101–1101–1402–2503
36	0100–2501–1401–0101–1101–1402–1701
38	0100–2501–1401–0101–1101–1402–0601
39	0100–2501–1401–0101–1101–1402–1103
41	0100–2501–1401–0101–1101–1402–1104
42	0100–2501–1401–0101–1101–1402–1104–0901
44	0100–2501–1401–0101–1101–1402–1104–0701
46	0100–2501–1401–0101–1101–1402–1104–1403
48	0100–2501–1401–0101–1101–1402–2503–2101
49	0100–2501–1401–0101–1101–1402–2503–2101–2201
51	0100–2501–1401–0101–1101–1402–0102–0401–0301–0501
53	0100–2501–1401–0101–1101–1402–1104–1404

paths are used instead and contain both the downward linking and branch information. The upward pointers and the reversed paths are used to guide the calculation of tolerance stackups for resultant dimensions and stock removal.

Our technique assumes that a model of a finished part has been created on a CAD system. Currently, CATCH interfaces only with UGII, but the principle is the same for other commercial CAD systems. A reference coordinate sys-

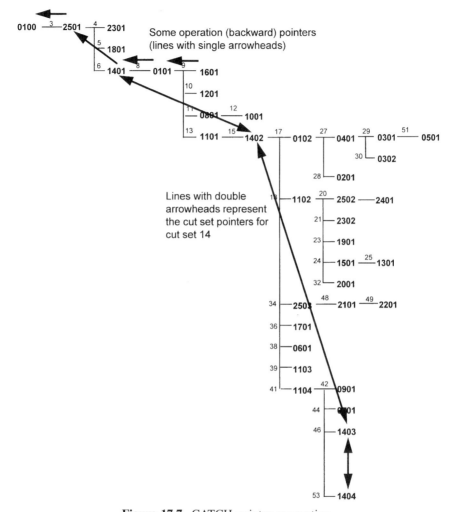

Figure 17.7. CATCH pointer convention.

tem (common datum) is established at the left-hand end of the model. The cut sets (machining surfaces or points) are identified by selecting critical points on the model. The point coordinates, relative to the common datum, are obtained directly from the CAD database. Design and other dimensions are obtained easily by taking the absolute differences between the coordinates of the endpoints defining the dimensions. The dimensions can be obtained at any stage in the manufacturing sequence. During the dimension calculations, the positions of all surfaces are computed independently relative to the common datum.

All practical manufacturing processes and geometric tolerances can be represented using our representational scheme. Some examples are given below.

- *Heat Treatment.* Heat treatment operations are represented in the same manner as normal matching operations.
- *Plating.* Plating is represented by an arrow and a small circle on the same surface. That is, the datum and cut faces come from the same cut set and the last two digits differ by one.
- *Geometric Tolerances, Radii Breakout Cuts, Angled Cuts, and so on.* Transformation rules are required to project the tolerances and stock removal allowances to the reference direction. Once this has been achieved, these operations can be processed in the normal manner. Gadzala (1959) and Wade (1967, 1983) provide a large number of transformation rules. We have not incorporated these rules in CATCH, as they are not required by Sundstrand.
- *Simultaneous Operations (Multitool Setups).* A multitool operation can be represented by a set of single machining operations under the following conditions:
  - The datum surfaces must be the same.
  - The tolerances for the operations cannot be changed independently as they are directly correlated (the cutting speed and feed rate are the same for all tools).
  - The operations must be added, deleted, and resequenced as a set during editing.

The details of tolerance stackup calculation have been covered fully in previous publications (Whybrew et al., 1990; Britton et al., 1992). The logic is straightforward. Two relevant surfaces in the tree are selected. The tolerances for the operations in the path between the two surfaces are added to obtain the stackup for the path.

Our current procedure for calculating working and inspection dimensions is as follows:

1. The positions of finished surfaces are set equal to the resultant dimensions plus or minus any offsets. These finished surfaces are the last surfaces in the cut sets.
2. The positions of the preceding surfaces in each cut set are calculated by working backward from the finished surfaces and adding or subtracting the appropriate stock removal allowances to the positions calculated previously.
3. The working dimensions are calculated by taking the absolute values of the differences between the positions of the surfaces defining the dimensions.

The algorithm for this procedure is simple and very fast, and it provides maximum flexibility to the process planner by allowing surfaces to be offset individually.

17.2.1   Procedure for Using CATCH

CATCH sits on top of UGII and is written in GRIP programming language, a proprietary language of EDS. The menus and other interfaces in CATCH are the same as for UGII, except that they have been customized for tolerance charting. CATCH is started from UGII using the UGII GRIP menu.

The overall procedure for tolerance planning and analysis using CATCH is shown in Figure 17.8. A full-size two-dimensional model must be created before CATCH is started. For new charts, a template of the chart is retrieved automatically and scaled to fit the part model. The user interface is via a menu structure, shown in Figure 17.9, which constrains the user to input data in the correct order. The program is fully interactive and on-screen editing allows the user to edit virtually all chart data, the exceptions being the design nominal dimensions (as these are obtained directly from the CAD database) and the tolerance stackup values.

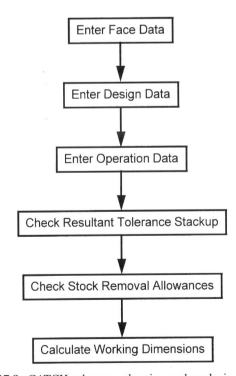

Figure 17.8. CATCH tolerance planning and analysis procedure.

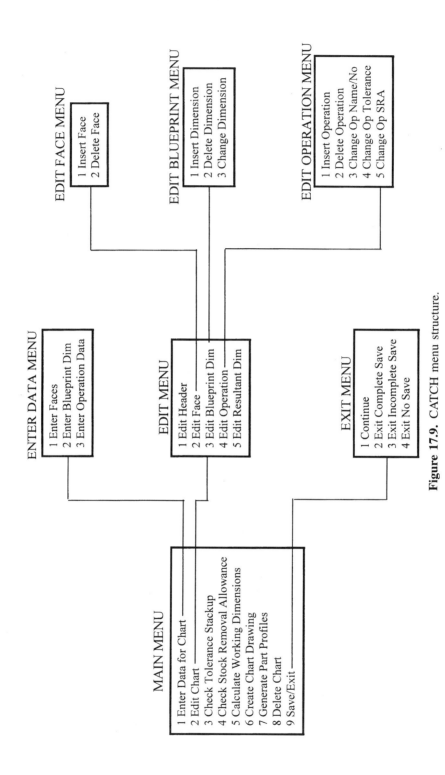

Figure 17.9. CATCH menu structure.

478

Machining surfaces and points are selected by clicking on points, lines, or arcs on the CAD model of the part using a mouse and a cursor. The direction of tool movement is input to set the sign of the stock removal allowances so that they are added and subtracted correctly during the working dimension calculations. When all the faces have been selected, the program orders and numbers them according to their coordinate position and displays the face lines and numbers in the display window on the screen.

Design specification dimensions are input by clicking on the appropriate face lines and entering the design tolerances. The program displays the design tolerances and symbols on the screen. It then calculates and displays the design nominal dimensions and sets the resultant nominal dimensions the same as the design nominal dimensions. These are then displayed on the screen.

An operation is input by selecting the type of operation, clicking on the face lines for the datum and cut faces, respectively, and inputting the following operation data: operation number, operation name, process tolerance, and stock removal allowance. The types of operations that can be entered are: machining, heat treatment, plating, first-operation part-off, stock issue, and blank line. The program checks that the operation is valid, generates its reversed path, and sets the relevant pointers. The data and operation symbol are displayed on the screen.

Once operation entry is complete, resultant tolerance stackups are calculated and displayed, and resultant lower and upper dimensions are compared with the design targets. If a design specification is not achievable, the relevant design specification dimension and all the operations that contribute to its resultant tolerance stackup are highlighted in yellow on the computer screen. It is not possible to progress to the next stage in the charting procedure until the process plan has been modified to satisfy the design specification.

Stock removal allowances are checked next, in a similar manner. If an operation has a tolerance stackup greater than the allowance, it (the complete line entry on the chart) is highlighted in green on the computer screen. There are two options available to the user during checking. First, the user can set the stock removal allowances equal to the tolerance stackups. This feature is sometimes used to generate initial allowances quickly for comparison with good machining practice. If this option is chosen, allowances do not need to be entered during input of the operations. Second, the user can override the check for any individual operation and accept a tolerance greater than an allowance. This is useful for grinding and similar operations. If this option is chosen, the allowance for that operation is highlighted in green to remind the user that the check has been overridden.

Working dimensions cannot be calculated until the process plan has been verified with regard to the design and stock removal allowance constraints. Resultant dimension offsets can be specified at this stage. The program will recompute resultant dimension upper and lower limits automatically and check these against the design specifications in case the user has made an error. Finally, a hard copy of the completed chart can be printed, if required.

17.2.2 Example

We now illustrate the procedure followed by CATCH with an example using the completed chart shown in Figures 17.1 and 17.2. We assume that the part sketch, face lines, and design specification dimensions and tolerances have been entered on the chart. In the following procedure, user inputs are highlighted in italics. All the other steps are performed automatically by CATCH.

1. Initialize the resultant dimensions by setting them equal to the design nominal dimensions.
2. *User enters the first operation* (*line 1 on the tolerance chart*). This is a special stock issue operation. It is not included in the rooted tree but is used to determine the size of the initial workpiece and to calculate the stock removal tolerance stackup for the following operation.
3. *User enters the next operation* (*line 3*). The datum surface is face 1 and the cut surface is face 25. It is the first operation in the rooted tree. CATCH calculates the reversed path, which is 0100–2501.
4. *User enters the next operation.* This is performed at the same work center as the preceding operation. The datum surface is face 25 and the cut surface is face 23. CATCH calculates the reversed path, which is 0100–2501–2301.
5. *Continue until all operations have been entered.* The reversed paths for all operations are shown in Table 17.1.
6. Determine the operations contributing to the design tolerances for all design specification dimensions and enter the Sundstrand index numbers in the appropriate columns. For example, consider the first design specification dimension (25–07). The operations contributing to its tolerance stackup lie in the path from 2503 and 0701 [i.e., operations 1402–2503 (line 34), 1402–1104 (line 41), 1104–0701 (line 44)]; refer to the rooted tree graph (Figure 17.6). Enter the index number of this design specification dimension (1) in index column 1 for each operation. The index numbers for all operations can be seen in Figure 17.2 and are found easily by inspection of the rooted tree graph.
7. Calculate the tolerance stackups for all design specification dimensions. For example, consider the first dimension. Its tolerance stackup is obtained by adding the operation tolerances for all operations with an index number of 1:

$$0.0015 + 0.0010 + 0.0005 = 0.0030 \text{ in.}$$

8. Compare the resultant dimensions against the design dimensions. For

example, consider the first design dimension (Figure 17.1):

lower limit of design dimension = 3.1810 − 0.0050 = 3.1760 in.

lower limit of resultant dimension = 3.1810 − 0.0030 = 3.1780 in.

upper limit of design dimension = 3.1810 + 0.0050 = 3.186in.

upper limit of resultant dimension = 3.1810 + 0.0030 = 3.184 in.

lower limit of resultant dimension ≧ lower limit of design dimension

upper limit of resultant dimension ≦ upper limit of design dimension

Therefore, these resultant dimensions are within limits. The other resultant dimensions are also within limits.

9. Calculate the stock removal tolerances. For example, consider face 14, which has four cuts made on it.
   - *Stackup for 1404* (*line 53*). Operations in path from 1403 to 1404 are 1104–1403 (line 46) and 1104–1404 (line 53).

     Stackup = 0.0005 + 0.0015 = 0.002 in.

   - *Stackup for 1403* (*line 46*). Operations in path from 1402 to 1403 are 1402–1104 (line 41) and 1104–1403 (line 46).

     Stackup = 0.0010 + 0.0005 = 0.0015 in.

   - *Stackup for 1402* (*line 15*). Operations in path 1401 to 1402 are 1401–0101 (line 8), 0101–1101 (line 13), and 1101–1402 (line 15).

     Stackup = 0.0015 + 0.0010 + 0.0010 = 0.0035 in.

   - *Stackup for 1401* (*line 6*). This is a first cut; the stackup equals the tolerance for the operation (0.001 in.).

10. Compare the stock tolerance stackups with the stock removal allowances. They are all satisfactory.

11. Calculate the working nominal dimensions working upward from the bottom of the chart. The positions of the surfaces are calculated first, then the nominal dimensions. We will illustrate for surfaces 7, 11, and 14. The finished surfaces 0701, 1104, and 1404 are set according to the dimensions of the finished workpiece. Their X-coordinate values, measured from the reference coordinate system, are 0.263, 0.924, and 1.314 in., respectively.
    - Surface 7 has only one cut, so no further calculation is required.

- The stock allowance for surface 11 must be subtracted to obtain the prior positions.
  - *For 1103:* stock allowance 1104 (line 41) = 0.003 in. Position of 1103 = 0.924 − 0.003 = 0.921 in.
  - *For 1102:* stock allowance 1103 (line 39) = 0.002 in. Position of 1102 = 0.921 − 0.002 = 0.919 in.
  - *For 1101:* stock allowance 1102 (line 18) = 0.007 in. Position of 1101 = 0.919 − 0.007 = 0.912 in.
- The stock allowances for surface 14 must be added to obtain the prior positions.
  - *For 1403:* stock allowance 1404 (line 53) = 0.001 in. Position of 1403 = 1.314 + 0.001 = 1.315 in.
  - *For 1402:* stock allowance 1403 (line 46) = 0.004 in. Position of 1402 = 1.315 + 0.004 = 1.319 in.
  - *For 1401:* stock allowance 1402 (line 15) = 0.012 in. Position of 1401 = 1.319 + 0.012 = 1.331 in.
- With these face positions known it is possible to determine the nominal dimensions between faces 7, 11, and 14 at any time during the manufacturing sequence.
  - Resultant nominal dimension 7–14 = $|0.263 - 1.314| = 1.051$ in.
  - Resultant nominal dimension 7–11 = $|0.263 - 0.924| = 0.661$ in.
  - Working nominal dimension for 1104–1404 (line 53) = $|0.924 - 1.314| = 0.390$ in.
  - Working nominal dimension for 1104–1403 (line 46) = $|0.924 - 1.315| = 0.391$ in.
  - Working nominal dimension for 1402–1104 (line 41) = $|1.319 - 0.924| = 0.395$ in.
  - Working nominal dimension for 1102–1402 (line 15) = $|0.912 - 1.319| = 0.407$ in.

12. *User adjusts any resultant dimensions, if necessary.* In our example some resultant dimensions (index numbers 6 and 10) have been offset. The first dimension, 7–6, has been offset 0.0005 in., by moving face 6. The second dimension, 14–13, has been offset 0.01 in. by moving face 13. CATCH automatically adjusts other resultant dimensions defined by these faces and the relevant working dimensions.

17.3   OPTIMIZATION OF TOLERANCES

Initial assignment of operation tolerances will often produce resultant tolerances that are smaller than the design tolerances. Based on the assumption that manufacturing costs decrease (somehow) with increase in operation tolerances, clearly it would be desirable to increase operation tolerances as much as possible subject to the design and stock allowance tolerance stackup constraints. Several computer-based approaches have been suggested for doing this.

Sack (1982) has developed an interesting approach to tolerance optimization. Maximum tolerances are initially assigned to each operation. If the design tolerances cannot be met, operation tolerances are reduced using a penalization function. Optimization is achieved when all the tolerance stackups are within the constraint limits. The CATC program developed by Ahluwalia and Karolin (1984) performs a very simple optimization. The user inputs initial tolerances. After the design and stock allowance tolerance stackup checks have proved satisfactory, the operation tolerances are increased starting from the last operation (last line) of the chart and working upward.

Some researchers (Irani et al., 1989; Ji, 1993; Tang et al., 1994) have used linear programming to optimize tolerances. The researchers differ in how they formulate the objective function and the constraint equations. A general formulation for linear programming is given below (modified from Ji, 1993).

The objective function is

$$\text{maximize} \sum_{j=1}^{m} w_j t_j$$

subject to:

(a) Process tolerance constraints

$$l_j \le t_j \le u_j \qquad j = 1,...,m$$

(b) Design tolerance constraints

$$\sum_{j \in DD_i} t_j \le dt_i \qquad i = 1,...,n$$

(c) Stock removal allowance constraints

$$sr_{jl} \le \sum_{k \in SR_j} t_k \le sr_{ju} \qquad j = 1,...,m$$

w_j = weight constant for operation j
t_j = tolerance assigned to operation j
l_j = lower tolerance bound for operation j
u_j = upper tolerance bound for operation j
dt_i = bilateral design tolerance for design specification dimension i
sr_{ju} sr_{jl} = lower bound of stock removal allowance for operation j
upper bound of stock removal allowance for operation j

t_k = tolerance assigned to operation k

m = total number of operations

n = total number of design specification dimensions

$j \in DD_i$ represents the set of operations that contribute to the tolerance stackup for design specification dimension i

$k \in SR_j$ represents the set of operations that contribute to the stock removal allowance tolerance stackup for operation j

The determination of the process sequence and the identification of the operations for the design tolerance and stock removal tolerance stackup constraint equations are performed externally and prior to optimization by the linear programming software. Operation tolerance assignment is performed within the software subject to the conditions specified by the objective function and constraint equations. Worst-case tolerancing is used because the constraint equations must be linear. The objective function may be linear or nonlinear. Nonlinear objective functions can be approximated by some linear programming software packages.

The working dimensions could be calculated within the linear programming software by adding further constraint equations. These additional equations are not required for tolerance assignment, and including them in the constraint equation set reduces significantly the efficiency of the linear programming software. Consequently, the researchers quoted above do not include the calculation of working dimensions in their optimization. They are easily calculated after optimization.

He and Gibson (1992) use a nonlinear optimizing technique to minimize the amount of scrap in a manufacturing sequence. There are three steps in the optimization: (1) the cost of scrap is minimized, (2) finished operation tolerances are maximized, and (3) roughing operation tolerances are maximized.

Based on our experience in implementing a practical tolerance charting system with Sundstrand Pacific Aerospace Pte. Ltd. (Singapore), it is our contention that existing computer-based methods for tolerance optimization in manufacturing are based on false assumptions. For instance:

• *Manufacturing sequence must exist prior to optimization.*

With the exception of Sack (1982) and Ahluwalia and Karolin (1984), all published optimization techniques require that the manufacturing sequence exists prior to optimization. The techniques only allow for change of tolerances on working dimensions, not for adjustment of the manufacturing sequence. That is, they are just optimization tools and not planning tools. In our experience the major gains in cost savings come from changing the sequence, not from tolerance adjustment. For example, removing a grinding operation and replacing it with a finishing operation on a lathe will yield large savings because of reduced setup and handling costs as well as the savings from using a cheaper operation. The generation of alternative process plans is the most critical feature of the whole process and is difficult to automate (Irani et al., 1989). In our opinion the current best approach is to use inter-active tolerance charting, which automates the planning procedure but also allows a process planner to input expert knowledge to generate alternative plans.

- *The design specification tolerances and nominal dimensions are taken as fixed constraints.*

In practice, nominal dimensions can be offset and design tolerance increases can be negotiated. In addition, by communicating the possibility of better than specified tolerances for a part it may be possible to gain savings from the manufacture of mating components in an assembly. Such savings may outweigh the savings from optimizing the tolerances on a single component. This is a main source of benefits from concurrent engineering. The objective function should be modified to include the design costs and benefits (Irani et al., 1989). This is another strong argument for using interactive tolerance charting; it is a tool for communication between design and manufacturing and permits easy and rapid adjustment of process plans in response to design changes.

- *A simple cost–tolerance relationship is assumed.*

In Chapter 2, Whybrew and Britton show that the relation between cost and tolerance is complex and cannot be reduced to a simple algebraic expression.

- *The stock removal allowances are set equal to the tolerance on the stock removal.*

The stock removal allowance for a process is not just a function of the tol-erance stackup of previous processes, it is a technological property of the process. The stock removal allowance is a recommended amount of material for removal for optimum performance of the process. It is normally related to depth of cut. If it is too large, the cutting forces will be too high and multiple roughing cuts may be required. If depth of cut is too small, the surface of the workpiece may deform locally without a chip being properly formed. In grinding this is known as *ploughing*. In turning and milling the surface is burnished by the tool instead of being cut. With work-hardening materials this will cause surface hardening and result in rapid tool wear and may also cause microcracks in the workpiece surface. Most modern cutting tools require a minimum depth of cut and feed for the chip breaking action to be effective, failure to satisfy these conditions will result in departure from the ideal surface finish.

· *Operation tolerances are assumed to be independent.*

With NC machining this may be a valid assumption, but in processes that involve gang milling, form tools, or combination tools, the dimensions and tolerances are not independent. Hence these processes cannot be included in the optimization.

For the foregoing reasons, we believe there is little benefit to be gained from tolerance optimization within a manufacturing sequence. However, manufacturing costs can be reduced significantly by comparing the costs of alternative process plans and selecting the lowest-cost plan. This requires the ability to generate and estimate the cost of alternative process plans rapidly (i.e., interactive process planning). Our rooted tree technique is an efficient and practical means for accomplishing this.

17.4   CONCLUSIONS

The importance of tolerancing in process planning has been discussed in Chapters 2 and 3. Currently, there is no commercial software available for performing tolerance charting, so many companies use manual charting methods, such as the one described in detail in Chapter 3. It is also possible to perform manual charting using the rooted tree algorithm described in this chapter, but the real power of the algorithm comes from its computerization. The implementation of the algorithm in CATCH is a proven practical tool satisfying the needs for tolerance analysis in precision manufacture.

Concurrent engineering requires the ability to analyze the manufacturing implications of any design changes quickly to assess their cost impacts. Because CATCH is interactive it can be used to generate a process plan and is not just a means of checking tolerance stacks in an existing plan. Thus it provides a means of developing and updating a process plan continuously in parallel with design development.

The rooted tree logic, which is the basis of CATCH, is capable of representing all feasible manufacturing sequences. All process planning data can be attached to the tree. Different rules can be developed for traversing the tree to provide application-specific requirements and to display the process plan from different viewpoints.

Given the power of our rooted tree technique the reader may well be wondering whether it can be used to analyze assembly tolerance stacks during design. We have given careful consideration to this and our conclusion is *no*! There are two reasons for this. First, in general, assembly trees are not rooted, they are cyclic. Second, tolerance charting is based on one-dimensional chains of dimensions which are defined by critical points. A designer must specify the critical points before dimension chains can be generated and checked. However, at the design stage, the designer does not know where these critical points are, nor how many are required to ensure that design functionality will be met. That is, the designer must envisage all possible combinations of tolerance variation in an assembly of parts to determine the worst cases for tolerance control. These worst cases are then used to dimension and tolerance the parts in the assembly using the critical point technique. Tolerance charting procedures are not appropriate either for generating the possible combinations or determining the worst cases because this requires three-dimensional analysis and synthesis (Roy et al., 1991). There are a number of three-dimensional computer-based techniques available for this purpose: for example, Wilhelm and Lu's (1992) technique or Chase's kinematic analysis method described in Chapter 5.

ACKNOWLEDGMENTS

The authors gratefully acknowledge the support given by Sundstrand Pacific Aerospace Pte. Ltd. (Singapore), especially Eric Tan Hung Heng, section manager of manufacturing engineering, and the process planners in manufacturing engineering. We have gained considerable practical working knowledge of tolerance charting and process planning from our association with Sundstrand.

We also wish to acknowledge Sermsuti-Anuwat Yongyooth, who collaborated with us at the University of Canterbury to develop the rooted tree graph technique. He also wrote the first computer program to use this technique for tolerance charting.

Thanks are also due to the following undergraduate students in the School of Mechanical and Production Engineering, Nanyang Technological University, for helping us to develop interactive tolerance charting prototype software: Kong Chien Ling, Teng Jen Pin, and Koh Kheng Yong, and to the technicians in the CAD/CAM laboratory for their technical assistance in the development of CATCH.

REFERENCES

Ahluwalia, R. S., and Karolin, A. V., 1984, CATC: A Computer Aided Tolerance Control System, *J. Manuf. Syst.,* Vol. 3, No. 2, pp. 153–160.

Bjorke, O., 1978, *Computer-Aided Tolerancing,* Tapir, Trondheim, Norway.

Britton, G. A., Whybrew, K., and Sermsuti-Anuwat, Y., 1992, A Manual Graph Theoretic Method for Teaching Tolerance Charting, *Int. J. Mech. Eng. Educ.* Vol. 20, No. 4, pp. 273–285.

Britton, G. A., Whybrew, K., and Tor, S. B., 1996, An Industrial Implementation of Computer-Aided Tolerance Charting, in *Int. J. Adv. Manuf. Technol.,* Vol. 12, No. 2, pp. 122–131.

Fainguelernt, D., Weill, R., and Bourdet, P., 1986, Computer Aided Tolerancing and Dimensioning in Process Planning, *Ann. CIRP,* Vol. 35, No. 1, pp. 381–386.

Gadzala, J. L., 1959, *Dimensional Control in Precision Manufacturing,* McGraw-Hill, New York.

He, J. R., and Gibson, P. R. 1992, Computer-Aided Geometrical Dimensioning and Tolerancing for Process-Operation Planning and Quality Control, *Int. J. Adv. Manuf. Technol.,* Vol. 7, pp. 11–20.

Irani, S. A., Mittal., R. O., and Lehtihet, E. A., 1989, Tolerance Chart Optimization, *Int. J. Prod. Res.,* Vol. 27, No. 9, pp. 1531–1552.

Ji, P., 1993, A Linear Programming Model for Tolerance Assignment in a Tolerance Chart, *Int. J. Prod. Res.,* Vol. 31, No. 3, pp. 739–751.

Koh, K. Y., 1993, *Development of an Interactive Tolerance Charting Software Package Cut Sequence,* Final Year Project Report 1992/1993, School of Mechanical and Production Engineering, Nanyang Technological University, Singapore.

Kong, C. L., 1992, *Development of Computerised Tolerance Charting System,* Final Year Project Report 1991/1992, School of Mechanical and Production Engineering, Nanyang Technological University, Singapore.

Li, J. K., and Zhang, C., 1989, Operational Dimensions and Tolerances Calculation in CAPP Systems for Precision Manufacturing, *Ann. CIRP,* Vol. 38, No. 1, pp. 403–406.

Ping, J., 1993, A Tree Approach for Tolerance Charting, *Int. J. Prod. Res.,* Vol. 31, No. 5, pp. 1023–1033.

Roy, U., Liu, C. R., and Woo, T. C., 1991, Review of Dimensioning and Tolerancing: Representation and Processing, *Comput.-Aid. Des.,* Vol. 23, No. 7, pp. 466–483.

Sack, C. F., Jr., 1982, Computer Managed Process Planning: A Bridge Between CAD and CAM, *Proceedings of AUTOFACT 4,* Philadelphia, PA, November–December, pp. 7.15–7.31.

Sermsuti-Anuwat, Y., 1992, *Computer-Aided Process Planning and Fixture Design* (CAPPFD), Ph.D. thesis, University of Canterbury, Christchurch, New Zealand.

Sermsuti-Anuwat, Y., Whybrew, K., and Britton, G. A., 1991, Some Recent Developments in Tolerance Charting and Its Role in CIM, *Proceedings of the First International Conference on Computer Integrated Manufacturing,* Singapore, pp. 285–288.

Sermsuti-Anuwat, Y., Whybrew, K., and McCallion, H., 1995, CAPPFD: A Tolerance Based Feature Sequencing CAPP System, *J. Syst. Eng.,* Vol. 5, No. 1, pp. 2–15.

Tang, G. R., Kung, R., and Chen, J. Y., 1994, Optimal Allocation of Process Tolerances and Stock Removals, *Int. J. Prod. Res.,* Vol. 32, No. 1, pp. 23–35.

Tang, G.-Y., Fuh, Y. M., and Kung, R., 1993, A List Approach to Tolerance Charting, *Comput. Ind.,* Vol. 22, pp. 291–302.

Tang, X., and Davies, B. J., 1988, Computer Aided Dimensional Planning, *Int. J. Prod. Res.,* Vol. 26, No. 2, pp. 283–297.

Teng, J. P., 1993, *Development of Interactive Tolerance Charting Software Package,* Final Year Project Report 1992/1993, School of Mechanical and Production Engineering, Nanyang Technological University, Singapore.

Wade, O. R., 1967, *Tolerance Control in Design and Manufacturing,* Industrial Press, New York.

Wade, O. R., 1983, Tolerance Control, Chapter 2 in *Tool and Manufacturing Engineers Handbook,* Vol. 1, T. J. Drozda and C. Wicks (eds.), Society of Manufacturing Engineers, Dearborn, MI.

Whybrew, K., Britton, G. A., Robinson, D. F., and Sermsuti-Anuwat, Y., 1990, A Graph-Theoretic Approach to Tolerance Charting, *Int. J. Adv. Manuf. Technol.,* Vol. 5, pp. 175–183.

Wilhelm, R. G., and Lu, S. C. Y., 1992, *Computer Methods for Tolerance Design,* World Scientific, Singapore.

Zhang, H. C., and Huq, M. E., 1992, Tolerancing Techniques: The State-of-the-Art, *Int. J. Prod. Res.,* Vol. 30, No. 9, pp. 2111–2135.

18

SOLID MODEL–BASED REPRESENTATION AND ASSESSMENT OF GEOMETRIC TOLERANCES IN CAD/CAM SYSTEMS

UTPAL ROY, XUZENG ZHANG, and YING-CHE FANG

Syracuse University
Syracuse, New York

18.1 INTRODUCTION

It is an established fact that all manufactured parts exhibit a variety of imperfections due to the uncertainties inherently present in any manufacturing process. Geometric tolerances are assigned in an attempt to control these imperfections. The objective of providing tolerancing information is to assure that as long as a part is manufactured within the allowable limits of the tolerance specification, the final product will be acceptable (i.e., the final product can be assembled and will meet the necessary functional and other geometric requirements). To achieve a desired part quality, the geometric characteristics of manufactured parts must therefore be evaluated against the specified design tolerances. Due to lack of proper definitions and guidelines, ambiguities arise from conflicting interpretations of the standard (ASME, 1982) on dimensioning and tolerancing, making the assessment criteria for different geometric characteristics very difficult. In this chapter, we address two major issues: (1) rigorous definitions of geometric dimensions and tol-

Advanced Tolerancing Techniques, Edited by Hong-Chao Zhang
ISBN 0-471-14594-7 © 1997 John Wiley & Sons, Inc.

erances, and (2) design of robust, postinspection, data analysis algorithms. It discusses how to define and establish geometric tolerancing, including the datum specification. It also outlines a computational geometry-based tolerance evaluation technique.

The starting point in the successful development of an advanced tolerance assessment module in the CAD/CAM system, aimed at discrete-parts manufacturing, is a complete and accessible database of computer-aided design (CAD) that helps provide an unambiguous definition of parts nominal shapes, and other technological and functional information, such as integrated dimensioning and tolerancing. Any engineering component can be viewed as a set of specifications regarding its geometrical and functional characteristics to reflect the design intent. The CAD model of the part must provide the dimensioning and tolerancing information for functional use, coupled with its usual geometric model. Solid modelers using the constructive solid geometry (CSG) boundary representation (B-Rep) technique has proved to be the most rational approach for the purpose (Roy and Liu, 1987). In this chapter a new tolerance representation scheme that exploits the advantages of CSG/B-Rep hierarchy and OOP (object-oriented programing) concept is presented.

18.2 REVIEW OF RELATED WORKS

18.2.1 Tolerance Representation

Based on the theory of *variational class,* Requicha (1977, 1984) was the first to advocate the representation of tolerances as properties or attributes of an object's features in a solid model and implemented the representation scheme in a CSG-based modeler, PADL-2 (Requicha and Chan, 1986). The tolerance (variational) information has been associated with the solid model by means of a graph, called a *VGraph* or *variational graph.* The VGraph has been linked with the nominal representation of PADL-2 via nominal faces (NFaces of an object) which are associated with the faces of primitives in the object's CSG representation.

Jayaraman and Srinivasan (1989a,b) have examined the issues of representing the geometric tolerances in solid models from the perspective of functional requirements related to the geometry of mechanical parts. Their research is concerned primarily with the positioning of parts with respect to each other in an assembly and with maintaining material bulk in critical portions of parts. They develop specific *virtual boundary requirements* (VBRs) to reflect the required functional conditions of the assembly and then discuss the theoretical basis of the interpretation of those virtual boundary requirements with the help of the theory of *solid model-based offsetting,* proposed by Rossignac and Requicha (1986).

In the GEOTOL system, Turner (1987) has attempted to associate the tolerancing information with the boundary representation of the part. All variations are applied to the part faces (currently limited to planar and cylindrical faces) of the nominal model. A prototype representational module has been built to provide IBM's geometric design processor (GDP) solid modeling system with a generalized CSG (GCSG) architecture.

Kimura et al. (1987) consider dimensioning and tolerances as geometric constraints and use first-order predicate logic to represent tolerances in their system.

Shah and Miller (1989, 1990) propose an object-oriented system for tolerance representation. A similar attempt has been made by Jacobsohn et al. (1990) to incorporate tolerances into their design system in an object-oriented programing environment.

Roy et al. (1989a,b) have reported a hybrid CSG/B-Rep representation scheme for tolerance representation. To exploit the benefits of both CSG and B-Rep systems, they link two systems and store information at different levels of the object creation process. On the CSG level, a feature tree (FT) is used to keep the object–feature relationship in a binary tree while a spatial relationship graph (SRG) is used to record the relationships of location and orientation of a feature with respect to other features at each level of the tree. A structured face–adjacency graph representation (SFAG) coupled with predefined datum reference frames (DRF) is also maintained at each level of this binary tree. A tolerance data structure (Roy et al., 1989b) is attached to both CSG and B-Rep data models via a global reference face list (RFL) to handle the dimensioning and tolerancing (D&T) information (which needs to be attached with either faces, edges, or vertices). A framelike data structure is used to store not only the D&T information, but also the technological information that is necessary in downstream manufacturing processes. Gossard et al. (1988) have used a similar hybrid CSG/B-Rep object modeling scheme; however, their system allows the representation of size tolerances only.

In most of the earlier works, attempts have been made to represent tolerance information in either a CSG-based or a B-Rep-based modeler. Due to the limitation of CAD data structure, most efforts are limited to a restricted domain of tolerance representation. Other works based on hybrid CSG/B-Rep schemes are memory intensive and are not suitable for representing complex objects. The schemes need to store a huge amount of data to maintain two databases simultaneously.

18.2.2 Tolerance Assessment

For assessment of geometrical tolerances, it is evident from the literature (Feng and Hopp, 1991; Shunmugam, 1986, 1987a,b; Lai and Wang, 1988) that several research attempts have been made to develop proper analytical techniques for tolerance assessment problem. Among these, the least squares techniques (Shunmugam, 1986, 1987b) are used most, but the assessment results obtained are not always the best. Some researchers therefore tried to derive the best values using exhaustive search techniques. The main problem with these methods is that the search techniques are computationally very expensive. It is therefore proposed in this chapter to use the computational geometry-based approach to improve the assessment methodologies. Lai and Wang (1988) have used this computational geometric technique in evaluating the straightness and roundness. Computational-based techniques have also been used by Traband et al. (1989) in the evaluation of straightness and flatness tolerance, although their application domain does not cover other types of geometric tolerances.

18.3   TOLERANCE REPRESENTATION IN CAD DATABASE

There are two types of tolerances, size and geometrical, that are normally used in product modeling. The tolerance representation is feature based (Roy and Liu, 1988) and it requires appropriate "geometric features" to be identified from the object's CAD data model for its application. Tolerance information is then attached to those features as attributes of the geometry of the object, as is done in the case of conventional engineering drawing.

A feature-based representation scheme, based on the hybrid CSG/B-Rep (constructive solid geometry/boundary representation) hierarchy and OOP concept, has been used in this chapter for object modeling to support the tolerancing information (Roy and Fang, 1995). The hierarchical organization of the features provides a multilevel representation of the part feature relation in a CSG-like data model. Tolerances (e.g., size, form, etc.) that needs higher-level features (i.e., prototype features) are represented in this data model. For other tolerances that need basic topological entities (e.g., vertices, edges, and faces) to be identified, the system provides a lower-level entity retrieval (LLER) mechanism to retrieve appropriate surfaces, edges, or vertices directly from the CSG model. In this way, tolerances are attached to different levels of geometrical entities according to specific needs and according to its required interactions with other features.

To represent all five types of geometric tolerances (e.g., form, orientation, profile, runout, and location tolerances), three types of features are necessary:

1. *Individual Feature:* a single surface, edge, or size feature that relates to a perfect geometric counterpart of itself as the desired form
2. *Related Feature:* a single surface or element feature that relates to a datum or datums
3. *Undetermined Feature:* a single surface or edge feature (whose perfect geometric profile is described) that may or may not relate to a datum or datums

Table 18.1 lists different types and characteristics of geometric tolerances and related features to which the tolerances are attached.

TABLE 18.1. Different Types and Characteristics of Geometric Tolerances (ASME, 1982)

	Type of Tolerance	Characteristic
For Individual Features	Form	Straightness
		Flatness
		Cylindricity
For Related Features	Orientation	Angularity
		Perpendicularity
		Parallelism
	Location	Position
		Concentricity
	Runout	Circular Runout
		Total Runout
For Undetermined Features	Profile	Profile of a line
		Profile of a Surface

As shown in the table, form tolerances are applied to individual features with no relationship with other features or datums. These can be represented as properties of a surface without considering the interaction with others. The orientation, runout, and location tolerances refer to a datum, a set of datums, or other features. In addition, these tolerances may be applied to a single surface, an edge, or a size feature, so its data structure is more complicated than that of the form tolerance. A datum reference frame is required as a default datum in case of related features for representing orientation, runout, and location tolerances.

Figure 18.1 shows a sample drawing of a typical part showing all the necessary geometric tolerances. Figure 18.2 demonstrated the data structure required to store the tolerancing information for the part. As shown in Figure 18.1, the part is consisted of two types of predefined features, base and hole. A reference frame acts as the datums of individual features. Users can input a set of datums as a default datum frame. For example, a user may specify surface A as the default primary datum, surface B as the default secondary datum, and surface C as the default tertiary datum. The position tolerance of the center hole in Figure 18.1 is then specified by using the default datum frame as its reference frame, while the other four small holes are specified by using the axis of the center hole as their secondary datum.

The size tolerances are related to nominal dimensions of the features only; therefore, it is stored at the feature (i.e., hole, base) level. The positional tolerances for the holes are specified against the datum surfaces A, B, and C (Figure 18.1). The position tolerance (location tolerance class) is then inherited by the holes as the superclass along with the required tolerance attributes.

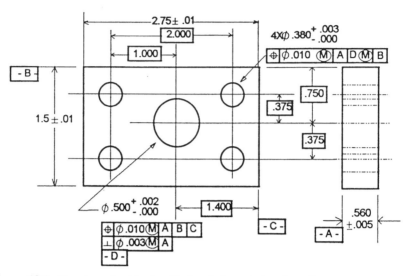

Figure 18.1. Drawing of a block-with-five-hole part showing all necessary geometric tolerances. (From Roy and Fang, 1987.)

In addition to inheriting attributes from the class of location tolerance, the center hole also inherits attributes of the orientation tolerance class for specifying the perpendicularity of its axis with respect to datums.

For profile tolerances, two types of profiles are normally considered: One is the profile of a surface and another is the profile of an edge. According to the design requirements, they may or may not relate to datums. The data structure of a profile tolerance can be decided according to whether it is relative to datums or not, and the methods presented above can be applied. The main idea behind this representational scheme is to capture knowledge about the topological and geometrical relationships among different kinds of features and to incorporate (both geometric and nongeometric) object attributes. Tolerance information is also attached to those features as attributes. The method of tolerance attachment is based on applying tolerances as the design constraints to the solid objects.

The CAD database thus created provides information regarding the design intent and part inspection requirements. Geometric dimensioning and tolerancing indicate how the part should be inspected and gaged to protect the design intent. Since functional data such as dimension text, feature control frame, and datum reference frame are essential for interpreting the part feature and generating inspection sequence, an appropriate reasoning module is provided to extract the required dimensioning and tolerancing information as required.

To automate the inspection of parts, it is then necessary to analyze the measurement data captured by coordinate measuring machines (CMMs) in order to detect out-of-tolerance conditions. CMMs were originally introduced

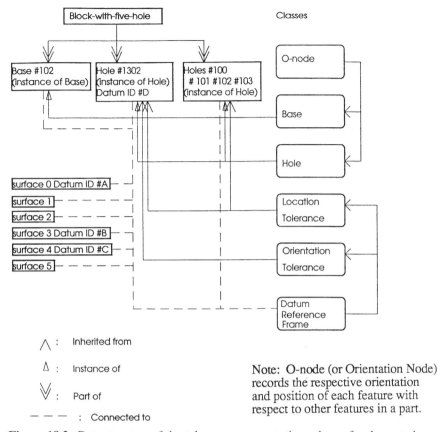

Figure 18.2. Data structure of the tolerance representation scheme for the part shown in Figure 18.1. *Note:* The O-node (or orientation node) records the respective orientation and position of each feature with respect to other features in a part. (From Roy and Fang, 1987.)

to industry as a flexible means of performing fast and accurate dimensional measurements on the manufactured objects. With the advancement of CMM technology, it is now possible to inspect complex parts containing many different features and tight tolerances. A high-level integration between CMM and CAD systems is the next objective toward the development of an intelligent manufacturing system. This dictates the needs for establishing procedures and algorithms for determining the geometric tolerances from the measured three-dimensional coordinates and for a systematic comparison of geometric variations of measured features with their specified geometric tolerances. We next discuss the assessment criteria for datums and other geometric tolerances to establish an unambiguous interpretation and a computational methodology to verify the tolerance per specification.

18.4   ASSESSMENT OF TOLERANCES

The metrology of CMMs is known as the coordinate measuring technique because they utilize the coordinates of points sampled from a part by a CMM to achieve the inspection results. The algorithms of the software used by CMMs are largely dependent on the assessment criteria of the manufactured features.

18.4.1   Establishment of Datums

The establishment of datums from the features of manufactured parts is most important in the assessment of tolerances because datums are used as the basis of establishing the geometric relationship between related features of a part. The ANSI standard (ASME, 1982) defines a datum as a theoretically exact point, axis, or plane derived from the true geometric counterpart of a specified datum feature (a manufactured part). Since the imperfectness of a manufactured part may result in different datum systems in reality, it is very important to define an explicit and unambiguous methodology for establishing a unique datum system. In the spirit of the standard's intent for datum specification, three important characteristics of a datum need to be specified when a datum is established from the manufactured features:

1. A datum is simulated as a basic feature, established from a nonperfect manufactured feature.
2. A datum established from a manufactured feature must be the *closest mating ideal feature* of the manufactured feature. This means that a datum for a manufactured (datum) feature is defined only from the effective points (the subset of the measured data points which possibly come in contact with another mating feature during manufacture, assembly, or inspection) of the manufactured feature rather than from *all* measured points of the feature.
3. A datum established from a manufactured feature is unique.

By definition, the datum is the reference for the measurements. If the datum is *not* unique, the qualification of a measured feature associated with the datum would become ambiguous. Therefore, to be established, a datum must have these three characteristics. We next discuss in detail how to establish primary, secondary, and tertiary datum planes from the measured data points. The procedure of establishing a datum frame from a planar surface is discussed as an example.

Establishment of the Primary Datum

1. Find the effective points of the planar surface from the measured data set. Effective points of the planar surface are those extreme points of a

convex hull of the surface, which are at the convex portions of the surface (Figure 18.3*a*).

2. Calculate the least squares plane from the set of effective points (Figure 18.3*b*).

3. Translate the least squares plane away from the surface such that it passes through the *farthest* effective points only (Figure 18.3*c*). Note that the least squares plane simulates an ideal closest plane that will come in contact with the manufactured surface during any assembly operation. The least squares plane at this condition yields the required datum surface from further manipulation.

Establishment of the Secondary Datum. Like a primary datum, a secondary datum must satisfy the three common characteristics of the datum along with the constraint of perpendicularity with the primary datum. This extra constraint dictates a slightly different technique for determining the effective points for a secondary datum.

1. Project all the data points required for establishing the secondary datum onto a plane parallel to the primary datum.

2. Determine the effective points of the projected planar point set as described in step 1 of the procedure for the establishment of a primary datum.

3. Calculate the least squares line from the set of effective points above by the least squares method (Figure 18.4*a*).

4. Translate the line obtained to the farthest point (a point of the effective point set farthest from the least squares line in Figure 18.4*b*).

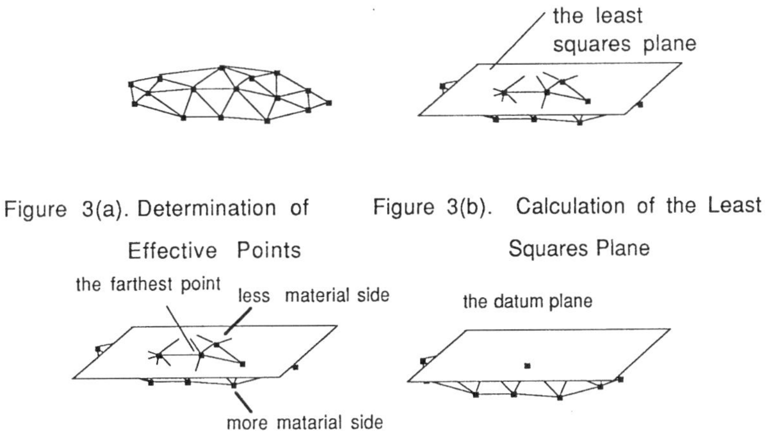

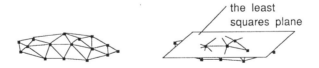

Figure 3(a). Determination of Effective Points

Figure 3(b). Calculation of the Least Squares Plane

Figure 18.3. Establishment of the primary datum.

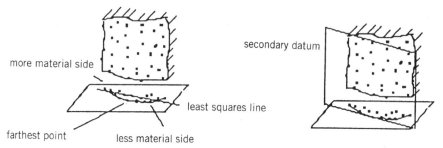

Figure 18.4. Establishment of the secondary datum.

5. Establish a plane passing through the line, obtained in step 4, and perpendicular to the primary datum surface (Figure 18.4). This plane is the desired secondary datum established from a planar surface.

Establishment of the Tertiary Datum. A tertiary plane requires that the three common characteristics of a datum be fulfilled and it needs to be normal to both a primary and a secondary datum. Suppose that we have a data set, S, to establish the tertiary datum plane. The following steps are required:

1. Project S twice on two planes, first onto a plane parallel to the primary datum plane, and obtain a set of S' of the projected points. Next, project the set S' onto another plane parallel to the secondary datum, and obtain the set S'' of the projected points. Notice that all points of set S'' lie on a line that is parallel to the primary datum plane.
2. Generate a plane perpendicular to the line and passing through an endpoint of S'' at the nonmaterial side. The plane obtained is the tertiary datum of the planar surface (Figure 18.5).

Other than the planar surfaces, abstract geometric entities (e.g., the median plane of a pair of planar surfaces, an axis of a cylinder, etc.) may also be used as the primary, secondary, or tertiary datums. The establishment of those datums has been discussed in detail by Zhang and Roy (1993).

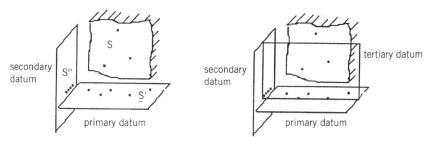

Figure 18.5. Establishment of the tertiary datum from a planar surface.

18.4.2 Geometric Tolerance Assessments

In this section we focus on the study of criteria for assessing geometric characteristics of manufactured parts. First we discuss size with respect to the manufactured features and the assessment criteria. A general definition for the sizes of manufactured features is developed based on the assembly function. We also propose the concepts of size feature and nonsize feature and then establish the procedures for calculating the size features. For assessing geometric characteristics other than the size characteristic, the concepts of tolerance zones (TZS) and minimum zones (MZs) as advocated by the standards (ISO, 1972) have been discussed, and the methodology for determining the parameters of those TZS and MZs from the design tolerance specification and the manufactured parts is outlined.

Size Tolerance. The ANSI standard (ASME, 1982) defines the size of a manufactured feature as a *measured size.* It is very ambiguous from an inspection standpoint because a nonideal, nonperfect manufactured feature does not have a fixed geometry that can be clearly measured. There must be a unique definition and measurement procedure of size assessment of a feature to ensure the mating condition in assembly. The size is the geometric characteristic of a manufactured feature, which in conjunction with the positional variation determines the assembly situation of the mating parts. The following two properties are important in establishing the size of a feature:

1. The size of a manufactured feature is a unique value.
2. The size of an actual feature is a geometric characteristic independent of other characteristics of the feature (e.g., form, orientation, and position characteristics).

The size of a manufactured feature is a geometric characteristic of a feature such that it determines the *clearance* in an assembly when the feature is mated with another feature of another part. It should be noted that the size of a manufactured feature can be determined by the size of a "perfect" feature that is the closest mating feature of the imperfectly manufactured feature. Not all manufactured features can always produce such unique-valued sizes from the closest mating features as is in the case with an imperfect cylinder or hole. For example, let a manufactured feature be two imperfect holes; the size (distance) between the imperfect holes may not be unique. When the two holes are oblong (Figure 18.6), not only is the closest mating feature (the pair of the largest inscribed cylinders) of the holes not unique, but the size (distance) between the holes is not unique. This size (distance) of the manufactured feature (the pair of oblong holes) does not satisfy the two properties listed earlier. We therefore define a manufactured feature that may not yield a unique value of size as a *nonsize feature,* and a manufactured feature that always yields a unique value of size as a *size feature.* We suggest that a nonsize feature should not be controlled by a size tolerance. By the definition,

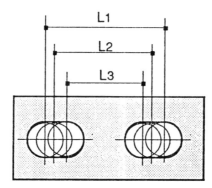

Figure 18.6. Distances between two oblong holes.

an imperfect cylinder, hole, boss, slot, spherical ball, and spherical hole are size features. On the other hand, a pair of parallel imperfect holes, a parallel imperfect plane and a hole, and a group of mounting holes with a fixed distance apart are nonsize features.

From the discussion above, we can now define the criteria for calculating sizes of typical size features.

The diameter of an external cylindrical surface (a cylinder) is the diameter (D) of the smallest circumscribed cylinder of the cylindrical surface (Figure 18.7). The diameter of an internal cylindrical surface (a hole) is the diameter (D) of the largest inscribed cylinder of the cylindrical surface (Figure 18.8).

The width of an external planar feature (a boss) is the distance (S) between two parallel planes that contact the planar surfaces from the nonmaterial sides with the least separation (Figure 18.9). The width of an internal planar feature (a slot) is the distance (S) between two parallel planes that contact the planar surfaces from the nonmaterial sides with the greatest separation (Figure 18.10). Like external and internal cylindrical surfaces, the sizes of external and internal spherical surfaces are the diameters of the smallest circumscribed ball and the largest inscribed ball of the spherical surfaces, respectively.

The proposed definitions of size and of size and nonsize features have two important characteristics. First, they avoid ambiguities regarding the size of

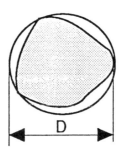

Figure 18.7. Diameter of a cylinder.

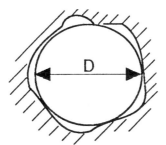

Figure 18.8. Diameter of a hole.

manufactured features as stated in the standard, which cause confusion in engineering practices, and they provide a guideline for developing the appropriate algorithms for CMM for assessing the size of manufactured features. Second, they give design engineers a criterion for identifying which features of manufactured parts cannot be controlled by a size tolerance.

Other Geometric Tolerances. For assessing other (form, orientation and location) geometric characteristics of manufactured parts, the standards (ASME, 1982; ISO, 1972) have introduced the concepts of tolerance zone (TZ) and minimum zone (MZ). The *tolerance zone* is used to describe the permissible range of variation of a geometric characteristic of manufactured parts for design purposes. On the other hand, the *minimum zone* defines a range of a deviation of the actual measured feature from its nominal feature, and the deviation needs to be checked with the prescribed design tolerance specification for the acceptability of the manufactured part.

A measured feature is said to be *acceptable* if it can be contained within its TZ or if it can be shown that the value of a MZ parameter (related to the specified design tolerance) of the manufactured feature is less than or equal to the tolerance specified. The inspection with TZs is carried out in the *inclusion principle,* while the inspection with MZs is the focus of the *minimum zone principle.* To assess a manufactured feature, either a TZ or MZ has to be established.

Our study (Roy and Zhang, 1993) reveals that the establishment of TZs and MZs (i.e., the determination of different parameters to establish the zones)

Figure 18.9. Width of a boss.

Figure 18.10. Width of a slot.

of a manufactured feature not only requires the design specification of the part tolerances (i.e., basic dimension, tolerance value, and datum specification if required, as stored in the CAD database) but also the actual shape, size, orientation, and location of the manufactured feature. For example, to establish a roundness TZ for an actual cylindrical feature, a pair of concentric circles with a certain radial separation (specified as the roundness tolerance value in the design) needs to be constructed. Besides the radial separation, two other parameters (the diameter of the inner circle or outer circle, and its center point), required to establish the pair of circles, are not supplied by the design specification; instead, they are determined from the actual feature. But the establishment of MZ for the roundness requires that we find *all* three parameters (i.e., diameter of either the inner or outer circle, its center point, and the separation between the circles) from the actual manufactured feature. Note that the design specification of the roundness dictates *only* the shape of the MZ (as a pair of concentric circles). The methods of establishing TZs and MZs (i.e., determining the parameters of shape, size, position, and orientation of TZs and MZs for the assessment of different geometric characteristics of the manufactured parts) have been discussed in detail (Roy and Zhang, 1993). The minimum zone principle has been shown to be easily implementable in CMMs.

18.5 COMPUTATIONAL TECHNIQUES FOR TOLERANCE EVALUATIONS: CASE STUDY FOR ROUNDNESS ERROR EVALUATION

After establishing the assessment criteria for different geometric tolerances, one needs to devise proper computational techniques to realize and implement these criteria in order to assess the manufactured part. In this section we study a faster algorithm for establishing the minimum zone (a pair of concentric circles with minimum radial separation) for assessing the roundness error (Roy and Zhang, 1992). The properties of convex hulls and Voronoi diagrams, the computational geometry-based techniques, have been exploited to establish the ideal pair of circles. The principles and procedures of this method are detailed in the work by Roy and Zhang (1992). The input to the system

TABLE 18.2. Data Points

					n			
	1	2	3	4	5	6	7	8
q	0°	45°	90°	135°	180°	225°	270°	315°
r	4.0	4.0	3.0	5.0	2.0	3.0	1.0	2.0
x_i	4.0	2.8284	0.0	−3.5355	−2.0	−2.1213	0.0	1.4142
y_i	0.0	2.8284	3.0	3.5355	0.0	−2.1213	−1.0	−1.4142

Source: Roy and Zhang (1992).

is a point set, S, obtained from the measured workpiece profile (on a cross-sectional plane that is perpendicular to the rotational axis, Table 18.2). Following is a brief discussion of the procedure.

1. Using the Graham scan technique, construct the convex hull, CH(S) from the simple polygon obtained from the measured data set, S.
2. Generate the farthest Voronoi diagram, FVor(S), from the established convex hull CH(S) above (Figure 18.11a). Also construct the nearest Voronoi diagram, NVor(S), from the point set, S (Figure 18.11b).
3. Establish the pair of concentric circles with minimum radial separation for each of the following three cases:

   Case 1. Compute the intersection of FVor(S) and NVor(S). Setting each of the intersecting points the center, draw a pair of concentric circles such that each circle passes through either two outermost or two innermost points (of the set S).

   Case II. For every Voronoi vertex of the NVor(S) diagram, draw an inner circle that passes through at least three points of S. The corresponding outer circle of the pair is determined by passing it through a point of S, which is farthest from the center of the inner circle. Thus for each point of the NVor(S) vertex set, a pair of concentric circles is generated.

   Case III. Like case II, instead of the NVor(S) diagram, we consider the FVor(S) diagram. Each vertex of FVor(S) is used to draw an outer

Figure 18.11. Nearest and farthest Voronoi diagrams. (From Roy and Zhang, 1992.)

circle that always passes through three outermost points of S. The corresponding inner circle is generated with its center on the same vertex point and passing through the point of S nearest the center. This case also results in a set of pairs of concentric circles centered at the FVor(S) vertices.

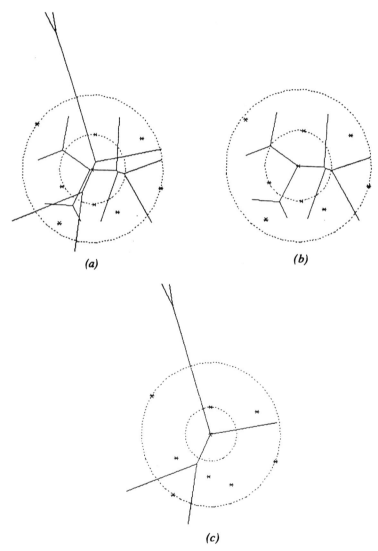

(a)

(b)

(c)

Figure 18.12. Calculation of radial separation in different cases: (a) 2.22433 for case I, (b) 2.23505 for case II, and (c) 2.6095 for case III. Therefore, the roundness error is the minimum one of {2.2433, 2.3505, 2.6095} = 2.2433. (From Roy and Zhang, 1992.)

4. Comparing the results from the three cases cited above, the roundness error is determined as the minimum radial separation of the concentric circles.

To illustrate the methodology, a sorted set of data points (Table 18.2) has been considered to establish the farthest and nearest Voronoi diagrams (Figure 18.11). The minimum separations of concentric circles are shown in Figure 18.12.

18.6   CONCLUSIONS

To develop a complete automated inspection system to determine conformity of a manufactured part to its design specification as stored in the CAD data model, a unique and unambiguous interpretation of geometric tolerances is an absolute requirement. The related issues have been discussed here and an outline of a uniform computational methodology (based on computational geometry techniques) has been established for assessing geometric character-istics of manufactured parts. It is valuable in closing the design–manufacturing–inspection loop in a modern intelligent manufacturing system. The extracted information on the shape errors of the manufactured parts can be fed directly back to the system control loop to improve the manufacturing process to produce precision parts.

REFERENCES

ASME, 1982, *Dimensioning and Tolerancing,* ANSI Y14.5M-1982, American Society for Mechanical Engineers, New York.

Feng, S. C., and Hopp, T. H., 1991, *A Review of Current Geometric Tolerancing Theories and Inspection Data Analysis Algorithms,* Technical Report NISTIR-4509, Manufacturing Engineering Laboratory, National Institute of Standards and Technology, Gaithersburg, MD.

Gossard, D. C., Zuffante, R. P., and Sakurai, H., 1988, Representing Dimensions, Tolerances, and Features in MCAE Systems, *IEEE Comput. Graph. Appl.,* Vol. 8, No. 2, March, pp. 51–59.

ISO, 1972, *Technical Drawing: Tolerances of Form and of Position,* ISO/DIS 1101, International Standardization Organization, Paris.

Jacobsohn, J. F., Radack, G. M., and Merat, F. L., 1990, Incorporating Knowledge of Geometric Dimensioning and Tolerancing into a Feature-Based CAD System, *Proceedings of the 2nd Rensselaer International Conference on CIM,* Rensselaer Polytechnic Institute, Troy, NY, May, pp. 152–159.

Jayaraman, R., and Srinivasan, V., 1989a, Geometric Tolerancing; I: Virtual Boundary Requirements, *IBM J. Res. Dev.,* Vol. 33; Manufacturing Research Department, IBM

Research Division, IBM T. J. Watson Research Center, Yorktown Heights, NY, August 1988.

Jayaraman, R., and Srinivasan, V., 1989b, Geometric Tolerancing, II: Conditional Tolerances, *IBM J. Res. Dev.,* Vol. 33; Manufacturing Research Department, IBM Research Division, IBM T. J. Watson Research Center, Yorktown Heights, NY, August 1988.

Kimura, F., Suzuki, H., and Wingrad, L., 1987, A Uniform Approach to Dimensioning and Tolerancing in Product Modeling, in *Computer Applications in Production and Engineering,* K. Bo, L. Estensen, P. Falster, and E. A. Warman (eds.), North-Holland, New York, pp. 165–178.

Lai, D., and Wang, J., 1988, A Computational Geometry Approach to Geometric Tolerancing, *Proceedings of the 16th North American Manufacturing Research Conference,* Urbana, Il, May 24–27, 1988, pp. 376–379.

Requicha, A. A. G., 1977, *Toward a Theory of Geometric Tolerancing,* Technical Memo 40, Production Automation Project, University of Rochester, Rochester, NY, March.

Requicha, A. A. G., 1984, Representation of Tolerances in Solid Modelling: Issues and Alternative Approaches, in *Solid Modeling by Computers: From Theory to Applications,* M. S. Pickett and J. W. Boyse (eds.), Plenum Press, New York, pp. 3–22.

Requicha, A. A. G., and Chan, S. C., 1986, Representation of Geometric Features, Tolerances and Attributes in Solid Modellers Based on Constructive Geometry, *IEEE J. Robot. Autom.,* Vol. RA-2, No. 3, September, pp. 156–166.

Rossignac, J. R., and Requicha, A. A. G., 1986, Offsetting Operations in Solid Modelling, *Comput.-Aid. Geom. Des.,* Vol. 3, pp. 129–148.

Roy, U., and Fang, Y., 1996, Tolerance Representation Scheme for a 3-Dimensional Product in Object Oriented Programing Environment, *IIE Transactions,* Vol. 28, pp. 809–819.

Roy, U., and Liu, C. R., 1987, Feature-Based Representational Scheme of a Solid Modeler for Providing Dimensioning and Tolerancing Information, *Proceedings of the Conference on Manufacturing Science and Technology,* MIT, Cambridge, MA. June, 1987. pp. 10–15. Also published in *Robot. Comput.-Integrat. Manuf.,* Vol. 4, No. 3–4, pp. 335–345.

Roy, U., and Zhang, X., 1992, Establishment of a Pair of Concentric Circles with the Minimum Radial Separation for Accessing Roundness Error, *Comput.-Aid. Des.,* Vol. 24, No. 3, March.

Roy, U., and Zhang, X., 1993, Criteria for Assessing Geometric Tolerances in Manufactured Part, *Proceedings of the 2nd International Conference and Exhibition on Computer Integrated Manufacturing, ICCIM'93,* Singapore, September 6–10.

Roy, U., Pollard, M. D., Mantooth, K., and Liu, C. R., 1989a, Tolerance Representation Scheme in Solid Model, Part I, *Proceedings of the 1989 Design Technical Conferences,* Montreal, Quebec, Canada, September, 17–21.

Roy, U., Mantooth, K., Pollard, M. D., and Liu, C. R., 1989b, Tolerance Representation Scheme in Solid Model, Part II, *Proceedings of the 1989 Design Technical Conferences,* Montreal, Quebec, Canada, September 17–21.

Shah, J. J., and Miller, D., 1989, A Structure for Integrating Geometric Tolerances with Form Features and Geometric Models, *Proceedings of the 1989 ASME Computers in Engineering Conference,* Orlando, Fla., July 30–August 3.

Shah, J. J., and Miller, D. W., 1990, A Structure for Supporting Geometric Tolerances in Product Definition Systems for CIM, *Manuf. Rev.,* Vol. 3, No. 1, March, pp. 23–31.

Shunmugam, M. S., 1986, On Assessment of Geometric Errors, *Int. J. Prod. Res.,* Vol. 24, No. 2, pp. 413–425.

Shunmugam, M. S., 1989a, Comparison of Linear and Normal Deviations of Forms of Engineering Surfaces, *Precis. Eng.,* Vol. 9, No. 2, April, pp. 96–102.

Shunmugam, M. S., 1987b, New Approach for Evaluating Form Errors of Engineering Surfaces, *Comput.-Aid. Des.,* Vol. 19, No. 7, September.

Traband, M. T., Joshi, S., Wysk, R. A., 1989, Evaluation of Straightness and Flatness Tolerances Using the Minimum Zone, *Manuf. Rev.,* Vol. 2, No. 3, September, pp. 189–195.

Turner, J. U., 1987, Tolerances in Computer-Aided Geometric Design, Ph.D. thesis, Rensselaer Polytechnic Institute, Troy, NY.

Zhang, X., and Roy, U., 1993, Criteria for Establishing Datums in Manufactured Parts, *J. Manuf. Syst.,* Vol. 12, No. 1, pp. 36–50.

19

NONTRADITIONAL TOLERANCING AND INSPECTION FOR PROCESS CONTROL

M. A. BUCKINGHAM

Renishaw Plc (Metrology Division)
Wotton Under Edge, Gloucestershire, England

19.1 INTRODUCTION

There is a problem with traditional tolerancing methods using BS 308 or ANSI Y14.5. No guidance is given to help chose an inspection method to establish part conformance or process control. Consequently, it has been demonstrated that good parts may be scrapped and bad parts accepted. It has been recognized that there needs to be a new standard for the specification of design intent which aligns with inspection methodology—in particular coordinate measuring machines. However, the international working parties are years away from producing standards that resolve this problem. Renishaw has developed its own in-house standard to address this problem, known as *gage point philosophy.*

Design intent is specified by means of *gage points* on the surface of features to be inspected together with tolerance fields that define the extent of allowable surface deviation. The uncertainty of surface position between gage points is a direct function of the capability of the process chosen to produce a given feature. Process capability is documented in a guidebook that details all allowable combinations of features, tools, and processes together with the

Advanced Tolerancing Techniques, Edited by Hong-Chao Zhang
ISBN 0-471-14594-7 © 1997 John Wiley & Sons, Inc.

minimum gage-point tolerance that can be applied to each combination. The design and tolerancing activity now becomes a dialogue between the designer and the production engineer who use the guidebook as a reference for all decisions concerning design intent and the manufacturing process.

The gage points specified for a given component are inspected to give direct feedback for process correction. Similar features produced with the same tool and process combination exhibit similar responses to process corrections and can therefore be controlled by a common gage point. The controlling feature that is inspected to give any necessary process corrections for the set of features with a given tool and process combination is known as a *critical feature*.

A three-dimensional solid model is used as the basis for all new product designs, alleviating the need to produce two-dimensional drawings, the computer numerically controlled (CNC) part programs and the gage-point inspection routines being generated automatically from the mathematical information stored in the CAD/CAM part file. Calibration and maintenance of the CNC machine tool is an essential element in ensuring part conformance. Laser calibration, ball bar traces, and "artifact comparison" techniques are used to establish and maintain the machine conformance to original performance specifications.

19.2   STATEMENT OF THE PROBLEM

Drawing standards such as BS 308 or ANSI Y14.5 give the engineer a powerful set of rules that allow him precise definition of what the part is allowed to be. However, they are not inspection standards and give no guidance whatsoever to an inspector, who must establish whether the manufactured part conforms to the drawing. In fact, there is no standard, national or international, for inspecting a part. Although the drawing defines the part, it will not tell you how to measure it, leading to ambiguity and confusion and ultimately, to different measured values for the same dimensions.

For the engineering industry to operate successfully (which it patently does!) under these conditions, assumptions are made by both the designer and the inspector. The inspector assumes an inspection methodology for each part based on knowledge of the part functionality and inspection custom and practice. The designer makes allowances for inspection variations and builds a safety margin into the drawing tolerances. The result is that good parts can be scrapped, bad parts accepted, and all parts carry the extra cost of a safety margin if the measurement uncertainties are not known accurately and treated as part of the tolerance.

The situation was brought into sharp focus recently by the issue of a GIDEP (Government/Industry Data Exchange Program) alert by an engineer at Westinghouse in the United States. This system is used in the United States to publicize data of national importance. He had clearly shown that equipment available to measure safety-critical aerospace components gave inconsistent results that would pass bad parts. The timing of this alert, coming as it did in the aftermath of the Space Shuttle disaster, made it impossible for the U.S. military to ignore. They insisted that all parts must conform to their drawings, without guidance as to how this was to be achieved.

The resulting impasse was addressed by a high-level meeting in Washington, DC, attended by recognized experts in the field. Working parties were set up in the United States to establish a mathematical definition of the drawing standard, allowing existing measuring equipment to be used. It will be many years before the necessary standards are in place to resolve this problem. Meanwhile, a concession has been put in place to overcome these impasses.

Next, we look at examples of how measured values can be misleading when determining the size of a part. Consider the shaft shown in Figure 19.1. An inspector measuring this component with a micrometer would find that the part would pass. Checking the part at the center or the edge would give the same result. If this part had been produced on a lathe, this form would be typical of a squareness error between the axis of rotation and the cross-slide and represents an extremely common situation.

Another example where gross errors can be experienced dependent on measurement technique is shown in Figure 19.2. A British 50-pence piece will

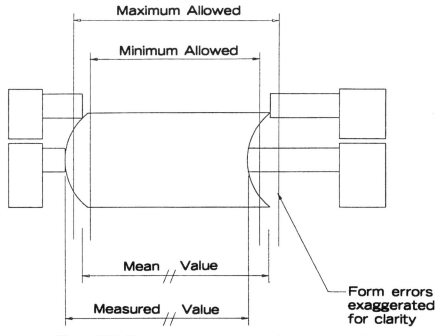

Figure 19.1. Form errors can cause misleading measured values.

give a consistent measurement of 30 mm when measured with a micrometer, irrespective of the part of the diameter measured. In reality, the outer diameter is 30.75 mm and the inner diameter is 29.25 mm. Clearly, the surface form or shape plays a major part in the size measured. This shape represents a surprisingly common result when attempting to grind a truly round object on a centerless grinder.

Of course, there are more sophisticated instruments for measuring such parts, in particular the coordinate measuring machine (CMM). This device can measure a number of discrete points on the surface, and using computer algorithms can produce a measured value for common features such as bores and planes. These algorithms make assumptions on the form of the surface between data points. These assumptions are not always published and are unlikely to be common between machines and manufacturers. Measurement decisions are therefore based on unknown assumptions.

To illustrate this point, consider the 50-pence piece again (see Figure 19.2). A conscientious inspector may decide that whereas three points are the minimum needed to define the diameter and center of a circle, he wishes to play safe and measures more points (e.g., seven). The actual number of points chosen is usually at his discretion. As can be seen from Figure 19.2, the result obtained on the CMM can vary from 30.75 to 29.25 mm in diameter, with a radial form error between 0 and 0.75 mm. The errors due to sampling are fundamental and inevitable. When measuring parts of an unknown form in precision engineering, they can be an alarming percentage of the tolerance.

The number and location of data samples, together with the processing algorithm, should be specified for each feature to be measured, to avoid the ambiguity that could cause bad parts to be passed or good components scrapped. However, this process must be understood in detail by the designer who must be satisfied that measurements resulting from this method will meet the design intent.

19.3   PRACTICAL SOLUTION

Instead of waiting for the outcome of the working parties and standards committees before implementing a solution, Renishaw has developed an in-house philosophy that integrates product design with inspection for process control and offers a workable framework for many companies faced with the difficulties of ensuring conformity of manufactured components. The problem was solved by a complete rethink on how the designer specifies the design intent or requirements.

By using the capability of a solid modeling CAD system to define the surface of the part mathematically, then specifying design intent by defining inspection points on the surface surrounded by tolerance "spheres" through which the surface must pass, we have developed an in-house standard for linking design intent to inspection. The number of data points is as many as

MICROMETER

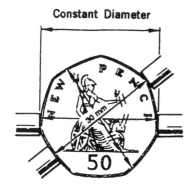

7 MEASUREMENTS
EACH GIVING SAME
RESULT - 30.00MM

C.M.M

7 POINTS ON FIRST CMM
EACH ON "HIGH SPOTS"
GIVES SIZE OF 30.75 MM

C.M.M

7 POINTS ON SECOND CMM
EACH ON "LOW SPOTS"
GIVES SIZE OF 29.25 MM

Figure 19.2. Measurement technique and data port selection can cause misleading measured values.

the designer needs to specify the design criteria adequately, and the maximum deviation between data points is a function of the manufacturing process, selected by the designer, on the basis of process capability data established using an in-house standard for defining and quantifying form errors resulting from particular manufacturing processes. The manufactured model defines the part for the design manufacture and inspection process. There are no detailed drawings, just the model complete with gage points and tolerance fields.

19.4   GAGE POINT PHILOSOPHY

The process control strategy known as gage point philosophy allows design intent to be conveyed in a language which is directly translated into process changes needed to ensure part conformance without analysis or interpretation of results, thereby removing the ambiguity and risk associated with traditional tolerancing and inspection techniques. Implementation of this philosophy necessitated a fundamental rethinking of the entire engineering function within the organization, from conceptual design to production, and become a vehicle for a cultural revolution which broke down all the traditional departmental barriers and forced the production engineers, who had in-depth knowledge of the processes, and the design engineers, who had in-depth knowledge of the design intent, to work together to create our new product designs. They could not work in isolation.

There are four basic building blocks within gage point philosophy:

1. Design intent specification
2. Understanding and specification of process capability
3. Process control and inspection
4. Calibration and maintenance

19.4.1   Design Intent

A standard engineering drawing using BS 308 or ANSI Y14.5 specifies the part in its nominal configuration as well as the allowable deviations of features from those nominal values in size position and form (see Figure 19.3). These drawings are then used to determine the manufacturing processes by which the part will be made, together with the measurement processes that will be used for both process control and part acceptance. It is the choice of processes that determine the accuracy to which the parts are manufactured.

If the manufacturing process is understood well enough, the specification and tolerancing of the component can be defined by the selection of the appropriate process for each feature, together with a process control strategy to correct for feature specific errors. Renishaw has developed a document entitled *Design for Manufacture Guidebook,* which details the process capa-

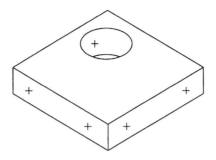

3D SOLID MODEL SHOWING "GAUGE POINTS"

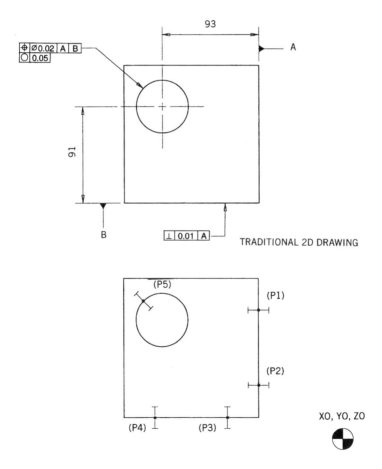

PLAN VIEW OF 3D SOLID SHOWING "GAUGE POINTS"

Figure 19.3. A comparison between "gage point philosophy" and traditional tolerance standards.

bility of its CNC machine tools to produce a range of features. Figure 19.4 details the Renishaw standard feature library. For each feature there is a corresponding range of standards tools that can be selected to produce it. Figure 19.5 details the Renishaw standard tool library.

For each combination of feature and tool there are a range of standard processes that can be used to produce the feature. Each combination of feature, tool, and process has a defined capability assigned in the form of an allowable tolerance which can be applied to the feature found by way of a rigorous process capability analysis developed in-house for the purpose. Figure 19.6 shows a typical extract from this section of the guidebook.

Design intent is specified by way of gage points. These are coordinates in a single reference frame (X_0, Y_0, Z_0, referenced to the machine tool "home" position). The coordinate represents a point on the surface of a feature that is to be measured. Allocated to each point is a tolerance band normal to the surface between which the points must lie to prove conformance (see Figure 19.3).

An error between the measured surface position and the nominal position is directly attributable to a correction in the tool offset active during the manufacturing process that produced the feature without inference or calculation. As long as the process selected to produce the feature has a capability that meets or exceeds the designer's requirements (i.e., the manufacturing errors are less than the allowable errors to achieve design intent) his single correction ensures that the errors are distributed about the nominal dimension and all subsequent errors are within the allowable band.

Renishaw uses a CAD/CAM system with three-dimensional solid modeling capability to design its new products. The component and assembly are created as mathematically defined entities with each surface and boundary in its nominal position. Traditionally, views of each component would be created, giving a two-dimensional representation of the model in various orientations. These views would then be dimensioned and toleranced using BS 308 or ANSI Y14.5 to produce a set of drawings from which the design intent is gathered and the manufacturing and inspection processes specified.

Using gage points, the design intent can be specified directly on the three-dimensional solid model of each component as a series of inspection points, with the tolerance attribute assigned to each point (see Figure 19.3). An automatic process by which the solid model is interrogated takes the gage point data and created a series of in-cycle-gaging routines which measure the specified points as part of the CNC part program. The solid model is also used within the CAM system to produce the CNC part program directly. There is no need to produce two-dimensional views as engineering drawings in an environment where gage point philosophy is used as the basis for process control.

19.4.2 Process Capability

Renishaw uses CNC vertical machining centers to produce a wide range of products for the metrology industry. An understanding of the CNC machine

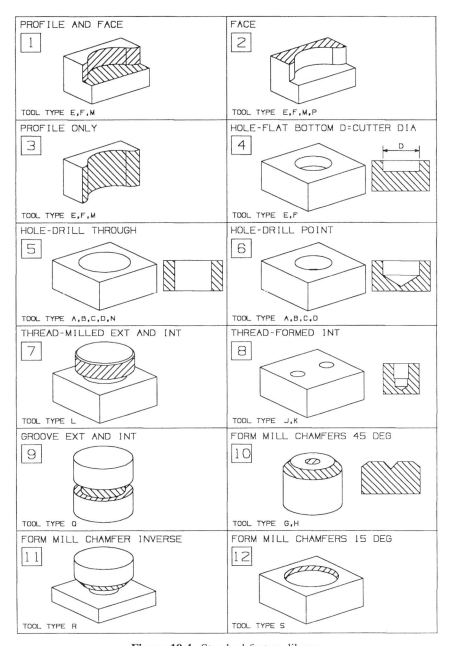

Figure 19.4. Standard feature library.

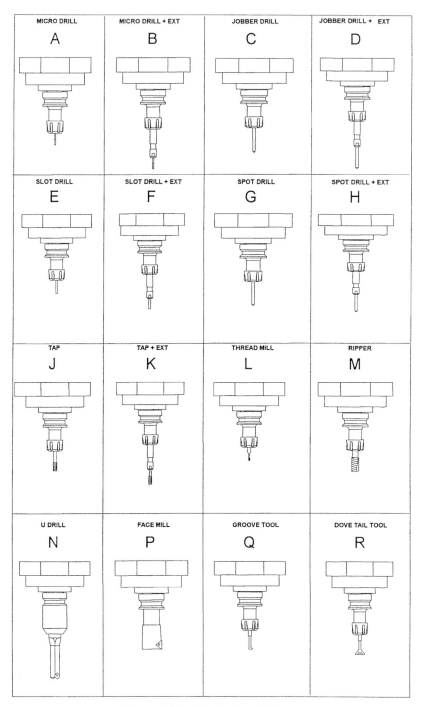

Figure 19.5. Standard tool library.

FEATURE	Tool type		PROJ/MM		SPEED/RP	FEED/MMP	SURFIN Ra	GP TOL
	Tool type	Micro drill	1.02		20	10000	400	0.8	0.07
HOLE DRILL THROUGH			1.8		20	10000	400	0.8	0.07
			2.05		20	10000	400	0.8	0.07
			2.53		20	10000	400	0.8	0.07
			3.05		20	10000	400	0.8	0.07
	Tool type	Micro drill with Ext	1.02		16/29	10000	400	0.8	0.07
			1.8		16/29	10000	400	0.8	0.07
			2.05		16/29	10000	400	0.8	0.07
			2.53		16/29	10000	400	0.8	0.07
			3.05		16/29	10000	400	0.8	0.07
	Tool type	Jobber drill	2.3		32	10000	0.1 MMP	1.6	0.15
			2.5		32	10000	0.1 MMP	1.6	0.15
			2.6		32	10000	0.1 MMP	1.6	0.15
			2.7		35	10000	0.1 MMP	1.6	0.15
			3		35	8000	0.1 MMP	1.6	0.15
			3.1		38	8000	0.1 MMP	1.6	0.15
			3.2		38	8000	0.1 MMP	1.6	0.15
			3.4		41	6000	0.1 MMP	1.6	0.15
			3.5		41	6000	0.1 MMP	1.6	0.15
			3.6		41	6000	0.1 MMP	1.6	0.15
			4		45	6000	0.1 MMP	1.6	0.15
			4.1		45	6000	0.1 MMP	1.6	0.15
			4.2		45	6000	0.1 MMP	1.6	0.15
			4.3		49	6000	0.1 MMP	1.6	0.15
			4.5		49	6000	0.1 MMP	1.6	0.15
			4.7		49	6000	0.1 MMP	1.6	0.15
			5		54	6000	0.1 MMP	1.6	0.15
			5.2		59	6000	0.1 MMP	1.6	0.15
			5.5		59	6000	0.1 MMP	1.6	0.15
			5.7		59	6000	0.1 MMP	1.6	0.15
			6		65	5000	0.1 MMP	1.6	0.15
			6.5		71	5000	0.1 MMP	1.6	0.15
			6.9		71	5000	0.1 MMP	1.6	0.15

Figure 19.6. Feature capability.

tool and the process by which components are made is an essential part of the control strategy we have developed. What is a component?

To a CNC machine tool, a component or piece part is merely a collection of features such as holes, faces, chamfers, slots, and radii in a block of material produced by interaction between the material and a tool (drill, end mill, slot drill, etc.). The interaction is controlled by spindle rotation and movement in three linear axes, designated X, Y, and Z. Each feature is produced independently by the machine.

The position, size, and form errors of each feature are primarily a function of:

1. *Positioning Errors Within the Machine Tool Control Circuit.* Actual position does not correspond to commanded position.
2. *Stiffness of the Tool.* Machining forces will cause deflection of the tool away from the material surface.
3. *Tool Size Errors.* Manufacturing tolerances and tool wear will cause the actual tool size to differ from the nominal size.
4. *Geometric Integrity of the Machine Structure.* This refers to the orthogonality of X-Y, X-Z, and Z-Y.

These errors fall into two categories. They are either global errors and affect the capability of all features or they are feature specific, caused by the specific combinations of axes commands and tools used to produce a particular feature. Global errors such as orthogonality, servo mismatch, and backlash are controlled by regular maintenance and calibration of machines (this is discussed later). Feature-specific errors are those normally controlled under production conditions. To match the needs of the process from a control viewpoint we need a system of independent dimensioning and tolerancing that would allow each feature within the component program to be controlled without affecting other dimensions.

19.4.3   Process Control and Inspection

There is a clear distinction between inspection and process control in a gage point environment. The purpose of a gage point is to facilitate process control, not to give a detailed report on the size and position of every feature. To this end, the number and positioning of these points is limited to those necessary to control the process adequately. Some features may not be inspected at all!

How is the process controlled? The design process is now a dialogue between the design engineer and the production engineer, who together select the appropriate manufacturing process for each feature in the component. When this exercise is complete, there will be a list of tools that are used to produce the component. Each tool will be used to form one or more features using the library of standard processes available in the guidebook.

By empirical experimentation Renishaw established that two similar features produced with the same tool using the same standard process (i.e., speed, feed, and depth of cut) exhibit similar errors and similar response to process corrections (i.e., tool offset corrections). Having established this underlying principle, it is possible to extrapolate this to state that all similar features produced by a given tool and process combination can be controlled by inspecting one of those features, the process correction made being valid for all such features. The feature selected is known as the *critical feature* and is nominally that feature in which the tolerance required to maintain design intent is smallest, or there being a number of features of equally criticality, that feature which produces ease of access for inspection. Once the features have processes allocated, gage points are applied to the critical features only, the process capability being relied upon to produce all other features with the same uncertainty.

During the "prove-out" phase of the process development, a detailed capability study is undertaken to establish the validity of the gage point set as a means of ensuring part conformance during which many additional information points are taken on noncritical features as a check that the process corrections made to the critical features are being applied. In production these information points can then be omitted. However, if during this phase a process anomaly is found (i.e., a particular feature is not being adequately controlled by the critical feature), it can be given its own gage point and be controlled as a separate critical feature with a separate offset.

19.4.4   Calibration and Maintenance

For this control strategy to be valid, the condition of the CNC machine tool needs to be monitored and corrected constantly to verify that the machine is performing to the specification that will allow the process uncertainty to be less than that needed to conform to design intent. How is this done?

The machine tool is calibrated regularly using a laser interferometer to ensure that each axis independently can be positioned within the uncertainty specification of the manufacturer. (Backlash and pitch error compensation features are normally provided to facilitate the elimination of these errors.) A Renishaw ball bar is then used to ensure that the interaction between axes (interpolation) can be performed within the uncertainty specification of the manufacturer. At this point we know that the machine is performing to specification; therefore, it will move to commanded positions with a known uncertainty. It will produce circles and profiles with a known uncertainty.

We then use a *calibrated artifact* to establish a baseline calibration for the machine tool as a measuring system. The artifact (see Figure 19.7) is a collection of features typical of those found on Renishaw products. It is made of the same material as that used in Rensihaw components (and therefore exhibits the same thermal characteristics). After normalization at 20°C the artifact is calibrated by means of inspection on a traceable CMM. A series

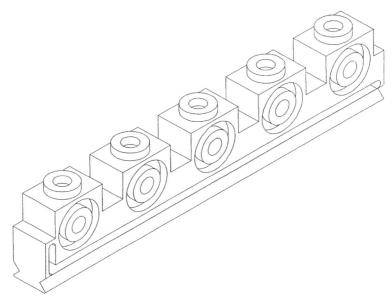

Figure 19.7. Calibrated artifact.

of predefined gage points is measured, the coordinates of which are the calibrated data. The artifact is then mounted in the working envelope of the machine tool in such a way as to allow it to occupy the same space as components when manufactured and be quickly moved automatically (see Figure 19.8). The machine tool is then used to measure the calibration gage points on the artifact using a spindle-mounted probe. These data are used to calibrate the machine tool measuring system and to characterize the machine in its "good" state. Subsequent remeasurement of the artifact during the manufacturing process allows the tracking of thermal errors and verification of the "health" of the machine in terms of accuracy, repeatability, and orthogonality, to ensure its ability to make components to its original capability. Any errors found form the basis of maintenance actions to repair or recalibrate the machine.

19.5 PRACTICAL PROBLEMS

No implementation of a radical new approach to the specification and verification of design intent can take place without problems being encountered. It is worth outlining two major areas that needed to be addressed before a successful system became a practical reality within Renishaw.

1. The knowledge level of the machining processes being specified was inadequate among the design engineers.

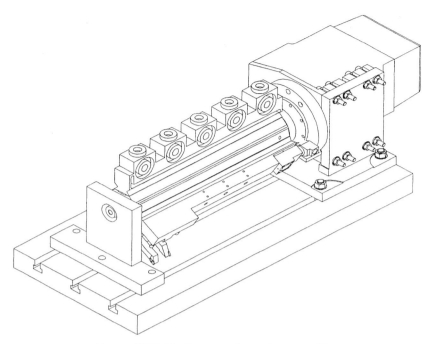

Figure 19.8. Working envelope of the machine

2. There was resistance to change among production engineers and design engineers.

19.5.1   Knowledge Level of Design About Manufacturing

Before the introduction of gage point philosophy Renishaw had separate functional groups for design and production engineering. Designers would produce drawings for piece parts which the production engineers would then turn into CNC programs. There was no need for a designer to understand how a machine tool works. However, in a concurrent engineering environment in which there are no drawings, the design interfaces directly with the process on which it will be produced. The quality of that design is very dependent on the level of detailed understanding of the manufacturing process possessed by the person designing the product.

There is no substitute for experience in this area. Consequently, key design engineers were given 12 months of seconding to the production engineering department to give them an understanding of CNC programming and the manufacturing processes used in their designs. The practical problems of managing the reduction in design capability and the additional training and support responsibilities within production engineering should not be underestimated, and this process of education takes many years to complete. The company has also determined that all new recruits who aim ultimately to become design engineers must start their careers in the production engineering discipline.

19.5.2   Resistance to Change

The cultural changes imposed by the needs of concurrent engineering and the breaking down of traditional departmental barriers are difficult for many people to accept. Going from a separate design and engineering function to an integrated team approach to the design of products has not been an easy transition. For example, design engineers now have to negotiate with production engineers concerning the size and tolerance of every feature to be produced. They no longer have the right to specify what *they* want. This change in relationship has produced many instances of disagreement, which had to be sorted out by mediation.

The *Design for Manufacture Guidebook,* which is used as the main reference for design and production engineering, was perceived within the production engineering fraternity as "deskilling" their job function and removing some elements of responsibility they had enjoyed beforehand. For example, the specification of tooling to machine the part was always the responsibility of the production engineer. Even though great care was taken in the specification of the guidebook to ensure ownership by those who would use it by agreeing to the standard tool library by consensus, there are always occasions when a production engineer believes that he or she could have specified a better tool for a particular job.

BIBLIOGRAPHY

McMurtry, D., 1993, *Metrology in the Field of Engineering,* Renishaw, Wotton Under Edge, Gloustershire, England.

Taylor, B. R., 1993, *Dimensional Control Through the Tolerancing of the Solid Model,* Renishaw, Wotton Under Edge, Gloustershire, England.

20

COMPARISON OF THE ORTHOGONAL LEAST SQUARES AND MINIMUM ENCLOSING ZONE METHODS FOR FORM ERROR ESTIMATION

MARY M. DOWLING, PAUL M. GRIFFIN, KWOK-LEUNG TSUI, and CHEN ZHOU

Georgia Institute of Technology
Atlanta, Georgia

20.1 INTRODUCTION

To satisfy designed functionalities, manufactured parts must conform to certain geometric constraints. These constraints are expressed in terms of ANSI Y14.5, the geometric dimensioning and tolerancing (GD&T) standard (ASME, 1994). In the GD&T standard, allowable variation of individual and related features is based on the envelope principle. In essence, the entire surface of the part feature of interest must lie within two envelopes of the ideal shape. The envelope principle has evolved from gaging technology. A *go gage* provides an envelope for checking the maximum material dimension (whether an internal feature is too small or an external feature is too large), while a *not go gage* provides an envelope for checking the minimum material dimension (whether an internal feature is too large or an external feature is

Advanced Tolerancing Techniques, Edited by Hong-Chao Zhang
ISBN 0-471-14594-7   1997 John Wiley & Sons, Inc.

too small). Hard gages have historically been the measurement tool of choice for tolerance verification, due to their precision and ease of use. However, they tend to be expensive and very inflexible.

Coordinate measuring machines (CMMs) have recently gained tremendous popularity in dimensional measurement due to their flexibility, accuracy, and ease of automation. A CMM is a computer-controlled device that uses a programmable probe to obtain measurements on a part surface, usually one point at a time. Despite the advantages offered by CMMs, their use has introduced a new problem in addition to the measurement errors (Harvie, 1986; Jones and Ulsoy, 1995, 1995). The design and measurement standards require knowledge of the entire surface, while a CMM measures only a small sample of the surface. In this chapter we focus on this CMM-specific problem. Satisfactory solutions to this problem are not easy to obtain, and both practitioners and researchers have shown great interest in the area.

A part is verified by comparing the part deviation range to the tolerance. The *deviation range* is a measure of the actual variation on the part surface. The tolerance is the allowable range assigned by the designer. Whenever a sample of observations is used to make inferences about a larger population (in this case the population of all measurements on a part surface), there is a chance of drawing incorrect conclusions. It is always possible that some unmeasured portions of a feature lie outside the estimated deviation range (i.e., the true deviation range is larger than the estimated deviation range). In such a case, acceptance of a defective part can occur. It is also possible for the estimated deviation range to exceed specifications while the true surface is, in fact, within the tolerable region. Clearly, estimation accuracy depends on sample size and estimation method. Selection of an appropriate sample size involves a trade-off between the cost of taking additional measurements and the cost of making incorrect decisions. The choice of estimation method involves a trade-off between the ease of use and the accuracy of the estimation.

The two most popular methods for form tolerance estimation are the least squares and minimum zone methods (Murphy and Abdin, 1981). Although least squares is the method most commonly used in practice, many researchers argue that the minimum zone method is more appropriate. Their argument is based on the fact that the minimum zone method has a close relationship with the envelope principle used in the ANSI tolerance standard definition, and further, the method asymptotically converges to the true feature's deviation range (with some restrictions on measurement error). The least squares method also converges asymptotically, but it may be to a value different from the true deviation range. However, it is important to keep in mind that only a small number of measurements are made in practice, due to time constraints, so the properties of each method should be studied in a small-sample context. Both methods tend to underestimate the true deviation range with small samples.

Many comparison studies in the literature advocate use of the minimum zone method for form evaluation because it results in estimates of the feature's deviation range that have the smallest values (Murthy and Abdin, 1981; Traband et al., 1989). The "smaller is better" viewpoint implicitly assumes that the sample is representative of the entire feature surface (including its extreme points). This is generally not the case in practice because it is usually feasible to collect relatively few measurements. The least squares approach treats the data as a sample rather than as the entire population of measurements. The empirical results of our study indicate that the resulting estimates of the deviation range for straightness and flatness have better statistical properties than do the minimum zone estimates.

In this chapter we compare the orthogonal least squares and minimum enclosing zone estimation techniques on straightness and flatness using real data collected with a CMM, as well as simulated data. In the next section, a brief overview of form tolerances is presented. The form tolerances used in the examples (straightness and flatness) are specified. In Section 20.3 we describe the orthogonal least squares and minimum enclosing zone estimation techniques. A comparison of these methods using simulation is presented in Section 20.4 along with an example provided by NIST. Conclusions from the comparison study are discussed in the Conclusion Section.

20.2   FORM TOLERANCES

As the name implies, form tolerances are used to control the shape or form of a feature (ASME, 1994). The most commonly used form tolerances are straightness, flatness, circularity, and cylindricity. In this chapter, we compare estimation methods for straightness and flatness.

The straightness form tolerance may be used for either surface control of axis control. In the case of surface control, all the line elements of the surface must be within the specified straightness tolerance. To meet this condition for straightness, each point on the surface must lie between two parallel lines of specified width. In the case of axis control, the straightness tolerance defines a diametrical zone within which each element of the axis must lie. The diameter of the zone is defined in the tolerance. It should be noted that no orientation need be specified.

The flatness form tolerance defines two parallel planes separated by a specified distance within which the entire surface must lie. Figure 20.1 shows how this tolerance is applied to a flat part. Note that the flatness tolerance is defined by the parallelogram in the left box of the feature control frame, and the width between the two parallel lines is defined by the number in the right box of the feature control frame.

20.3   ESTIMATION TECHNIQUES

20.3.1   Orthogonal Least Squares Technique

A common estimation technique for straightness and flatness is orthogonal least squares. For straightness, use a set of n measurements $(x_1, y_1),...,(x_n, y_n)$ to fit a straight line:

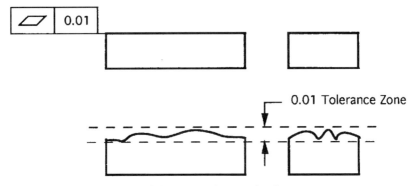

Figure 20.1. Form tolerance for flatness.

$$y = \beta_0 + \beta_1 x$$

The β_i values are chosen to minimize the sum of the orthogonal deviations. That is,

$$\min_{\beta} S_L(\beta) = \sum_i [e_i(\beta)]^2$$

where

$$e_i(\beta) = \frac{y_i - \beta_0 - \beta_1 x_i}{\sqrt{1 + \beta_1^2}}$$

The deviation range of the feature is estimated from the fitted line by

$$\hat{h}_L = \max_i e_i(\hat{\beta}) - \min_i e_i(\hat{\beta})$$

where $\hat{\beta}$ is the estimate of β from minimizing $S_l(\beta)$. For straightness the solution for $\hat{\beta}$ is (Bowker and Lieberman, 1972)

$$\beta_1 = \frac{\sum_i (x_i - \bar{x})(y_i - \bar{y})}{\sum_i x^2 - 1/n(\sum_i x_i)^2}$$

$$\beta_0 = \bar{y} - \beta_1 \bar{x}$$

If $\hat{h}_L$ is less than the specified tolerance, the feature is judged acceptable. Otherwise, the feature is determined to be out of tolerance.

It is important to notice here that orthogonal least squares minimizes the sum of the *orthogonal* deviations. Other estimators that rely on linear (vertical) deviations have been studied in the literature. The use of orthogonal deviations is more intuitively appealing in a physical sense (since the straightness and flatness tolerances are measured by orthogonal deviations) and the concept generalizes to more complex nonlinear features. However, solution of the orthogonal least squares problem generally requires an iterative search technique. If the feature has been aligned properly, linear least squares will give similar results with much less effort (Shunmugan, 1987). Our purpose here is not to compare these methods, but rather, to show how a least squares approach compares to the minimum zone estimation technique. We use orthogonal least squares to be more general.

In flatness, we wish to use n measurements $(x_1, y_1, z_1), \ldots, (x_n, y_n, z_n)$, to fit a plane:

$$z = \beta_0 + \beta_1 x + \beta_2 y$$

The objective function used is similar to that of straightness. However, in the case of flatness, the orthogonal deviation for the ith measurement is

$$e_i(\beta) = \frac{z_i - \beta_0 - \beta_1 x_i - \beta_2 y_i}{\sqrt{1 + \beta_1^2 + \beta_2^2}}$$

20.3.2 Minimum Enclosing Zone Technique

The minimum enclosing zone technique is an alternative method for straightness and flatness estimation. In this case the orthogonal minimax estimator is used. That is, the maximum orthogonal deviation from the line (or plane) is minimized:

$$\min_{\beta} S_M(\beta) = \max_i |e_i(\beta)|$$

where $e_i(\beta)$ is the orthogonal deviation as above. The objective can be solved numerically, but exact solutions are possible for the cases of straightness and flatness using well-known results in computational geometry (Preparata and Aromog, 1985; Traband et al., 1989). For straightness, the deviation range is estimated by finding the minimum enclosing rectangle for the set of measurements. This rectangle will have at least one line that corresponds to one of the edges of the convex hull of the set of points. A simple algorithm for straightness then is:

1. Compute the convex hull on the set of measurements.
2. For each edge of the convex hull, determine the minimum rectangle defined by this edge and the remaining measurements.
3. Pick the rectangle that has the smallest width, $\hat{h}_M$.

A simple decision rule says that if $\hat{h}_M$ is less than the specified tolerance of straightness, the feature is acceptable. Otherwise, the feature is out of tolerance. The algorithm stated above has a worst-case complexity of $O(n^2)$. A more efficient algorithm [$O(n \log n)$] that makes use of antipodal pairs is discussed by Traband et al. (1989) and Preparata and Stramos (1985). However, this algorithm is slightly more difficult to code.

A similar result from computational geometry is used to determine the deviation for flatness. In this case the minimum enclosing box for the set of measurements either has a side that contains at least one facet of the convex hull of the measurements, or has at least two parallel sides each of which contains an edge of the convex hull. A simple algorithm for flatness then is:

1. Compute the convex hull on the set of measurements.
2. For each facet of the convex hull, determine the minimum box defined by this facet and the remaining measurements.
3. For each nonorthogonal pair of edges on the convex hull, determine the two parallel planes that contain these edges. If all the points are contained in between these planes, determine the minimum enclosing box defined by these planes and the remaining measured points.
4. Pick the box that has the smallest width, $\hat{h}_M$.

Flatness is then verified by comparing $\hat{h}_M$ to the specified flatness tolerance. The algorithm given above determines the minimum enclosing box with worst-case complexity $O(n^2)$.

It should be noted that while finding the orthogonal minimax estimator is relatively easy for straightness and flatness, the only other form tolerance that has an exact solution is circularity. In this case, more difficult algorithms involving the minimum- and maximum-point Voronoi diagrams are used to determine the estimator. There are no known exact methods for cylindricity or sculptured forms. In both these cases, numerical methods must be used.

20.3.3   Example

We illustrate the differences in the orthogonal least squares and minimum enclosing zone with an example provided by the National Institute of Standards and Technology (NIST). Figure 20.2 shows a set of 400 measurements taken over a 2-in. length. This was considered to be a sufficiently dense sample to characterize the feature's profile. The orthogonal least squares line is the dashed line shown in this figure. The estimated deviation range determined by this method is 0.065 mm. The convex hull on the set of points is shown as the polygon enclosing the points. The minimum enclosing box contains the long lower edge of this convex hull. The deviation range estimated from the minimum enclosing zone is 0.053 mm. If this set of measurements

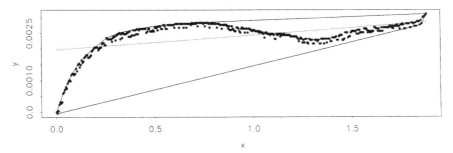

Figure 20.2. Data for the NIST example. The orthogonal least squares line is shown by the dashed line, and the convex hull is shown around the data.

completely characterizes the feature of interest, orthogonal least squares over-estimated the true tolerance range by just over 23%. However, rarely would an inspector have the time to take 400 measurements to verify a straightness tolerance. For this reason we are interested in studying how these methods differ as a function of sample size.

It should be pointed out that in this example the difference between regular least squares and orthogonal least squares is quite large. Further, we could have used this cxample in our study by repeated sampling from the 400 points to test the properties of estimators. However, this would not have included any kind of process variability since it is limited to a single part. This is the reason for using the simulated data in our study discussed in the next section.

20.4   SIMULATION STUDY

In this section we present the results from a simulation study on straightness and flatness estimation. We choose simulation study because we can get a variety of error magnitude combinations quickly. Both orthogonal least squares and minimum enclosing zone techniques are compared. Most previous work comparing methods has been done by taking repeated measures on a single true part (Elmaraghy et al., 1990; Weckenmann et al., 1991). Therefore, process variability is not accounted for. In our simulation we generate populations of test parts based on physical properties. In this way we include fixturing, surface variation, process variation, and measurement variation in the study.

20.4.1   Straightness

Simulated features were generated according to a model that represents face milling of a flat surface in a vise. Figure 20.3 illustrates a rectangular part

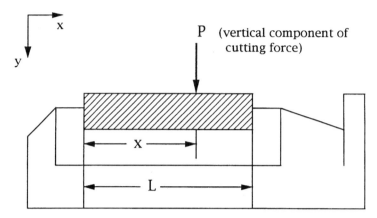

Figure 20.3. Rectangular part fixtured in a vise with cutting force shown.

fixtured in a vise. In this case the cutting tool has a vertical component of cutting force that deflects the workpiece. the deflection at a point P, y_P, is defined by (Beer and Johnston, 1977).

$$y_P = \frac{P}{3L^2EI} [x^3(L - x)^2]$$

where E is Young's modulus and I is the second moment of inertia about the centerline. The maximum deflection, R, occurs when $x = L/2$. In this case

$$R = \frac{P}{192EI} L^3$$

If R is specified for a workplace, the deflection as a function of x is

$$y_P = \frac{64}{L^6} R[x^3(L - x)^2]$$

In addition to deflection, we include surface variation, process variation, and measurement error in the model. The surface variation in face milling is due to both high-frequency variation of the cutting tool from vibration and to the commonly used up- and-down tool path followed on a rectangular part. The surface variation is assumed to follow a sine-wave pattern. There is empirical evidence of this effect in face milling. The process variation is due to random noise, and it is assumed to be normally distributed. Finally, there is measurement error due to the precision of the CMM. The distribution of the measurement error was determined empirically to follow a multinomial distribution. Test lines were generated according to the model

$$Y = \frac{64}{L^6} R[x^3(L - x)^3] + A \sin\left(\frac{2\pi}{\lambda} x\right) + N(0,\sigma^2)$$

where L = length of the line
    A = amplitude of the sine wave
    λ = wavelength
    σ = standard deviation of random error
    R = range of deflection

Three combinations of the line model were used in the simulation to represent a range of typical circumstances. In the first, process variation dominated the other sources of error. Deflection dominated the other sources of error in the second combination, and surface variation dominated in the third combination. The parameter values for each combination are shown below. In each case the length of the feature was L = 100 mm and the wavelength was λ = 3 mm. The units for A, R, and σ are millimeters.

Dominant Error Source	A	R	σ
Process error	0.003	0.005	0.003
Deflection	0.02	0.180	0.001
Surface variation	0.06	0.060	0.003

Figure 20.4 shows representative sets of 1000 points from each combination.
    For the simulation, a sample of n points is measured on the generated line model. The sampling is done using a randomized grid (i.e., stratified sample). That is, the line is broken into n equal-sized widths and a random point is measured within each width. The randomized grid sampling scheme is more robust than uniform sampling since it is insensitive to periodic variation. The steps in the simulation for each test case are as follows:

1. Generate 1000 points (enough to be a good approximation for the entire population).
2. Determine the "true" deviation range, h^*, from the 1000 points.
3. For n = 5,6,...,10, 15, 20,...,50, select a sample of size n from the 1000 points (using a randomized grid sampling procedure). Generate and add measurement error to each point measured.
4. For each sample i, determine $\hat{h}_L(i)$ and $\hat{h}_M(i)$.

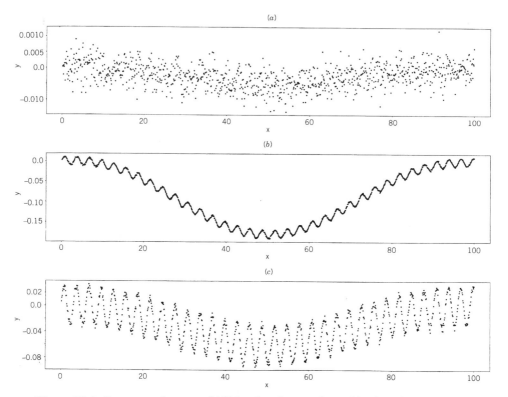

Figure 20.4. Representative sets of 1000 points from each combination. Combinations 1, 2, and 3 are shown in (a), (b), and (c), respectively.

5. Calculate the ratios $\hat{h}_L(i)/h^*$ and $\hat{h}_M(i)/h^*$.
6. Repeat this process 100 times.
7. Calculate absolute bias and root mean-squared error for each sample size.

The performance measures used in the study are absolute bias (AB) and root-mean-squared error (RMSE). These measures are defined as

$$AB = \left| h^* - \frac{1}{100} \sum_{i=1}^{100} \hat{h}(i) \right|$$

$$RMSE = \sqrt{\frac{1}{100} \sum_{i=1}^{100} \left[\frac{\hat{h}(i)}{h^*} - 1 \right]^2}$$

The absolute bias is a measure of how well an estimator compares with the true values. The root-mean-squared error is a measure that takes into account both the bias and the variance of an estimator. Obviously, we would like an estimation scheme that has low AB and RMSE. We would note that there are related measures, such as mean-squared error and variance. However, RMSE and AB are reported because they are the ones usually reported by statisticians.

Results for the two estimation methods at the various sample sizes on the three scenarios are illustrated with boxplots in Figures 20.5 to 20.7. For each box in the figure, location of the maximum value, 75% quantile, median, 25% quantile, and minimum value are given from top to bottom, respectively. The dashed line across the figure shows the value for the true deviation range. Figures 20.8 to 20.10 show the absolute bias and the root-mean-squared error for each case. All units are millimeters. Several important observations may be drawn from these figures.

1. The minimum zone method suffers from bias more than the orthogonal least squares method in all cases. However, orthogonal least squares may overestimate the true minimum deviation of the surface (as shown in Figure 20.4(c)).
2. The RMSE of the orthogonal least squares method is strictly less that the RMSE of the minimum zone method.
3. The variance of the orthogonal least squares method is higher than that of the minimum zone method. Notice, however, that even though the variance tends to be higher for the orthogonal least squares method, the RMSE is made up almost entirely by the absolute bias. This is seen by the fact that the RMSE and absolute bias lines are so close together. Therefore, variance does not appear to be nearly as important as bias.
4. The minimum zone method appears to converge less rapidly to the true minimum deviation when the process variation is large relative to the other sources of error. This is illustrated in Figure 20.7.

20.4.2 Flatness

As in Section 20.4.1, the simulated features were generated according to a model that represents face milling of a flat surface in a vise. In this case we assume that a rectangular plate is fixtured in such a way that it is simply supported upward around four edges with a concentrated force downward at the center. This is an approximation for the distortion when the fixturing force dominates. The deflection is modeled by (Timoshenko, 1959).

$$W = -\sum_{m=1}^{8} [\sinh(c1_m c2_y) + a_m \tanh(a_m) \sinh(c1_m c2_y)$$

$$-c1_m c2_y \cosh(c1_m c2_y)] \frac{\sin(mpa_m\zeta)(mpa_m x)}{m^3 \cosh(a_m)}$$

where

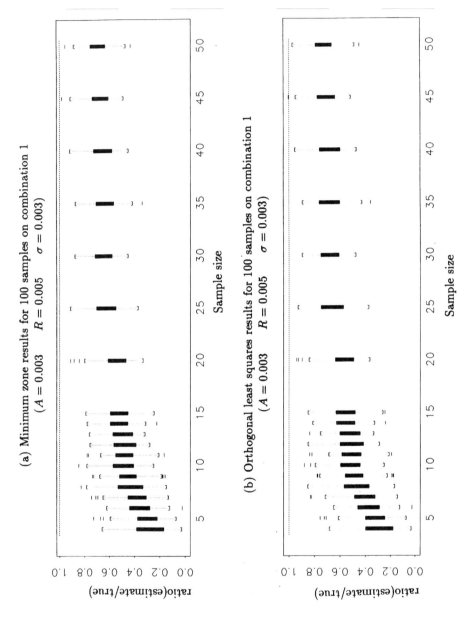

Figure 20.5. Boxplots for combination 1: (*a*) minimum zone; (*b*) orthogonal least squares.

538

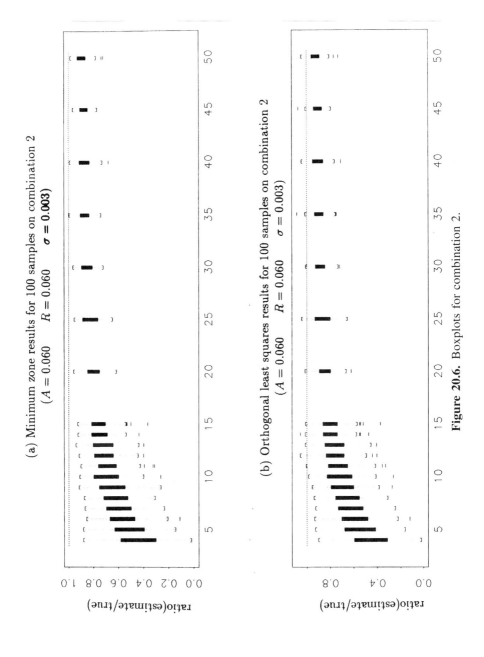

Figure 20.6. Boxplots for combination 2.

539

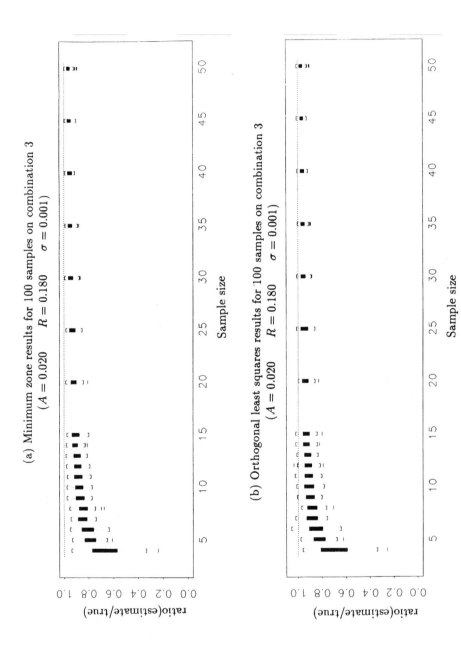

Figure 20.7. Boxplots for combination 3.

540

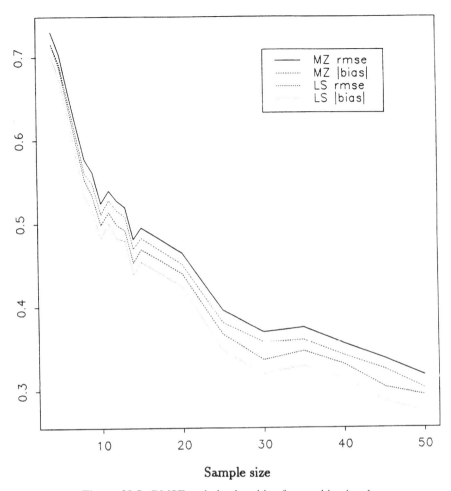

Figure 20.8. RMSE and absolute bias for combination 1.

$$a_m = 0.5\,\pi m$$

$$c1_m = 0.025 a_m$$

$$c2_y = b - 2y$$

$$mpa_m = 0.025\,\pi m$$

As in the case of straightness, we include surface variation, process variation, and measurement error in the model. The overall model is then

$$Z = W + a_1 \sin(v_1 x) + a_2 \sin(v_2 y) + N(0,\sigma)$$

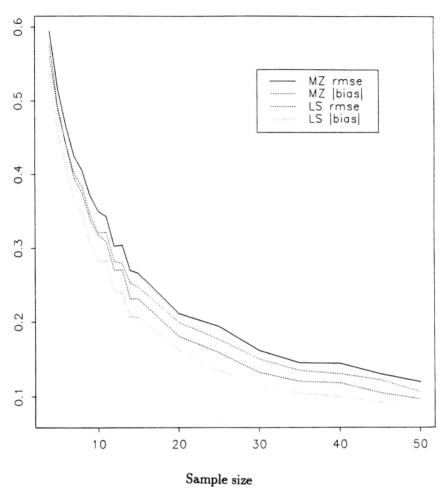

Sample size

Figure 20.9. RMSE and absolute bias for combination 2.

For this model, a_1 and a_2 define the amplitude of the surface waviness, ζ the range of the deflection, and σ the amount of process variation.

For the simulation study, a 40×40 grid of points generated using the surface model just described was assumed to be representative of the surface. The simulation was performed using the same set of steps as in the straightness case, with a generalization of the sampling method. For flatness we use the Latin hypercube sampling as described by McKay et al. (1979). A sample of n points is taken as follows:

1. Divide the length into n equal-sized sections and the width into n equal-sized sections. The result is an $n \times n$ grid of equal area rectangles. A rectangle in the grid is defined by position (i,j).

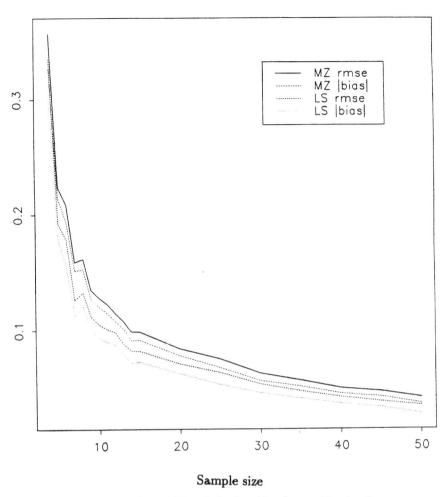

Figure 20.10. RMSE and absolute bias for combination 3.

2. Generate n random numbers and sort them from highest to lowest. Store the positions (p_i) of the random numbers. (For example, if $n = 4$, the random numbers 0.536, 0.691, 0.023, and 0.848 will have the positions $p_1 = 2$, $p_2 = 3$, $p_3 = 1$, $p_4 = 4$.)
3. Pick the set of rectangles $S = \{i, p_i\}$ for $i = 1$ to n.
4. For each rectangle in S, take a measurement at a random point within it.

This sampling scheme works very well since it tends to spread the measurements over the feature surface while maintaining a fair amount of randomness. Notice that there is exactly one measurement in every row and every column. Figure 20.11 illustrates the method for $n = 8$.

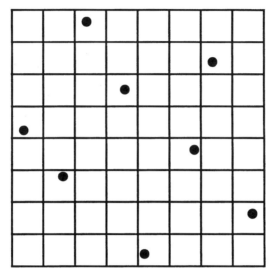

Figure 20.11. Latin hypercube sampling method for $n = 8$.

Three combinations of the planar model parameters were used in the simulation study. In the first, the deflection dominated the other sources of error. The dominant sources of error in the three cases were deflection, surface variation, and process variation, respectively. The parameter values for each combination are shown below. In each case, the size of the feature was 100 mm by 100 mm.

Dominant Error Source	a_1	a_2	ζ	σ
Deflection	0.010	0.010	4.000	0.015
Surface variation	0.040	0.040	1.000	0.015
Process variation	0.010	0.010	1.000	0.040

Figure 20.12 shows a surface plot for half the surface for the first combination. The process variation error was left out in this figure to better illustrate the deflection and surface variation. Figures 20.13 to 20.15 use boxplots to summarize the results for the two estimation methods on the three cases at various sample sizes. Figure 20.16 to 20.18 show the absolute bias and the root-mean-squared error for each model. There are several important observations that may be drawn from Figures 20.13 to 20.18.

1. All of the observations made for the straightness comparison study hold true for flatness as well.

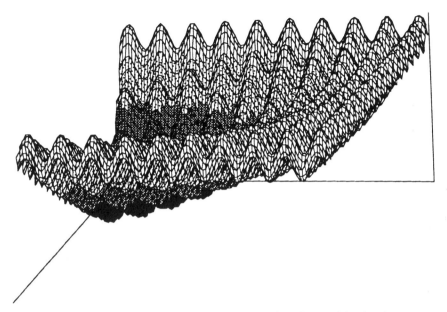

Figure 20.12. Surface plot for half the surface for combination 3.

2. As in the case of straightness, the RMSE of the orthogonal least squares method is strictly less than the RMSE of the minimum zone method. In fact, the RMSE for flatness from the minimum zone method can be two or three times higher than the RMSE from the orthogonal least squares method.

3. The difference between the minimum zone and orthogonal least squares method is much more pronounced with flatness than with straightness.

We should note that Figures 20.8 to 20.10 and 20.16 to 20.18 would be smoother if many more than 100 simulations were done for each sample size. However, the points made from the comparison are still quite clear from these simulations.

20.5 CONCLUSIONS

In the past, estimation techniques have been compared based on how small the estimated feature deviation is for a given sample. However, what is more important is how well the technique estimates the true feature deviation range. Further, most studies of estimation techniques in the past were based on one or two parts. Therefore, process variation was not accounted for.

In this paper we compared the minimum zone and orthogonal least squares methods through the use of a simulation study for the straightness and flatness

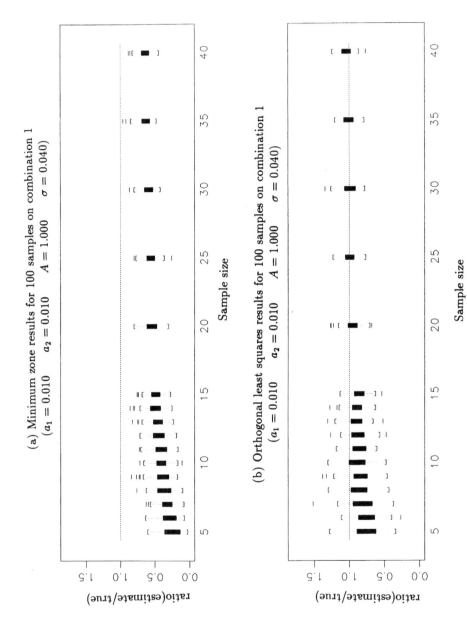

Figure 20.13. Boxplots for combination 1: (*a*) minimum zone; (*b*) orthogonal least squares.

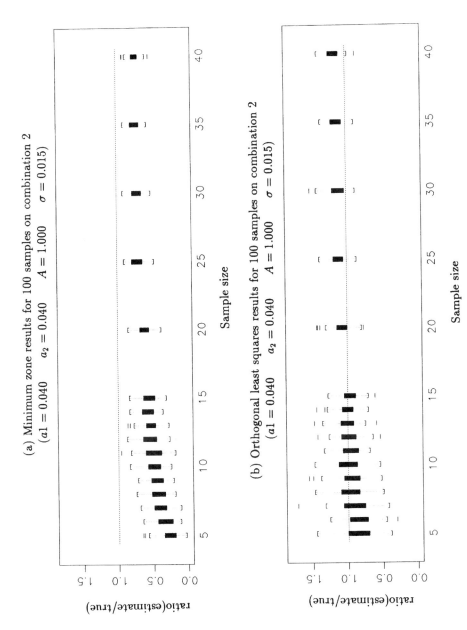

Figure 20.14. Boxplots for combination 2.

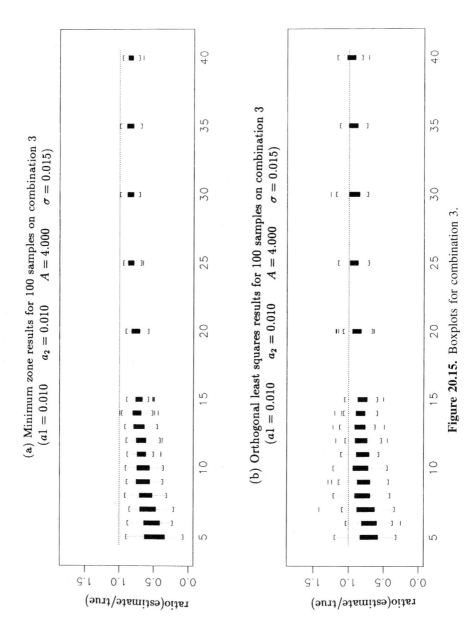

Figure 20.15. Boxplots for combination 3.

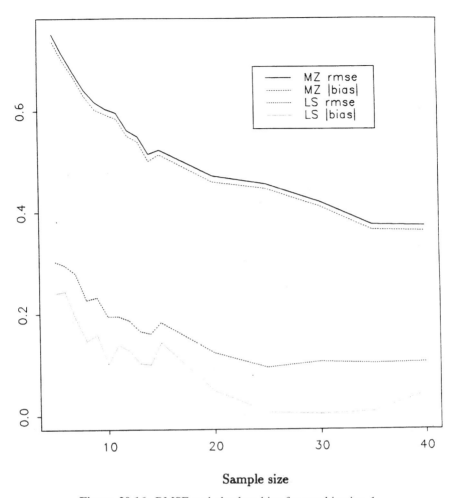

Figure 20.16. RMSE and absolute bias for combination 1.

form tolerances. The study included errors from the following sources: deflection, surface variation, and process variation. We found that for all cases, the orthogonal least squares method had less mean-squared error than did the minimum zone method. At times, this difference was quite large, particularly so for small samples (i.e., the practical case).

There are other advantages that the orthogonal least squares method has over the minimum zone method. First, in the absence of measurement error, the minimum zone method will always underestimate the true feature deviation (sometimes by a large amount). Second, the minimum zone method has not been developed for all the form tolerances. The method exists for the two cases considered here as well as circularity (Roy and Zhang, 1992); however, there has not yet been developed an efficient method for cylindricity or sculp-

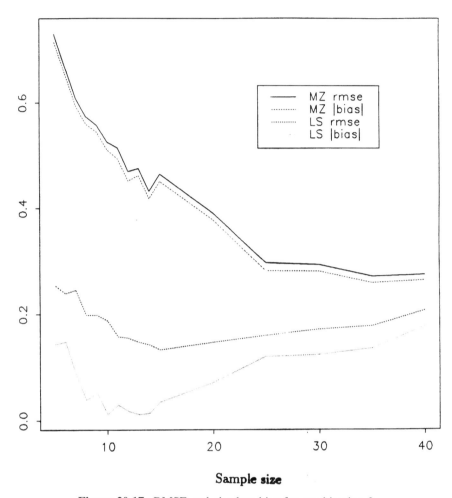

Figure 20.17. RMSE and absolute bias for combination 2.

tured forms. Finally, the computational effort for the minimum zone method tends to be larger than that required for the orthogonal least squares method.

In light of these findings, it appears that the orthogonal least squares is a preferred method over the minimum zone method. However, further studies will need to be done for other tolerances, such as circularity. It is worth mentioning that the argument over which method is best is largely academic. Practitioners use least squares because of computational ease anyway. There seems to be widespread opinion, however, among researchers that the minimum zone method would be best if it were only computationally practical. One of the points we want to make is that even though the minimum zone method is a combinatorial algorithm, it is being applied to a *sample* of points and so it is an *estimate*. It is therefore very important to consider the properties of that estimate.

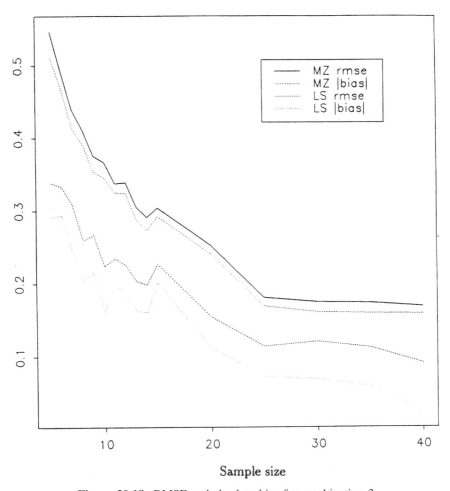

Figure 20.18. RMSE and absolute bias for combination 3.

ACKNOWLEDGMENTS

This work was supported in part by the National Science Foundation, Grants DDM-9203054 and DDM-9257918. The authors gratefully acknowledge Carl Zeiss Inc. for providing us with coordinate measuring equipment, software, and technical support.

REFERENCES

ASME, *Dimensioning and Tolerancing,* American Society of Mechanical Engineers, 1994, ANSI Y14.5M-1994, New York.

Beer, F. P., and Johnston, E. R., 1977, *Engineering Statistics,* 2nd ed., Prentice Hall, Upper Saddle River, NJ.

Dowling, M., Griffin, P. M., Tsui, K., and Zhou, C., Statistical Issues in Geometric Tolerance Verification Using Coordinate Measuring Machines, *Technometrics,* Vol 39; No. 1, pp. 3–17, 1997.

Elmaraghy, W. H., Elmaraghy, H. A., and Wu, Z., 1990, Determination of Actual Geometric Deviations Using Coordinate Measuring Machine Data, *Manuf. Rev.,* Vol. 3, pp. 32–39.

Harvie, A., 1986 Factors Affecting Component Measurement on Coordinate Measuring Machines, *Precis. Eng.,* Vol. 8, pp. 13–18.

Hopp, T. H., 1993, Computational Metrology, *Manuf. Rev.* Vol. 6, pp. 295–304.

Jones, S. D., and Ulsoy, A. G., 1995, Optimization Strategy for Maximizing Coordinate Measuring Machine Productivity, Part 1: Quantifying the Effects of Measurement Quality, *J. Eng. Ind.,* Vol. 117, No. 4, pp. 601–609.

Jones, S. D., and Ulsoy, A. G., 1995, Optimization Strategy for Maximizing Coordinate Measuring Machine Productivity, Part 1: Problem Formulation, Solution, and Experimental Results, *J. Eng. Ind.* Vol. 117, No. 4, pp. 610–618.

Lai, K., and Wang, J., 1988, A Computational Geometry Approach to Geometric Tolerancing, *Proc. XVI NAMRC,* pp. 376–379.

McKay, M. D., Conover, W. J., and Bechman, R. J., 1979, A Comparison of Three Methods for Selecting Values of Input Variables in the Analysis of Output from a Computer Code, *Technometrics,* Vol. 21, pp. 29–245.

Murthy, T. S. R., 1982, A Comparison of Different Algorithms for Cylindricity Evaluation, *Int. J. Mach. Tool Des. and Res.,* Vol. 22, pp. 283–292.

Murthy, T. S. R., and Abdin, S. Z.,. 1981, Minimum Zone Evaluation of Surfaces, *Int. J. Mach. Tool Des. and Res.,* Vol. 20, pp. 123–136.

Preparata, F. P., and Shamos, M. I., 1985, *Computational Geometry: An Introduction,* Springer-Verlag, New York.

Roy, U., and Zhang, X., 1992, Establishment of a Pair of Concentric Circles with the Minimum Radial Separation for Assessing Roundness Error, *Comput.-Aid. Des.,* Vol. 24, pp. 161–168.

Shunmugan, M. S., 1987, New Approach for Evaluating Form Errors of Engineering Surfaces, *Comput.-Aid. Des.,* Vol. 19, pp. 368–374.

Shunmugan, M. S., 1990, Criteria for Computer-Aided Form Evaluation, *J. Eng. Ind.,* Vol. 113, pp. 233–238.

Timoshenko, S., 1959, *Theory of Plates and Shells,* McGraw-Hill, New York.

Traband, M. T., Joshi, S., Wysk, R. A., and Cavalier, T. M., 1989, Evaluation of Straightness and Flatness Tolerances Using the Minimum Zone, *Manuf. Rev.,* Vol. 2, pp. 189–195.

Weckenmann, A., Heinrichowski, M., and Mordhorst, H. J., 1991, Design of Gauges and Multipoint Measuring Systems Using Coordinate-Measuring Machine Data and Computer Simulation, *Precis. Eng.,* Vol. 13, pp. 203–207.

21

INDIRECT TRANSFER OF TOLERANCE AND SURFACE FINISH TO MANUFACTURING[1]

SAMPATH KUMAR and SHIVAKUMAR RAMAN

University of Oklahoma
Norman, Oklahoma

21.1 INTRODUCTION

A key factor in the success of mass production of mechanical components is *interchangeable manufacturing,* which can be defined as a means by which parts can be made in widely separated localities and brought together for assembly and the parts will fit together properly (Giesecke et al., 1980). An economic way of ensuring interchangeable manufacturing is by prudently specifying a tolerance on the various geometric attributes on a component. *Tolerance* can be defined as "the total permissible variation from design size, form or location" (ASME, 1982). Tolerance specifications also determine the range of looseness/tightness (fit) between mating parts. Thus tolerances are very important in meeting the desired functionality. Interesting though it might seem, there is very little, if any, published literature that has mathematically related the tolerances on certain features on a component to the machining parameters (speeds, feeds) used to fabricate those features.

Another factor that depends directly on machining parameters and that affects the functionality of a component is its *surface finish* or *surface texture,*

---

[1]Parts of this work have been published at the 1993 ASME Winter Annual Meeting within the reference noted as Kumar and Raman (1993).

Advanced Tolerancing Techniques, Edited by Hong-Chao Zhang
ISBN 0-471-14594-7 © 1997 John Wiley & Sons, Inc.

defined as "the geometric irregularities generated by a machining method" on a machined surface (*Machining Data Handbook,* 1980). Different machining processes have different characteristic surface finish capabilities. Intelligent selection of machining processes and their parameters is essentially to insure economy of manufacture.

A multitude of CAD/CAM packages have been available on the market for several years now. With the exception of a few expensive and advanced software packages, most of these packages do not "understand" tolerance or surface finish specifications. That is, they cannot interpret, analyze, calculate, or make decisions about the tolerance/surface finish information stored in them (Varghese and Atkinson, 1987; Truslove, 1988; Weill, 1988). In the context of integrated CAD/CAM, this is a major handicap, hindering total data exchange between the design and manufacturing domains. In practice, transfer of these specifications is done manually: through drawings, process specification sheets, and the like. The importance of total exchange of component information between design/manufacturing and the utilization of this information for automatic decision making has been emphasized by several researchers, including Black (1986), Truslove (1988), and Weill (1988). Other relevant tolerancing literature that has influenced this work includes the papers by Requicha (1983a,b), Fainguelernt et al. (1986), Xiaoquing and Davies (1988), and Roy and Liu (1988).

Ideally, for a total exchange of data between design and manufacturing, there has to be a geometric transfer of dimensions and tolerances from design to manufacture. Implementation of such transfer is somewhat complicated by the ambiguities that exist in the definition of design and manufacturing features, inadequate tools for tolerance representation in CAD systems, and the relative absence of documented correlations between tolerances and machining conditions. This leads to deviations from the ideally desired and a textual transfer of dimensional entities between design and manufacturing. In this chapter, simple methodology is developed for tolerance transfer by highlighting the relationship between size tolerances and surface finish. Since this research aims to transfer size tolerances for numerical control (NC) code generation through the route of surface finish related to the tolerance, an "indirect" or nongeometric (textual) transfer is achieved. A personal computer (PC) environment and inexpensive software are used to demonstrate the methodology. A major objective was to create a framework and test its feasibility and reliability, even if it caters to specific families of parts and a few operations only. Further, the methodology is explained through the software implementation performed.

21.2   TOLERANCE SPECIFICATION

Component functionality is typically the almighty specification in manufacturing. The terms *quality, performance,* and *requirement* are closely related and the translation of the "voice of the customer" is a mammoth task. Specifications and expectations of products are subjective at best, and a derivation of objective measures from these subjective characteristics for design, manufacture, and inspection is highly critical. For example, an automobile that runs smoothly, consumes less fuel, and has no squeaks may appeal to one consumer, whereas the speed pickup in a certain time frame is important to another. How such product expectation translates into the automobile design parameters, and consequently, how it affects critical dimensions such as wheelbase, center of gravity, overall dimensions, and the dimensions of the individual parts that comprise the grand assembly, are matters typically addressed by the design engineer. With the notion of tolerances, a fact of real-life manufacturing, this problem is compounded. The tolerance specifications relate to a product's functional capability as well as, in most instances, the entire assembly's performance. Further, tolerance specification is typically cost-driven.

Tolerances are specified on components with the understanding that perfect parts are infinitely expensive to manufacture in large quantities. The functionality of a component is, however, realized as the most important goal in tolerance design. Based on the documented relationships between size tolerances and cost, and the process capability of different operations, economy of tolerance selection is achieved. Usually, a tighter tolerance specification results in a higher cost of manufacturing and inspection, and a loose tolerance specification may be easier to achieve, although functionally limited.

Tolerance specification research is divided into two main categories: tolerance analysis and tolerance allocation (Chase and Greenwood, 1988; Wu et al., 1988). *Tolerance analysis* deals with the aggregation of component tolerances to determine the assembly tolerance and does not specifically involve cost. Common methods employed for tolerance analysis include Monte Carlo simulation and the method of moments. *Tolerance allocation,* on the other hand, deals with the distribution of tolerances of assemblies to discrete component tolerances. Cost-based tolerance allocation methods typically minimize a manufacturing cost function subject to assembly constraints. Some research has also concentrated on joint tolerance allocation and process selection among a limited range of manufacturing capabilities (Chase et al., 1990; Nagarwala et al., 1995). There also exist some methods that suggest the use of loss functions for tolerance allocation. Here the manufacturing cost and loss are proposed to be minimized simultaneously subject to assembly constraints (Kapur et al., 1990). Several automatic systems for tolerance specification are available, important ones discussed by Bjorke (1989), Dong and Soom (1986), and Lokhandwala et al. (1992). It is to be noted that only some of the publications in this well-researched area of tolerance allocation and analysis are cited here, and the reader is referred to Kumar and Raman (1992) and other recent surveys for more details on these subjects.

The subject of most discussions until now, as well as the subject of this paper involves only size tolerances. However, dimension and tolerance specifications typically consist of the size, profile, form, orientation, position, and runout specifications. Although CAD systems attempt to provide total and unambiguous representation of geometric dimensions and tolerances, many of the present-day definitions and interpretations of tolerances are based on engineering drawings that employ projection methods and baseline or chain dimensioning. Dimension specifications are typically annotational or textual and left to the interpretation of the design and manufacturing engineers, although standards such as ANSI Y14.5M attempt to avoid ambiguities. With inspection and quality control thrown in, and with the advent of concurrent engineering, some interpretation issues of these standards are themselves questioned by occasionally contradicting measurement, inspection, and tooling (specifically, coordinate measuring machines) standards.

In engineering drawings, these dimensions and tolerances are specified with the use of datum features and feature control frames that have tried to alleviate the ambiguities arising from local notes. The definitions emphasize the role of individual and related features, a matter highly critical to process selection. In fact, a well-prepared design drawing automatically supplies precedence information for process selection and sequencing (Irani et al., 1995; Mani and Raman, 1996). Process planning and fixturing activities are affected by, as well as affect, tolerance application. A related area is tolerance charting (Irani et al., 1989), where issues of stock removal and cut selection are handled. Intermediate surface generation and the tolerance specification for these surfaces during manufacture are compounded further when statistical distributions of processes and cuts are considered.

21.3   SURFACE ROUGHNESS AND FINISH

Just as tolerances, the surface finish is an important characteristic of a component that governs the functionality of the component as well as its appearance. The tribological boundary conditions of mating surfaces is affected strongly by the finishes of the mating pair, which in turn affect friction, wear, and lubrication. Further, their specification is dependent on the capability of machining processes and process parameters. Surface finish is typically defined in terms of surface roughness, waviness, lay, and flaws. These characteristics are usually compounded with the form and size tolerance specifications, making inspection difficult. The most common indicator of surface texture is the arithmetic average of the surface with respect to a centerline (R_a), with the root-mean-square value (R_q) applied sparingly.

Standard handbooks provide conservative guideline charts for selection of operations based on their surface roughness generating capabilities. Sandcasting and hot-forming processes produce coarser surfaces, while coldshaping and traditional machining operations such as turning and milling produce finer finishes. Finishing processes such as grinding, lapping, honing, and polishing produce finer details. It is also observed that a particular surface finish or range of finishes can be obtained by more than one process. Hence selection of a process or sequence of processes to achieve a desired finish depends on several factors, including cost, machinery and tool availability, and manufacturing time. In practice this decision is made by experienced process engineers with adequate knowledge of shop constraints. Assuming the absence of these constraints, it is fairly easy to come up with a sequence of operations that will optimize a particular consideration, for example, cost. *Machining Data Handbook* (1980) provides a plot identifying the cost–finish relationship and states that "the cost of producing a machined surface increases with increasing requirements for finer finishes." Hence coarser finishes must be used whenever allowed by functionality/appearance demands.

Among the more prominent factors that determine the surface finish within a specified machining operation are speed, feed, depth of cut, tool geometry, tool surface preparation, cutting time, and machine tool rigidity. For instance, in grinding, the peripheral wheel speed, the speed of traverse, the rate of feed, the grit size, the bonding material, and the state of wheel dressing are all parameters that affect the finish. Some calculations of the *theoretical surface roughness* as a function of the feed, the tool radius, the end cutting edge angle, and the side cutting angle have been illustrated (Figure 21.1) in the *Metals Handbook* (1989). This roughness is the best finish commonly produced by that particular tool and thus provides an indication of the minimum surface roughness possible with a given tool geometry and feed rate. The actual roughness is typically poorer, due to effects of other cutting parameters, built-up formations on tools, and such. In some situations surface finish is better than that predicted by the theoretical surface roughness, when a wear land that develops on the tool provides a wiping action, tending to smooth out the theoretical roughness (*Metals Handbook,* 1989).

Many roughness studies are empirical in nature, some recommending the use of higher speeds, larger nose radii, and smaller feeds for obtaining better finishes. For economy of manufacture, a coarser finish is preferred, with the

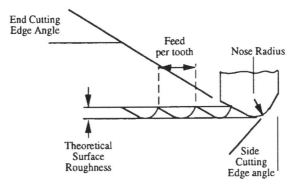

Figure 21.1. Theoretical surface roughness. (Adapted from *Metals Handbook,* 1989.)

choice of the best operation sequence and the appropriate selection of machining and material parameters that have the maximum controllable effect on the finish.

21.4   SURFACE FINISH AND SIZE TOLERANCES

Intuitively, it can be realized that there must be some relation between the surface roughness and dimensional tolerance. Figure 21.2 (and Table 21.1) is a typical representation of such a relationship (Trucks, 1974). Trucks (1974) states that "it is not feasible to hold a tolerance of 0.0001 inch on a part which is machined to an average roughness of 125 microinch rms." The measurement of roughness involves determination of the average linear deviation of the actual surface from the nominal surface. Hence a requirement for the accurate measurement of a dimension is that the variations introduced by the surface roughness should not exceed the tolerance placed on that di-

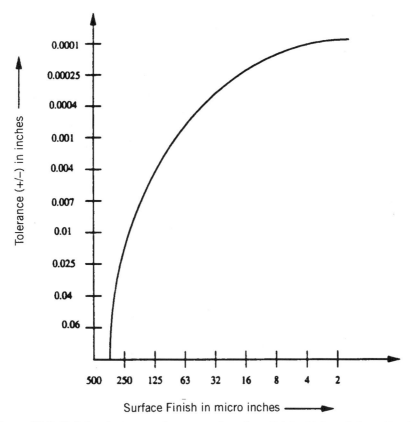

Figure 21.2. Relation between tolerance and surface finish. (Adapted from Trucks, 1974.)

TABLE 21.1. Relation Between Roughness and Corresponding Suitable Tolerance Range (Trucks, 1974)

Roughness (μin.)	Suitable Tolerance Range ($\pm$ in.) if Roughness Expressed as:	
	R_a	R_q
4	0.0005	0.00015
8	0.0005	0.00027
16	0.001	0.004
32	0.002	0.001
63	0.003	0.005
125	0.004	0.01
250	0.007	0.04
500	0.013	N.A.[a]
1000	0.025	N.A.
2000	0.05	N.A.

[a]N.A., not applicable.

mension. If this is not the case, measurement of the dimension will be subject to an uncertainty greater than the required tolerance, as illustrated by Figure 21.3 (*Machinery's Handbook,* 1984). If the tolerance zone for this surface were less than or equal to the profile height, the latter dominates the former and the former cannot be measured reliably.

To state otherwise, a smaller roughness/larger tolerance must be specified. This implies that on parts where very tight dimensional tolerances are specified, a suitably fine finish has to be specified (even if there is no need), so that useful dimensional measurements can be made. However, whenever a

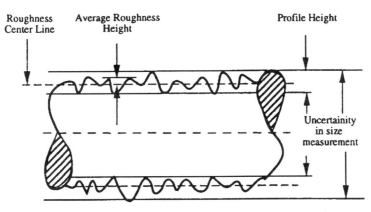

Figure 21.3. Uncertainty in dimensional measurement due to surface roughness. (Adapted from *Machinery's Handbook,* 1984.)

high surface finish may be specified for appearance, it may not be accompanied by a tight tolerance specification. The problem in reality is more complicated, with additional considerations of waviness, form tolerances (such as straightness, flatness, roundness, and cylindricity), orientation tolerances, profile, runout, and position tolerances. To maintain simplicity, these aspects are not considered at the present time.

In many instances the capability of a manufacturing process/machine to produce a tight tolerance coincides with its ability to produce a fine finish. Although the tolerances themselves may not be easily related to cutting parameters, the surface roughness can be so related and it might be advantageous to relate the cutting parameters with tolerances by highlighting the tolerance–roughness relationship. Tolerances can be allocated with or without cost considerations. Using standard tables or using a cost function optimization, tolerances can be allocated to each component in an assembly. One such tolerance allocation system is attempted by Pulat and Raman (1993), whereby size tolerances are allocated to different dimensions of parts. A sample output such as the one shown in Table 21.2 can be prepared based on the foregoing system. Note that the dimensional entities and their tolerances may also be derived using several known and published techniques.

21.5   PROCESS PLANNING AND INTEGRATED CAD/CAM SYSTEMS

Process planning involves the coordinated selection and sequencing of manufacturing activities to convert a raw material into a finished product. To simplify the representation of the goals, constraints, and variables involved in feature identification and process planning, computer-aided process planning (CAPP) is employed. Since this research involves the development of meth-

TABLE 21.2. Typical Output from Any Tolerance Allocation Procedure

RECORD #	TYPE	DIMN	TOLNL1
1	DIA	4.50	.01
2	DIA	3.50	.005
3	DIA	2.00	.002
4	LINEAR	3.50	.2
5	LINEAR	4.75	.15
6	LINEAR	2.00	.09

RECORD #	TYPE	DIMN	TOLNL1
1	DIA	1.00	.003

odologies for process planning and NC code generation, a brief review of relevant articles is included here. Two major approaches exist in process planning: variant and generative process planning. Variant planning methods use principles of group technology (GT) and parts classification and coding to prepare process plans. Generative process planning creates process plans of a component based on logical procedures, similar to the ones that a human process planner would use. Alting and Zhang (1989), Wang and Wysk (1987a,b), Joshi et al. (1987), and Shah and Bhatnagar (1989) are all important papers in these areas. Assembly process planning, a related area, involves the identification of the assembly sequence to put together (or take apart) an assembly from its individual components, as illustrated by Baldwin et al. (1991) and Kurup et al. (1995).

In the context of an integrated manufacturing system, the computer assists in the generation of tool path and NC code after creating the process plan from the CAD drawing. This integration effort requires taking the digital data generated by the CAD drawing and translating them into the input format required by CAPP, then taking the output of CAPP and converting it into the data format required by CAM. Owing to the multitude of standards followed by the various CAD/CAPP/CAM software, many of these data conversions are very difficult. Standard interfaces such as IGES attempt to alleviate these difficulties. Knutilla and Park (1990) describe research to overcome some of these interfacing problems.

21.6   REQUIREMENTS FOR AN INTEGRATED SYSTEM

The ability to achieve and maintain a specified size/form tolerance depends on factors such as how well (firmness and accuracy) the workpiece is held, the quality of the jigs and fixtures used, and the rigidity of the machine. Cutting parameters such as the speed and feed have little, if any, quantified influence on the ability to maintain a dimensional tolerance, although tolerance charting suggests the influence of the depth of cut in tolerancing. This is to say that a surface can be held within specifications even when the cutting parameters are varied within a wide range. Hence the direct effect of cutting parameters on tolerance is not considered.

On the other hand, surface finish is very much dependent on the cutting parameters—in particular, the feed. The other factors remaining constant, an increase in speed can result in a better finish; an increase in feed, a poorer surface finish; and an increase in depth of cut, a poorer surface finish. The other critical variables affecting finish include the tool material and tool geometry. Hence it becomes important to identify the best cutting parameters that will result in the desired surface finish. The use of computer-assisted cutting parameter selection has been recognized as a critical area by both industry and research institutes. Most existing systems use one of the following techniques for parameter selection (Wang and Wysk, 1986): data retrieval methods, optimization mathematical models, and empirical equation methods. Many of these have been designed to prove the feasibility of computer-assisted parameter selection, not to be an integral part of CAD/CAM system. That is, the input/output of such systems demand significant human intervention to make data exchange meaningful.

Before selecting the cutting parameters it might be important to verify that the specified surface finish requirement is compatible with the tolerance specification. If this check is not built-in as an integral part of the process plan generating mechanism, chances are that any incompatibility might go unnoticed until at a later stage. Any downstream correction may not only be very expensive, but may also upset the entire production plan.

Many commercial (NC Polaris, Mastercam, Smartcam) and experimental software (Sakal and Chow, 1991) packages are available that convert a component drawing into NC code needed to machine. Many of these are good at determining the cutter/tool path but lack a sound system that recommends the cutting parameters—they need to be input by the user. Even in systems that recommend cutting parameters, little consideration is given to surface finish requirements. Based on these considerations, it can be concluded that there is a need to develop an integrated system that will:

1. Verify the tolerance/surface finish compatibility and suggest a finish, if not compatible.
2. Generate a preliminary process plan.
3. Select and suggest machining parameters based on the surface finish requirements.
4. Generate the NC code for machining the component by using the corresponding machining parameters for each surface.
5. Simulate the cutterpath to verify the NC code.

21.7 OVERVIEW OF METHODOLOGY

A simple and unique system has been developed that considers the issues described in Figures 21.4 and 21.5. It must be recognized that the focus is to create a framework and study the feasibility of such a system and not to create a perfect commercial software. All the same, the methodology is better explained by considering the software implementation. Considering the constraints of time and resources, the scope of the system has been restricted to concentric cylindrical components with steps (only) and no tapers. Only rough, semifinish, finish turning, and facing operations for external surfaces and centering, drilling, rough, semifinish and finish boring for internal surfaces are considered, and it is assumed that the bar stock is fed continuously.

The CAD module implemented in this work to demonstrate the methodology is an earlier version of AutoCAD that uses Autolisp. Note, however,

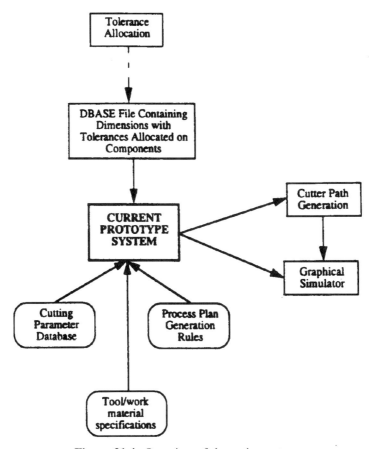

Figure 21.4. Overview of the entire system.

that this module can easily be replaced at any stage. The importance of the contribution lies not in feature extraction done for this stage and the programming package used but the suggestions made for the downstream activities performed thereafter. Further, the tolerance allocation is not tied to the CAD module at present and can be integrated as desired, at a later time. As mentioned earlier, the framework is developed based on engineering drawings and not based on geometric CAD modeling, thus achieving only a textual transfer of size tolerance entities. It is hoped that this software development will inspire small-parts manufacturers to develop more comprehensive integrated systems for preparing process plans and NC codes based on important annotational information. The contribution of the present chapter lies mainly in consideration of multiple functions toward integrated system development and the use of the tolerance–surface finish compatibility for generating machining sequence and machining parameters.

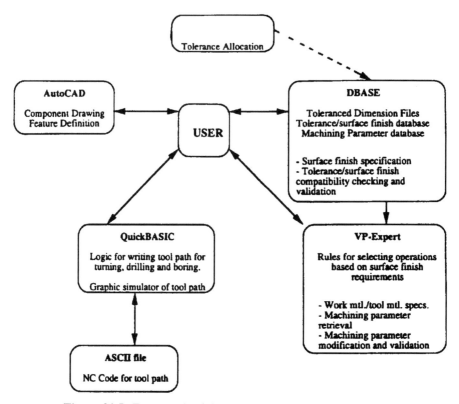

Figure 21.5. Framework of the current software implementation.

Entity information from a prepared drawing and tolerance design output are used as input data for textual tolerance transfer and integrated system development. The size tolerance data are checked for compatibility with surface roughness. If not compatible, a prudent surface finish is assigned. In instances where no surface finish is specified, an economical one is assigned. For instance, if the tolerance specification is tight and the surface roughness specification is very loose, to ensure compatibility, the surface roughness is tightened. The modified surface roughness and the stock to be removed are considered for deciding the combination of rough pass(es), semifinish pass, and finish pass machining for external operations. Based on the materials of the tool and work selected, the machining parameters for these operations are estimated based on surface finish specifications. Using these machining parameters and the geometric information derived earlier, the cutting path for converting the raw stock into the finished part is determined. This is converted into NC code-relevant information, which can be modified and used as such. A graphical verification of the code generated is performed to ensure proper machining.

21.8 DESCRIPTION OF PROTOTYPE SOFTWARE DEVELOPMENT

An AutoLISP program (Lisp.lsp) is used for feature definition that prompts the user to go over the outside and inside features and obtain vertex points. These points recorded in ASCII files define the contours of the external (Ptlsto) and internal (Ptlsti) surfaces. These files are used in subsequent modules.

A database module developed in Dbase houses the various databases and the consultation mechanism for surface finish specification and validation. For machining of external surfaces of cylindrical components, turning is most commonly employed. Turning is usually employed when the surface finish requirements (R_a) range from 16 to 250 μin. (0.4064 to 6.3500 μm) and in some special cases up to 2 μin. (0.0508 μm). From Figure 21.2, these finish values translate to a tolerance range of ± 0.001 to ± 0.007 in. (0.0254 to 0.1778 mm) for normal applications and up to ± 0.0005 in. (0.0127 mm) for special cases. These values cover a common range of finish/ tolerances in shop practice, and finer requirements may demand advanced operations, such as lapping and polishing. Depending on the final surface finish required, the suitable combinations of rough, semifinish, and/or finish operations need to be selected in a way that minimizes the time of machining as well as maintains specification. For purposes of this experimental system, the values recommended by the *Machining Data Handbook* (1980) for turning tools are adopted. These are good starting-point recommendations and are conservative estimates. Optimum values particular to a shop may be used in the database instead of the ones used, as deemed appropriate. For example, for a HSS tool and work material hardness of 125 to 175 BHN (Brinell hardness number), the following recommendations apply:

- Speed: 110 to 150 ft/min (33 to 45 m/min)
- Feed: 0.015 to 0.007 in./rev (0.3810 to 0.1778 mm/rev)
- Depth of cut: 0.150 to 0.025 in. (3.8100 to 0.6350 mm)

The lower range of speed (110 ft/min), higher range of feed (0.015 in./ rev), and higher range of depth of cut (0.150 in.) are taken to represent the recommendations for rough turning. The higher range of speed (150 ft/min), lower range of feed (0.007 in./rev), and lower range of depth of cut (0.025 in.) are taken as recommendations for finish turning. Each of these ranges is interpolated linearly to yield the corresponding recommendations for semifinish turning. This results in a value of 130 ft/min (39 m/min) for speed, 0.010 in./rev (0.2540 mm/rev) for feed, and 0.09 in. (2.286 mm) for depth of cut. A similar procedure is employed for each tool material/work material/work hardness combination to obtain the databases Rfturn, Semiturn, and Finturn. A similar logic applies for the selection of boring for machining internal surfaces and for values found in the files Rfbore, Semibore, Finbore, and Drill. To add a new work material, tool material, or hardness range, one may append the data to these files appropriately.

A database program (Into.prg) retrieves the dimension and tolerance specifications for the external diametral features, from "tempor.dbf" (prepared file from tolerance allocation). Another program, "Tolsfin.dbf," contains the tolerance–surface finish relationship shown in Table 21.1. It is to be noted that the actual database implemented modifies information contained in Table 21.1 to cover surface roughness ranges corresponding to tolerances rather than discrete values [refer to Kumar (1992b) for more information]. This database is searched to find the range of surface finish specification (R_a) that would be most suitable for the specified tolerance to be maintained and measured. The user is asked to input a surface finish specification within the recommended range. If there is no requested specific finish requirement, the highest default value in that range is selected, so as to ensure reduced costs. This process is repeated for each external surface and the output stored in a file "compreo.dbf" which contains the dimension, tolerance, and surface roughness value for every external surface. A similar program (Inti.prg) is used for each internal surface.

An expert module houses the rules for selecting the operations that would result in the surface finish specified. This module also serves as a user interface for selecting work and tool materials, work material hardness, and for modifying and validating cutting parameters selected. VP-Expert (*VP-Expert,* 1987) is a popular expert system shell capable of being interfaced with dBase III Plus (Datapro, 1988) and of communicating with external ASCII files; the knowledge is represented in the form of "If-Then" rules, and consultation is performed through a backward-chaining mechanism.

A knowledge-base file (Begin.kbs; logic shown in Figure 21.6) helps in selecting the work material, tool material, and hardness range by reading them out of the database file (Rfturn.dbf). Additions or deletions made to the database are reflected here automatically. Since the raw material is assumed to be power-fed, the only input required is the diameter of the bar (in inches). The dimension and surface finish are retrieved and stored into arrays dimn [] and surfin [] for each feature in file "compreo.dbf." A program (Begin.kbs) takes in the basic information on the tool and the work material and chains it a program "Process.kbs," the logic of which is shown in Figure 21.7. The latter contains the knowledge to determine the processes needed for each external feature. Further, it has the knowledge base to recommend and modify the cutting parameters for each process selected. The logic for process determination can be described as follows:

1. If the surface finish $R_a \leq 63$ μin. (1.6002 μm), finish turning is needed; if $63 < R_a < 250$ μin. (1.6002 $< R_a <$ 6.3500 μm), semifinish is needed, and if $R_a > 250$ μin. (6.350 μm), only a rough turning operation is sufficient. These are conservative selections based on assumptions and can be modified as desired by the user.

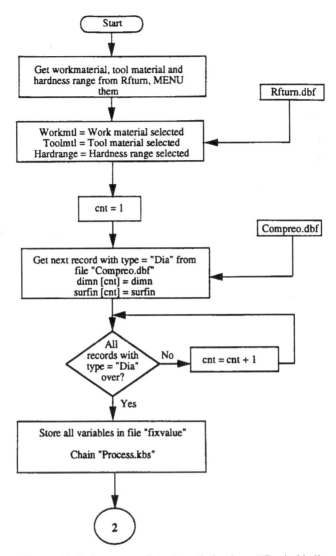

Figure 21.6. Structure of the knowledge-base "Begin.kbs".

2. A finish turning operation is always preceded by semifinish turning, which in turn is preceded by rough turning.

The databases Rfturn, Semiturn, and Finturn are searched, depending on whether just a rough turn, a rough and semi-finish turn, or a rough, semifinish, and finish turning are required, for each dimension, for the specified work material, tool material, hardness range, and surface finish needed. The values of cutting parameters matching the conditions recommended are retrieved and displayed. The user is given an option of changing any of these. The number

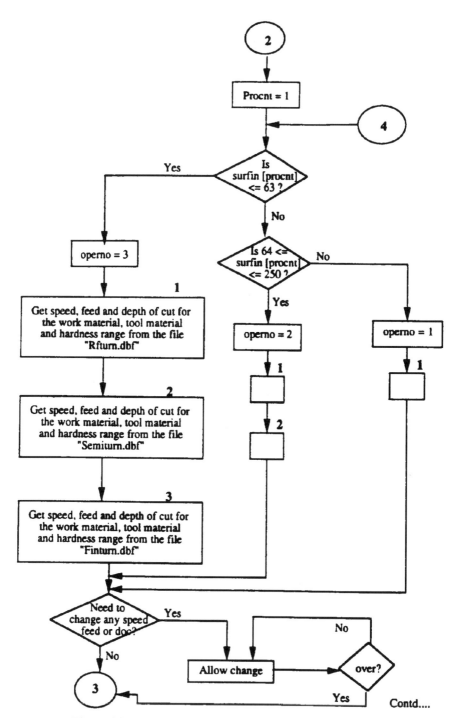

Figure 21.7. Structure of the knowledge-base "Process.kbs".

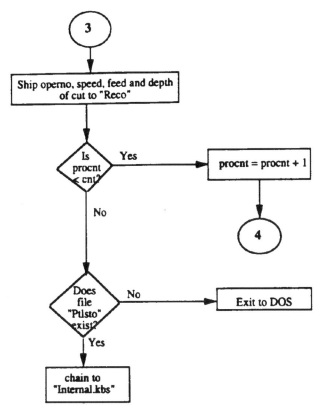

Figure 21.7. (*Continued*)

of operations, the diametral dimension, and the speed, feed, and depth-of-cut values for each operation are written into a file "Rec." The knowledge base then checks for the existence of a file "Ptlsti," an absence of which indicates the absence of any internal surface to be machined. Upon the presence of the file, the knowledge base chains program "Internal.kbs" for internal features (with internal machining operations drilling and rough, semifinish, and finish boring). The procedures needed for internal operations are then performed.

A QuickBASIC module has the programs necessary for determining the cutter path and generating the NC code necessary to machine the component. "Relptso" and "Relptsi" contain scaled coordinates of vertex points of the external and internal surfaces, respectively; written by the QuickBASIC module. "Dia" is an ASCII file that has the diameter of the raw stock, written by the VP-Expert module.

A line drawn through the center of the machine spindle is the Z-axis: a negative movement is toward the headstock. Negative X movement signifies movement of the cross-slide toward the centerline of the spindle. Operations are assumed to be performed on an NC machining center with an indexable

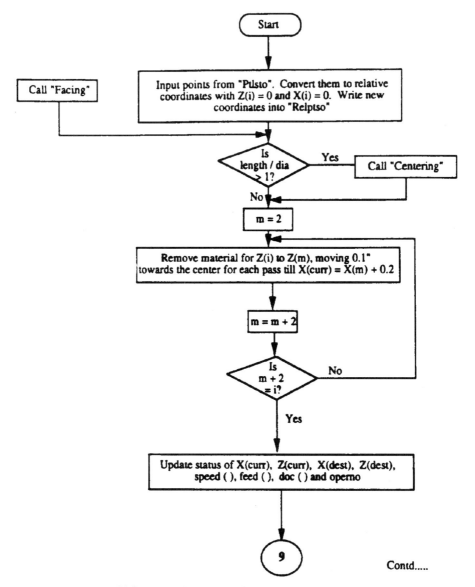

Figure 21.8. Logic for determining cutter path during turning.

turret. Absolute programming is selected and the following standard words of the word address format are used: N, G, X/Z, S, F, M. Other typical words are not used in this experimental system and may be added as desired at a later stage. Figure 21.8 illustrates the logic for determining the cutter path for external machining. This logic controls the cutter motion, updating the direction, depth of cut, speed, feed, and surface being machined until all external surfaces are machined. After each surface is machined, a subroutine is called

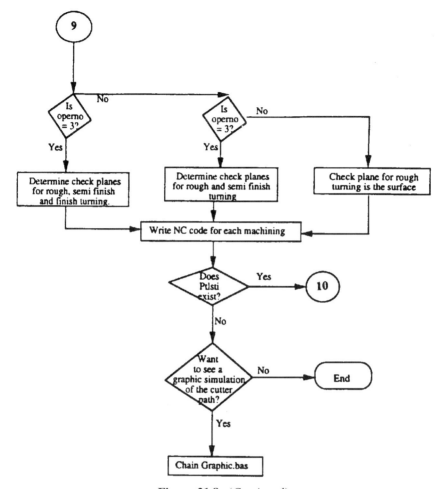

Figure 21.8. (*Continued*)

that updates the variables to those of the next surface to be machined. Program logic is also developed for boring, drilling, and centering.

A graphical simulator that will emulate the movement of the cutter is also programmed, seeking to confirm the NC code generated. Any unmachined raw stock, unnecessary machining, or tool/work collision can hence be identified. Different depths of cut and feed rates corresponding to different machining operations also show up.

21.9 EXAMPLE

Figure 21.9 is an illustration of the chosen component that is input into AutoCAD (Autodesk, 1989). Three steps and the internal cylinder must be

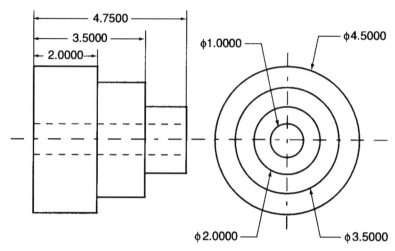

Figure 21.9. Example part with three steps and one hole.

machined, to machine this component from a rough cylindrical stock. The dimensions shown are nominal dimensions. Using an appropriate allocation procedure, tolerances may be specified to each feature of a component, based on usual considerations of functionality, cost and so on. An input such as shown in Table 21.2 is assumed for this part. Note that this may not be an accurate allocation of tolerances and has been used here only for demonstration purposes. Further work is also needed in developing a sturdy input module for inputting dimension/tolerance information with the right format into the present system developed. The feature definition program stores the vertex points of the external and internal features into specific files "Ptlsto" and : "Ptlsti". Files "Compreo" and "Comprei" can be written as shown in Table 21.3. "Compreo" is the result of the Dbase program "Into.dbf". For surfaces with tolerance values of 0.01 in. (0.2540 mm) and 0.005 in. (0.1270 mm), the user has specified surface finish requirements of 160 μin. (4.0640 μm)

TABLE 21.3. Output of the Dbase Module (Files "Compreo" and "Comprei")

RECORD #	TYPE	DIMN	TOLERANCE	SURFIN
1	DIA	4.50	.01	160
2	DIA	3.50	.005	90
3	DIA	2.00	.002	48

RECORD #	TYPE	DIMN	TOLERANCE	SURFIN
1	DIA	1.00	.003	82

and 90 μin. (2.2860 μm), respectively; for the surface with a tolerance 0.002 in. (0.0508 mm), the user had no surface finish specification, so the system defaulted to the highest surface finish possible for this tolerance specification, namely 48 μin. (1.2192 μm). Similarly, the output of "Inti.prg" is the file "Comprei", containing the user-specified surface finish for the surface with tolerance 0.003 in. (0.0762 mm).

Using the VP-Expert consultation screens, data are input into the various knowledge bases and output files are generated. The QuickBASIC module uses these files to determine the cutter path and write the NC code information, a segment of which is shown in Figure 21.10. The reader is referred to Kumar (1992a,b) for the source code (containing databases), specific data input, and expert system consultations.

21.10 SUMMARY AND CONCLUSIONS

A simple methodology for converting size tolerance information to determine cutting parameters and information for NC code generation is demonstrated using software implementation. Tolerance specifications are suitably modified into surface roughness specifications to achieve a transfer into the manufacturing domain. The surface finish specifications are hence used for machining parameter determination, machining sequence determination, and NC code generation. A software implementation is achieved, mainly to prove the validity of the proposal. The software is developed as a consultation-type system, enabling significant user interaction, which is desired in such activities. This work is preliminary, and significant research must be conducted to make this software cater to various types of tolerances and surface attributes. Moreover, the integrated framework is developed based on textual output obtainable from computerized engineering drawings and not on geometric output obtainable from true CAD models, thus achieving only an indirect transfer of size tolerance entities. All the same, it is hoped that this software development will inspire small-parts manufacturers to develop more comprehensive integrated systems for preparing process plans and NC codes based on important annotational information whenever advanced CAD/CAM systems are not available. The principal contribution of the present paper lies in the consideration of multiple functions in integrated system development and in the use of the tolerance–surface finish compatibility for generating machining sequence and machining parameters.

ACKNOWLEDGMENTS

The authors wish to acknowledge the support provided by S. Pulat, M. Nagarwala, and M. Lokhandwala during the development of this work. The authors also wish to thank the Technology Utilization Office of NASA, George C. Marshall Space Center,

```
"N",59,"G","01","Z",3.6,"X",1.93,"S",340,"F",.015,"D",.15
"N",60,"G","01","Z",2,"X",1.93,"S",340,"F",.01,"D",.09
"N",61,"G","00","Z",2,"X",2.02,"S",340,"F",.01,"D",.09
"N",62,"G","01","Z",3.6,"X",2.02,"S",340,"F",.01,"D",.09
"N",63,"G","01","Z",3.6,"X",1.84,"S",340,"F",.01,"D",.09
"N",64,"G","01","Z",2,"X",1.84,"S",340,"F",.01,"D",.09
"N",65,"G","00","Z",2,"X",1.93,"S",340,"F",.01,"D",.09
"N",66,"G","01","Z",3.6,"X",1.93,"S",340,"F",.01,"D",.09
"N",67,"G","01","Z",3.6,"X",1.75,"S",340,"F",.01,"D",.09
"N",68,"G","01","Z",2,"X",1.75,"S",340,"F",.01,"D",.09
"N",69,"G","00","Z",2,"X",1.84,"S",340,"F",.01,"D",.09
"N",70,"G","00","Z",3.6,"X",1.84,"S",340,"F",.01,"D",.09
"N",71,"G","00","Z",2.1,"X",2.6,"S",340,"F",.015,"D",.15
"N",72,"G","01","Z",2.1,"X",2.43,"S",340,"F",.015,"D",.15
"N",73,"G","01","Z",0,"X",2.43,"S",340,"F",.01,"D",.09
"N",74,"G","00","Z",0,"X",2.52,"S",340,"F",.01,"D",.09
"N",75,"G","01","Z",2.1,"X",2.52,"S",340,"F",.01,"D",.09
"N",76,"G","01","Z",2.1,"X",2.34,"S",340,"F",.01,"D",.09
"N",77,"G","01","Z",0,"X",2.34,"S",340,"F",.01,"D",.09
"N",78,"G","00","Z",0,"X",2.43,"S",340,"F",.01,"D",.09
"N",79,"G","01","Z",2.1,"X",2.43,"S",340,"F",.01,"D",.09
"N",80,"G","01","Z",2.1,"X",2.25,"S",340,"F",.01,"D",.09
"N",81,"G","01","Z",0,"X",2.25,"S",340,"F",.01,"D",.09
"N",82,"G","00","Z",0,"X",2.34,"S",340,"F",.01,"D",.09
"N",83,"G","00","Z",2.1,"X",2.34,"S",340,"F",.01,"D",.09
"N",84,"G","00","Z",.1,"X",.35,"S",340,"F",.015,"D",.15
"N",85,"G","00","Z",4.95,"X",.125,"S",40,"F",.004,"D",.25
"N",86,"G","81","Z",-.1,"X",.125,"S",40,"F",.004,"D",.25
"N",87,"G","000","Z",4.75,"X",.125,"S",40,"F",.004,"D",.25
"N",88,"G","00","Z",4.85,"X",.025,"S",60,"F",.007,"D",.1
"N",89,"G","1","Z",4.85,"X",.225,"S",60,"F",.007,"D",.2
"N",90,"G","1","Z",0,"X",.225,"S",60,"F",.007,"D",.2
"N",91,"G","0","Z",0,"X",.125,"S",60,"F",.007,"D",.2
"N",92,"G","0","Z",4.85,"X",.125,"S",60,"F",.007,"D",.2
"N",93,"G","1","Z",4.85,"X",.3,"S",60,"F",.007,"D",.2
"N",94,"G","1","Z",0,"X",.3,"S",60,"F",.007,"D",.2
"N",95,"G","0","Z",0,"X",.2,"S",60,"F",.007,"D",.2
"N",96,"G","0","Z",4.85,"X",.2,"S",60,"F",.007,"D",.2
"N",97,"G","1","Z",4.85,"X",.4,"S",60,"F",.007,"D",.1
"N",98,"G","1","Z",0,"X",.4,"S",60,"F",.007,"D",.1
"N",99,"G","0","Z",0,"X",.3,"S",60,"F",.007,"D",.1
"N",100,"G","0","Z",4.85,"X",.3,"S",60,"F",.007,"D",.1
"N",101,"G","1","Z",4.85,"X",.4,"S",60,"F",.005,"D",.05
"N",102,"G","1","Z",0,"X",.4,"S",60,"F",.005,"D",.05
"N",103,"G","0","Z",0,"X",.35,"S",60,"F",.005,"D",.05
"N",104,"G","0","Z",4.85,"X",.35,"S",60,"F",.005,"D",.05
"N",105,"G","1","Z",4.85,"X",.45,"S",60,"F",.005,"D",.05
"N",106,"G","1","Z",0,"X",.45,"S",60,"F",.005,"D",.05
"N",107,"G","0","Z",0,"X",.4,"S",60,"F",.005,"D",.05
"N",108,"G","0","Z",4.85,"X",.4,"S",60,"F",.005,"D",.05
"N",109,"G","1","Z",4.85,"X",.5,"S",60,"F",.005,"D",.05
"N",110,"G","1","Z",0,"X",.5,"S",60,"F",.005,"D",.05
"N",111,"G","0","Z",0,"X",.45,"S",60,"F",.005,"D",.05
"N",112,"G","0","Z",4.85,"X",.45,"S",60,"F",.005,"D",.05
"N",113,"G","1","Z",4.85,"X",.5,"S",60,"F",.005,"D",.05
"N",114,"G","1","Z",0,"X",.5,"S",60,"F",.005,"D",.05
"N",115,"G","0","Z",0,"X",.45,"S",60,"F",.005,"D",.05
"N",116,"G","0","Z",4.75,"X",.45,"S",60,"F",.005,"D",.05
"N",117,"G","00","Z",4.85,"X",.4,"S",60,"F",.007,"D",.1
```

Figure 21.10. Segment of the NC code output.

for its partial financial support of this project, and L. Adams and R. Burton of Entec Corporation for their practical viewpoints that have helped in various developmental aspects of this system.

SYMBOLS USED IN THE FLOW CHARTS/ALGORITHMS

cnt	counter for external operations
cnti	counter for internal operations
dimn []	array containing the diameters of all external surfaces
dimni []	array containing the diameters of all internal surfaces
doc ()	array containing the depth of cut for each operation for the current surface
feed ()	Array containing the feeds for each operation for the current surface
gcode	preparatory function code used in NC programs
operno	number of operations needed to achieve the desired surface finish Example: 2 = rough turning (boring) followed by semifinish turning (boring)
procnt	counter used in the VP-Expert module
speed ()	array containing the speeds for each operation for the current surface
surfin []	array containing the surface finish of each external surface
surfini []	array containing the surface finish requirements of all internal surfaces
X(curr)	X-coordinate of the current point
X(dest)	X-coordinate of the destination point
Z(curr)	Z-coordinate of the current point
Z(dest)	Z-coordinate of the destination point

REFERENCES

Alting, L. and Zhang, H. C., 1988, Computer Aided Process Planning: The State of the Art Survey, *Int. J. Prod. Res.,* Vol. 27, No. 4, pp. 553–588.

ASME, 1982, *Dimensioning and Tolerancing,* ANSI Y14.5-1982, American Society of Mechanical Engineers, New York.

Autodesk, 1989, *Reference Manual, AutoCAD Release 10,* Autodesk, Inc., Sausolito, CA.

Baldwin, D. F., Abell, T. E., Lui, M. M., De Fazio, T. L., and Whitney, D. E., 1991, An Integrated Computer Aid for Generating and Evaluating Assembly Sequences for Mechanical Products, *IEEE Trans. Robot. Autom.,* Vol. 7, No. 1, pp. 78–94.

Bjorke, O., 1989, *Computer-Aided Tolerancing,* 2nd ed., ASME Press, New York.

Black, O., 1986, Assuring Confident Re-use of Production Drawing Information by the NC Programming Activity, *Comput. Aid. Eng.,* pp. 159–163.

Chase, K. W., and Greenwood, W. H., 1988, Design Issues in Mechanical Tolerance Analysis, *Manuf. Rev.,* Vol. 1, No. 1, pp. 50–59.

Chase, K. W., Greenwood, W. H., Loosli, B. G., and Haglund, L. F., 1990, Least Cost Tolerance Allocation for Mechanical Assemblies with Automated Process Selection, *Manuf. Rev.,* Vol. 3, No. 1, pp. 49–59.

Datapro, 1988, *dBASE III and Advanced dBASE III,* 2nd ed., Datapro Information Technology, Bombay, India.

Dong, Z., and Soom, A., 1986, Automatic Tolerance Analysis from a CAD Database, *Proceedings of the ASME Design Engineering Technical Conference,* Columbus, Ohio, Oct. 5–8, pp. 1–8.

Fainguelernt, D., Weill, R., and Bourdet, P., 1986, Computer Aided Tolerancing and Dimensioning in Process Planning, *Ann. CIRP,* Vol. 35, No. 1, pp. 381–386.

Giesecke, F. E., Mitchell, A., Spencer, H. C., Hill, I. L., and Dygdon, J. T., 1986, *Technical Drawing,* 7th ed., Macmillan, New York.

Irani, S. A., Mittal, R. O., and Lehtihet, E. A., 1989, Tolerance Chart Optimization, *Int. J. Prod. Res.,* Vol. 27, No. 9, pp. 1531–1552.

Irani, S. A., Koo, H. Y., and Raman, S., 1995, Feature-Based Operation Sequence Generation in CAPP, *Int. J. Prod. Res.,* Vol. 33, No. 1, pp. 17–39.

Joshi, S., Vissa, N. N., and Chang, T. C., 1987, Expert Process Planning System with Solid Model Interface, *Int. J. Prod. Res.,* Vol. 26, No. 5, pp. 863–885.

Kapur, K. C., Raman, S., and Pulat, S., 1990, Methodology for Tolerance Design Using Quality Loss Function, *Comput. Ind. Eng.,* Vol. 19, No. 1–4, pp. 254–257.

Knutilla, A. J., and Park, B. C., 1990, Computer Managed Process Planning for Cylindrical Parts, *Proceedings of the 5th International Conference on CAD/CAM, Robotics and Factories of the Future,* Norfolk, VA, Springer-Verlag, New York, pp. 153–165.

Kumar, S., 1992a, *A Prototype for Tolerance/Surface Finish Transfer from CAD to NC,* M.S. thesis, University of Oklahoma, Norman, OK.

Kumar, S., 1992b, *Source Code Manual,* NN-KU-1, IE Library, University of Oklahoma, Norman, OK.

Kumar, S., and Raman, S., 1992, Computer Aided Tolerancing: The Past, Present and the Future, *J. Des. Manuf.,* Vol. 2, No. 1, pp. 29–41.

Kumar, S., and Raman, S., 1993, A Simple Framework for Interpreting Tolerance/ Surface Finish Specifications, in *Manufacturing Science and Engineering,* K. F. Ehmann (ed.), PED-Vol. 64, American Society of Mechanical Engineers, New York, pp. 253–266.

Kurup, K., Raman, S., and Pulat, S., 1995, Computer Aided Mechanical Disassembly Sequence Generation, *Transactions of the North American Manufacturing Research Institute (NAMRI) of SME,* Vol. 23, pp. 247–254.

Lokhandwala, M., Raman, S., and Pulat, S., 1992, An Integrated System for Tolerance Allocation and Tolerance Analysis, in *Concurrent Engineering,* D. Dutta, A. C. Woo, et al. (eds.), PED-Vol. 59, American Society of Mechanical Engineers, New York, pp. 195–209.

Machinery's Handbook, 1984, 22nd ed., Industrial Press, New York.

Machining Data Handbook, 1980, 3rd ed., Metcut Research Associates, Cincinnati, OH.

Mani, J., and Raman, S., 1996, A Methodology for Manufacturing Precedence Representation and Alternate Task Sequencing in CAPP, *Transactions of the North*

American Manufacturing Research Institute (NAMRI) of SME, Vol. 24, pp. 251–256.

Metals Handbook, 1989, Vol. 16, *Machining,* ASM International, Metals Park, OH.

Nagarwala, M., Pulat, S., and Raman, S., 1995, A Slope-Based Method for Tolerance Allocation in Minimum Cost Assembly, *Concurrent Eng.: Res. Appl.,* Vol. 3, No. 4, pp. 319–238.

Pulat, S. P., and Raman, S., 1993, *A Decision Support System for Tolerance Allocation,* Final Project Report submitted to NASA, George C. Marshall Space Center, Technology Utilization Office, Huntsville, Alabama.

Requicha, A. A. G., 1983a, Toward a Theory of Geometric Tolerancing, *Int. J. Robot. Res.,* Vol. 2, No. 4, pp. 45–60.

Requicha, A. A. G., 1983b, Representation of Tolerances in Solid Modeling: Issues and Alternative Approaches, *GM Solid Modeling Symposium,* Warren, MI, pp. 3–22.

Roy, U., and Liu, C. R., 1988, Feature-Based Representational Scheme of a Solid Modeler for Providing Dimensioning and Tolerancing Information, *Robot. Comput. Integrat. Manuf.,* Vol. 4, No. 3, pp. 335–345.

Sakal, R. L., and Chow, J. G., 1991, Design and Development of an Integrated CAD/CAM/CAPP Machining System for Prismatic Parts, *Transactions of the North American Manufacturing Research Institute (NAMRI) of SME,* Vol. 19, pp. 267–273.

Shah, J. J., and Bhatnagar, A. S., 1989, Group Technology Classification from Feature-Based Geometric Models, *Manuf. Rev.,* Vol. 2, No. 2, pp. 204–213.

Trucks, H. E., 1974, *Designing for Economical Production,* Society of Manufacturing Engineers, Dearborn, MI.

Truslove, K. C. E., 1988, The Implications of Tolerancing for Computer-Aided Mechanical Design, *Comput. Aid. Eng.,* April, pp. 79–85.

Varghese, M., and Atkinson, J., 1987, Automated Dimensional Tolerancing on a Turnkey CAD System, *Proceedings of the 2nd International Conference on Computer-Aided Production Engineering,* Edinburgh, April, pp. 43–48.

VP-Expert: Rule-Based Expert System Development Tool, 1987, Paperback Software, Berkeley, CA.

Wang, H. P., and Wysk, R. A., 1986, An Expert System for Machining Data Selection, *Comput. Ind. Eng.,* Vol. 10, No. 2, pp. 99–107.

Wang, H. P., and Wysk, R. A., 1987a, Turbo-CAPP: A Knowledge-Based Computer Aided Process Planning System, *Proceedings of the 19th CIRP International Seminar on Manufacturing Systems,* State College, PA, June 1–2, pp. 161–167.

Wang, H. P., and Wysk, R. A.,., 1987b, A Knowledge-Based Approach for Automated Process Planning, *Int. J. Prod. Res.,* Vol. 26, No. 6, pp. 999–1014.

Weill, R., 1988, Integrating Dimensioning and Tolerancing in Computer Aided Process Planning, *Robot. Comput. Integrat. Manuf.,* Vol. 4, No. 1–2, pp. 41–48.

Wu, Z., Elmaraghy, W. H., and Elmaraghy, H. A., 1988, Evaluation of Cost-Tolerance Algorithms for Design Tolerance Analysis and Synthesis, *Manufac. Rev.,* Vol. 1, No. 3, pp. 168–179.

Xiaoqing, T., and Davies, B. J., 1988, Computer Aided Dimensional Planning, *Int. J. Prod. Res.,* Vol. 26, No. 2, pp. 283–297.

INDEX